Principles and Practices of Geoenvironmental Mapping

Principles and Practices of Geoenvironmental Mapping

Editor:

Dr. Tanjina Nur, Ph.D.

Delve Publishing

www.delvepublishing.com

Principles and Practices of Geoenvironmental Mapping

Editor: Dr. Tanjina Nur, Ph.D.

708 3rd Avenue, 6th Floor
New York, NY 10017
United States of America
www.delvepublishing.com

ISBN: 978-1-68095-821-8

Library of Congress Control Number: 2016951563

About the Editor

Dr. Tanjina Nur, Ph.D.

Tanjina finished her PhD in Civil and Environmental Engineering in 2014 from University of Technology Sydney (UTS). Now she is working as Post-Doctoral Researcher in the Centre for Technology in Water and Wastewater (CTWW) and published about eight International journal papers with 80 citations. Her research interest is wastewater treatment technology using adsorption process.

Contents

	List of Contributors	*xi*
	Preface	*xvii*
	Introduction	*xix*
Chapter 1	**Landslide Susceptibility Mapping in Vertical Distribution Law of Precipitation Area**	**1**
1.	Introduction	1
2.	Methodology	2
3.	Study Area and Data	3
4.	Results and Discussion	8
	Conclusions	12
	Acknowledgments	13
	References	13
Chapter 2	**A Combination of Geographically Weighted Regression, Particle Swarm Optimization and Support Vector Machine for Landslide Susceptibility Mapping**	**17**
1.	Introduction	17
2.	Related Techniques	18
3.	The Proposed GWR-PSO-SVM Model	21
4.	Study Area and Data	24
5.	Results	28
6.	Discussion	37
	Conclusions	40
	Acknowledgments	40
	References	40
Chapter 3	**Evaluation of Different Soil Salinity Mapping Using Remote Sensing Techniques in Arid Ecosystems**	**45**
1.	Introduction	45
2.	Materials and Methods	46
3.	Results and Discussion	47
	Conclusions	50
	Acknowledgment	50
	References	50
Chapter 4	**Soil Clay Content Mapping Using a Time Series of Landsat TM Data in Semi-Arid Lands**	**53**
1.	Introduction	53
2.	Study Area and Data	54
3.	Methods	56
4.	Results	58
5.	Discussion	61
	Conclusions	63
	Acknowledgments	63
	References	63
Chapter 5	**Mapping and Modelling Spatial Variation in Soil Salinity Based on Remote Sensing Indicators and Regression Techniques**	**67**
1.	Introduction	67
2.	Materials and Methods	68
3.	Results	70

4. Discussion ... 74
Conclusions ... 76
Acknowledgments ... 77
References ... 77

Chapter 6 Operational Mapping of Soil Moisture Using Synthetic Aperture Radar Data ... 81
1. Introduction ... 81
2. Description af Databases ... 82
3. Soil Moisture Mapping ... 86
Conclusions ... 93
Acknowledgments ... 94
References ... 94

Chapter 7 Land Surface Water Mapping Using Multi-Scale Level Sets and a Visual Saliency Model From Sar Images ... 97
1. Introduction ... 97
2. Proposed Method ... 98
3. Experimental Data Set ... 102
4. Results and Discussion ... 102
Conclusions ... 109
Acknowledgments ... 110
References ... 110

Chapter 8 Automated Subpixel Surface Water Mapping from Heterogeneous Urban Environments Using Landsat 8 Oli Imagery ... 111
1. Introduction ... 111
2. Study Areas and Data Sources ... 112
3. Methodology ... 113
4. Results ... 116
5. Discussion ... 119
Conclusions ... 120
Acknowledgments ... 120
References ... 120

Chapter 9 Operational Surface Water Detection and Monitoring Using Radarsat 2 ... 123
1. Introduction ... 123
2. Materials ... 124
3. Methods ... 125
4. Results ... 128
5. Discussion ... 133
Conclusions ... 134
Acknowledgments ... 134
References ... 135

Chapter 10 Flood Hazard Mapping by Using Geographic Information System and Hydraulic Model ... 137
1. Introduction ... 137
2. Study Area ... 138
3. Application and Results ... 139
Conclusions ... 143
References ... 143

Chapter 11 **Urban Flood Mapping Based on Unmanned Aerial Vehicle Remote Sensing and Random Forest Classifier** **145**
1. Introduction 145
2. Study Area 146
3. Methods 146
4. Results 150
5. Discussion 153
Conclusions 153
Acknowledgments 154
Author Contributions 154
References 154

Chapter 12 **Flood Mapping and Flood Dynamics of the Mekong Delta: Envisat-Asar-Wsm Based Time Series Analyses** **157**
1. Introduction: Remote Sensing Based Flood Monitoring 157
2. Study Area: The Mekong Delta 158
3. Available Satellite and Ancillary Data 160
4. Methods and Data Analyses 161
5. Results 163
6. Discussion 169
Conclusion 170
Acknowledgments 171
References 171

Chapter 13 **Flood Mapping Based On Multiple Endmember Spectral Mixture Analysis and Random Forest Classifier** **175**
1. Introduction 175
2. Study Area and Datasets 176
3. Results 181
4. Discussion 185
Conclusions 187
Acknowledgments 187
Author Contributions 187
Conflicts of Interest 187
References 187

Chapter 14 **Flood Risk Mapping Using Lidar** **191**
1. Introduction 191
2. Methods 193
3. Results 195
4. Discussion and Conclusions 200
Acknowledgements 200
References 200

Chapter 15 **Land Cover Mapping Using the HC-MMK Approach** **205**
1. Introduction 205
2. Study Area 206
3. Methodology 207
4. Results 212
5. Discussions 216
Conclusions 218
Acknowledgments 219

Author Contributions 219
References 219

Chapter 16 **Designing an Experiment to Investigate Subpixel Mapping as an Alternative Method to Obtain Land Use/Land Cover Maps 223**
1. Introduction 223
2. Experimental Design 224
3. Case Study 226
4. Samspm 228
5. Discussion 233

Conclusions 235
Acknowledgments 235
Author Contributions 235
Conflicts of Interest 235
References 235

Chapter 17 **An Automatic Procedure for Early Disaster Change Mapping Based on Optical Remote Sensing 237**
1. Introduction 237
2. The Procedure and Methods 238
3. Experiments Results and Analysis 242
4. Acknowledgments 255

References 256

Chapter 18 **Development of Mobile Mapping System For 3D Road Asset Inventory 259**
1. Introduction 259
2. Background 260
3. Methodology 261
4. Camera Calibration [39–41] 263
5. Discussion 269

Conclusions 270
References 270

Index 273

List of Contributors

Chen Cao
College of Construction Engineering, Jilin University, Changchun 130026, China

Qing Wang
College of Construction Engineering, Jilin University, Changchun 130026, China

Jianping Chen
College of Construction Engineering, Jilin University, Changchun 130026, China

Yunkai Ruan
College of Construction Engineering, Jilin University, Changchun 130026, China

Lianjing Zheng
College of Construction Engineering, Jilin University, Changchun 130026, China

Shengyuan Song
College of Construction Engineering, Jilin University, Changchun 130026, China

Cencen Niu
College of Construction Engineering, Jilin University, Changchun 130026, China

Xianyu Yu
Faculty of Information Engineering, China University of Geosciences, Wuhan 430074, China
Institute of Geophysics and Geomatics, China University of Geosciences, Wuhan 430074, China

Yi Wang
Institute of Geophysics and Geomatics, China University of Geosciences, Wuhan 430074, China

Ruiqing Niu
Institute of Geophysics and Geomatics, China University of Geosciences, Wuhan 430074, China

Youjian Hu
Faculty of Information Engineering, China University of Geosciences, Wuhan 430074, China

Mohamed Elhag
Department of Hydrology and Water Resources Management, Faculty of Meteorology, Environment & Arid Land Agriculture, King Abdulaziz University, Jeddah 21589, Saudi Arabia

Marouen Shabou
CESBIO/UMR 5126, 18 av.Edouard Belin, bpi 2801, F-31401 Toulouse Cedex 9, France
Rural engineering, water and forest department, INAT/University of Carthage, 43, Avenue Charles Nicolle 1082 Tunis-Mahrajène, Tunisia

Bernard Mougenot
CESBIO/UMR 5126, 18 av.Edouard Belin, bpi 2801, F-31401 Toulouse Cedex 9, France

Zohra Lili Chabaane
Rural engineering, water and forest department, INAT/University of Carthage, 43, Avenue Charles Nicolle 1082 Tunis-Mahrajène, Tunisia

Christian Walter
Agrocampus Ouest, INRA, UMR 1069 SAS, 65 rue de st Brieuc- CS 84215, F-35042 Rennes Cedex, France

Gilles Boulet
CESBIO/UMR 5126, 18 av.Edouard Belin, bpi 2801, F-31401 Toulouse Cedex 9, France

Nadhira Ben Aissa
Rural engineering, water and forest department, INAT/University of Carthage, 43, Avenue Charles Nicolle 1082 Tunis-Mahrajène, Tunisia

Mehrez Zribi
CESBIO/UMR 5126, 18 av.Edouard Belin, bpi 2801, F-31401 Toulouse Cedex 9, France

Amal Allbed
Department of Ecosystem Management, School of Environmental and Rural Science, University of New England, Armidale, NSW 2351, Australia

Lalit Kumar
Department of Ecosystem Management, School of Environmental and Rural Science, University of New England, Armidale, NSW 2351, Australia

Priyakant Sinha
Department of Ecosystem Management, School of Environmental and Rural Science, University of New England, Armidale, NSW 2351, Australia

Nicolas Baghdadi
French Geological Survey (BRGM), BP 36009, 45060 Orléans Cedex 2, France

Maelle Aubert
French Geological Survey (BRGM), BP 36009, 45060 Orléans Cedex 2, France

Olivier Cerdan
French Geological Survey (BRGM), BP 36009, 45060 Orléans Cedex 2, France

Laurent Franchistéguy
Météo-France, 42 avenue G. Coriolis, 31057 Toulouse Cedex, France

Christian Viel
SCHAPI, 42 avenue Coriolis, 31057 Toulouse Cedex, France

Eric Martin
Météo-France, 42 avenue G. Coriolis, 31057 Toulouse Cedex, France

Mehrez Zribi
CETP/CNRS, 10/12, avenue de l'Europe, 78140 Velizy, France

Jean François Desprats
French Geological Survey (BRGM), BP 36009, 45060 Orléans Cedex 2, France

Chuan Xu
State Key Laboratory of Information Engineering in Surveying, Mapping and Remote Sensing, Wuhan University, Wuhan 430079, China

Haigang Sui
State Key Laboratory of Information Engineering in Surveying, Mapping and Remote Sensing, Wuhan University, Wuhan 430079, China

Feng Xu
National Disaster Reduction Center of China, NO.6 GuangBaiDongLu, Chaoyang District, Beijing 100124, China

Huan Xie
College of Surveying and Geo-informatics, Tongji University, Shanghai 200092, China

Xin Luo
College of Surveying and Geo-informatics, Tongji University, Shanghai 200092, China

Xiong Xu
College of Surveying and Geo-informatics, Tongji University, Shanghai 200092, China

Haiyan Pan
College of Surveying and Geo-informatics, Tongji University, Shanghai 200092, China

Xiaohua Tong
College of Surveying and Geo-informatics, Tongji University, Shanghai 200092, China

Sandra Bolanos
Science and Risk Assessment Directorate, Environment and Climate Change Canada, 351 Boulevard St-Joseph, Gatineau, QC K1A 0H3, Canada

Doug Stiff
National Hydrological Services, Environment and Climate Change Canada, 373 Sussex Dr., Ottawa, ON K1A 0H3, Canada

Brian Brisco
Canada Centre for Mapping and Earth Observation, Natural Resources Canada, 560 Rochester St, Ottawa, ON K1S 5K2, Canada

Alain Pietroniro
National Hydrological Services, Environment and Climate Change Canada, 11 Innovation Blvd., Saskatoon, SK S7N 3H5, Canada

Vahdettin Demir
Civil Engineering Department, Canik Basari University, Samsun, Turkey

Ozgur Kisi
Civil Engineering Department, Canik Basari University, Samsun, Turkey

Quanlong Feng
[1]State Key Laboratory of Remote Sensing Science, Institute of Remote Sensing and Digital Earth Chinese Academy of Sciences, No. 20 Datun Road, Chaoyang District, Beijing 100101, China

Jiantao Liu
[1]State Key Laboratory of Remote Sensing Science, Institute of Remote Sensing and Digital Earth Chinese Academy of Sciences, No. 20 Datun Road, Chaoyang District, Beijing 100101, China

Jianhua Gong
State Key Laboratory of Remote Sensing Science, Institute of Remote Sensing and Digital Earth Chinese Academy of Sciences, No. 20 Datun Road, Chaoyang District, Beijing 100101, China
Zhejiang-Chinese Academy of Sciences (CAS) Application Center for Geoinformatics, No. 568 Jinyang East Road, Jiashan 314100, China

Claudia Kuenzer
[1]German Remote Sensing Data Center, DFD, German Earth Observation Center (EOC), German Aerospace Center (DLR), Oberpfaffenhofen, D-82234 Wessling, Germany

Huadong Guo
Center for Earth Observation & Digital Earth (CEODE), Chinese Academy of Science, Beijing 100094, China

Juliane Huth
German Remote Sensing Data Center, DFD, German Earth Observation Center (EOC), German Aerospace Center (DLR), Oberpfaffenhofen, D-82234 Wessling, Germany

Patrick Leinenkugel
German Remote Sensing Data Center, DFD, German Earth Observation Center (EOC), German Aerospace Center (DLR), Oberpfaffenhofen, D-82234 Wessling, Germany

Xinwu Li
Center for Earth Observation & Digital Earth (CEODE), Chinese Academy of Science, Beijing 100094, China

Stefan Dech
German Remote Sensing Data Center, DFD, German Earth Observation Center (EOC), German Aerospace Center (DLR), Oberpfaffenhofen, D-82234 Wessling, Germany

Quanlong Feng
State Key Laboratory of Remote Sensing Science, Institute of Remote Sensing and Digital Earth Chinese Academy of Sciences, No.20, Datun Road, Chaoyang District, 100101 Beijing, China

Jianhua Gong
State Key Laboratory of Remote Sensing Science, Institute of Remote Sensing and Digital Earth Chinese Academy of Sciences, No.20, Datun Road, Chaoyang District, 100101 Beijing, China
Zhejiang-CAS Application Center for Geoinformatics, No.568, Jinyang East Road, 314100 Jiashan, China

Jiantao Liu
State Key Laboratory of Remote Sensing Science, Institute of Remote Sensing and Digital Earth Chinese Academy of Sciences, No.20, Datun Road, Chaoyang District, 100101 Beijing, China

Yi Li
State Key Laboratory of Remote Sensing Science, Institute of Remote Sensing and Digital Earth Chinese Academy of Sciences, No.20, Datun Road, Chaoyang District, 100101 Beijing, China

Tim L. Webster
Applied Geomatics Research Group, Nova Scotia Community College, Middleton, Nova Scotia B0S 1M0, NS, Canada

Guangbin Lei
Institute of Mountain Hazards and Environment, Chinese Academy of Sciences, Chengdu 610041, China
University of Chinese Academy of Sciences, Beijing 100049, China

Ainong Li
Institute of Mountain Hazards and Environment, Chinese Academy of Sciences, Chengdu 610041, China

Jinhu Bian
Institute of Mountain Hazards and Environment, Chinese Academy of Sciences, Chengdu 610041, China
University of Chinese Academy of Sciences, Beijing 100049, China

Zhengjian Zhang
Institute of Mountain Hazards and Environment, Chinese Academy of Sciences, Chengdu 610041, China

Huaan Jin
Institute of Mountain Hazards and Environment, Chinese Academy of Sciences, Chengdu 610041, China

Xi Nan
Institute of Mountain Hazards and Environment, Chinese Academy of Sciences, Chengdu 610041, China

Wei Zhao
Institute of Mountain Hazards and Environment, Chinese Academy of Sciences, Chengdu 610041, China

Jiyan Wang
Institute of Mountain Hazards and Environment, Chinese Academy of Sciences, Chengdu 610041, China
University of Chinese Academy of Sciences, Beijing 100049, China

Xiaomin Cao
Institute of Mountain Hazards and Environment, Chinese Academy of Sciences, Chengdu 610041, China
University of Chinese Academy of Sciences, Beijing 100049, China

Jianbo Tan
Institute of Mountain Hazards and Environment, Chinese Academy of Sciences, Chengdu 610041, China
University of Chinese Academy of Sciences, Beijing 100049, China

Qiannan Liu
Institute of Mountain Hazards and Environment, Chinese Academy of Sciences, Chengdu 610041, China
University of Chinese Academy of Sciences, Beijing 100049, China

Huan Yu
College of Earth Sciences, Chengdu University of Technology, Chengdu 610059, China

Guangbin Yang
School of Geographic and Environmental Sciences, Guizhou Normal University, Guiyang 550001, China

Wenlan Feng
College of Resources and Environment, Chengdu University of Information Technology, Chengdu 610225, China

Yong Ge
State Key Laboratory of Resources and Environmental Information System, Institute of Geographic Sciences & Natural Resources Research, Chinese Academy of Sciences, Beijing 100101, China

Jiangsu Center for Collaborative Innovation in Geographical Information Resource Development and Application, Nanjing 210023, China

Yu Jiang
State Key Laboratory of Resources and Environmental Information System, Institute of Geographic Sciences & Natural Resources Research, Chinese Academy of Sciences, Beijing 100101, China
University of Chinese Academy of Sciences, Beijing 100049, China

Yuehong Chen
State Key Laboratory of Resources and Environmental Information System, Institute of Geographic Sciences & Natural Resources Research, Chinese Academy of Sciences, Beijing 100101, China
University of Chinese Academy of Sciences, Beijing 100049, China

Alfred Stein
Faculty of Geo-Information Science and Earth Observation (ITC), University of Twente, Enschede 7500, The Netherlands

Dong Jiang
State Key Laboratory of Resources and Environmental Information System, Institute of Geographic Sciences & Natural Resources Research, Chinese Academy of Sciences, Beijing 100101, China

Yuanxin Jia
State Key Laboratory of Resources and Environmental Information System, Institute of Geographic Sciences & Natural Resources Research, Chinese Academy of Sciences, Beijing 100101, China
University of Chinese Academy of Sciences, Beijing 100049, China

Yong Ma
Institute of Remote Sensing and Digital Earth, Chinese Academy of Sciences, Beijing 100094, China
University of Chinese Academy of Sciences, Beijing 100049, China

Fu Chen
Institute of Remote Sensing and Digital Earth, Chinese Academy of Sciences, Beijing 100094, China

Jianbo Liu
Institute of Remote Sensing and Digital Earth, Chinese Academy of Sciences, Beijing 100094, China

Yang He
Institute of Remote Sensing and Digital Earth, Chinese Academy of Sciences, Beijing 100094, China
University of Chinese Academy of Sciences, Beijing 100049, China

Jianbo Duan
Institute of Remote Sensing and Digital Earth, Chinese Academy of Sciences, Beijing 100094, China

Xinpeng Li
Institute of Remote Sensing and Digital Earth, Chinese Academy of Sciences, Beijing 100094, China
University of Chinese Academy of Sciences, Beijing 100049, China

Nivedita Sairam
Department of Civil, Environmental and Geomatics Engineering, Florida Atlantic University;777 Glades Road, Boca Raton, FL 33431, USA

Sudhagar Nagarajan
Department of Civil, Environmental and Geomatics Engineering, Florida Atlantic University;777 Glades Road, Boca Raton, FL 33431, USA

Scott Ornitz
Keith and Schnars P.A, 6500 North Andrews Avenue, Fort Lauderdale, FL 33309, USA

Preface

This book illustrates the range of Principles and Practices of Geoenvironmental Mapping which presently carries out around the world. Specialists in several countries have contributed a number of sub disciplinary and thematic topics including landslides, soil salinity, land surface water, land cover and use, flood risk etc. At beginning, we discuss the landslide susceptibility mapping in vertical Distribution law of precipitation area of the hydropower station reservoir and a combination of geographically weighted regression, optimization and support vector machine for landslide susceptibility mapping. Then we present the evaluation of different soil salinity mapping using remote sensing techniques in arid ecosystems and soil clay content mapping using a time series of Landsat data in semi-arid lands. Mapping and modelling spatial variation in soil salinity using remote sensing indicators and regression techniques and operational mapping of soil moisture using synthetic aperture radar data are also discussed.

In the middle part of the book, we focus on land surface water mapping using multi-scale level sets and automated subpixel surface water mapping from heterogeneous urban environments using Landsat imagery. Then operational surface water detection and monitoring using Radarsat 2 and flood hazard mapping by using geographic information system and hydraulic model are also discussed. We also try to present urban flood mapping based on unmanned aerial vehicle remote sensing and random forest classifier and flood mapping and flood dynamics of the Mekong delta.

The last chapters of the book describe flood mapping based on multiple endmember spectral mixture analysis and random forest classifier and flood risk mapping using LiDAR for Nova Scotia. We also try to find out land cover mapping in Southwestern China using the HC-MMK approach and designing of an experiment to investigate subpixel mapping as an alternative method to obtain land use/land cover maps. At the end, we describe an automatic procedure for early disaster change mapping based on optical remote sensing and development of mobile mapping system for 3D road asset inventory.

Editor
Dr. Tanjina Nur, Ph.D.

INTRODUCTION

Geologic mapping is a highly interpretive, scientific process that can produce a range of map products for many different uses, including assessing ground-water quality and contamination risks; predicting earthquake, volcano, and landslide hazards; characterizing energy and mineral resources and their extraction costs; waste repository siting; land management and land-use planning; and general education. The value of geologic map information in public and private decision-making (such as for the siting of landfills and highways) has repeatedly been described anecdotally, and has been demonstrated in benefit-cost analyses to reduce uncertainty and, by extension, potential costs.

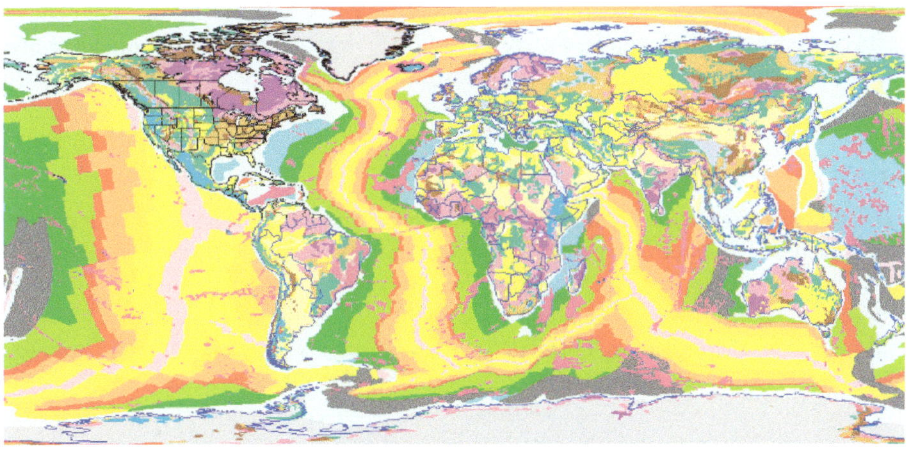

A map is a visual representation of an area - a symbolic depiction highlighting the relationships between elements of that space such as objects, regions and themes. A map showing the distribution of rock units and structures across a region, usually on a plane surface, is thus a geological map. Geological field mapping is the process of selecting an area of interest and identifying all the geological aspects of that area with the purpose of preparing a detailed geological report and a map to summarize the report. A geological map will thus show the various rock types of the region, the structures, geological formations, geothermal manifestations, age relationships, distribution of mineral ore deposits and fossils etc. and all these features may be super imposed over a topographic map or a base map. Basically, the quality of a geologic map will depend upon the accuracy and the precision of the field work. Further still quality depends on the completeness with which certain data, both geologic and geographic are presented on the maps; and on the care with which scale, colors, conventions, etc. With the development of technology however, geological maps today are more precise than ever as a combination of accurate satellite imagery, aerial photographs, high tech geological equipment and Geographic Information Systems (GIS) advancements are applied. The interpretation of a geological map depends on one's training, interest and the techniques used. It is also fundamental to be able to visualize scenarios that may have been involved during the formation processes of the features displayed as this forms an essential foundation for the analysis of geological maps. The ability to form a three- dimensional image from a two dimensional plan is in real sense the major part of geologic map interpretation.

Objectives of Geological Field Mapping

Principally there are various reasons why geological field mapping is carried out all of which entail collecting variable amounts of field data. These reasons can, however be summarized into three broad objectives which may sometimes be interconnected; natural resources, government mandate and academic objectives.

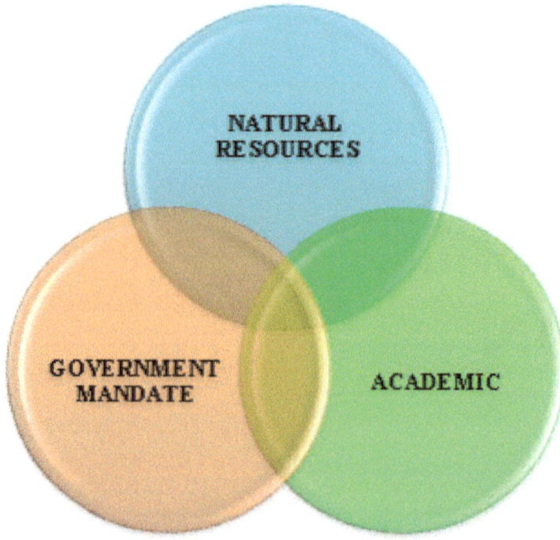

Natural Resources

These are naturally occurring materials that exist within the earth's crust and are extracted mostly for economic purposes. Exploitation of all natural resources requires the appreciation of basic geology and optimum utilization of a potential area requires that the resource is mapped out. Understanding the spatial distribution and deformation of rock units at the surface is critical in order to develop 3-dimensional models of the subsurface geology. Mineral and oil exploration proceeds in this way, particularly when considering the siting of boreholes. In geothermal fields, the geology of the area has to be thoroughly investigated and understood using a combination of geological, geophysical and geochemical methods before a decision is made on the productivity of that area. Geological mapping is usually the first task in any reconnaissance study as the question of whether or not a geothermal system exists can be answered using geological studies. Geophysical investigations answer the question of the extent of the system while geochemical investigations are used to estimate parameters such as the temperature of the system.

Government Mandate

In order to better understand an unknown area, geological mapping is carried out by learning and/or government institutions. The maps created are thereafter used as references for planning purposes in development projects, prevention of adverse destruction during natural disasters such as earthquakes (hazard preparedness) impact assessment studies, proper utilization of resources etc. Most government structures have a geology ministry/department that ensures that geological maps are available, that they are accurate and updated whenever changes arise. Ultimately all governments have an important role to collect and administer geological data, information and knowledge while protecting and preserving the area in question at the same time.

Academic

The third reason is to shed some light on the processes that have molded the earth to what it is today. Processes in the subsurface are not fully understood and the in mapping geological features on the surface we get some insight into the more complex systems. Geologists are forced to leave no stone unturned in acquiring the basic data for developing models of geologic history and process, and, in contrast with many other sciences, to use a healthy dose of imagination in interpreting what is almost always an incomplete data set. Many earth science institutes work tirelessly in creating models and proposing theories that may have explained various geological formations, structures and lithologies, where geological maps have been the foundation of these studies.

Phases of Geological Mapping

Field mapping projects are carried out in three phases which have a stepwise relationship.

Planning/Research

This phase of mapping is mostly carried out in the office although a short reconnaissance field trip may be included. Once a field has been identified for mapping, it is only rational to learn everything in its regard so as to make a workable field programme. A field programme is a step-by-step guide that outlines the time to be spent in the field and the objectives of each day thus ensuring successful and satisfactory results. Compilation of all available and relevant data is implemented, in an effort to avoid duplication of work and most importantly simplify the field study. The number of days spent during the research phase may vary based on the purpose of the study, the detail required and availability of funds. All possible geological reports and data, including aerial and satellite photographs, maps whether topographical, base maps, legislative boundary maps etc. must be obtained. The most important document however is the official approval by a government office and/or a local community. Without this one may be met with hostility and lack of cooperation by the locals and this may lead to an unsuccessful field study. A budget must also be prepared indicating the number of people to be involved in mapping, their daily allowance, fuel costs if vehicles are to be used, consumables and a miscellaneous kitty should be included. It is wise to involve the local community when employing casuals or language translators as they understand their environment better than anyone else and will thus be quick to pinpoint areas that maybe of interest. Furthermore, they are in the best suited to relate historical events where necessary. Issues that may hinder optimum productivity such as bad weather, wild animals and events of sudden sickness must be taken into consideration and planned for appropriately.

Observing/Mapping/Collecting

This phase is focal in any geological field mapping and is carried out in the field for the sole purpose of collecting data. This data may be collected in the form of photographs, measurements, notes and physical samples. Therefore, one must be fully equipped with all the necessary tools, be physically psyched and mentally prepared to make note of not only geological features but of the entire surrounding. For example, phreatophytes may be used to locate structures especially when they grow in a certain alignment. When good strategy is involved during the planning phase, unfortunate incidences in the field are addressed promptly with no major disruptions of the field program. In tropical countries such as Kenya, it is advisable to plan for field studies during the dry seasons. Thus it is in the best interest of everyone to commence mapping work very early in the morning so as to accomplish a substantial amount of work before the temperatures rise too high. In colder climates such as Iceland mapping is usually planned for the summer and may begin at anytime of the day and extend well into the night as daylight is prolonged to about twenty hours. Thus there should be no strict working hours in the field and this should be agreed upon among the field crew. Teamwork should be encouraged given that more observations are likely to be made and geological contentious issues discussed. A mapping project must be qualitative as well as quantitative. Accuracy whilst taking readings should be emphasized and no amount of information is too much therefore all possible data should be collected.

Reporting

A report is as good as its data and thus the need to collect good accurate data cannot be overemphasized. Ultimately when all possible available data has been collected it is taken back to the office or laboratory for sorting, interpretation and analysis. This phase is the most challenging of all three as wrong analysis or misinterpretation of data can lead to an inaccurate report and in consequence misinformation. The resulting map or model is in most incidences drawn by persons of another section (for example cartography) who may have had no involvement in the field whatsoever. Thus it is crucial that the data collector, in this case the geologists, work closely with those involved in actual drawing of the map, since they can select the data that is relevant for final presentation in the map and that which is not.

Tools and Equipment used in Geological Field Mapping

Maps: During the research phase all existing data and maps of the field in question are collected. All suitable maps available whether physical, political, relief, road, physical, and topographic should be carried to the field as it is possible that details in one may not be present in another. Most importantly for geological fieldwork, a geological map is expected to be handy especially as a reference. Depending on the areal extent of the field and the detail required, the scale of the map is an important aspect to consider.

Aerial photographs and stereoscope: This 3D imagery tool is very vital for all geological field work especially where large features such as volcanoes, calderas craters etc. are involved. In addition photos are also used to pinpoint thermally anomalous areas in geothermal fields when infrared thermography is applied. These photographs are studied with the aid of a stereoscope and are put to use before and during the field study; for planning in the case of the former and for confirmation purposes in case of the latter. Where aerial photographs may not exist, satellite imagery may be used although it may not possess the fine detail as seen on the aerial photographs.

Compass/Clinometer: A compass is an instrument used for determining direction and has recently been supplanted by modern devices such as the Global Positioning System (GPS). However when acquiring a clinometer it is advisable to purchase one that has a built in compass in it. This is for the reason that a GPS can at times malfunction or it may not locate satellites in areas of thick/dense forest cover. A clinometer is an instrument used for measuring inclination, tilt and elevation of rock outcrops. This is particularly important in areas that have been subjected to tectonic movement. These days, however technology advancement ensures quick and easy measurement of strike and dip of a surface (bedding plane, foliation, fault plane etc.) in a single action.

GPS(Geographic Positioning System): It is a satellite based navigation system comprising three basic parts; the satellites in space, monitoring stations on earth and the GPS receivers. This equipment is used in geological field mapping for finding ones position, mapping lithology, tracking structures, measuring elevation, storing sampling points and descriptions of formations when samples are collected. The GPS functions and capabilities are improving rapidly with advancement in technology and as such it is important to purchase one that is relatively modern.

Geological hammer: It is basic equipment for any geologist as it is the tool used for collecting samples. The best geological hammers are the ones with a piece chisel head made of hardened steel and a rubber coated shock reduction handle. For harder formations, a crack geological hammer forged from fine grain carbon steel with fiber glass shaft fitted with comfortable shock-absorbing rubber grip, is recommended.

Hand held lens: It is used to make the first analysis of rock samples in the field before further analysis is performed in the laboratories. The analysis needs to be detailed and descriptive giving all properties of the sample: rock type, color, texture, identifiable mineralogy, alteration as well as the physical properties such as folding, foliation, intrusions, layering etc.

Sample bags which best suit geological samples are canvas in fabric have a sewn in tie tape and a label tag on the outside to insert the sample number and location point. Plastic bags may be used where the sample is soft, disintegrated or wet.

Measuring tape is important for taking actual measurements of lithology and structures. For instance in lithology that exhibit layering, it is necessary to measure the thickness of each layer precisely, as this can be used as a correlation tool with other similar sequences that may be encountered.

Field notebook, masking tape, marker pens: All important observations must be written down in a concise, orderly and legible manner. The field note book should be hardcover for ease when writing and should easily fit into a pocket. Masking tape and marker pens are used for labelling samples before they are put into the sample bags.

Field camera is the only back up memory in addition to the notebook. It is very important to take photographs of all interesting features ensuring a scale is used in each instance; video clips can also be of use. One must also remember to take a GPS reading for each photograph taken. Photographs are important for descriptive purposes especially during report writing and for making presentations.

Safety clothing: These include sturdy shoes, clothes that are tough in fabric preferably jeans or khaki, hat and sunglasses to protect one from the sun a bag pack, plus a raincoat and wellington boots just in case of wet environments. Safety glasses and gloves are important especially when hammering rock samples. Safety clothing should aid one to be more effective by being comfortable and safe at the same time.

First aid kit: Safety in the field is always a priority; nevertheless accidents cannot be ruled out thus the need to be prepared with a first aid kit. Furthermore, it is crucial that at least one person in the field crew be trained in basic first aid techniques.

Parameters to Consider in Geological Mapping

Geology: When mapping rock types, it is important to have a carefully laid out plan for the entire location so that traverses can be made without leaving out some areas. A careful study of aerial photographs and topographical maps can aid one to avoid areas that are impractical to traverse on foot or impossible to climb because they are too steep. In areas where there is thick soil cover, riverbeds, valleys or exposures on hill tops are sort after for outcrops.

Collecting fresh samples is a process performed carefully, ensuring that the sample is not altered. The sample size depends on its intended use, be it for binocular, petrographic, X-ray analysis, whole rock analysis or just for reference. However, a sample should not be too big as to become a problem to carry while in the field. When acquiring a sample for whole rock analysis, it is advisable to break it using another rock from the same formation to ensure that it is not contaminated by metals from the geological hammer.

Strike and dip are a technique of relating the orientation of a plane in three dimensional spaces. The technique is usually applied to the orientation of tilted layers of rock and thus one can easily see the disarticulation a formation has undergone. As mentioned earlier a clinometer is used for these measurements which should be carried out more than once with utmost precision.

Geological landforms: Geological landforms in this paper refer to features such as mountains/hills, calderas craters, canyons etc. The magnitude of these formations (usually large) can be estimated from aerial photographs and maps prior to the actual field work. The dimensions of these formations must therefore be obtained by actual measurement. The recommended method is to traverse the formation on foot while using a GPS to track the movements. For example, to map the dimensions of a caldera it is in best practice to first measure the rim by walking around it ensuring the GPS is in tracking mode. Since the elevation will be known by walking up the caldera one should then walk into it for further measurements. All lithology should be mapped both in the interior and exterior of such formations. In some instances landforms such as calderas and craters may not be obvious on the surface due to erosion and weathering which may leave only a few exposures on the surface. If such outcrops are ignored it would be difficult to realize the connection if another exposure of the same kind is found in another location. Needless to say, a structure can easily be left out if work is carried out hurriedly.

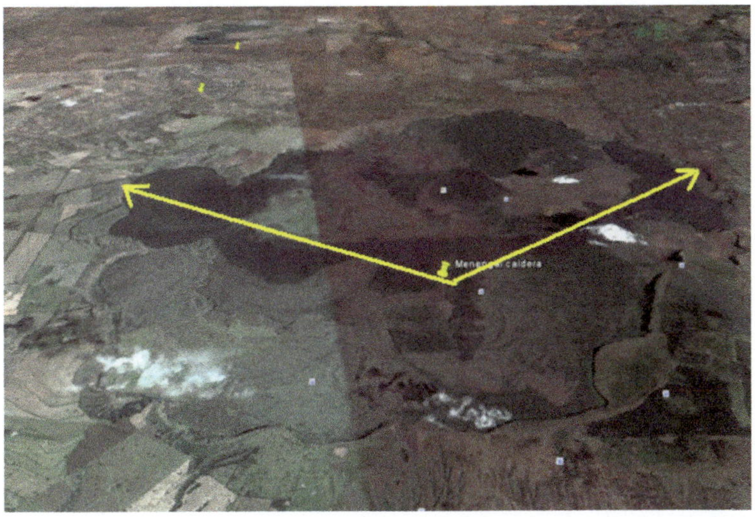

Geological structures (Faults and fractures): Faults and fractures always indicate movement within the earth's crust. They are distinguished from each other by whether displacement (lateral or vertical) has taken place; faults show displacement whilst fractures do not. Differences aside, the mapping of these structures needs to be done meticulously due to the fact that these structures may easily be confused with simple erosional surfaces. Erosional surfaces may sometimes be an indication of an underlying structure. The disparity is whether any observable disturbance can be seen in the bedrock. Structures like geological features may not always be obvious on the surface. In volcanic areas, they may manifest as volcanic centers or fumaroles oriented in a linear direction which become obvious once a map is compiled. Thus it is imperative to be accurate in measuring all formations and structures with well calibrated equipment so that if extrapolations are to be made for modelling purposes the error margin is very slim.

Geothermal manifestations: Geothermal manifestations, whether hot springs, fumaroles, altered or hot ground, geysers, steam jets or sulphur deposition, must be mapped and their extent measured. All observable characteristics should be fully described and photographs taken. Intrusive rocks must be analyzed in detail making note of their composition, quantity, orientation and if any sequences can be identified they should be numbered accordingly. All new mineralization should be sampled regardless whether the same formation has been sampled elsewhere.

Compiling Geological Maps

With the aid of GIS (Geographic Information Systems), maps of good quality and great accuracy are produced. Evidently it is important to consult with the GIS department before field work to calibrate the GPS in format/units that can easily be used with their software. Nonetheless, it is important for a geologist to draw a simple sketch map during field work for comparison purposes.

Interpreting Geological Maps

A map is basically a visual summary of an entire report and the two should complement each other. When the interpretation process is at hand, it is crucial to have group discussions with all those involved in the actual fieldwork. It is also deemed important to have discussions with those who are experienced in field of geology as well as other related scientific fields.

CHAPTER 1

LANDSLIDE SUSCEPTIBILITY MAPPING IN VERTICAL DISTRIBUTION LAW OF PRECIPITATION AREA

Chen Cao, Qing Wang, Jianping Chen, Yunkai Ruan, Lianjing Zheng, Shengyuan Song and Cencen Niu

Keywords: Geographic information system (GIS), rapid bedrock uplift, analytical hierarchy process, information content, spatial analysis

INTRODUCTION

The interaction between triggering mechanisms and natural conditions directly determines the occurrence and frequency of landslides [1,2,3,4,5]. To understand these natural hazards and predict potential landslide hazard areas, landslide susceptibility mapping (LSM) is considered to be an effective method to reduce the hazard impacts [6]. Many approaches can be used to predict the occurrence of slope failures, such as physically-based and statistical approaches [7,8,9,10]. The physically-based method is appropriate for analyzing the specific event. Combined with the field, GIS technology and the nonlinear method are utilized for LSM, which is more appropriate due to their flexibility.

Landslides are one of the most common natural hazards in the Three Gorges Reservoir [11]. The water level is an important factor to activate old landslides and trigger new landslides [12]. Thus, attention should be given to the reservoir area of the hydropower station, where landslides are more probable and frequently occur [11]. Landslide susceptibility mapping is necessary to mitigate and even avoid natural or secondary hazards. In fact, human field investigation in a large region, especially in mountainous areas, is very difficult. Hence, the prediction of hazards in a certain area based on limited data is difficult and demands a practical approach to achieve this goal. Some researchers have recently applied different classification methods to LSM prediction. Many statistical methods exist for LSM, such as the logistic regression (LR) method [13,14,15,16,17], statistical index (SI) [18,19], discriminant analysis (DA) [20] and bivariate statistical analysis (BSA) [21]. However, the statistical approaches have very strict mathematical reasoning, and their application must meet strict demands. In fact, the prediction samples usually cannot pass the hypothesis test in the assessment process. Some other approaches have also been developed, according to the methodologies of decision tree (DT) [22], genetic algorithm (GA) [23], artificial neural network (ANN) [15,24,25,26,27,28,29], and support vector machine (SVM) [30,31,32,33]. These objective statistical methods were used for evaluating the relationships between various influencing factors and landslide inventories. However, some methods have a few limitations in terms of LSM. Tehrany [34] stated that the DT method requires enhancement because some difficulties were encountered while defining the rules. During the process of ANN modeling, Tiwari and Chatterjee [35] stated that the length of the dataset could cause errors. Therefore, solely relying on objective methods has some limitations and can easily lead to misleading information. Hence, the introduction of a subjective method is necessary.

The pairwise comparison of the analytic hierarchy process (AHP) is based on expert opinions and thus introduces a degree of subjectivity for the criteria of significance [36]. However, it is sometimes criticized for its subjectivity. Along with traditional applications, a new trend uses AHP in conjunction with others methods [37]. An accurate objective method, the information content model (ICM), can be integrated with AHP to provide a framework for LSM in this study. The combined method makes use of the advantages of both the subjective method (AHP) and the objective method (ICM) to assess

Citation: Cao, C.; Wang, Q.; Chen, J.; Ruan, Y.; Zheng, L.; Song, S.; Niu, C. Landslide Susceptibility Mapping in Vertical Distribution Law of Precipitation Area. Water 2016, 8, 270. doi:10.3390/w8070270.

criteria and improve the accuracy of the results. The proposed method avoids the inherent disadvantages of using subjective or objective methods in isolation. Meanwhile, the geographical information system (GIS), which has quick access to the obtained data, global positioning technique, and remote sensing techniques, has been widely used and integrated with the aforementioned methods.

Approaches should be applicable for a special region. In this study, the Xulong hydropower station is located in a rapidly uplifting region in Southwestern China. Landslide susceptibility mapping should be considered from a new perspective based on the characteristics of this area. Based on GIS technology and field investigations, we obtained a database of 69 landslides along the Jinsha River at the Xulong hydropower station reservoir. The study area is located in the rapidly uplifting region of the Tibetan Plateau that is caused by the neotectonic movement. Furthermore, under the influence of the southwest and southeast monsoons, the study area is dry, with low values of rainfall, and the foehn effect is significant. Due to the foehn effect and topographical enclosure, the valley along the Jinsha River has a special dry and hot climate. According to the special climate and different influencing factors, this study selected nine factors, including vertical distributed annual precipitation, as the assessment factors, and established a landslide susceptibility assessment factor system. Meanwhile, this work chose the pixel unit as the evaluation unit to extract the information of each factor and discussed the intervals of each factor based on GIS technology. The analytic hierarchy process (AHP) and the information content model (ICM) were combined to establish a landslide susceptibility assessment model. Finally, the historic landslide data and evaluation accuracy analysis [38] were used for validation.

METHODOLOGY

The Analytic Hierarchy Process

The analytic hierarchy process (AHP) is a multi-criteria decision analysis method proposed by Saaty [39]. The weights of these criteria are defined after they are ranked according to their relative importance. Thus, once all the criteria are sorted in a hierarchical manner, a pairwise comparison matrix for each criterion is created to enable a significance comparison. The relative significances of the criteria are evaluated on a scale of 1–9, indicating less importance to greater importance. Weighting by AHP is widely used in many applications [40,41] and it is recommended to be used for regional studies. The steps of the AHP for weighting are as follows:

1. The first step is to build the hierarchical structure of the target problem.
2. Saaty proposed a scaling method to score the parameters in each layer. By comparing the importance of the parameters in each level, a_{ij} is used to present the ratio of x_i and x_j, which builds the judgment matrix $A = (a_{ij})$:

$$A = (a_{ij})_{n\times n} = \begin{bmatrix} a_{11} & a_{12} & \cdots & a_{1n} \\ a_{21} & a_{22} & \cdots & a_{2n} \\ \cdots & \cdots & \cdots & \cdots \\ a_{n1} & a_{n2} & \cdots & a_{nn} \end{bmatrix} \quad (1)$$

3. The judge matrix needs to meet the following equation:

$$A\omega = \lambda_{max}\omega \quad (2)$$

where ω is maximum characteristic vector corresponding feature vector of judge matrix A. The weight value can be obtained after normalizing the feature vector.

4. The consistency check is needed to analyze whether the weights distribution is reasonable or not. The consistency indicator (CI) can be obtained by the following equation:

$$CI = \frac{\lambda_{max} - n}{n - 1} \quad (3)$$

The random consistence ratio (CR) can be obtained by the following equation:

$$CR = CI/RI \quad (4)$$

where n is the order of RI, RI is a random indicator (RI). A different order of the corresponding RI values can be obtained listed in Table 1.

Table 1: Consistency random factor (RI)

m	3	4	5	6	7	8	9	10	11	12
RI	0.58	0.90	1.12	1.24	1.32	1.41	1.45	1.49	1.52	1.54

If CR < 0.1, the judgment matrix has met the consistency test standards. This means that the weight of the factors is reasonable. It is necessary to adjust the matrix until the CR meets the requirement.

The Information Content Model

The information content model (ICM) is evaluated from the theory of communication proposed by C.E Shannon. He firstly proposed the concept of information and the calculation equation of information entropy. Recently, more and more researchers applied the ICM on geological hazard assessment and environmental quality evaluation [42,43,44]. This method overlays the information content provided by landslide influencing factors and sets the total information content as the quantitative expression factor of landslide susceptibility assessment. The calculation equation is as follows:

$$I(y, x_1x_2...x_n) = \ln\frac{P(y, x_1x_2...x_n)}{P(y)} \quad (5)$$

Where I (y, $x_1x_2 \ldots x_n$) is the information content provided by factor combinations x_1 $x_2 \ldots$ x_n for landslide hazard, P (y, $x_1x_2 \ldots x_n$) is the probability of landslide under the condition of factors combinations x_1 $x_2 \ldots$ x_n,P(y) is the probability of landslide.

The procedure of the GIS-based information content model is as follows:

1. Calculate the information content I (x_i, H) of each influencing factor x_i:

$$I(x_i, H) = \ln\frac{P(x_i/H)}{P(x_i)} \tag{6}$$

Where P (x_i/H)is the probability of x_i under the condition of landslide, P (x_i) is the occurrence probability of x_i in the study area. This equation is a theoretical model; frequency is used for probability calculation in the actual application:

$$I(x_i, H) = \ln\frac{N_i/N}{S_i/S} \tag{7}$$

where S is the total amount of pixels in the study area, N is total landslide pixels amount in the study area, S_i is the pixels amount of factor x_i in the study area, N_i is landslide pixels amount among the factor x_i.

2. Calculate the amount information content of each evaluation factor:

$$I_i = \sum_{i=1}^{n} I(x_i, H) = \sum_{i=1}^{n} \ln\frac{N_i/N}{S_i/S} \tag{8}$$

Where I_i is the amount information content of each evaluation factor, n is total number of factors.

3. The total information content I_i is used as the comprehensive factor. The higher the value is, the greater the landslide susceptibility.

The Landslide Susceptibility Assessment

Based on the information content model, this study considered the weight of each influencing factor. Combined with the AHP weighting method, this study overlays the related factor information content to calculate the information weight values and the total information weighs of each evaluation unit:

$$I(x_i, H)_w = W_i \times I(x_i, H) = W_i \times \ln\frac{N_i/N}{S_i/S} \tag{9}$$

$$I_{iw} = \sum_{i=1}^{n} I(x_i, H) = \sum_{i=1}^{n} W_i \times \ln\frac{N_i/N}{S_i/S} \tag{10}$$

The information weight value of each evaluation factor and each evaluation unit are calculated by the following equations: where I_{iw} is the comprehensive information weight value of evaluation unit for the evaluation factors except for precipitation, $_{Wi}$ are the weights of the evaluation factors calculated by the AHP method, S is the total amount of pixels in the study area, N is total amount of the number of landslide pixels amount in the study area, S_i is the amount of pixels of factor xi in the study area, and N_i is the number of landslide pixels among the factor xi pixels.

STUDY AREA AND DATA

Study Area

The study area is located in the mountains separating the Sichuan and Yunnan provinces of China, along the upper reaches of the Jinsha River (Figure 1). The Jinsha River flows from south to north and is controlled by regional tectonics. In this area, the topography is characterized as a steep and deep valley within the area of the high and very high mountains. The Xulong hydropower station reservoir is located in the middle-high tectonic erosion mountains, with erosion accumulation landform types. The elevation difference is mainly greater than 1000 m, and the maximum elevation difference is 2800 m.

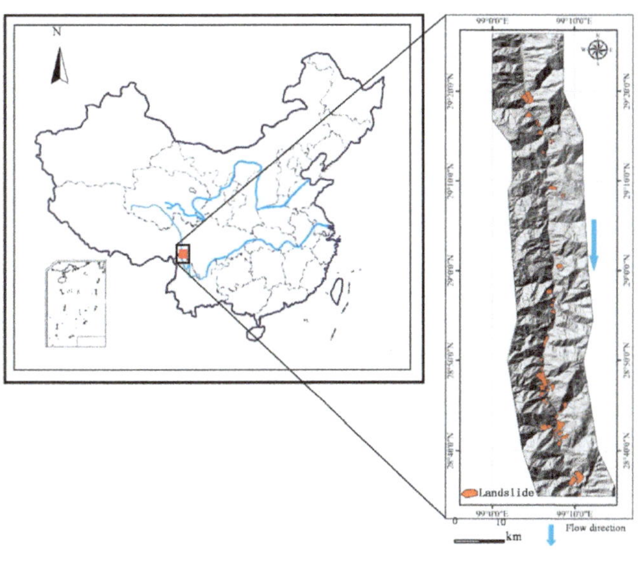

Figure 1: Geographical position of the reservoir area of Xulong hydropower station.

The reservoir area is located along the southeastern margin of the Tibetan Plateau. Since the Quaternary, neotectonic movement has made the study area a rapidly uplifting region [45]. Due to the landform characteristics and intensity differences of the neotectonic movement, landslides have widely occurred [46]. The deformation rate of the Jinsha River belt is 5 mm a year [47]. The regional neotectonic movement zoning is shown in Figure 2. The structural movement of the strong uplift caused a rapid river incision, yielding secondary disasters, such as landslides and collapses. Meanwhile, this area is influenced by the southwest and southeast monsoons, which contribute to the foehn effect in this area. Due to the foehn effect and the topographical enclosure, the characteristics are complex. The climate is very dry and has sufficient sunlight. Additionally, the diurnal temperature variation is quite big. The annual temperature is 13.8–19.2°C. The mean annual precipitation ranges from 354.2 mm to 648 mm in the low and middle elevation area. However, because of the unique landforms and high mean elevation, this area follows a significant vertical distribution law of precipitation. The annual precipitation at the high elevation may reach more than 1000 mm. The Jinsha River and its tributaries mainly form a V shape. The flow velocity and river discharge are both high, which has a great impact on the landslides along the river.

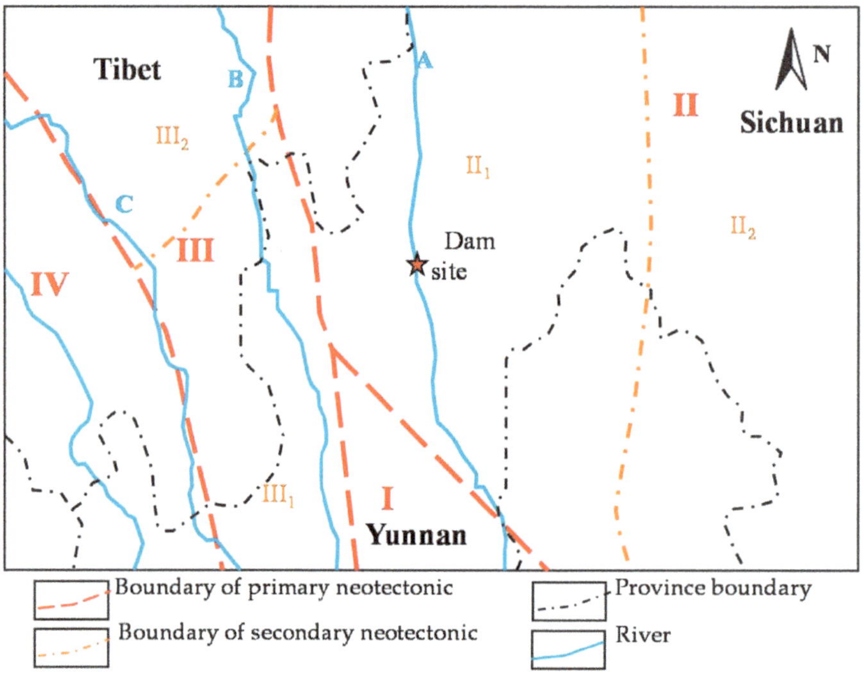

Figure 2: Regional neotectonic movement zoning map: I: Zhongdian-Yulong-jokul rapid uplift zone; II: The western of Sichuan rapid uplift zone; II_1: Gajinjokul strong rapid uplift zone; II_2: Daocheng-Gongga mountain tilted uplift zone; III: Qiangtang-Changdu rapid uplift zone; III_1: Bijiang-Baoshan tilted uplift zone; III_2: Changdu fault block uplift zone; IV: Nianqingtanggula-Gaoligong mountain fault block uplift zone; A: The Jinsha River; B: The Lancang River, C: The Nu River.

The exposed strata in the reservoir are from the Middle Proterozoic, Paleozoic, Mesozoic and Cenozoic. Magmatite and metamorphic rocks also exist. The Middle Proterozoic strata include the Xiongsong formation (Pt_2X). The lithology of Pt_2X is mainly composed of phyllite, marble, limestone, schist, and slate, which are materials that are prone to landslides. The stratigraphic groups in the Paleozoic include the Gerong formation (D_1g), Qiongcuo formation (D_1q), and Ophiolite formation (DTJ). The lithology of the Jinsha River Ophiolite formation is mainly composed of basic-ultrabasic rocks, spilite keratophyre, and radiolarian cherts. The stratigraphic groups in the Mesozoic are mainly Triassic (T) and include the Bulun formation (T_1b), the Qugasi formation (T_3q), the Waigucun formation (T_3w), the Jiabila formation (T_3j), the Sanhedong formation (T_3sh), the Waluba formation (T_3wl), and the Maichujing formation (T_3m). The lithology of the Triassic formations is mainly composed of mafic volcanic rocks, carbonate, limestone, basalt, and esite, conglomerate, and sandstone. The Cenozoic strata include the Relu formation (E_2r) and Quaternary sediment. The Relu formation mainly includes gravelly sandstone and gritstone. The Quaternary Period sediment includes alluvial and lateral moraine material, in addition to lacustrine accumulation, chemical accumulation, and residual accumulation. There are many deep and major faults in the study area, which are affected by tectonic movement. Most of the faults formed as active faults on the basis of the old faults.

Extensive field investigations and observations were identified and mapped in the Xulong hydropower reservoir, which was used to produce a detailed and reliable landslide inventory map. A total of 69 landslides were identified and mapped in the study area by evaluating aerial photos supported by field investigation (Figure 1). A series of field investigations were undertaken to identify the relationship between the occurrence of landslides and the environmental factors. The landslide types of the 69 landslides were various, mainly including rock slope deformation, rock planar slide, and rock flexural topple [48]. Figure 3 gives some examples of the landslides.

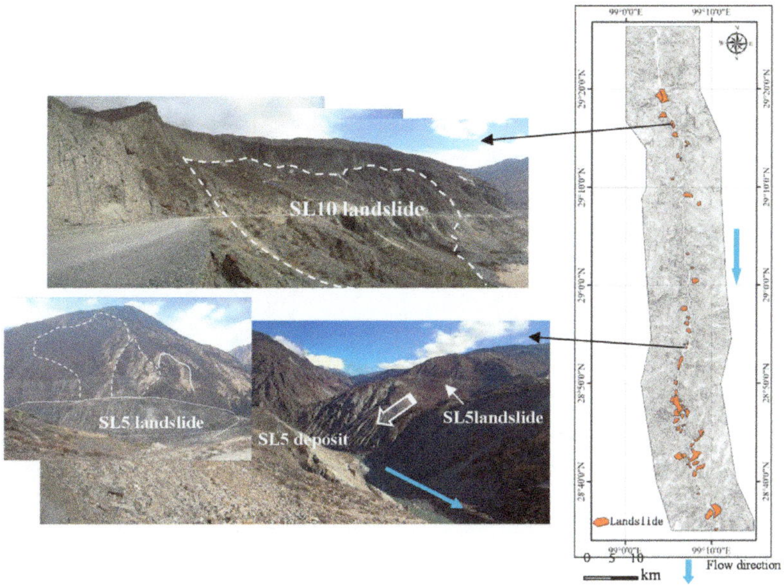

Figure 3: Landslide inventory in the study area

Influencing Factors

Various causative factors data, such as the slope angle, slope aspect, curvature, geology, distance-to-fault, distance-to-river, vegetation, bedrock uplift, and annual precipitation, were selected for the LSM of the Xulong hydropower station. These variables were selected because they have been successfully used in previous studies [31,49,50,51,52,53,54,55]. Note that there are few people living in the study area. Hence, the traffic is not significant and people's activities have little influence on landslide occurrences. Precipitation follows a vertical distribution law due to the foehn effect; the higher the elevation is, the higher the precipitation is. Precipitation-induced landslides have been the main topic in recent years [56,57,58]. Evidence has shown that precipitation follows a vertical distribution law. Higher precipitation zones should, thus, have more landslides. Precipitation and landslides show a very close relationship over large spatial scales, suggesting that precipitation drives landslide variations [59]. Considering the distribution of the investigated landslides, solely using the ICM cannot reflect this discipline. Precipitation is the main predisposing factor of landslides, which should be given more attention. Thus, the annual precipitation should not be treated as the other predisposing factors; it was necessary to analyze the annual precipitation factor separately for landslide susceptibility assessment.

In fact, the selection of regional LSM influencing factors should be applicable and practical for the Xulong hydropower station. Meanwhile, the data obtained by GIS should be reliable. Elevation is controlled by some geologic and geomorphologic processes [60]. The geological map has a scale of 1:10,000, and the digital elevation model (DEM) has a resolution of 5 m × 5 m, covering an area of 1413 km^2 (Figure 4). The topography of the reservoir area is highland with elevation ranging from 2100 m to 5160 m, with 20 m interval contours in the geological map. Slope angle is an important factor influencing the slope stability. It was extracted from the DEM through the GIS software at a resolution of 5 m × 5 m. The slope angle varies between flat and 73.8°. The slope map was reclassified into eight classes: (1) 0–10°, (2) 10°–20°, (3) 20°–30°, (4) 30°–40°, (5) 40°–50°, (6) 50°–60°, (7) 60°–70°, and (8) >70° (Figure 5a). The slope aspect map was also extracted from the DEM through the GIS software at a resolution of 5 m × 5 m. Some microclimatic parameters, such as exposure to sunlight and winds, precipitation intensity, and soil moisture are controlled by slope aspect [29]. The slope aspect was divided into nine types, including flat, north (337.5°–22.5°), northeast (22.5°–67.5°), east (67.5°–112.5°), southeast (112.5°–157.5°), south (157.5°–202.5°), southwest (202.5°–247.5°), west (247.5°–292.5°), and northwest (292.5°–337.5°) (Figure 5b). Curvature describes the slope shape. The slope shape affects the landslide development, which provides space for slope sliding. Based on the slope shape, the slope can be divided into three types, including concave, convex, and flat (Figure 5c). The influence of curvature on the slope erosion processes is the convergence or divergence of water during downhill flow [61].

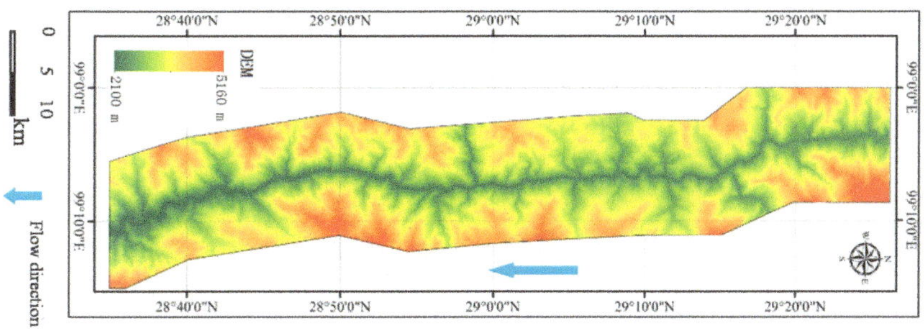

Figure 4: The digital elevation model of reservoir in Xulong hydropower station.

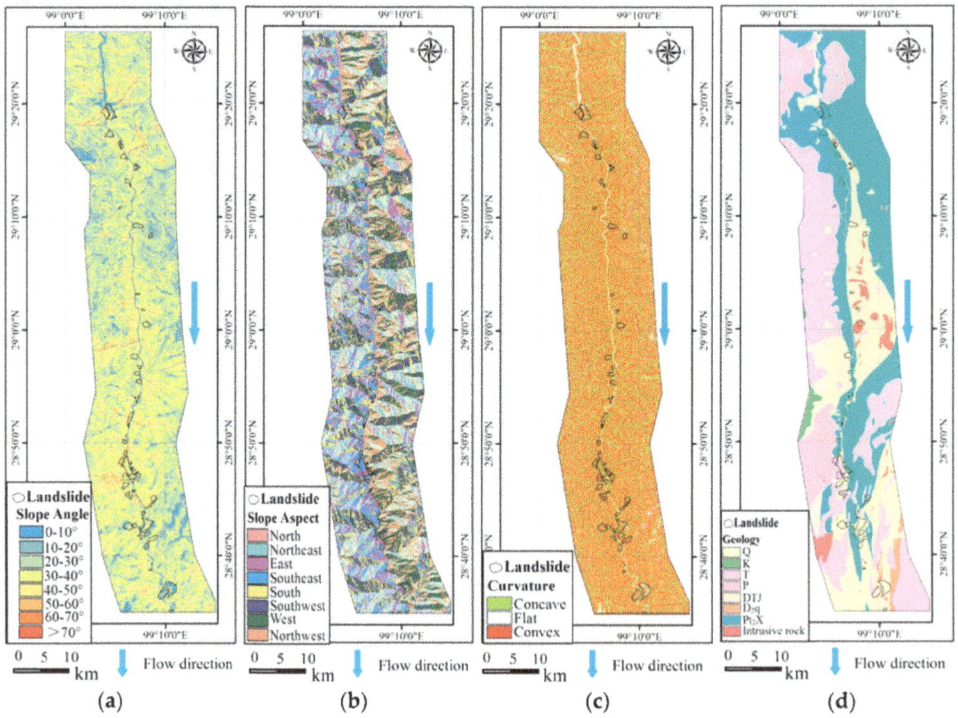

Figure 5: Influencing factors maps of the study area: (**a**) slope angle; (**b**) slope aspect; (**c**) curvature; and (**d**) geology.

The study area is covered with various types of geologic formations. According to the geological map of scale 1:50,000 and field investigation, this study mainly groups the geology into eight types. The general geological setting of the area is shown in Figure 5d. Field surveys show that faults have a great influence on landslide occurrence. In the reservoir of the Xulong hydropower station, deep and long faults developed substantially. The distance-to-fault is an extremely important evaluation, factor and the distance-to-fault map was calculated in 200 m intervals (Figure 6a). The landslides in the reservoir are mostly distributed within 1000 m of faults. The slopes on the banks of the river often suffer river erosion. In general, at a closer distance to the river, the erosion is stronger and the probability of the occurrence of landslides is higher. The distance-to-river map was calculated in 200 m intervals (Figure 6b).The dominant vegetation species are grasses and shrubs. The vegetation distribution is obtained by SPOT5 images. The vegetation distribution is shown in Figure 6c, and the vegetation is divided into vegetation cover and no cover.

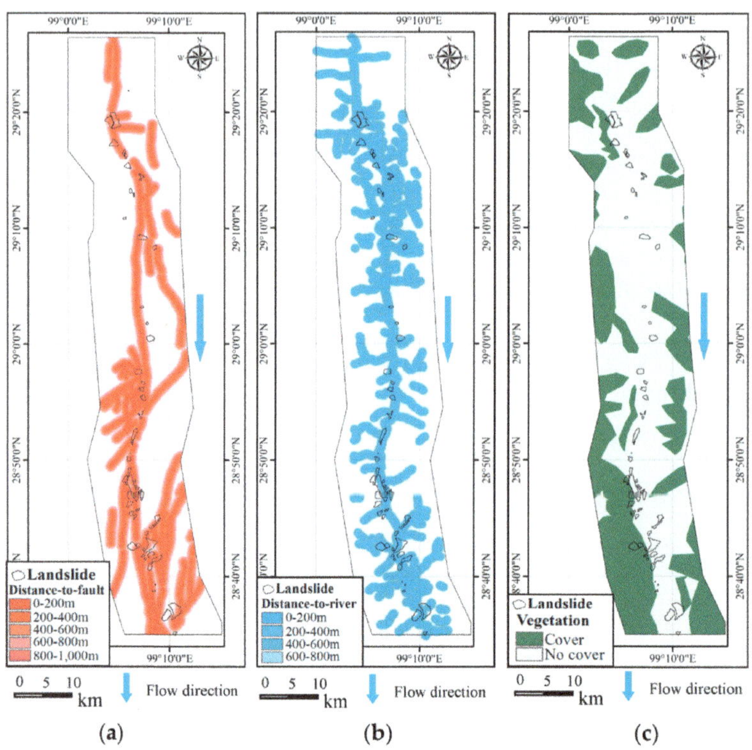

Figure 6: Influencing indicators maps of the study area: (**a**) distance-to-fault; (**b**) distance-to-river; and (**c**) vegetation.

The reservoir belongs to the Daocheng-Gongga Mountain tilted uplift zone (Figure 2), which has been experiencing rapid bedrock uplift since the Quaternary Period. According to the observation data collected from 1970 to 2012, the rate of bedrock uplift could have reached 5.8 ± 1.0 mm a year in this study area [62]. The increase of landslides is proportional to the increase rates of bedrock uplift [63,64]. It is apparent that landslides mainly occur along the banks of the Jinsha River (Figure 1), which means that the interaction of bedrock uplift and river incision contributes to the landslide occurrence. In the active landslide period, the landslides were impacted by the interaction between rapid tectonic uplift and the Jinsha River incision. In this period, the angle of slope increased along the river, as well as the slope potential energy [63]. Burbank [46] considered that the equilibrium was maintained between bedrock uplift and river incision, with landslides allowing hillslopes to adjust efficiently to the rapid river incision. The evidence reflected that a relationship exists between the landslide occurrence and rapid bedrock uplift. The average slope angle among all areas suggested that a common threshold controlled the occurrence of landslides [46]. Additionally, it has often been argued that high rates of bedrock uplift and denudation should be correlated with steep slope angles [65]. This study used the slope angle to represent the influence of the bedrock uplift.

There is a vertical distribution law of precipitation in this area. According to previous studies [66,67], precipitation is proportional to the elevation. The coefficients of tendency increase with increasing elevation, and it is particularly evident in areas above an elevation of 2500 m. The higher elevation enhances the increasing trend of precipitation erosion. The precipitation erosion force change trend coefficient increases by 0.05 for every 500 m increase in elevation. The elevation of the reservoir ranges from 2100 m to 5160 m. In this study, the elevation is divided into six zones based on Liu [67]: (1) 2100 m–2600 m, (2) 2600 m–3100 m, (3) 3100 m–3600 m, (4) 3600 m–4100 m, (5) 4100 m–4600 m and (6) 4600 m–5160 m.

Most of the precipitation stations are distributed from Benzilan to Batang county, rather than along the Jinsha River. Four precipitation stations are distributed along the Jinsha River, as listed in Table 2 [68,69,70]. This study established a linear relationship between the annual precipitation and the elevation (Figure 7). In Figure 7, the blue points represent the data from the four precipitation stations in Table 2. The linear equation is shown as follows:

$$y = 0.244\,e - 157.7 \quad (2100\text{m} < e < 5160\text{m}) \tag{11}$$

where y (mm) is the annual precipitation and e (m) is the elevation. The precipitation gradient is 24.4 mm/100 m, in terms of the spatial distribution. The annual precipitation was divided into six zones based on Equation (11):(1) the annual precipitation range in the elevation range of 2100 m to 2600 m is 355 mm to 477 mm; (2) the annual precipitation range in the elevation range of 2600 m to 3100 m is 477 mm to 599 mm; (3) the annual precipitation range in the elevation range of 3100 m to 3600 m is 599 mm to 720 mm; (4) the annual precipitation range in the elevation range of 3600 m to 4100 m is 720 mm to 843 mm; (5) the annual precipitation range in the elevation range of 4100 m to 4600 m is 843 mm to 965 mm; and (6) the annual precipitation range in the elevation range of 4600 m to 5160 m is 965 mm to 1101 mm. The annual precipitation distribution map is shown in Figure 8. Moreover, the landform of the Xulong reservoir is similar to Gongwang Mountain [71], which is located in the north of Yunnan Province along the Jinsha River. The precipitation of Gongwang Mountain is very low, and the precipitation gradient is 24 mm/100 m, which is close to that observed at the study site, that is, 24.4 mm/100 m. Therefore, 24.4 mm/100 m is a reasonable precipitation gradient of the Xulong hydropower station reservoir.

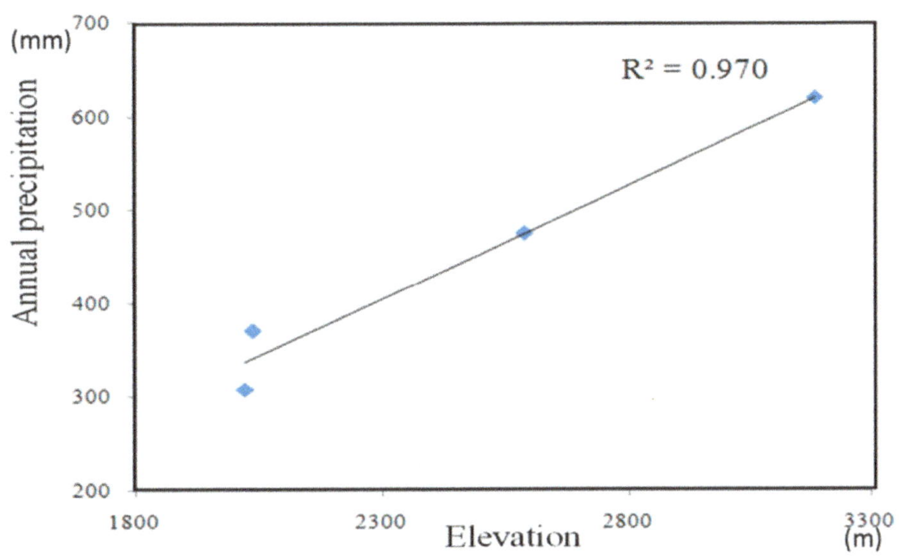

Figure 7: The relationship between elevation and annual precipitation based on the four meteorological stations.

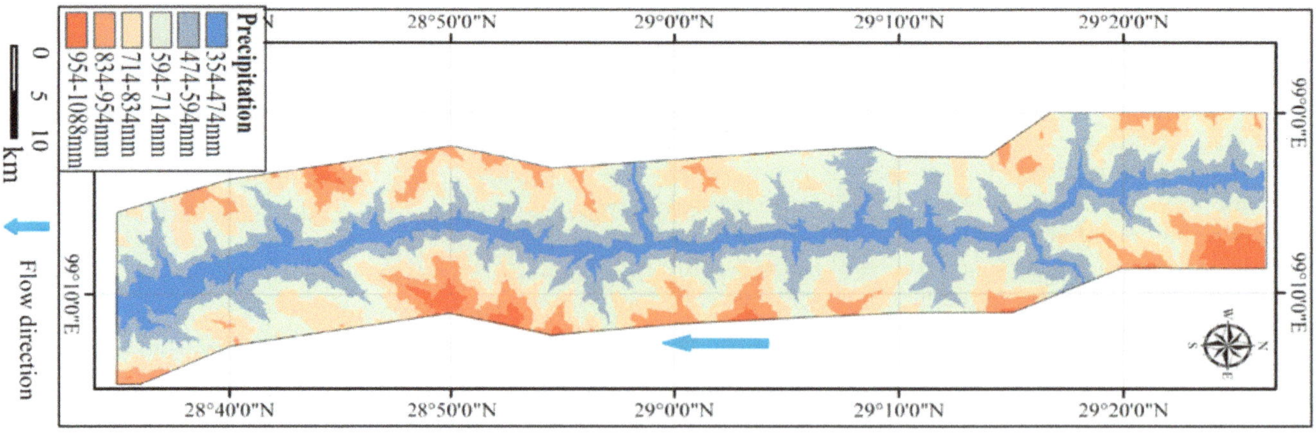

Figure 8: The annual precipitation classification of the study area.

Table 2: Annual precipitation measured by precipitation stations along the upstream of the Jinsha River at different elevation

Precipitation Station	Longitude	Latitude	Elevation/m	Annual Precipitation/mm	Data Resources/year
Benzilan	99°17′	28°17′	2023	308	1965–1988
Shangqiaotou	99°24′	28°10′	2040	369.68	1961–2004
Batang	99°06′	30°00′	2590	474.4	1960–2012
Dege	98°35′	31°48′	3184	619.81	1960–2012

Data Processing

Note that the AHP–ICM method for the landslide susceptibility mapping is only used for the eight factors: slope angle, slope aspect, curvature, geology, distance-to-fault, distance-to-river, vegetation and bedrock uplift (represented by slope angle). Since the eight-factor comprehensive information content (CIC) and annual precipitation have different ranges, normalizing the eight-factor CIC and the annual precipitation is necessary. This study normalizes the data to (0, 1) on the basis of the following equation:

$$A = (CIC - CIC_{min})/(CIC_{max} - CIC_{min}) \qquad (12)$$

where A is the landslide susceptibility value of the eight factors, CIC is the different unit information content, andCIC_{max} and CIC_{min} are the maximum and minimum comprehensive information contents, respectively.

In this study, the elevation is divided into six zones based on Liu [67]: (1) 2100 m–2600 m, (2) 2600 m–3100 m, (3) 3100 m–3600 m, (4) 3600 m–4100 m, (5) 4100 m–4600 m, and (6) 4600 m–5160 m. The six divisions of annual precipitation are: (1) 355 mm–477 mm, (2) 477 mm–599 mm, (3) 599 mm–720 mm and (4) 720 mm–843 mm, (5) 843 mm–965 mm, and (6) 965 mm–1101 mm. The assignment value is based on the following equation:

$$V_i = \frac{(r_i + r_{i+1})(e_{i+1} - e_i)}{(r_{max} + r_{min})(e_{max} - e_{min})} \; (i = 1,2,\ldots,6) \qquad (13)$$

where V_i is the assignment value of the annual precipitation pixel; r_i is the annual precipitation interpolation; e_i is the corresponding elevation; r_{max} and r_{min} are maximum and minimum annual precipitation in the study area, respectively; and e_{max} and e_{min} are maximum and minimum elevation in the study area, respectively. As discussed in Section 3.1, the annual precipitation has a linear relationship with the elevation.

The normalization of annual precipitation is calculated by the following equation:

$$N = (V_i - V_{min})/(V_{max} - V_{min})(i = 1,2,\ldots,6) \qquad (14)$$

where N is the normalization of annual precipitation, V_i is the assignment value of the annual precipitation pixel, and V_{max} and V_{min} are the maximum and minimum assignment values of the annual precipitation, respectively. Eventually, the landslide susceptibility value is calculated by the following equation:

$$S = (1 - W_r) \times A + W_r \times N \qquad (15)$$

where S is the eventual landslide susceptibility value, A is the eight-factor assessment value of landslide susceptibility, W_r is the weight of the annual precipitation calculated by AHP, and N_i is the annual precipitation normalization value of the six zones, respectively.

RESULTS AND DISCUSSION

Many of the factors influencing landslide occurrence could be collected to systematically assess the other areas that might have slope failures. For some areas, this analysis is essential so that people can consult the landslide susceptibility maps to avoid areas of higher landslide risk. In this study, landslide susceptibility maps had been constructed using the analytic hierarchy process and the information content method for the Xulong hydropower station reservoir. A landslide susceptibility map is provided for this area. In the future construction of this hydropower station, this map may help avoid losses in human lives and property.

Determination of Analytic Hierarchy Process (AHP)Weights

Based on the landslide susceptibility assessment factor system, the hierarchical structure was established. According to previous studies about the relationships between influencing factors and landslides, the judgment matrix of landslide susceptibility assessment is shown in Table 3. Note that this study took nine influencing factors into consideration for landslide susceptibility. However, the influence of the rapid bedrock uplift factor was represented by slope angle (Figure 5a). This study used the slope angle, slope aspect, curvature, geology, distance-to-fault, distance-to-river, vegetation, and annual precipitation for weighting with the AHP method. The feature vector (the weight) was calculated (Table 3). Based on Equations (3) and (4), the CR was calculated; CR = 0.0403, which was less than 0.1. This means that the judgment matrix met the consistency check and weight allocation was reasonable.

Table 3: The analytic hierarchy process judgment matrix and influencing factor weights

Heading	X1	X2	X3	X4	X5	X6	X7	X8	Weights
X1	1	1	2	4	4	6	7	8	0.2803
X2	1	1	2	3	3	5	6	7	0.2452
X3	1/2	1/2	1	3	3	4	5	6	0.1800
X4	1/4	1/3	1/3	1	1	3	4	5	0.0948
X5	1/4	1/3	1/3	1	1	3	4	5	0.0948
X6	1/6	1/5	1/4	1/3	1/2	1	2	3	0.0482
X7	1/7	1/6	1/5	1/4	1/3	1/2	1	2	0.0329
X8	1/8	1/7	1/6	1/5	1/4	1/3	1/2	1	0.0237

Notes: X1: precipitation; X2: lithology; X3: slope angle; X4: distance-to-river; X5: distance-to-fault; X6: vegetation; X7: slope curvature; X8: slope aspect.

The Information Content (IC) of Eight Factors

Based on Figures 5 and 6, the eight factors were calculated and listed in Table 4. According to Equations (7) and (8), the information contents of units of influencing factors were calculated. If $I_i < 0$, the landslide occurrence probability of the influencing factor is lower than that of the average probability. When $I_i=0$, the landslide occurrence probability of the influencing factor is equal to that of the average probability. If $I_i > 0$, the landslide occurrence probability of the influencing factor is higher than that of the average probability.

Table 4 shows that for most landslides (72.81%), the slope angle is distributed between 20° and 50°. However, the IC values of 0° to 10°, 10° to 20°, 50° to 60° and 60° to 70° were 0.8997, 0.3498, 0.4180, and 0.2850, respectively. This indicates that the slopes with slope angles of 0° to 20° and 50° to 70° were prone to failure in this study area. The substitution of steeper slopes for gentler slopes is the reason for landslide occurrence, which is a process driven by the rapid bedrock uplift and river incision. This was the most advantageous slope angle condition of landslides. For the slope aspect factor, the highest probabilities of landslide occurrence were mainly in the east, southeast, and south directions. The IC values were 0.5566, 0.3843, and 0.3583, respectively. The north, southwest, and northwest did not develop. Because the slope body of the south is easily loosened and broken under water and heat, the critical slope angle was lower than that of the northern slopes. The curvature of landslides was mainly concave (43.9%) and convex (43.19%), but the IC values were not high. For the landslides in the reservoir, most of the geological formations were composed of Quaternary Period (Q) loess, mud, and gravel, Xiongsong Formation (Pt_2) schist, phyllite and marble, and the Devonian to Triassic ophiolite suite. The IC values were 1.3652, 0.3278, and 0.3023, respectively. The landslides along the Jinsha River were widely developed. The Xiongsong Formation (Pt_2) geologic suite had a layered fracture structure. Their rock strength was not high, which provided a precondition for the occurrence of landslides. Meanwhile, the ophiolite suite had a complicated geologic combination, and its anti-weathering ability was poor. The landslides were mainly distributed within 200 m–600 m of the faults and 0 m–400 m of the river. It could be observed that the farther distance from the river, the smaller the probability of landslides occurrence. The plants on the ground could improve the shear strength of rock and soil mass. The IC values of vegetation cover and no cover were −0.7883 and 0.2879, respectively. This suggests that the ground with no vegetation was more prone to landslides. The eight-factor LSM result is displayed in Figure 9a.

Table 4: Distribution of the training pixels

Factor	Class	Landslide Not Occurred		Landslide Occurred		Total Count	Information Content
		Count	Ratio/%	Count	Ratio/%		
Slope Angle (Rapid Bedrock Uplift)/°	0–10	1,858,723	3.36	110,024	8.56	1,968,747	0.8997
	10–20	5,788,727	10.48	192,876	15.01	5,981,603	0.3498
	20–30	14,993,603	27.14	326,073	25.38	15,319,676	–0.0656
	30–40	22,210,924	40.21	389,651	30.33	22,600,575	–0.2763
	40–50	9,069,973	16.42	219,551	17.09	9289,524	0.0391
	50–60	1,195,655	2.16	42,750	3.33	1,238,405	0.4180
	60–70	118,723	0.21	3700	0.29	122,423	0.2850
	>70	1572	0	0	0	1572	0
Slope Aspect	Flat	499,251	0.01	12,648	0.01	511,899	0.1861
	North	6,890,597	0.12	90,223	0.07	6,980,820	−0.5644
	Northeast	5,660,124	0.10	138,129	0.11	5,798,253	0.0470
	East	6,459,419	0.12	266,725	0.21	6,726,144	0.5566
	Southeast	6,699,468	0.12	231,321	0.18	6,930,789	0.3843
	South	6,775,601	0.12	227,748	0.18	7,003,349	0.3583
	Southwest	6,890,429	0.12	14,262	0.01	6,904,691	−2.3973
	West	7,870,201	0.14	168,562	0.13	8,038,763	−0.0806
	Northwest	7,492,796	0.14	135,002	0.11	7,627,798	−0.2501
Curvature	Concave	23,966,881	43.39	563,969	43.90	24,530,850	0.0115
	Flat	5,947,546	10.77	165,781	12.90	6,113,327	0.1766
	Convex	25,323,475	45.84	554,878	43.19	25,878,353	−0.0582
Geology	Q	2,451,097	4.44	239,012	18.61	2,690,109	1.3652
	K	4,718,96	0.85	0	0	471,896	0.0000
	T	16,883,871	30.57	160,526	12.5	17,044,397	−0.8791
	P	7,619,949	13.79	19,629	1.53	7,639,578	−2.1783
	DTJ	12,172,394	22.04	395,719	30.8	12,568,113	0.3278
	D2q	491,512	0.89	0	0	491,512	0.0000
	Pt2X	13,753,469	24.90	435,508	33.9	14,188,977	0.3023
	Intrusive rock	1,393,883	2.52	34,142	2.66	1,428,025	0.0527
Distance-to-Fault/m	0-200	5,488,207	24.54	170,371	16.07	5,658,578	−0.4073
	200–400	5,129,121	22.93	298,303	28.14	5,427,424	0.1945
	400–600	4,465,698	19.97	300,739	28.37	4,766,437	0.3326
	600–800	3,871,123	17.31	181,969	17.17	4,053,092	−0.0077
	800–1000	3,412,901	15.26	108,548	10.24	3,521,449	−0.3837
Distance-to-River/m	0–200	7,180,378	28.70	424,595	37.73	7,604,973	0.2602
	200–400	6,451,048	25.79	345,492	30.70	6,796,540	0.1665
	400–600	6,005,606	24.01	211,146	18.76	6,216,752	−0.2368
	600–800	5,378,631	21.50	144,181	12.81	5,522,812	−0.5000
Vegetation	Cover	21,230,458	38.44	221,769	17.26	21,452,227	−0.7883
	No cover	34,006,789	61.56	1,063,500	82.74	35,070,289	0.2879

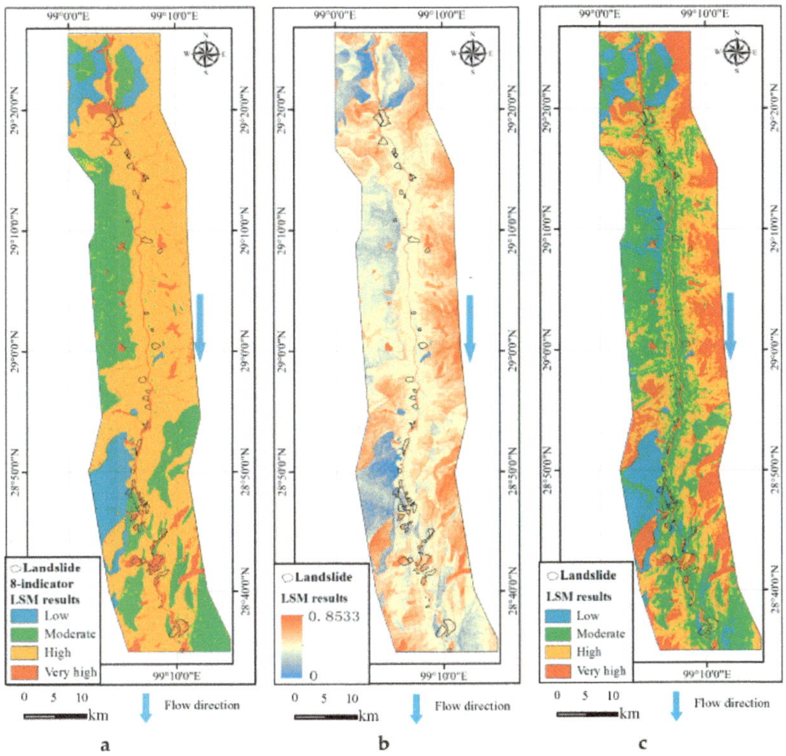

Figure 9: Landslide susceptibility map results using the new approach: (a) eight-factor LSM results; (b) LSM discrete results; and (c) LSM classification results.

Landslides Susceptibility Mapping

The weights of the influencing factors were calculated by the AHP. The method for the eight-factor LSM result was based on the ICM, whereas the annual precipitation factor was analyzed separately. Then, the eight-factor LSM was combined with the annual precipitation factor based on Equation (15) for the eventual landslide susceptibility mapping. The eventual landslide susceptibility map was generated through the GIS software and the value scope is 0–0.8533. The discrete result is shown in Figure 9b. The natural break classification scheme was applied for the classification of the landslide susceptibility zones. Recently, the classification method mainly used the natural break classification scheme [72,73,74,75,76]. The landslide susceptibility maps were reclassified into four classes: low (0–0.3323), moderate (0.3323–0.4593), high (0.4593–0.5699), and very high (0.5699–0.8533).

The eventual landslide susceptibility results are shown in Figure 9c. The area of the very high zone was 252.23 km^2, accounting for 15.17% of the whole study area (Table 5). This area is mainly distributed in the high elevation areas of Quaternary-aged material. The very high susceptibility zone is not stable and is prone to large-scale landslides. The high susceptibility zone is 479.99 km^2, accounting for 34.11% of the whole reservoir. This sub-class basically represents the landslide susceptibility of the whole study area. It is mainly distributed on the left bank of the Jinsha River in the middle elevation areas. The geology of this area was mostly Xiongsong Formation (Pt_2X) schist, phyllite, marble, and the Devonian to Triassic ophiolite suite. Middle-large landslides were prone to occur in this zone. The moderate zone was 493.46 km^2, accounting for 35.06% of the area. This zone mainly had small-middle scale landslides. However, the moderate zone was mainly distributed in the low elevation area, where the annual precipitation was not as high as the higher elevation area. The rest of the 181.58 km^2 was of low susceptibility, that is, landslides were not likely to occur. Overall, approximately 87.1% of the reservoir area may have experienced landslides. It was found that 89.8% of the landslides occurred in areas allocated to the moderate, high and very high areas. The low, moderate, high, and very high zones accounted for 10.2%, 48.92%, 30.12% and 10.46% of the landslide occurrence, respectively. This suggests that the susceptibility mapping of the study area was consistent with the actual consequences. The high and very high susceptibility zones accounted for 52.03% of the Xulong hydropower station reservoir.

Table 5: Statistical results of LSM in the Xulong hydropower station reservoir

Susceptibility	Landslide Occurred			Total Study Area		
	Count	Ratio (%)	Area (km^2)	Count	Ratio (%)	Area (km^2)
Low	131,027	10.20	3.28	7,263,251	12.90	181.58
Moderate	628,504	48.92	15.71	19,738,352	35.06	493.46
High	390,774	30.12	9.77	19,199,915	34.11	479.99
Very High	134,422	10.46	3.36	10,089,277	17.92	252.23

Discussions and Validation

Most of the aforementioned literature used objective methods for LSM, which were mostly based on objective data from field investigation aerial photos. However, regarding the implemented special area, these approaches had limitations, such as obtaining unclear data [35] and the strict demands of the methods. Any incorrect results can easily be conveyed into the weighting assignments. Therefore, solely relying on objective methods has some limitations and can easily lead to misleading information. Based on expert opinion, AHP introduced a degree of subjectivity when used to make comparison judgments. The expert experiences combined with ICM permitted a better flexibility in the landslide susceptibility analysis. The proposed approach retained many advantages that AHP had, especially its hierarchical structure, reduced inconsistency from the pairwise comparison, and the priority vectors generated [77]. The AHP-ICM approach can be applied as a quantitative solution for LSM considering both the priority for landslide susceptibility factors and the objective investigated information. The results using the combination method were superior to those from using the alternative method alone.

It was found that most of the study area was rocky and poorly vegetated. Due to the effects of the vertical distribution law of precipitation, the vegetation also had a vertical distribution law along with precipitation. There was almost no vegetation in the low elevation area, whereas the vegetation at high elevation was widely distributed. The vegetation coverage increased with the elevation because of the higher precipitation.

Precipitation commonly occurred as snowfall during the winter in the high elevation area. Matsuura [78] found that the maximum daily displacements of each year were observed not in snow melting periods, but immediately before or at the beginning of snow cover periods. According to [78], the displacement of a landslide that has a shallow sliding surface in a snowy region was found to be affected by snow accumulation conditions. A slight effect was still observed for landslides in the snow cover period. Snow accumulation is not annual. In warm seasons, the snow melts. In this study, precipitation is considered throughout the whole year. Slope instability appears to be mainly forced by the snow melt accumulated during the winter season, which, in turn, promotes rock and soil water saturation and landslide occurrence in the following spring-summer generated by the snow melt [79]. Yamasaki [80] also considered that precipitation was presented as snowfall, suggesting a possible increase in slope instability because of the more frequent rock water saturation in high elevation areas.

This study introduced an approach for landslide susceptibility mapping considering the vertical distribution law of precipitation in the study area. Figure 9a shows the eight-factor LSM results. It shows that the very high susceptibility zones were mainly distributed in the area covered by Quaternary material. The high susceptibility zones were mainly distributed in the areas of the Xiongsong Formation (Pt_2) geologic suite and Ophiolite formation (DTJ). It seemed that the lithology played a dominant role in the LSM. However, based on Table 3, the precipitation was the most important factor influencing landslide occurrence. When taking the vertical distribution of precipitation into consideration, the LSM results became totally different. It is shown in Figure 9c that the susceptibility values of the high elevation areas were higher than those of the low elevation areas. The susceptibility values increased with increasing elevation. It was obvious that the vertical precipitation distribution had a great contribution to the landslide distribution. It should be noted that there were some red belts at the top of the mountain (Figure 9c). The red belts represent the very high susceptibility zones. This pattern followed the vertical distribution law of precipitation.

To validate the accuracy of the results, this study used the evaluation accuracy analysis method as the validation [38]. The equation is as follows:

$$P = \frac{M_1}{N_1} \times \left(\frac{M - M_1}{N - N_1}\right)^{1/3} \tag{16}$$

where P is the accuracy of the prediction, N is the total pixel number of the study area, N_1 is the pixel number of landslides, M is the pixel number above the critical value of the study area, and M_1 is the pixel number of landslides above the critical value. Since all of the moderate, high and very high susceptibility zones yield different sized landslides, this study sets the moderate susceptibility zone as the critical value. The parameters are shown in Table 6.

Table 6: The evaluation results accuracy analysis

Susceptibility Degree	*M*	*N*	M_1	N_1	*P*
Moderate	19,738,352	56,290,795	628,504	1,284,727	85.74%
High	19,199,915		390,774		
Very High	10,089,277		134,422		
Counts Number	49,027,544		1,153,700		

Based on Equation (16), the prediction accuracy was calculated as P = 85.74%. The prediction accuracy was very high and the landslide susceptibility map using the AHP-ICM evaluation results was reasonable and reliable in terms of the vertical distribution law of precipitation.

CONCLUSIONS

The preparation of the landslide susceptibility map is one of most important planning agencies for hazard prediction in the reservoir of the Xulong hydropower station. It should be implemented in the process of the hydropower station construction. Based on the field investigation data, this study provided a landslide distribution and data of their basic influencing factors. Among the landslide-related factors, the slope angle, slope aspect, curvature, geology, distance-to-river, distance-to-fault, vegetation, bedrock uplift, and annual precipitation were used for landslide susceptibility mapping.

It has often been argued that the high rates of bedrock uplift and denudation should be correlated with steep slope angles. The rapid bedrock uplift of this area also plays an important role in the slope angle distribution, which directly influences the occurrence of landslides. With the analytic hierarchy process, this study collected the observations and suggestions of many experts and references and established a weighting model of the related factors. The eight factors were analyzed with the information content method, except for annual precipitation.

The evidence shows that a vertical distribution law of precipitation exists in the study area. The precipitation directly determines the occurrence probability of landslides. The ICM cannot reflect the vertical distribution law of precipitation, which represents the characteristics of the implemented area. Analyzing the annual precipitation factor separately can satisfy this condition. The eight-factor landslide susceptibility map was combined with the annual precipitation map for the eventual landslide susceptibility mapping. The landslide susceptibility map was generated and the value scope was 0–0.8533.

Objective methods have been widely used for landslide susceptibility analysis. However, most objective methods have their limitations, which would easily introduce some mistakes in assigning factor weights. The AHP method was useful for many cases because of its ease of applicability and the structure of AHP. This study proposed an approach for integrating an accurate objective method (ICM) with an overall guidance subjective method (AHP) to develop a landslide susceptibility map. The combined method with the advantages of the two methods can obtain accurate results. Moreover, the proposed method permits better flexibility in LSM. GIS was applied to obtain the landslide susceptibility map.

The landslide susceptibility map comprises four classes, including low, moderate, high and very high. These susceptibility zone ratios were 12.9%, 35.06%, 34.11% and 17.92% of the study area, respectively. The precipitation had a great effect on the occurrence of landslides. It was necessary to consider this factor correctly. Since there was a vertical distribution law of precipitation in the study area, the top of the mountain displayed a series of very high susceptibility pixels. The validation results showed that the prediction accuracy was 85.74%, which meant that the susceptibility map was confirmed to be reliable and reasonable. This study could serve as an effective guide for the further construction of the Xulong hydropower station. It also verifies that the method for obtaining the vertical distribution law of precipitation in the study area was appropriate.

ACKNOWLEDGMENTS

This work was supported by the State Key Program of National Natural Science of China (Grant No.41330636). Natural Science Foundations of China (Grant No. 41402243). Graduate student innovation fund project of Jilin University (No. 2015013). Thanks to anonymous reviewers for their valuable feedback on the manuscript.

REFERENCES

1. Roering, J.J.; Kirchner, J.W.; Dietrich, W.E. Evidence for nonlinear, diffusive sediment transport on hillslopes and implications for landscape morphology. Water Resour. Res. 1999, 35, 853–870.
2. Malamud, B.D.; Turcotte, D.L.; Guzzetti, F.; Reichenbach, P. Landslides, earthquakes, and erosion. Earth Planet. Sci. Lett. 2004, 229, 45–59.
3. Kawabata, D.; Bandibas, J. Landslide susceptibility mapping using geological data, a DEM from ASTER images and an artificial neural network (ANN). Geomorphology 2009, 113, 97–109.
4. Guzzetti, F.; Mondini, A.C.; Cardinali, M.; Fiorucci, F.; Santangelo, M.; Chang, K.-T. Landslide inventory maps: New tools for an old problem. Earth Sci. Rev. 2012, 112, 42–66.
5. Peruccacci, S.; Brunetti, M.T.; Luciani, S.; Vennari, C.; Guzzetti, F. Lithological and seasonal control on rainfall thresholds for the possible initiation of landslides in central Italy. Geomorphology 2012, 139, 79–90.
6. Feizizadeh, B.; Blaschke, T. Landslide risk assessment based on GIS multi-criteria evaluation: A case study in Bostan-Abad County, Iran. J. Earth Sci. Eng. 2011, 1, 66–71.
7. Montgomery, D.R.; Dietrich, W.E. A physically based model for the topographic control on shallow landsliding. Water Resour. Res. 1994, 30, 1153–1171.
8. Van Westen, C.; Van Asch, T.W.; Soeters, R. Landslide hazard and risk zonation—Why is it still so difficult?Bull. Eng. Geol. Environ. 2006, 65, 167–184.
9. Van Westen, C.J.; Castellanos, E.; Kuriakose, S.L. Spatial data for landslide susceptibility, hazard, and vulnerability assessment: An overview. Eng.Geol. 2008, 102, 112–131.
10. Corominas, J.; van Westen, C.; Frattini, P.; Cascini, L.; Malet, J.P.; Fotopoulou, S.; Catani, F.; Van Den Eeckhaut, M.; Mavrouli, O.; Agliardi, F.; et al. Recommendations for the quantitative analysis of landslide risk. Bull. Eng. Geol. Environ. 2014, 73, 209–263.
11. Bi, R.N.; Schleier, M.; Rohn, J.; Ehret, D.; Xiang, W. Landslide susceptibility analysis based on arcgis and artificial neural network for a large catchment in three gorges region, China. Environ. Earth Sci. 2014, 72, 1925–1938.
12. Li, D.; Yin, K.; Leo, C. Analysis of baishuihe landslide influenced by the effects of reservoir water and rainfall. Environ. Earth Sci. 2010, 60, 677–687.
13. Can, T.; Nefeslioglu, H.A.; Gokceoglu, C.; Sonmez, H.; Duman, T.Y. Susceptibility assessments of shallow earthflows triggered by heavy rainfall at three catchments by logistic regression analyses. Geomorphology 2005,72, 250–271.
14. Lee, S.; Evangelista, D. Earthquake-induced landslide-susceptibility mapping using an artificial neural network. Nat. Hazards Earth Syst. Sci. 2006, 6, 687–695.

15. Pradhan, B.; Lee, S. Landslide susceptibility assessment and factor effect analysis: Backpropagation artificial neural networks and their comparison with frequency ratio and bivariate logistic regression modelling.Environ. Model. Softw. 2010, 25, 747–759.
16. Akgun, A. A comparison of landslide susceptibility maps produced by logistic regression, multi-criteria decision, and likelihood ratio methods: A case study at Izmir, Turkey. Landslides 2012, 9, 93–106.
17. Oh, H.-J.; Park, N.-W.; Lee, S.-S.; Lee, S. Extraction of landslide-related factors from aster imagery and its application to landslide susceptibility mapping. Int. J. Remote Sens. 2012, 33, 3211–3231.
18. Regmi, A.D.; Devkota, K.C.; Yoshida, K.; Pradhan, B.; Pourghasemi, H.R.; Kumamoto, T.; Akgun, A. Application of frequency ratio, statistical index, and weights-of-evidence models and their comparison in landslide susceptibility mapping in Central Nepal Himalaya. Arab. J. Geosci. 2014, 7, 725–742.
19. Bui, D.T.; Lofman, O.; Revhaug, I.; Dick, O. Landslide susceptibility analysis in the Hoa Binh province of Vietnam using statistical index and logistic regression. Nat. Hazards 2011, 59, 1413–1444.
20. Dong, J.J.; Tung, Y.H.; Chen, C.C.; Liao, J.J.; Pan, Y.W. Discriminant analysis of the geomorphic characteristics and stability of landslide dams. Geomorphology 2009, 110, 162–171.
21. Wang, L.J.; Guo, M.; Sawada, K.; Lin, J.; Zhang, J.C. Landslide susceptibility mapping in Mizunami City, Japan: A comparison between logistic regression, bivariate statistical analysis and multivariate adaptive regression spline models. Catena 2015, 135, 271–282.
22. Wan, S. A spatial decision support system for extracting the core factors and thresholds for landslide susceptibility map. Eng. Geol. 2009, 108, 237–251.
23. Liu, S.-H.; Lin, C.-W.; Tseng, C.-M. A statistical model for the impact of the 1999 Chi-Chi earthquake on the subsequent rainfall-induced landslides. Eng. Geol. 2013, 156, 11–19.
24. Lee, S.; Ryu, J.-H.; Won, J.-S.; Park, H.-J. Determination and application of the weights for landslide susceptibility mapping using an artificial neural network. Eng. Geol. 2004, 71, 289–302.
25. Gomez, H.; Kavzoglu, T. Assessment of shallow landslide susceptibility using artificial neural networks in Jabonosa River Basin, Venezuela. Eng. Geol. 2005, 78, 11–27.
26. Caniani, D.; Pascale, S.; Sdao, F.; Sole, A. Neural networks and landslide susceptibility: A case study of the urban area of Potenza. Nat. Hazards 2008, 45, 55–72.
27. Melchiorre, C.; Matteucci, M.; Azzoni, A.; Zanchi, A. Artificial neural networks and cluster analysis in landslide susceptibility zonation. Geomorphology 2008, 94, 379–400.
28. Lin, H.-M.; Chang, S.-K.; Wu, J.-H.; Juang, C.H. Neural network-based model for assessing failure potential of highway slopes in the Alishan, Taiwan area: Pre-and post-earthquake investigation. Eng. Geol. 2009, 104, 280–289.
29. Conforti, M.; Pascale, S.; Robustelli, G.; Sdao, F. Evaluation of prediction capability of the artificial neural networks for mapping landslide susceptibility in the Turbolo River catchment (Northern Calabria, Italy). Catena 2014, 113, 236–250.
30. Samui, P. Slope stability analysis: A support vector machine approach. Environ. Geol. 2008, 56, 255–267.
31. Marjanović, M.; Kovačević, M.; Bajat, B.; Voženílek, V. Landslide susceptibility assessment using SVM machine learning algorithm. Eng. Geol. 2011, 123, 225–234.
32. Xu, C.; Dai, F.; Xu, X.; Lee, Y.H. Gis-based support vector machine modeling of earthquake-triggered landslide susceptibility in the Jianjiang River Watershed, China. Geomorphology 2012, 145, 70–80.
33. Peng, L.; Niu, R.; Huang, B.; Wu, X.; Zhao, Y.; Ye, R. Landslide susceptibility mapping based on rough set theory and support vector machines: A case of the Three Gorges area, China. Geomorphology 2014, 204, 287–301.
34. Tehrany, M.S.; Pradhan, B.; Jebur, M.N. Spatial prediction of flood susceptible areas using rule based decision tree (DT) and a novel ensemble bivariate and multivariate statistical models in GIS. J. Hydrol. 2013, 504, 69–79.
35. Tiwari, M.K.; Chatterjee, C. Uncertainty assessment and ensemble flood forecasting using bootstrap based artificial neural networks (BANNs). J. Hydrol. 2010, 382, 20–33.
36. Kazakis, N.; Kougias, I.; Patsialis, T. Assessment of flood hazard areas at a regional scale using an index-based approach and analytical hierarchy process: Application in Rhodope-Evros region, Greece. Sci. Total Environ.2015, 538, 555–563.
37. Ho, W. Integrated analytic hierarchy process and its applications—A literature review. Eur. J. Oper. Res. 2008,186, 211–228.
38. Jade, S.; Sarkar, S. Statistical models for slope instability classification. Eng.Geol. 1993, 36, 91–98.
39. Saaty, T.L. Modeling unstructured decision problems—The theory of analytical hierarchies. Math. Comput. Simul. 1978, 20, 147–158.
40. Junior, R.V.; Varandas, S.; Fernandes, L.S.; Pacheco, F. Environmental land use conflicts: A threat to soil conservation. Land Use Policy 2014, 41, 172–185.
41. Oikonomidis, D.; Dimogianni, S.; Kazakis, N.; Voudouris, K. A GIS/remote sensing-based methodology for groundwater potentiality assessment in Tirnavos area, Greece. J. Hydrol. 2015, 525, 197–208.
42. Lin, M.-L.; Tung, C.-C. A GIS-based potential analysis of the landslides induced by the Chi-Chi earthquake.Eng. Geol. 2004, 71, 63–77.
43. Gao, K.; Cui, P.; Zhao, C.; Wei, F. Landslide hazard evaluation of wanzhou based on GIS information value method in the Three Gorges reservoir. Chin. J. Rock Mech. Eng. 2006, 25, 991–996.
44. Du, J.; Yang, Q.; Yan, J. Hazard evaluation of secondary geological disaster based on GISand information value method. Earth Sci. J. China Univ. Geosci. 2010, 35, 324–330.

45. Chen, Y.; Booth, D.C. The Wenchuan Earthquake of 2008: Anatomy of a Disaster; Springer Science & Business Media: New York, NY, USA, 2011.

46. Burbank, D.W.; Leland, J.; Fielding, E.; Anderson, R.S.; Brozovic, N.; Reid, M.R.; Duncan, C. Bedrock incision, rock uplift and threshold hillslopes in the northwestern Himalayas. Nature 1996, 379, 505–510.

47. Wang, E.; Burchfiel, B.C. Late cenozoic to holocene deformation in southwestern Sichuan and Adjacent Yunnan, China, and its role in formation of the southeastern part of the Tibetan Plateau. Geol. Soc. Am. Bull.2000, 112, 413–423.

48. Hungr, O.; Leroueil, S.; Picarelli, L. The varnes classification of landslide types, an update. Landslides 2014, 11, 167–194.

49. Guzzetti, F.; Reichenbach, P.; Ardizzone, F.; Cardinali, M.; Galli, M. Estimating the quality of landslide susceptibility models. Geomorphology 2006, 81, 166–184.

50. Kanungo, D.; Arora, M.; Sarkar, S.; Gupta, R. A comparative study of conventional, ann black box, fuzzy and combined neural and fuzzy weighting procedures for landslide susceptibility zonation in Darjeeling Himalayas. Eng.Geol. 2006, 85, 347–366.

51. Yalcin, A. Gis-based landslide susceptibility mapping using analytical hierarchy process and bivariate statistics in Ardesen (Turkey): Comparisons of results and confirmations. Catena 2008, 72, 1–12.

52. Akgun, A.; Sezer, E.A.; Nefeslioglu, H.A.; Gokceoglu, C.; Pradhan, B. An easy-to-use MATLAB program (Mamland) for the assessment of landslide susceptibility using a Mamdani fuzzy algorithm. Comput. Geosci. UK 2012, 38, 23–34.

53. Mohammady, M.; Pourghasemi, H.R.; Pradhan, B. Landslide susceptibility mapping at Golestan Province, iran: A comparison between frequency ratio, Dempster–Shafer, and weights-of-evidence models. J. Asian Earth Sci.2012, 61, 221–236.

54. Pradhan, B.; Chaudhari, A.; Adinarayana, J.; Buchroithner, M.F. Soil erosion assessment and its correlation with landslide events using remote sensing data and GIS: A case study at Penang Island, Malaysia. Environ. Monit. Assess. 2012, 184, 715–727.

55. Roodposhti, M.S.; Rahimi, S.; Beglou, M.J. PROMETHEE II and fuzzy AHP: An enhanced GIS-based landslide susceptibility mapping. Nat. Hazards 2014, 73, 77–95.

56. Chue, Y.S.; Chen, J.W.; Chen, Y.R. Rainfall-induced slope landslide pontential and landslide distribution characteristics assessment. J. Mar. Sci. Technol. Taiwan 2015, 23, 705–716.

57. Kritikos, T.; Davies, T. Assessment of rainfall-generated shallow landslide/debris-flow susceptibility and runout using a GIS-based approach: Application to western Southern Alps of New Zealand. Landslides 2015, 12, 1051–1075.

58. Lee, M.J.; Park, I.; Won, J.S.; Lee, S. Landslide hazard mapping considering rainfall probability in Inje, Korea. Geomat. Nat. Hazard Risk 2016, 7, 424–446.

59. Kirschbaum, D.; Adler, R.; Adler, D.; Peters-Lidard, C.; Huffman, G. Global distribution of extreme precipitation and high-impact landslides in 2010 relative to previous years. J. Hydrometeorol. 2012, 13, 1536–1551.

60. Ayalew, L.; Yamagishi, H. The application of Gis-based logistic regression for landslide susceptibility mapping in the Kakuda-Yahiko Mountains, Central Japan. Geomorphology 2005, 65, 15–31.

61. Oh, H.-J.; Pradhan, B. Application of a neuro-fuzzy model to landslide-susceptibility mapping for shallow landslides in a tropical hilly area. Comput. Geosci. UK 2011, 37, 1264–1276.

62. Hao, M.; Wang, Q.; Shen, Z.; Cui, D.; Ji, L.; Li, Y.; Qin, S. Present day crustal vertical movement inferred from precise leveling data in eastern margin of Tibetan Plateau. Tectonophysics 2014, 632, 281–292.

63. Deng, L.-Q.; Wamg, X.-P. Relationship between neotectonism and landslides in Reservoir area of three-gorges project on Yangtze River. J. Eng. Geol. 2000, 8, 136–141.

64. Zhang, X.-J.; Zeng, Q.-L.; Ma, Y.-S. Basic characteristics of the differential uplift of the yulong-haba block and its geological hazard effects. Geol. China 2006, 5, 1075–1082.

65. Simons, M. The Morphological Analysis of Landforms: A New Review of the Work of Walther Penck (1888–1923); JSTOR: New York, NY, USA, 1962.

66. Gouvas, M.; Sakellariou, N.; Xystrakis, F. The relationship between altitude of meteorological stations and average monthly and annual precipitation. Stud. Geophys. Geod. 2009, 53, 557–570.

67. Liu, B.-T.; Tao, H.-P.; Song, C.-F. Temporal and spatial variations of rainfall erosivity in southwest China from 1960 to 2009. Adv. Earth Sci. 2012, 27, 499–509.

68. Huang, Y.; Li, Z.-S. Analysis of little rainfall of dry valley in low latitude area. Res. Water Resour. 2006, 27, 1–3.

69. Shan, B. Reservoir Bank Geological Disaster Susceptibility Assessment and Study on Zoning of Disaster Risk in Benzilan Water-Source Reservoir Area Based on 3s Technology. Ph.D. Thesis, Jilin University, Changchun, China, 2014.

70. Li, L.-Q.; Xiong, L.-H.; Jiang, C.; Zhang, H.-G. Impact of air temperature on annual runoff of batang station in the headstream of Yangtze River. Resour. Environ. Yangtze Basin 2015, 24, 1142–1149.

71. Wu, L.-Q.; Li, X.-H. Analysis on change of annual precipitation with altitude in high mountain areas. Yunnan Geogr. Environ. Res. 2004, 2, 4–7.

72. Falaschi, F.; Giacomelli, F.; Federici, P.R.; Puccinelli, A.; Avanzi, G.D.; Pochini, A.; Ribolini, A. Logistic regression versus artificial neural networks: Landslide susceptibility evaluation in a sample area of the Serchio River Valley, Italy. Nat. Hazards 2009, 50, 551–569.

73. Bednarik, M.; Magulova, B.; Matys, M.; Marschalko, M. Landslide susceptibility assessment of the kralovany-liptovsky mikulas railway case study. Phys. Chem. Earth 2010, 35, 162–171.

74. Constantin, M.; Bednarik, M.; Jurchescu, M.C.; Vlaicu, M. Landslide susceptibility assessment using the bivariate statistical analysis and the index of entropy in the Sibiciu Basin (Romania). Environ. Earth Sci. 2011,63, 397–406.

75. Pourghasemi, H.R.; Mohammady, M.; Pradhan, B. Landslide susceptibility mapping using index of entropy and conditional probability models in GIS: Safarood Basin, Iran. Catena 2012, 97, 71–84.

76. Pourghasemi, H.R.; Jirandeh, A.G.; Pradhan, B.; Xu, C.; Gokceoglu, C. Landslide susceptibility mapping using support vector machine and GIS at the Golestan Province, Iran. J. Earth Syst. Sci. 2013, 122, 349–369.

77. Feizizadeh, B.; Roodposhti, M.S.; Jankowski, P.; Blaschke, T. A GIS-based extended fuzzy multi-criteria evaluation for landslide susceptibility mapping. Comput. Geosci. UK 2014, 73, 208–221.

78. Matsuura, S.; Asano, S.; Okamoto, T.; Takeuchi, Y. Characteristics of the displacement of a landslide with shallow sliding surface in a heavy snow district of Japan. Eng. Geol. 2003, 69, 15–35.

79. Moreiras, S.; Lisboa, M.S.; Mastrantonio, L. The role of snow melting upon landslides in the central Argentinean Andes. Earth Surf. Proc. Landf. 2012, 37, 1106–1119.

80. Yamasaki, S.; Nagata, H.; Kawaguchi, T. Long-traveling landslides in deep snow conditions induced by the 2011 Nagano Prefecture Earthquake, Japan. Landslides 2014, 11, 605–613.

CHAPTER 2

A COMBINATION OF GEOGRAPHICALLY WEIGHTED REGRESSION, PARTICLE SWARM OPTIMIZATION AND SUPPORT VECTOR MACHINE FOR LANDSLIDE SUSCEPTIBILITY MAPPING

Xianyu Yu, Yi Wang, Ruiqing Niu and Youjian Hu

Keywords: Landslide susceptibility mapping, geographically weighted regression, support vector machine, particle swarm optimization, Three Gorges Reservoir

INTRODUCTION

It is known that the area in the Three Gorges Reservoir along the Yangtze River is characterized by many active and reactivated landslides caused by the periodic water level fluctuation of the reservoir [1], which poses a serious threat to the security of life and property. Up to 2009, more than 3800 landslides have been recorded in this region [2]. Thus, it is crucial to predict slope failures in the Three Gorges area.

Landslide susceptibility evaluation is a complex task [3]. Compared to the traditional geological survey methods, such as landslide field reconnaissance, landslide spatial prediction is more convenient and efficient, due to the integration of geographical information systems (GIS) technology and statistical analysis principles. The spatial prediction of landslide susceptibility mapping is considered as one of the most important steps for landslide hazard mitigation and management [4], which has encouraged research towards knowledge-driven and data-driven models [5]. Knowledge-driven models, such as analytic hierarchy process (AHP) and fuzzy mathematics [5,6], are based on the analysis of landslide formation mechanism(s), and expert experience and knowledge are used to choose the most important environmental factors of landslides and quantitative weight values. On the other hand, data-driven models include logistic regression (LR) [7,8,9], artificial neural network (ANN) [10,11,12,13], SVM [14,15,16,17] and geographically weighted regression (GWR) [18,19], etc. These models are based on overlay analysis to calculate quantitative relationship between various environmental factors and the known distributions of landslides. Therefore, they are always used to determine weights of predictors, i.e., values/indices of landslide susceptibility.

Since support vector machine (SVM) can demonstrate satisfactory classification accuracies when a limited number of training samples is available, and it has been widely utilized to perform landslide susceptibility mapping [14,15,16,17,20,21]. However, the proper selection of a kernel function and its corresponding parameters is still an open problem, which can greatly influence the final prediction accuracy. To obtain the optimal parameters for SVM, some researchers worked on combining the particle swarm optimization (PSO) algorithm with the classical SVM model [22,23,24]. PSO is a population-based stochastic optimization technique developed by Eberhart and Kennedy [25], inspired by social behavior of bird flocking or fish schooling. This technique has many similarities with evolutionary computation techniques such as Genetic Algorithms (GA) [26]. For instance, the system is initialized with a population of random solutions and searches for optima by updating generations. However, unlike GA, PSO has no

Citation: Yu, X.; Wang, Y.; Niu, R.; Hu, Y. A Combination of Geographically Weighted Regression, Particle Swarm Optimization and Support Vector Machine for Landslide Susceptibility Mapping. Int. J. Environ. Res. Public Health 2016, 13, 487.

evolution operators such as crossover and mutation. In PSO, the potential solutions, called particles, fly through the problem space by following the current optimum particles. Compared to GA, the advantages of PSO are that it is easy to implement and there are few parameters to adjust [27]. To better perform landslide prediction, this technique can estimate optimum parameters for the SVM prediction model. For instance, Huang and Dun [22] proposed a PSO–SVM model to improve classification accuracies with an appropriate feature subset. One year later, Zhao and Yin [23] integrated the SVM, PSO and numerical analysis techniques to intelligent displacement back analysis in geomechanical parameter identification. More recently, Ren et al. [24] presented a landslide prediction method for the Shuping landslide by using a PSO-SVM model and wavelet analysis. However, the drawbacks of these techniques are threefold: first, the PSO algorithm always falls into a local optimum, especially in a very large area. Second, spatial autocorrelation in study areas is not taken into account. Finally, these methods applied a global model in a certain area and considered that the impacts of environmental factors are equal for the entire region, so they cannot describe the local characteristics of spatial landslide occurrences.

This paper presents an effective PSO-SVM model based on GWR for landslide susceptibility mapping. It should be noted that in practice different degrees of impact may occur at a local scale for study areas [18]. Moreover, the impacts of environmental factors always vary with spatial locations. It is well-known that most variables in real-world applications tend to be moderately spatially autocorrelated because of the way phenomena are geographically organized [28,29]. Therefore, spatial autocorrelation is always used to measure the degree to which a set of spatial features and their associated data values tend to be clustered together in space or dispersed [30,31]. Recently, many contributions have been devoted to using GWR to account for spatial autocorrelation and these have validated that GWR can be an effective estimator of spatial autocorrelation [32,33,34,35]. Inspired by previous works, we utilize the GWR technique to segment the study area into several prediction regions with a proper size. To this end, each computing unit in the study area is assigned a GWR coefficient by exploiting an appropriate kernel type and selection criteria. Meanwhile, each environmental factor is divided into several classes by the natural breaks method. By superposing these classification maps, different degrees of impacts at a local scale for these environmental factors are taken into account as well. As a consequence, the GWR coefficients in each prediction region are similar, while they make a great difference in different regions, i.e., spatial autocorrelations of environmental factors between them are greatly suppressed. Secondly, the PSO-SVM model is used in each prediction region for landslide susceptibility mapping. The PSO algorithm is utilized for the SVM model to search for optimal parameters in each prediction region. In this way, the problem of local optimum can be effectively overcome. In addition, the SVM model can be locally applied to each prediction region for accurate landslide susceptibility maps.

The remainder of this paper is organized as follows: Section 2 reviews the related techniques on GWR, PSO and SVM. Section 3 presents the proposed GWR-PSO-SVM model. Section 4 describes the study area and data used in this work. Section 5 reports experiments including comparative results between the traditional SVM-based prediction models and ours. Section 6 presents some discussions of our model and the last section states our concluding remarks.

RELATED TECHNIQUES

Geographically Weighted Regression

Geographically Weighted Regression (GWR) is a fairly recent contribution to modelling spatially heterogeneous processes [28,29,36,37] that has attracted much attention for its elegant performance when exploring local variations in a study area [18,38,39]. GWR is implemented by obtaining regression equations for each spatial zone separately [40] and its basic model can be written as:

$$y_i = \beta_0(u_i, v_i) + \sum_{k=1}^{Q} \beta_k(u_i, v_i) x_{ik} + \varepsilon_i \tag{1}$$

where (u_i,v_i) denotes the coordinates of the ith sample in space (e.g., latitude and longitude), $i = 1,2,\cdots,L$, L and Qare the number of samples and regression coefficients, respectively. y_i is the dependent variable at location i, x_{ik} is the value of the kth explanatory variable at location i, $\beta_k(u_i,v_i)$ is the local regression coefficients for the kth explanatory variable at location i, and $\beta_0(u_i,v_i)$ is the intercept parameter at location i. Then, the least square estimate of β_i can be defined as follows:

$$\hat{\beta}_i = \left(X^T W_i X\right)^{-1} X^T W_i Y \tag{2}$$

and its variance is:

$$\mathrm{var}(\hat{\beta}) = \left(X^T W_i^{-1} X\right)^{-1} \tag{3}$$

where W_i is n × n diagonal matrix, whose diagonal elements are the geographical weights:

$$W_i = \begin{bmatrix} W_{i1} & 0 & \cdots & 0 \\ 0 & W_{i2} & \cdots & 0 \\ \vdots & \vdots & \ddots & \vdots \\ 0 & 0 & \cdots & W_{in} \end{bmatrix} \tag{4}$$

the choice of W_i depends on the selected kernel function, which can be fixed (i.e., fixed bandwidth) or adaptive kernels (i.e., varying bandwidths) in [41]. In practical, it is found that GWR is not sensitive to the choice of Gaussian function and bi-square function, but rather the bandwidth of the specific weight function. Based on the maximum likelihood principle, Akaike [42] proposed a general model selection criterion, called the Akaike Information Criterion (AIC), which is shown as follows:

$$AIC = -2\ln L\left(\hat{\theta}_L, x\right) + 2q \tag{5}$$

where $L\left(\hat{\theta}_L, x\right)$ is the maximized likelihood of the parameter vector θ, x is a random sample, $\hat{\theta}_L$ is the maximum likelihood estimate of θ, q is the number of the unknown parameters. The larger the likelihood function, the better the estimator. In this work, a minimum AIC model is selected as the "optimal" model.

Support Vector Machine

The support vector machine (SVM) incorporates mainly two learning techniques [43], i.e., Vapnik–Chervonenkis (VC) dimensional and statistical learning theories. One of the most important applications of SVM is classification. Because of its satisfactory performance and capabilities of fault-tolerance, SVM has recently attracted increasing attention and is widely used in machine learning, data mining and knowledge discovery [44,45], as well as landslide susceptibility assessment [14–17]. The SVM method is briefly introduced as follows [46,47]: assuming that a set of linear separable training vectors x_i (i = 1,2,¨ ¨ ¨ ,R, R is the total number of vectors) consists of two classes $y_i = \pm 1$, which denote as landslide occurrence or not. The aim of SVM is to find an n-dimensional hyperplane to split two classes by the maximum gap, as shown in Figure 1. The n-dimensional hyperplane can be minimized as:

$$\begin{cases} \min \frac{1}{2} \| w \|^2, \\ s.t., \ y_i (w \cdot x_i + b) \geqslant 1 \end{cases} \tag{6}$$

where ||w|| is the two-norm of w, b is used to increase the interval to ensure that the hyperplane does not cross the origin, x_i is the point of the hyperplane, and w is a vector perpendicular to the hyperplane. By embedding a non-negative Lagrange multiplier (λ_i), the cost function can be obtained as follows:

$$L(w, b, \lambda) = \frac{1}{2} \| w \|^2 - \sum_{i=1}^{n} \lambda_i \left(y_i (w \cdot x_i + b) - 1 \right) \tag{7}$$

The solution can be obtained by dual minimizing Equation (7) with respect to w and b. In the non-separable case, one can complete the constraints by introducing a non-negative ξ_i, then Equation (7) can be produced as follows:

$$\begin{cases} \min \frac{1}{2} \| w \|^2 + C \sum_{i=1}^{n} \xi_i, \\ s.t., \ y_i ((w \cdot x_i) + b) \geqslant 1 - \xi_i \end{cases} \tag{8}$$

where $\xi_i(\xi_i \geq 0)$ is the slack variable, C is a penalty variable of the error term, which denotes the distance from a wrong point to its correct position.

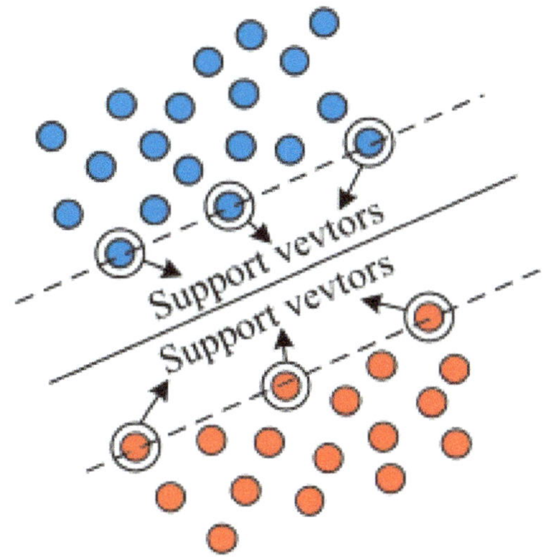

Figure 1: Illustration of the SVM principle.

In addition, the Gaussian Radial Basis Function (RBF) is used as a kernel function introduced by Vapnik [43] to account for the nonlinear decision boundary:

$$K(x_i, x_j) = \exp(-\gamma \| x_i - x_j \|^2) \tag{9}$$

where γ is a positive variable to measure the width of the Gaussian kernel in RBF. This function is robust and can account for the nonlinear decision boundary.

Particle Swarm Optimization

The PSO algorithm is an evolutionary computation technique [25], which is derived from the complex adaptive system (CAS). The algorithm was originally inspired by the regularity of the activity of birds, and then a simplified model was established based on swarm intelligence. In PSO, the solution of each optimization problem is a bird in the search space, called a "particle". PSO is initialized to a group of random particles and used to search the optimal solution by iterative evolution. In each iteration, the particles update themselves by tracking extremes of velocity and position. The above-mentioned behavior of the ith particles can be mathematically expressed as follows [48]:

$$\begin{cases} V_i^{n+1} = t \cdot V_i^n + c_1 \cdot r_1 \cdot (p_i^n - x_i^n) + c_2 \cdot r_2 \cdot (p_g^n - x_i^n) \\ x_i^{n+1} = x_i^n + V_i^n \end{cases} \tag{10}$$

where i = 1, 2, ···, K, K is the total number of particles, n is the current number of iteration. t is the inertia weight, p_i^n and p_g^n are the individual optimal position of the ith particle and the optimal position of all particles at the iteration of n, respectively. Vni and xni are the velocity and the current position of the ith particle, respectively. V_i^{n+1} and x_i^{n+1} are the updated velocity and position of the ith particle at the iteration of n + 1, respectively. c_1 and c_2 are learning factors, r_1 and r_2 are two random numbers, ranging from 0 to 1. The process of the PSO algorithm is displayed in Figure 2.

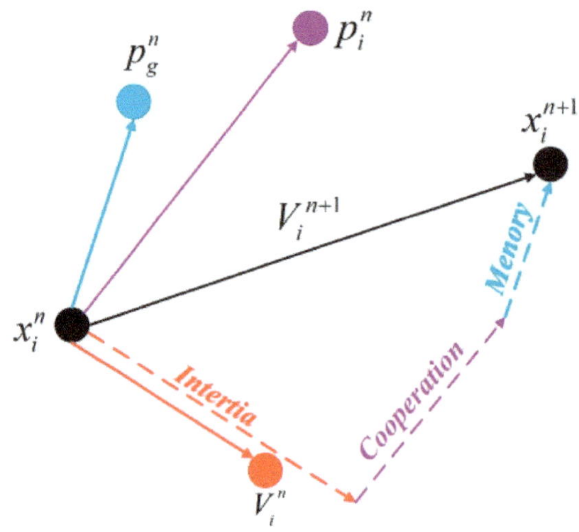

Figure 2: The process of the PSO algorithm.

The PSO-SVM Model

In order to improve the performance of the SVM model, the key issue is the selection of the parameters. Although the introduction of a kernel function can achieve the same purpose, the problem of selecting parameters of a kernel function still exists [22]. Combination of the PSO algorithm and SVM model can effectively solve this problem. Taking the RBF function as the kernel function, we demonstrate the flowchart of the PSO-SVM algorithm in Figure 3. To make this algorithm clearer, the details of this algorithm is briefly introduced in Table 1 as follows [22,49]:

Table 1: Procedures of the PSO-SVM algorithm

Input: Training and Verification Samples. **Output: The Result of the PSO-SVM Model.**	
1.	**Initialization parameters:** Generate initial particles comprised of C and γ of the SVM model. And set the PSO parameters including number of iterations, population size, maximum iteration number, the learning factors, the initial particle swarm location, the random flight velocity and two random number in range [0,1]. Initial the iteration = 0, and perform the training process from step **3–9** for each particle.
2.	**Data set:** Selection the training and verification samples.
3.	Set iteration $n = n + 1$.
4.	**SVM model training:** Conduct 10-fold cross validation (CV) on the training samples, and calculate the average CV accuracy based on the (C, γ).
5.	Evaluate its fitness by the average CV accuracy which is obtained in step **4**.
6.	Update the global and local optimal solution according to the result of the fitness evaluation.
7.	Each particle moves to its new location x_i^n by velocity V_i^n according to Equation (10).
8.	Until this iteration, the local optimal solution of the ith particle p_i^n, are compared with the new location x_i^n, the better will be the new p_i^n in the iteration n+1. And the same way on the p_g^n, which is the global optimal solution of all particles until iteration n.
9.	**Stop condition checking:** If the maximum iterations predefined are met, go to step **3**. Otherwise, go to step **10**.
10.	To avoid overtraining, stop training when the iteration has the best CV accuracy.
11.	Build the SVM model on the verification samples based on the SVM model optimal parameters (C, γ), which are obtained with the stopping iteration determined in the step **10**.
12.	End the training and verification procedure and get the result of the PSO-SVM model.

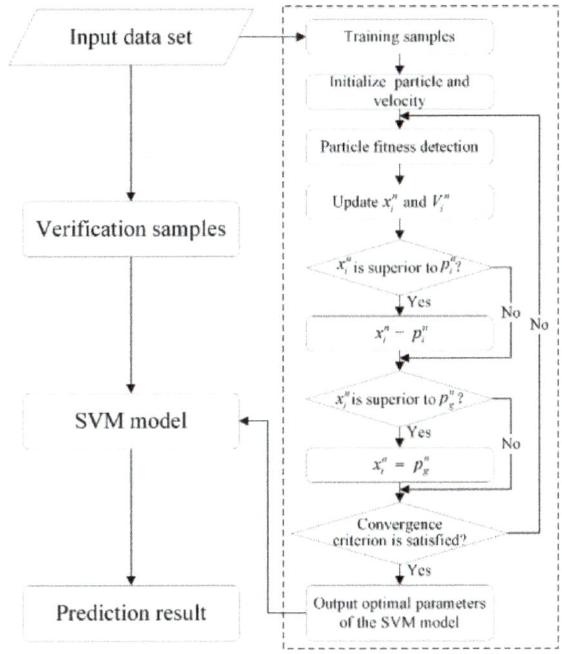

Figure 3: The flowchart of the PSO-SVM model.

THE PROPOSED GWR-PSO-SVM MODEL

In this work, we present a coupled model by combining the techniques of GWR, PSO and SVM. The flowchart of our method is summarized in Figure 4. In the following, each step of our method is briefly introduced.

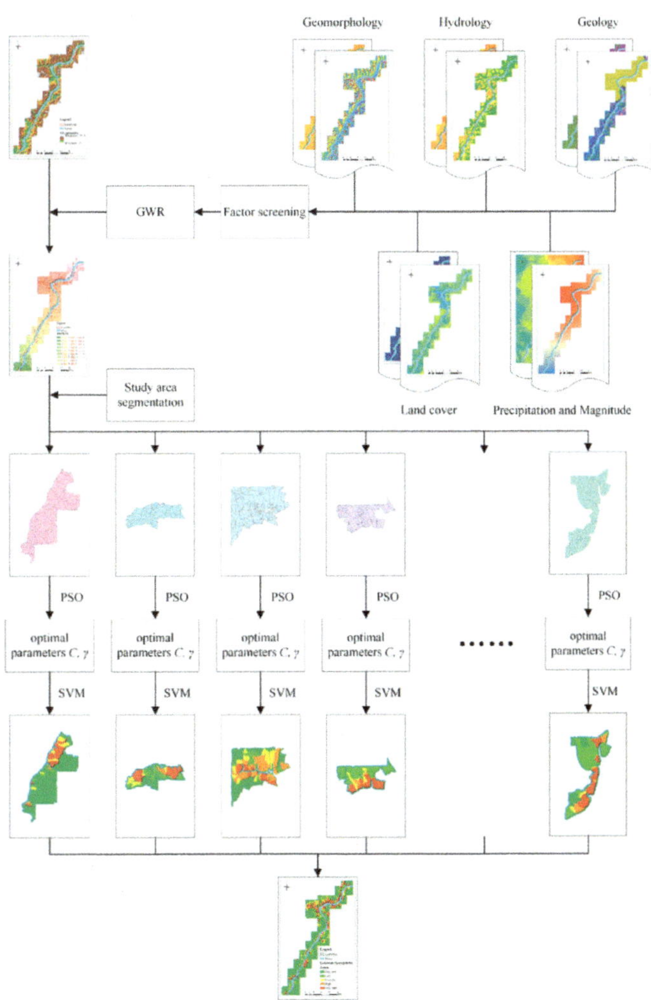

Figure 4: Flow-chart of our proposed method.

Factor Screening

It is well-known that some environmental factors have very high correlations. If our coupling model is constructed by using these factors, it may cause errors and cannot effectively improve prediction accuracies. Therefore, it is necessary to screen environmental factors. Correlation analysis is one of commonly used methods for the selection of environmental factors and is considered in our method. In addition, the required environmental factors are further screened based on their importance values. Finally, the remaining environmental factors are used for the subsequent landslide prediction.

Study Area Segmentation

It is well-known that GWR allows different relationships to exist at different points in the study area and improves the modeling performance by reducing spatial autocorrelations [50]. Based on Tobler's theory about nearness and similarity, observations which are nearer a certain location should have a greater weight in the estimation than observations which are further away [51]. Therefore, we can utilize this technique to estimate parameters for a model at some locations. To segment the study area, we produce and map GWR coefficient values to explore the spatial variability of relationships between the study area and the environmental factors.

The natural breaks method is a typical classification method, which is based on the inherent nature of the packet data [52]. Meanwhile, GWR coefficient values can be used to characterize the spatial autocorrelation of factors. Therefore, we prefer to cluster the study area into several classes in which the GWR coefficient values are greatly similar, with respect to each environmental factor. Meanwhile, it should be noted that the total class number makes great impact on the resultant segmentation maps. Specifically, if the value of N is very large, there are too many small partitions in the segmentation map, which causes the difficulties of constructing samples for training and verification and obtaining satisfactory prediction accuracies, as discussed in Section 6.3. In addition, spatial dependency cannot effectively reduced since the region centers are very close. Otherwise, if the value of N is too small, there are very few large partitions in the segmentation map, which means that spatial autocorrelations cannot be effectively alleviated in each region and greatly influence prediction results. Furthermore, our method cannot achieve regional scale landslide prediction due to very few prediction regions in the entire study area. To make it clearer, the influence of prediction regions is detailed discussed in Section 6.2.

To further weaken spatial autocorrelations, we prefer to superpose classification maps of the selected environmental factors, as shown in Figure 5. Meanwhile, the required environmental factors can be chosen according to importance values of all the environmental factors, measured by the SVM model. It can be observed that the superposition process is a simple intersection of all classes obtained from the most important environmental factors. In addition, the process always results in over-segmentation of the study area, though the GWR coefficient values in each region are consistent for individual environmental factor. As a result, spatial autocorrelations cannot be thoroughly removed since the Euclidean distance between a pair of prediction region centers is too close. In addition, it is very difficult to select training and verification samples for landslide prediction due to quite small regions in the study area. Therefore, it is necessary to merge these small regions in the superposed map. For this aim, the distribution of landslides in the study area should be considered, i.e., (i) prediction regions which separate landslides should be merged as one prediction region; (ii) adjacent small regions including landslides, which are far from other landslides area, should be merged into one prediction region; (iii) a large region without landslide should not be merged with regions containing landslide, as shown in Figure 6.

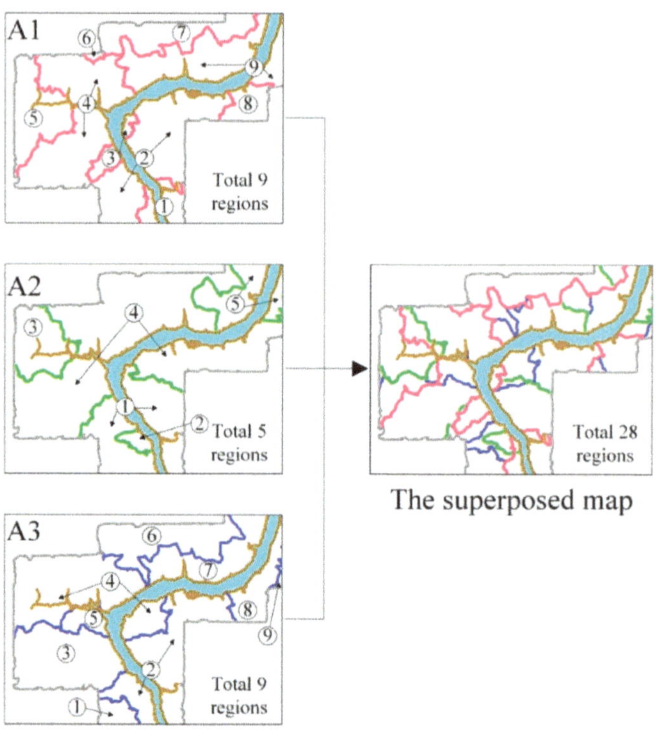

Figure 5: The superposition of classification maps of three environmental factors A1, A2 and A3.

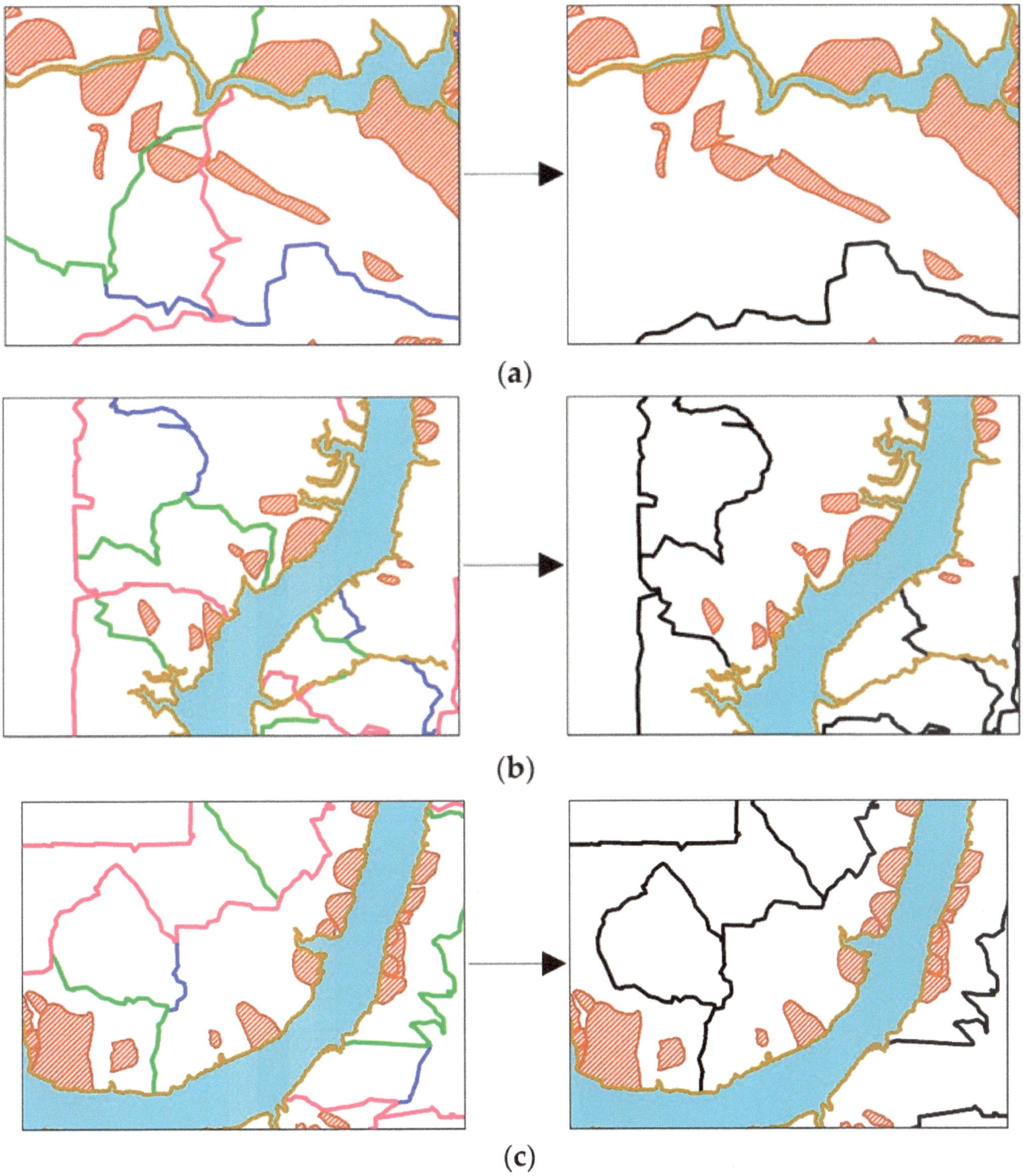

Figure 6: The merging of prediction regions. (**a**) The merging of prediction regions which separate landslides; (**b**) the merging of adjacent small regions including landslides, which are far from other landslides area; (**c**) the preservation of large regions without landslide.

The GWR-PSO-SVM Model

Once the study area is divided into several prediction regions by clustering GWR coefficients, the SVM model with the kernel function of RBF is used as the prediction component of the coupling model. Moreover, to improve the performance of prediction, the PSO algorithm is embedded into the SVM model to obtain the optimal parameters C and γ for each prediction region. The details of the GWR-PSO-SVM model for landslide prediction are shown in Table 2 as follows:

Table 2: Procedures of the GWR-PSO-SVM algorithm

Input: Ancillary Data of the Study Area.
Output: The Landslide Susceptibility Map.

Step 1: Extract environmental factors

✓ Extract environmental factors from ancillary data, including digital elevation models, geological maps, topographical maps and remote sensing images, *etc.* Note that all data should be resampled to the same spatial resolution.

✓ To each computing unit, a value is assigned to represent its corresponding environmental factor.

Step 2: Environmental factors screening

✓ Calculate the Pearson product-moment correlation coefficient (PPMCC) between any pair of environmental factors and exclude the environmental factors with high correlations. If the PPMCC value is greater than a predefined threshold T_1, the corresponding environmental factors are excluded according to the actual situation of the study area and previous research works.

✓ Calculate the importance value in the SVM model for each remaining environmental factor. In this work, the importance values, which are greater than a predefined threshold T_2, are preserved as the final ones for the corresponding environmental factors. Finally, these environmental factors are used for the subsequent landslide prediction.

Step 3: Study area segmentation

✓ Select an appropriate kernel function and information criterion method according to Equations (4) and (5), respectively.

✓ Calculate a GWR coefficient for each computing unit of each environmental factor according to Equations (1)–(3) by inputting the geographic coordinates of each center point and the values of all computing unit mentioned in **Step 1**.

✓ Divide each environmental factor into N classes using the natural breaks method based on GWR coefficient values. In this work, M environmental factors, which are determined in **Step 1**, are chosen for study area segmentation. As a result, M classification maps are produced.

✓ Superpose all the classification maps to obtain a superposed map and merge very small regions in this map to generate a final prediction region map according to Figure 6.

STUDY AREA AND DATA

General Characteristics

The Three Gorges span from the western Sichuan Basin upstream to the eastern Jianghan Basin downstream [53]. Wanzhou is a district of Chongqing Municipality, bordering Sichuan Province to the northwest and Hubei Province to the southeast. It is one of the main ports of the Yangtze River basin and the important industrial, cultural, trade and transportation center in Yudong. The site covers an area of 3457 km^2 and lies between longitudes of 107°52'22"–108°53'25" and latitudes of 30°24'25"–31°14'58", belonging to the subtropical moist climate zone, with a mild climate and abundant rainfall. The annual average precipitation is 1191.3 mm and around 70% of the annual precipitation falls from May to September. Our study area is located in the center of Wanzhou district, distributed along the 80 km-long Yangtze River, with an area of 552 km^2 and its elevation is between 21 m and 1015 m, as shown in Figure 7.

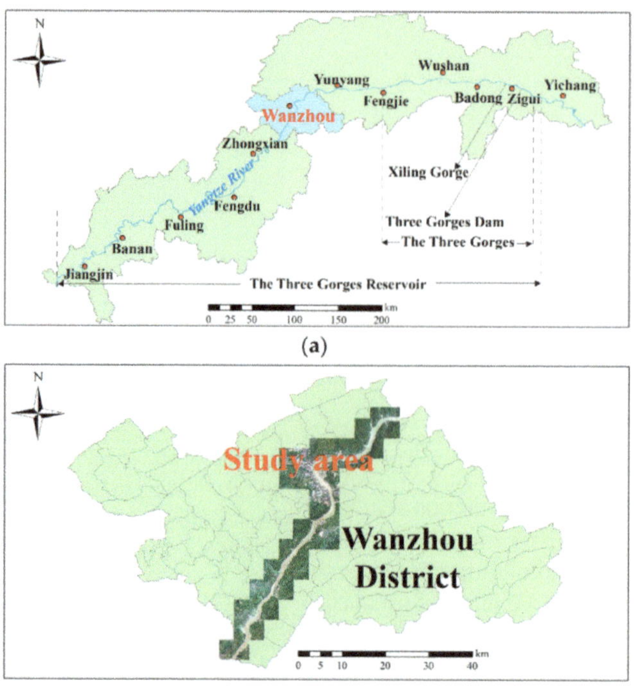

(b)

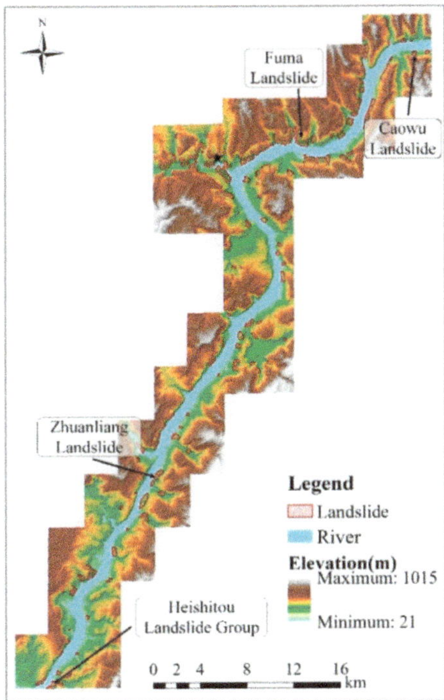

(c)

Figure 7: Location map of the study area. (**a**) Site map of the Three Gorges Reservoir; (**b**) site map of our study area; (**c**) digital elevation mode (DEM) overlaid with previously investigated landslides. The red hatched regions represent previously investigated landslides in the study area.

Geological Setting

The Wanzhou district is located at the two wings of the Wanxian synclinorium of the Eastern Sichuan fold belt. Meanwhile, anticline and syncline exist alternately in this area and construct a typical ejective fold structure [54]. The geological and tectonic framework map and a schematic geologic cross-section of the study area are shown in Figure 8a,b, respectively [55].

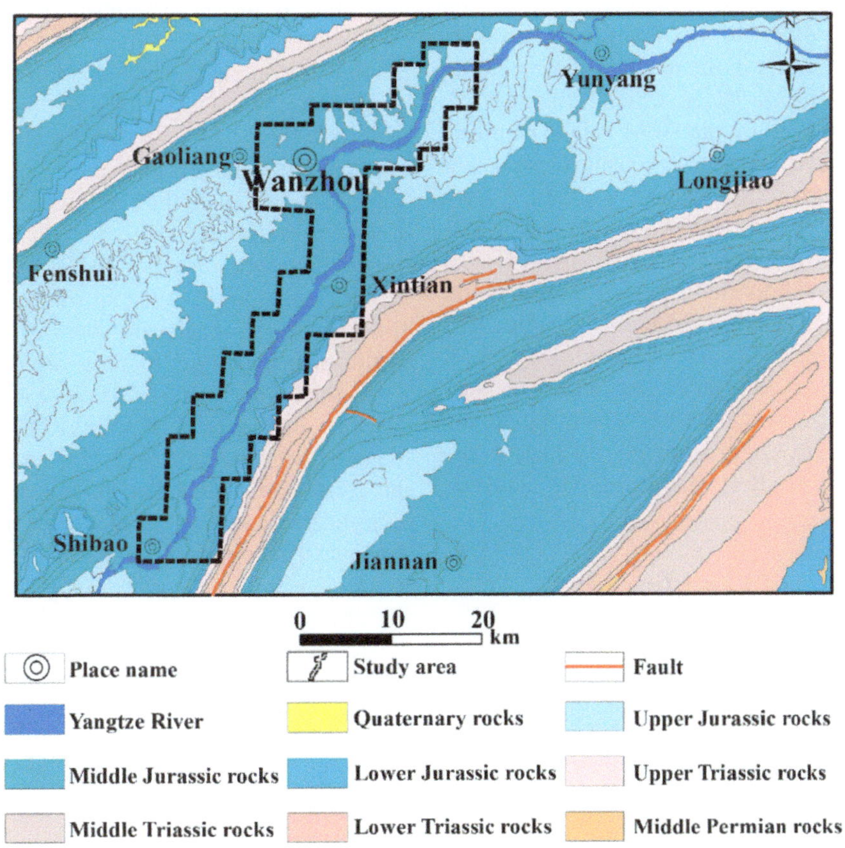

(a)

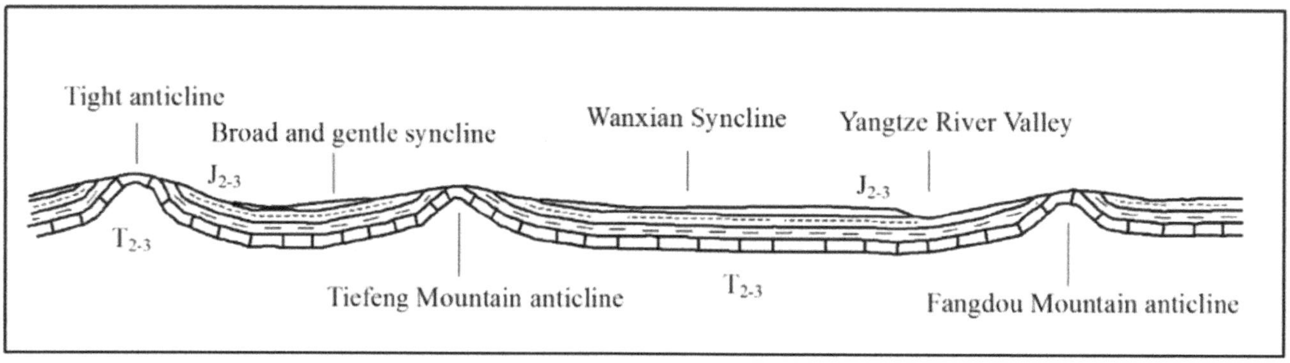

(b)

Figure 8: Geological maps of the study area. (**a**) Geological and tectonic sketch; (**b**) a schematic geological cross-section.

Description of Landslides

In the study area, the accurate sizes and shapes of previously investigated landslides can be extracted from the Headquarters of Prevention and Control of Geo-Hazards in Area of Three Gorges Reservoir [56]. In addition, high-resolution aerial photographs are used to detect neogenic landslides which are caused by the impoundment of the Three Gorges Project from 2003, while historical and literature data are employed to identify previous landslides, which were activated during Holocene and/or Pleistocene age, before the impoundment of the Three Gorges Project. In this work, 233 landslides were mapped in the study area.

Note that we cannot obtain terrain data under the Yangtze River, since there are no such information recorded in topographic maps or Advanced Spaceborne Thermal Emission and Reflection Radiometer (ASTER) G-DEM data. As a result, DEM values always vary greatly at the junction between both sides and surface of the Yangtze River, which influences the environmental factors produced by the DEM data. Therefore, we excluded the Yangtze River from the study area. For prediction, computing units are automatically obtained from high-quality digital terrain models (DTMs) by the slope-units method, which is used to partition the territory into hydrological regions between drainage and divide lines [57]. In this work, our study area is divided into 1909 slope-units, including 416 for landslides with total areas of 24.06 km^2, covering 4.36% of the study area. It can be observed from Figure 7c that the sizes of landslides in this area are very different. For instance, the Fuma landslide with an area of approximately 1.12 km^2 is the biggest landslide, while the smallest Xianjia 6 group landslide has an area of 3539.77 m^2.

Environmental Factors of Landslides

In this work, ancillary data used for extraction of environmental factors are the following:

- High-resolution aerial photographs;
- 1:50,000-Scale geological maps [55];
- ASTER G-DEM data with a spatial resolution of 30 m;
- Landsat-8 OLI+ sensor data, acquired on 24 February 2013, with the Path/Row number of 127/39 and its spatial resolution of 30 m for the extraction of land-use and calculation of Normalized Difference Vegetable Index (NDVI) and Normalized Difference Water Index (NDWI);
- Precipitation and seismic data from the China Meteorological Administration and the China Earthquake Administration for obtaining the precipitation and seismic factors.

Many researchers have verified the correlations between various environmental factors and landslide occurrence [58]. Based on these contributions and the characteristics of the study area, 29 environmental factors are selected to predict the potential distribution of landslides, including geomorphological, geological, hydrological, land cover, meteorological and geophysical factors. The selected environmental factors and their original values are listed in Table 1. In particular, the classification for the bedding structure is shown in Table 3. This factor is based on the topography bedding intersection angle (TOBIA) index [59] using the slope aspect, slope angle, bed dip direction and bed dip angle in space. In addition, the numbers of landslides corresponding to different bedding structures are demonstrated in Figure 9. From this figure, landslide failure can be caused by any type of slope in Figure 9. It should be mentioned that there are many horizontal strata landslides in the study area [60]. Since the formation mechanism of this type of landslides is very complicated and beyond the scope of this article, the gently dipping structure is not addressed in this work. Meanwhile, the figure depicts that there are very strong relationships between the different types of slope and the occurrence of landslides. Therefore, this factor is an important indicator of landslide and should be taken into account for prediction.

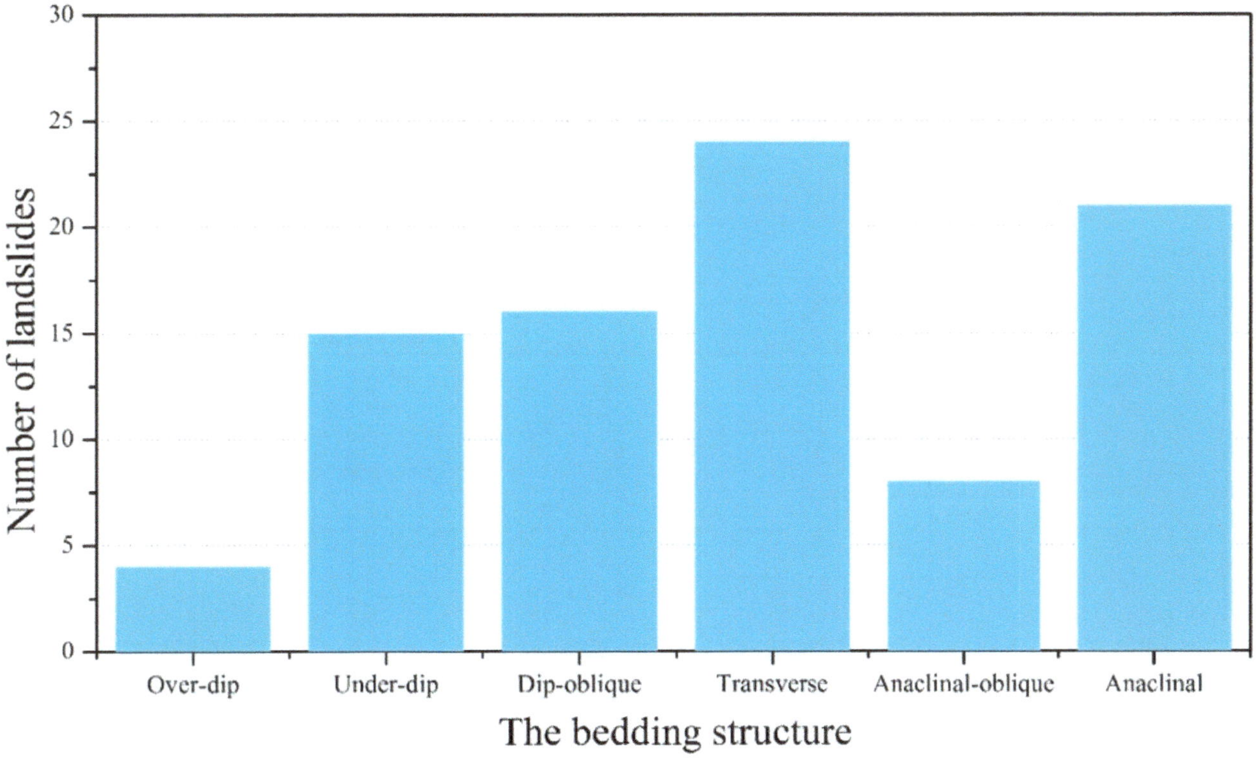

Figure 9: Statistical results of the numbers of landslides corresponding to different bedding structures.

Table 3: Classification of the bedding structure

Type	Definition
Over-dip slope	$\|\alpha-\beta\| \in [0°,30°)$ *or* $\|\alpha-\beta\| \in [330°,360°), \gamma > 10°$ *and* $\delta > \gamma$
Under-dip slope	$\|\alpha-\beta\| \in [0°,30°)$ *or* $\|\alpha-\beta\| \in [330°,360°), \gamma > 10°$ *and* $\delta < \gamma$
Dip-oblique slope	$\|\alpha-\beta\| \in [30°,60°)$ *or* $\|\alpha-\beta\| \in [300°,330°)$
Transverse slope	$\|\alpha-\beta\| \in [60°,120°)$ *or* $\|\alpha-\beta\| \in [240°,300°)$
Anaclinal-oblique slope	$\|\alpha-\beta\| \in [120°,150°)$ *or* $\|\alpha-\beta\| \in [210°,240°)$
Anaclinal slope	$\|\alpha-\beta\| \in [150°,210°)$

α: Slope aspect; β: bed dip direction; γ: bed dip angle; δ: slope angle.

It is known that the slope-units method is different from the grid-cells one, because the former is irregular, which means that the resultant areas by the slope-units method are different from each other. Therefore, the first problem of the slope-units method is that how to assign a normalized value to each slope-unit. If the original value of an environmental factor in Table 4 is a continuous variable, such as elevation, slope angle, terrain surface convexity and so on, the mean value of this factor is computed as the normalized value of the corresponding slope-unit, while if the original value of an environmental factor is a discrete variable, such as slope form, lithology, bedding structure and land-use, the most frequently occurring value of this factor is used as the value of this slope-unit. By using this idea, the 1909 slope-units are assigned to a unique value of each factor. To obtain landslide susceptibility of the study area, this value is used in all prediction models in this work.

Table 4: Landslide environmental factors and their respective values

	Environmental Factors	Value
Geomorphology	Elevation (m)	124.2727–922.3077
	Slope angle (°)	3.2045–36.2898
	Slope aspect (°)	28.4827–321.5051
	Terrain surface convexity (°/100m)	0.5979–0.2449
	Plane curvature (°/100m)	−0.4023–0.4832
	Profile curvature (°/100m)	−1.2441–1.2856
	Slope form	(1) V/V; (2) GE/V; (3) X/V; (4) V/GR; (5) GE/GR; (6) X/GR; (7) V/X; (8) GE/X; (9) X/X
	Slope height (m)	374.6390–3.6325
	Mid-slope position	0.1272–0.9491
	Terrain surface texture	0.8495–0.3018
	Terrain roughness index	1.1589–16.4521
	Terrain convergence index	−27.6027–19.7669
	Terrain curvature (°/100m)	−1.5762–1.4682
	Terrain position index	−14.6285–9.5591
Geology	Lithology	(1) mudstone, shale and Quaternary deposits; (2) sandstones and thinly bedded limestones; (3) limestones and massive sandstones
	Bedding structure	(1) over-dip slope; (2) under-dip slope; (3) dip-oblique slope; (4) transverse slope; (5) anaclinal-oblique slope; (6) anaclinal slope
Hydrology	Catchment area (m^2)	1156.0378–105,783.4666
	Catchment slope (°)	0.0485–0.5675
	Flow path length (m)	50.1196–2352.5587
	Valley depth (m)	3.4642–258.2873
	Stream power index	−617,299.4571–281,486.9383
	Distance from drainage (m)	18.4328–5637.6471
	Topographic wetness index	8.2193–14.7816
	Vertical distance to channel network (m)	−184.3475–461.4196
Land cover	Land-use	(1) water; (2) residential; (3) forest; (4) agriculture; (5) grassland
	NDVI	−0.4856–0.8337
	NDWI	0.0206–0.69411
Meteorology	Precipitation (mm)	1134.0551–1192.7400
Geophysics	Magnitude (Ms)	1.2617–2.1209

RESULTS

Experimental Results of the GWR-PSO-SVM Model

As mentioned in Section 3.3, the classical PPMCC is used to weaken the correlations of the selected environmental factors and $T_1 = 0.5$. For simplicity, correlations of geomorphological and hydrological factors are listed in Table 5 andTable 6 and 10 factors are excluded for all the models used here. As a result, the remaining 19 environmental factors are relatively independent and can be further screened based on their importance values ranging from 0 to 0.205, as illustrated in Figure 10, obtained using SPSS Clementine 12 software (IBM, Armonk, NY, USA). To this end, we set T_2= 0.02 and exclude the environmental factor whose importance value is less than T_2. Finally, 12 environmental factors are selected for the construction of the coupling model, i.e., catchment slope, distance from drainage, NDVI, bedding structure, slope angle, topographic wetness index, precipitation, lithology, NDWI, vertical distance to channel network, land-use and elevation.

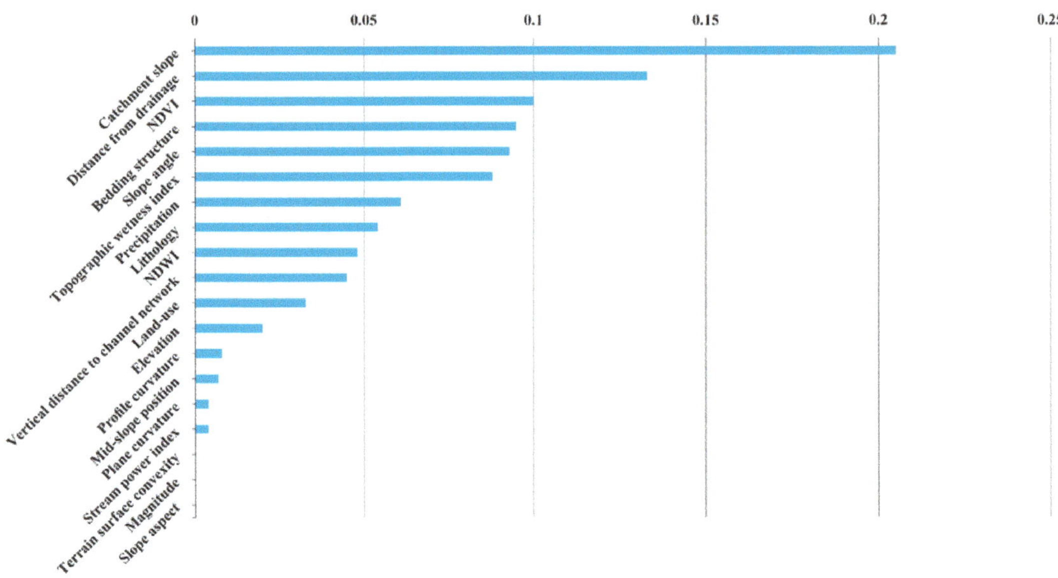

Figure 10: Importance values of the remaining 19 environmental factors.

Table 5: Correlations of geomorphological factors

Environmental Factor	ELE	SLAN	SLAS	SLHE	SLFO	TST	TRI	TPI	TCI	MSLP	PLCU	PRCU	TCU	TSC
ELE	1													
SLAN	0.198	1												
SLAS	0.022	−0.099	1											
SLHE	0.321	0.581	0.03	1										
SLFO	−0.013	0.093	0.16	0.206	1									
TST	−0.255	−0.739	0.045	−0.562	−0.022	1								
TRI	0.188	0.995	−0.105	0.579	0.091	−0.735	1							
TPI	0.125	0.138	0.122	0.338	0.761	−0.062	0.133	1						
TCI	0.117	0.054	0.221	0.241	0.787	−0.013	0.047	0.810	1					
MSLP	0.08	0.007	0.015	0.176	−0.163	−0.143	0.025	−0.15	−0.16	1				
PLCU	−0.103	0.112	0.187	0.162	0.735	−0.052	0.114	0.601	0.641	−0.093	1			
PRCU	−0.172	−0.103	−0.08	−0.224	−0.564	0.017	−0.095	−0.809	−0.661	0.14	−0.3	1		
TCU	0.071	0.131	0.155	0.243	0.782	−0.04	0.127	0.889	0.804	−0.15	0.728	−0.872	1	
TSC	0.083	0.155	−0.015	0.356	0.169	0.172	0.142	0.204	0.165	0.021	0.034	−0.2	0.161	1

ELE = elevation, SLAN = slope angle, SLAS = slope aspect, SLHE = slope height, SLFO = slope form, TST = terrain surface texture, TRI = terrain roughness index, TPI = terrain position index, TCI = terrain convergence index, MSLP = mid-slope position, PLCU = plane curvature, PRCU = profile curvature, TCU = terrain curvature, TSC = terrain surface convexity.

Table 6: Correlations of hydrological factors

Environmental Factor	DISD	CMA	FPL	TWI	VADE	CMSL	SPI	VDCN
DISD	1							
CMA	0.011	1						
FPL	−0.109	0.551	1					
TWI	−0.026	0.607	0.545	1				
VADE	−0.112	0.678	0.675	0.65	1			
CMSL	−0.007	0.327	0.41	0.411	0.638	1		
SPI	−0.055	−0.013	0.004	−0.112	−0.052	−0.004	1	
VDCN	−0.368	0.259	0.424	0.222	0.475	0.292	0.045	1

DISD = distance from drainage, CMA = catchment area, FPL = flow path length, TWI = topographic wetness index, VADE = valley depth, CMSL = catchment slope, SPI = stream power index, VDCN = vertical distance to channel network.

According to the selection criterion mentioned in Section 3.2, the most important environmental factors, i.e., catchment slope, distance from drainage and NDVI, are selected as the regional division factors, whose GWR coefficients are obtained by exploiting an adaptive bi-square kernel and AIC in the GWR method. The GWR coefficient values of catchment slope are shown in Figure 11. It can be easily observed from the figure that different clusters with respect to GWR are spatially developed. Based on the relationship between GWR and spatial autocorrelation mentioned in Section 1, we can easily infer that the GWR coefficients in each cluster are very close. Consequently, spatial dependency are greatly reduced if each cluster is considered as a spatial variable. Therefore, it is possible that the study area can be partitioned into different prediction regions while spatial autocorrelations are very limited.

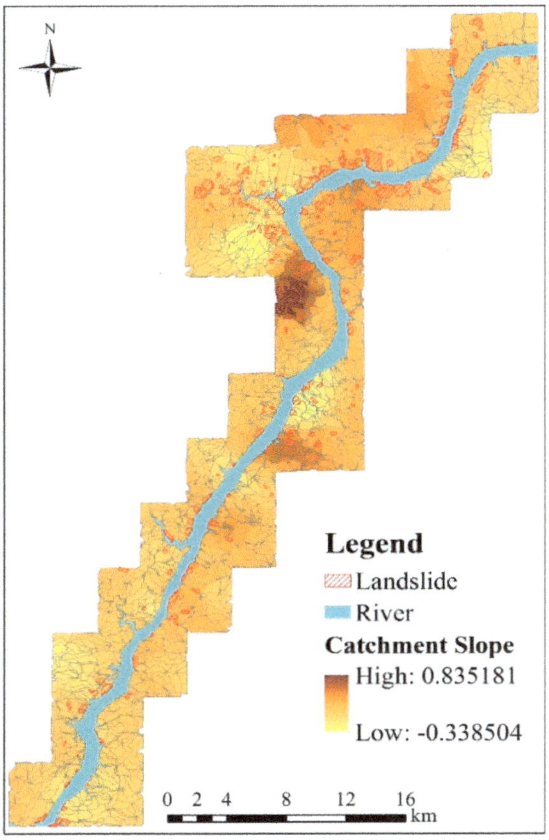

Figure 11: The GWR coefficient map of the catchment slope.

In this work, we set N = 3, i.e., these selected environmental factors are clustered into three classes by the natural breaks method and the corresponding classification maps are shown in Figure 12a–c. For convenience, the slope-unit without landslide is named as the non-landslide slope-unit, while the slope-unit including landslide is named as the landslide slope-unit. The result of simple superposition is shown in Figure 13a. According to the three rules for merging regions mentioned in Section 3.2, the study area is finally divided into 34 prediction regions by superposing all classification maps. For simplicity, each prediction region is assigned to a unique label, as shown in Figure 13b. It can be observed from this figure that 25 regions contain landslides in the study area. The numbers of the slope-units and the landslide slope-unit are listed in Table 7.

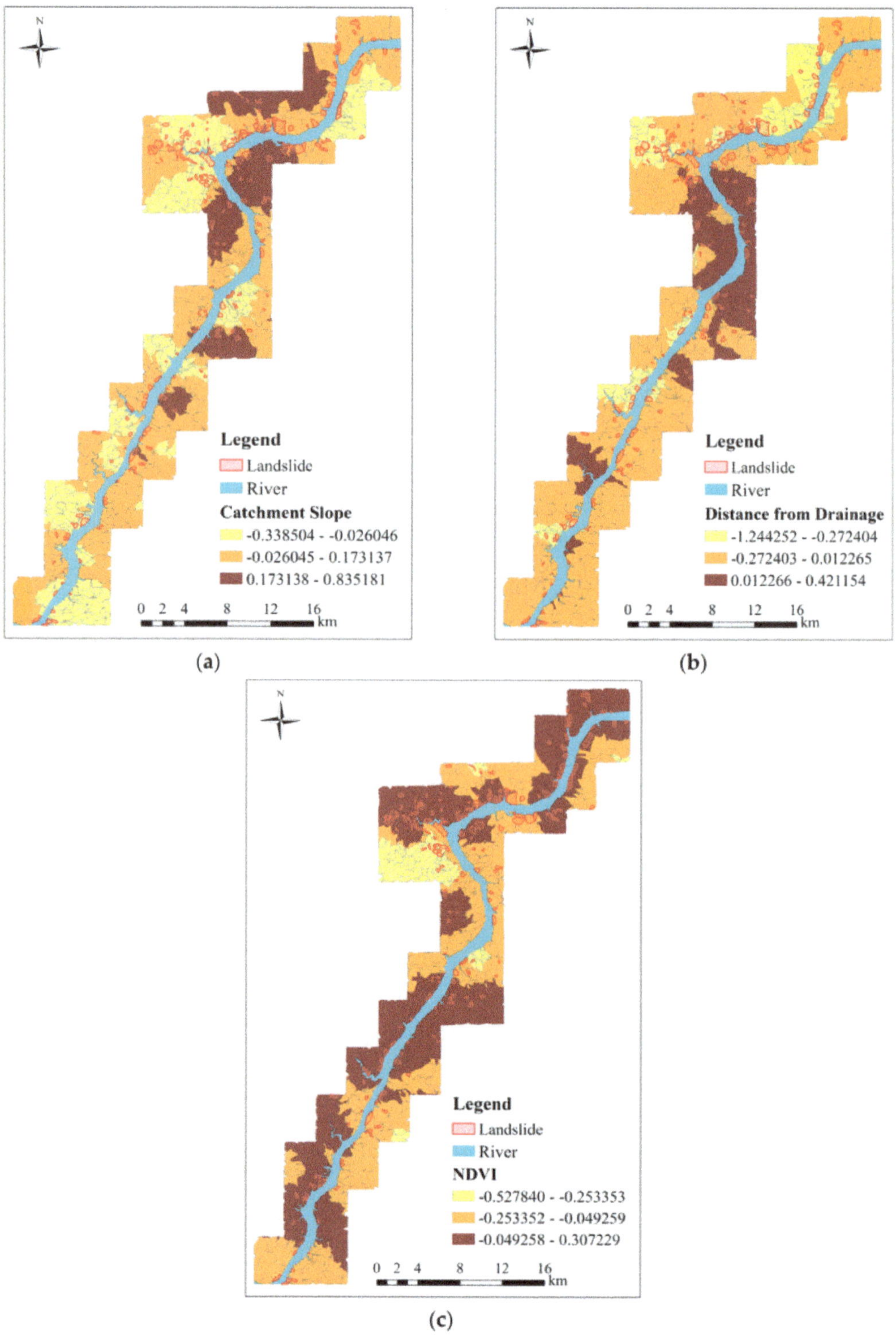

Figure 12: The GWR coefficient values and classification maps of environmental factors. (**a**) Catchment Slope; (**b**) the distance from drainage; (**c**) NDVI.

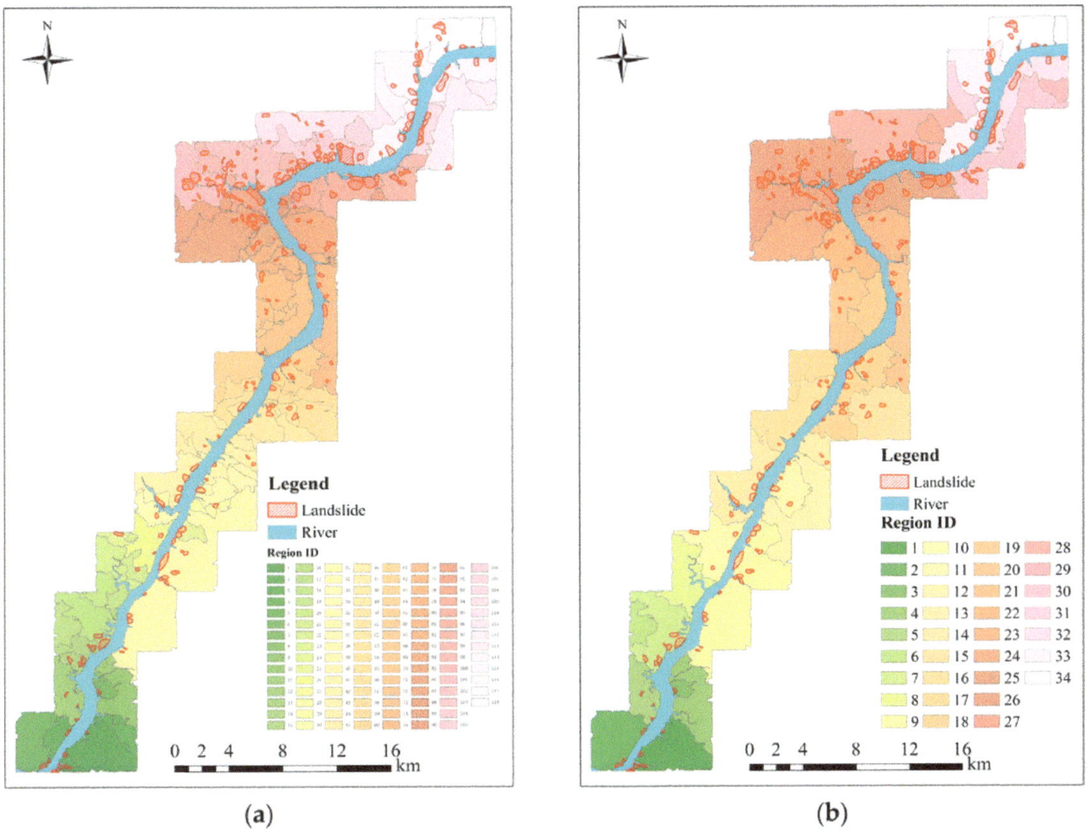

Figure 13: The resultant prediction regions of the study area. (**a**) The resultant prediction region map after the superposition process; (**b**) the final prediction region map after the merging process.

Table 7: The numbers of the total slope-units and the landslide slope-units for each prediction region

Region ID	Number of Slope-Units	Number of Landslide Slope-Units	Region ID	Number of Slope-Units	Number of Landslide Slope-Units
1	59	9	18	75	18
2	51	5	19	40	0
3	8	2	20	63	14
4	59	0	21	52	12
5	52	5	22	54	12
6	17	0	23	52	13
7	61	19	24	57	15
8	61	0	25	71	24
9	138	29	26	134	60
10	57	9	27	10	0
11	38	12	28	80	36
12	80	0	29	9	0
13	21	2	30	76	12
14	90	23	31	7	0
15	64	8	32	47	22
16	77	14	33	70	31
17	42	0	34	37	10

For the GWR-PSO-SVM prediction model, all of prediction regions must be sampled as input variables. For each prediction region in Figure 13b, the label of the landslide slope-unit is assigned as "1", while the label of the non-landslide slope-unit is assigned as "0". In our experiment, we use the same number of landslide slope-units and non-landslide slope-units in each prediction region to form training and verification samples. It can be observed from Figure 13 that the total number of non-landslide slope-units in each prediction region is always more than that of the landslide slope-units. Therefore, all of the landslide slope-units and the same number of the randomly selected non-landslide slope-units form the required samples. Meanwhile, the proposed GWR-PSO-SVM model is a local model, which generates the optimal C and γ of the SVM model for each prediction region by using the PSO algorithm, as shown in Table 8. It should be noted that the prediction regions without landslides are not included in this table. Meanwhile, we perform the SVM classifier to estimate the likelihood that each slope-unit contains the existing landslides and demonstrate the corresponding probability maps in Figure 14. The probability value in the map ranging from 0% to 100% represents the different degrees of landslide susceptibility.

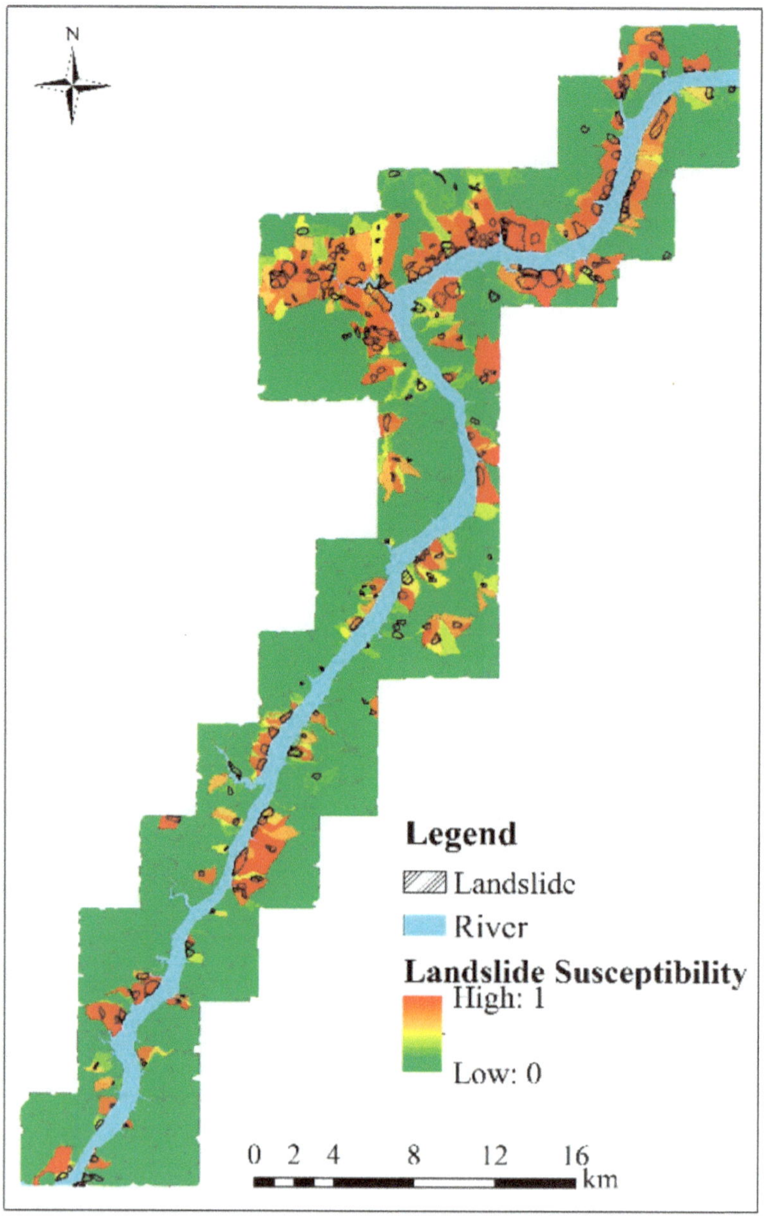

Figure 14: The landslide susceptibility map by the GWR-PSO-SVM model.

Table 8: The parameter settings of C and γ calculated by the PSO algorithm for the GWR-PSO-SVM model

	Region ID	*C*	*γ*	Region ID	*C*	*γ*
	1	6.1826	0.13879	20	5.9453	0.29134
	2	1.2965	0.32455	21	5.3659	0.38439
	3	2.4682	0.31596	22	3.3548	0.17105
	5	1.4832	0.36957	23	5.8234	0.36851
	7	8.6235	0.51243	24	2.1629	0.47592
GWR-PSO-SVM Model	9	4.1356	0.67572	25	3.2592	0.45665
	10	2.3659	0.49986	26	6.5359	0.67853
	11	2.6971	0.33645	28	6.2157	0.47935
	13	4.3651	0.42631	30	7.2853	0.63428
	14	5.8652	0.42375	32	6.4075	3.35874
	15	1.4964	0.56916	33	5.3364	0.47516
	16	4.7569	0.32793	34	4.8435	0.67203
	18	1.4259	0.47157		-	

Methods to Assess Models Performance

To objectively evaluate the performance of the models considered, three methods are utilized. The first measure is overall prediction accuracy, which is used to evaluate prediction correctness and can be defined as:

$$p = \frac{a+b}{S} \cdot 100\% \tag{11}$$

where a and b are the numbers of correctly predicted landslide and non-landslide slope-units in the landslide susceptibility maps, respectively. S is the total number of slope-units in the study area. According to (11), this measure can be appropriately applied to evaluate the global models, such as the SVM, PSO-SVM, RS-SVM models, by taking into account the entire study area. While it is used for the GWR-based models, the measure can be computed in each prediction region. In this work, the final measure of overall prediction accuracy is defined as follows:

$$p = \frac{\sum_{i=1}^{n_{pr}} (a_i + b_i)}{\sum_{i=1}^{n_{pr}} S_i} \cdot 100\% \tag{12}$$

where $i = 1,2,\ldots,n_{pr}$ (n_{pr} is total number of prediction regions), a_i and b_i are the numbers of correctly predicted landslide and non-landslide slope-units in the ith prediction region, respectively. S_i is the number of slope-units involved in the current prediction region.

The second measure is exploited to evaluate prediction accuracy of landslide areas in each class of landslide susceptibility maps obtained by the mentioned models according to the distribution of our study area. This measure is named as class-specific accuracy and is defined as follows:

$$p_j = \frac{A_j}{B_j} \cdot 100\% \tag{13}$$

where $j = 1,2,\cdots,M$ (M is total number of landslide susceptibility zones), A_j and B_j are the numbers of landslide slope-units and total slope-units in the jth landslide susceptibility zone, respectively. To perform this measure, our study area is classified into M landslide susceptibility zones. In this work, the fixed interval method is used to achieve this aim and it is based on previous studies to segment study areas by the predefined thresholds, which is widely used for comparison of multiple models [7,46,61].

The third measure is the classical receiver operation characteristic (ROC) curve and its area under curve (AUC). In a ROC curve the true positive rate (sensitivity) is plotted in function of the false positive rate (100-specificity) for different cut-off points. Each point on the ROC curve represents a sensitivity/specificity pair corresponding to a particular decision threshold. A test with perfect discrimination (no overlap in the two distributions) has a ROC curve that passes through the upper left corner (100% sensitivity, 100% specificity). Therefore, the closer a curve is to the upper left corner, the better are the prediction results [62].

Comparison with Further Models

To better demonstrate the performance of our model, several models are compared to our method, including: (1) the SVM model, in which the study area are globally used for sampling and prediction; (2) the PSO-SVM model, in which the PSO algorithm is used to obtain the optimal C and γ to improve prediction accuracies; (3) the landslide susceptibility mapping method based on rough set (RS) and SVM proposed by Peng et al. [46]. RS theory is an effective tool introduced by Pawlak [63] and discussed in many review papers [64,65,66,67,68,69,70]. This technique can deal with vagueness and uncertainty information and identify cause-effect relationships in databases as a form of data mining and knowledge discovery [46,63,71]. Meanwhile, it has been widely used in various disciplines of science [72], including remote sensing [73], geographic information science [74], and landslide susceptibility mapping [71],etc. In the work of [46], it was employed to select key environmental factors for landslide prediction; (4) the GWR-SVM model, which is a local model and similar to our coupling model, without the PSO step to obtain the optimal C and γ.

For a fair comparison, the same mapping unit and original environmental factors are used for all models used here. It should be noted that the RS-SVM model is different from the other models due to the fact that its input environmental factors are determined based on the RS theory after the PPMCC analysis. In our experiments, all of the remaining 12 factors are used for input variables for the SVM, PSO-SVM, GWR-SVM and our models, while 14 factors are selected based on the RS theory in the RS-SVM model, excluding land-use, mid-slope position, plane curvature, stream power index, terrain surface convexity from the remaining 19 factors.

It is well-known that the selection of samples for training and verification is a key step for the SVM prediction model. As mentioned above, the classical SVM, PSO-SVM and RS-SVM models can be considered as global ones due to the fact that the entire study area is taken into account for selecting samples, i.e., all of the landslide slope-units in the study area and the same number of the randomly selected non-landslide slope-units are used for training their respective SVM models, while all of the slope-units in the study area are utilized for verification. Nevertheless, the selection scheme of the remaining GWR-based models is performed for each prediction region, instead of the entire study area, as mentioned in Section 5.1. Therefore, the sample size of each model in this work is measured using the number of slope-units in the study area or each prediction region. Table 9 depicts the training and verification sample sizes of all the models. In addition, the PSO algorithm is used for the PSO-SVM and GWR-PSO-SVM models to obtain the optimal C and γ to improve prediction performance of the SVM model.

Table 9: The training and verification sample of the five models

Model	Region ID	Training Sample	Verification Sample	Region ID	Training Sample	Verification Sample
GWR-PSO-SVM and GWR-SVM	1	18	59	20	28	63
	2	10	51	21	24	52
	3	4	8	22	24	54
	5	10	52	23	26	52
	7	38	61	24	30	57
	9	58	138	25	48	71
	10	18	57	26	120	134
	11	24	38	28	72	80
	13	4	21	30	24	76
	14	46	90	32	44	47
	15	16	64	33	62	70
	16	28	77	34	20	37
	18	36	75			
SVM		832	1909			
PSO-SVM		832	1909			
RS-SVM		832	1909			

To make probability maps more readable, we can divide probability values by using fixed interval method in ArcGIS software into five susceptibility categories, i.e., very low, low, medium, high and very high, corresponding thresholds are fixed to 0.1, 0.35, 0.75 and 0.9, respectively, as shown in Figure 15. It can be observed from Figure 15that all of the models can achieve the purpose of landslide prediction. Meanwhile, the very high-susceptibility zones are apparently mapped in the main urban area of Wanzhou district in all the susceptibility maps, which accords with the fact that the previously investigated landslides are mainly distributed in this area. The distribution of high and very high-susceptibility zones is greatly different for each model.

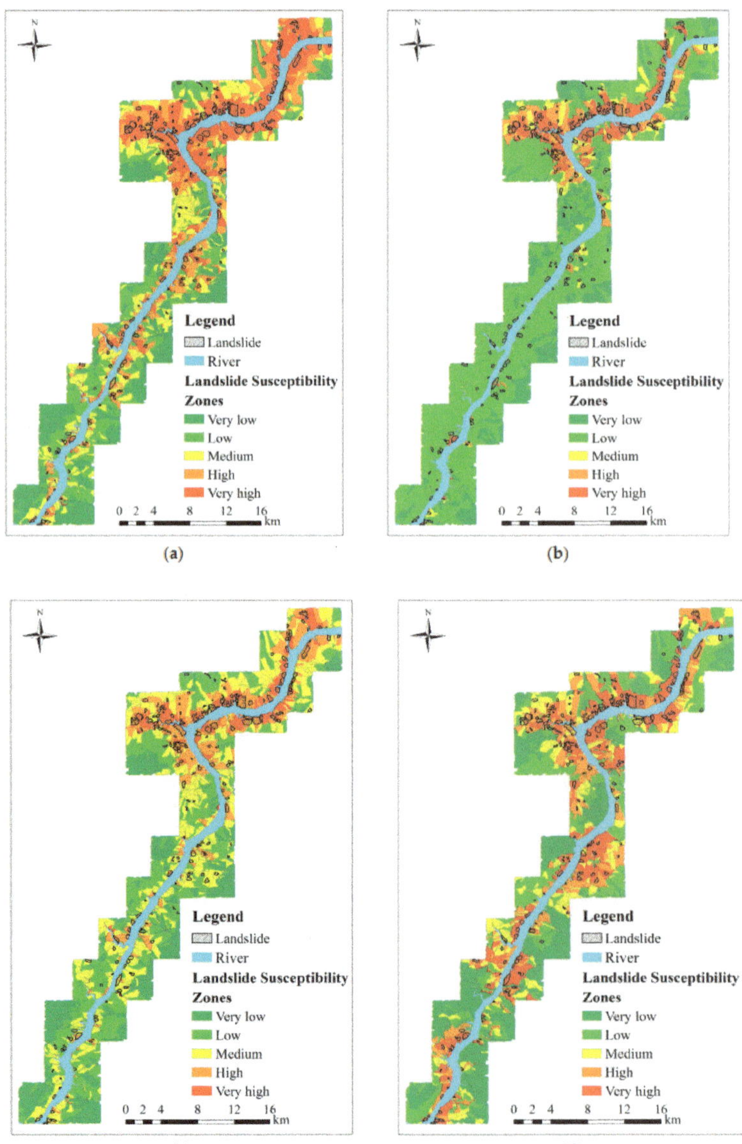

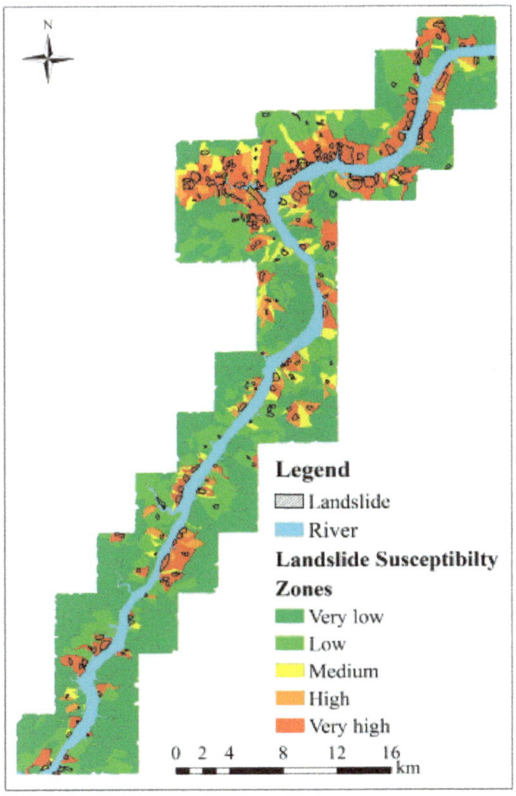

(e)

Figure 15: Landslide susceptibility zoning using the fixed interval method. (**a**) SVM; (**b**) PSO-SVM; (**c**) RS-SVM; (**d**) GWR-SVM; (**e**) GWR-PSO-SVM.

For instance, most of the previously investigated landslides are located in high or very high-susceptibility zones in the maps of the SVM, RS-SVM and GWR-SVM models. However, a large number of slope-units are unreliably classified by these models as high or very high-susceptibility zones as well. Landslides are typically a minority class in the study area, the PSO algorithm always results in local optima of the SVM model, when it is applied to the entire study area. As a consequence, the previously investigated landslides in the southwest of the study area cannot effectively be predicted by the PSO-SVM model. In contrast, the map by our model is consistent to the ground truth of landslide distribution. Although the PSO algorithm is used in our method to optimize the parameters in the SVM model, the division of our study area into prediction regions with appropriate sizes can greatly overcome trapping in local optimum. The high and very high-susceptibility zones mainly concentrate in the previously investigated landslide areas, while most of non-landslide areas are classified as low and very low-susceptibility zones, which guarantee the reliability of prediction results of landslide susceptibility. The overall accuracies of landslide susceptibility mapping by all the models used here are listed in Table 10.

Table 10: Overall accuracies by all the prediction models

Model	Correct	Total	Accuracy
SVM	1415	1909	74.12%
PSO-SVM	1590	1909	83.29%
RS-SVM	1427	1909	74.75%
GWR-SVM	1140	1584	71.97%
GWR-PSO-SVM	1443	1584	*91.10%*

In this table, the item of "Correct" indicates the number of slope-units that are correctly predicted in prediction regions, while the item of "Total" means the number of slope-units in prediction regions. It should be noted that this "total" number in the GWR-SVM and GWR-PSO-SVM models are calculated using the prediction regions including landslides. It is obvious that the GWR-PSO-SVM model can achieve the best prediction accuracy of 91.10%, which is 7.8%–19.1% higher than the traditional SVM-based models. To further compare the performance of all the models, the class-specific accuracies are shown in Figure 16. It can be clearly seen that the class-specific accuracy of the very high-susceptibility zone achieved by our model is highest (96.27%) when compared with the other models, which means that our model can detect the very high-susceptibility zones mainly including the previously investigated landslides.

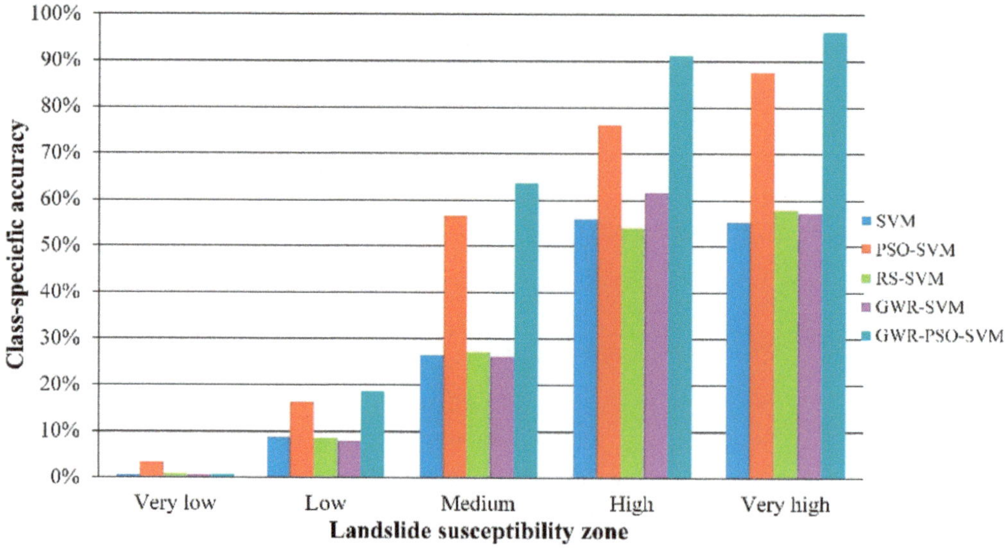

Figure 16: The class-specific accuracies by different prediction models using the fixed interval method.

The ROC curves of all the methods are plotted in Figure 17. It is known that the closer the ROC curve is to the upper left corner, the higher the overall accuracy of the test is. As can be observed from Figure 17, we can obtain similar conclusions as for the two previous evaluation measures, i.e., the GWR-PSO-SVM model can achieve the best prediction result. Meanwhile, the ROC plots of the GWR-SVM and the RS-SVM models are pretty close to each other. Since the PSO algorithm is not very robust when it is applied to the whole study area, the ROC plot of the PSO-SVM model is not continuous and is close to the upper left corner when the value (of the 1-specificity) is 0.2, but worse than the RS-SVM model, GWR-SVM and our models when the value is larger than 0.2. In addition, the corresponding AUC is listed in Table 11. The larger the value of AUC, the better the performance of the prediction model. As shown in this table, our model can produce the largest area of 0.971, when compared with the other models.

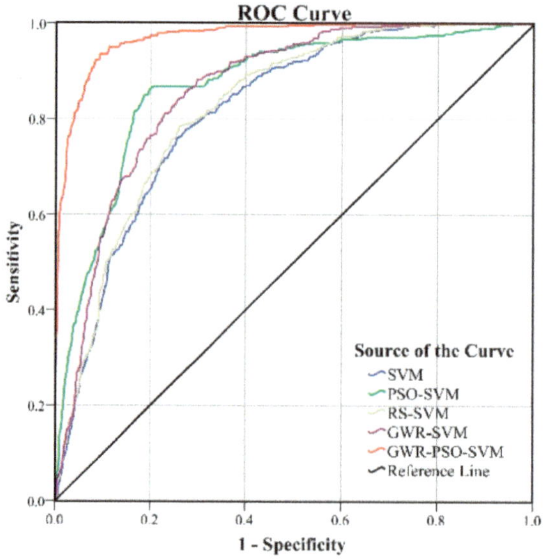

Figure 17: The ROC curve of all the prediction models.

Table 11: The AUC of four models

Model	Area	Std. Error	Asymptotic Sig.	Asymptotic 95% Confidence Interval	
				Lower Bound	Upper Bound
SVM	0.817	0.011	0.000	0.796	0.837
PSO-SVM	0.869	0.010	0.000	0.850	0.889
RS-SVM	0.825	0.010	0.000	0.804	0.845
GWR-SVM	0.860	0.009	0.000	0.842	0.878
GWR-PSO-SVM	0.971	0.004	0.000	0.963	0.978

Std. = Standard; Sig. = Significant.

It should be noted that there are a few non-landslide regions in the prediction region map (Figure 12b), since landslides are typically a minority class in the study area. To compare the performance of our model with the global models, we assume in this work that the overall prediction accuracies of these non-landslide regions are 100%, which may improve the overall accuracy of the entire study area. Meanwhile, our experiments not reported here confirm that the AUC value of our model can still reach 0.962 by removing these non-landslide regions from the study area. Furthermore, all the prediction models were applied to Zigui to Badong section in the Three Gorges Reservoir for landslide susceptibility mapping. The experimental results demonstrated that the GWR-PSO-SVM model can obtain the best prediction result as well and the AUC value of 0.965, which is highest among all the models. Therefore, the universality of our model can be validated. Finally, to objectively compare our model with the other models, we select the same number of landslide slope-units and non-landslide slope-units in each prediction region. Although the number of training samples is relatively small in certain prediction regions, the influence on the overall prediction accuracy is very limited.

DISCUSSION

Impact of Environmental Factors

It should be noted that the global and regional prediction results of the study area are always different, mainly due to two reasons. The first one is the prediction model. Since the SVM model has been used as a universal model and can obtain satisfactory results, it is exploited by all the models used here for landslide susceptibility mapping. The second one is the impact of environmental factors. There are several crucial environmental factors for landslide prediction, such as elevation, slope angle and so on. However, the most crucial factors are different in different parts of the study area. For instance, the environmental factor of distance from drainage is greatly significant for landslide failures in the area along the Yangtze River, while slope angle may be the most important environmental factor in the areas far away from the Yangtze River. Therefore, the introduction of the GWR technique into landslide susceptibility mapping may avoid these two problems and improve the prediction accuracy. The importance values of all the environmental factors in each prediction region, obtained using SPSS Clementine 12 software, are displayed in Figure 18. It can be observed that the importance values of the final 12 environmental factors produced in Section 5.1 at each prediction region are different. Meanwhile, in each prediction region, the rank of each environmental factor in terms of the important value is greatly different.

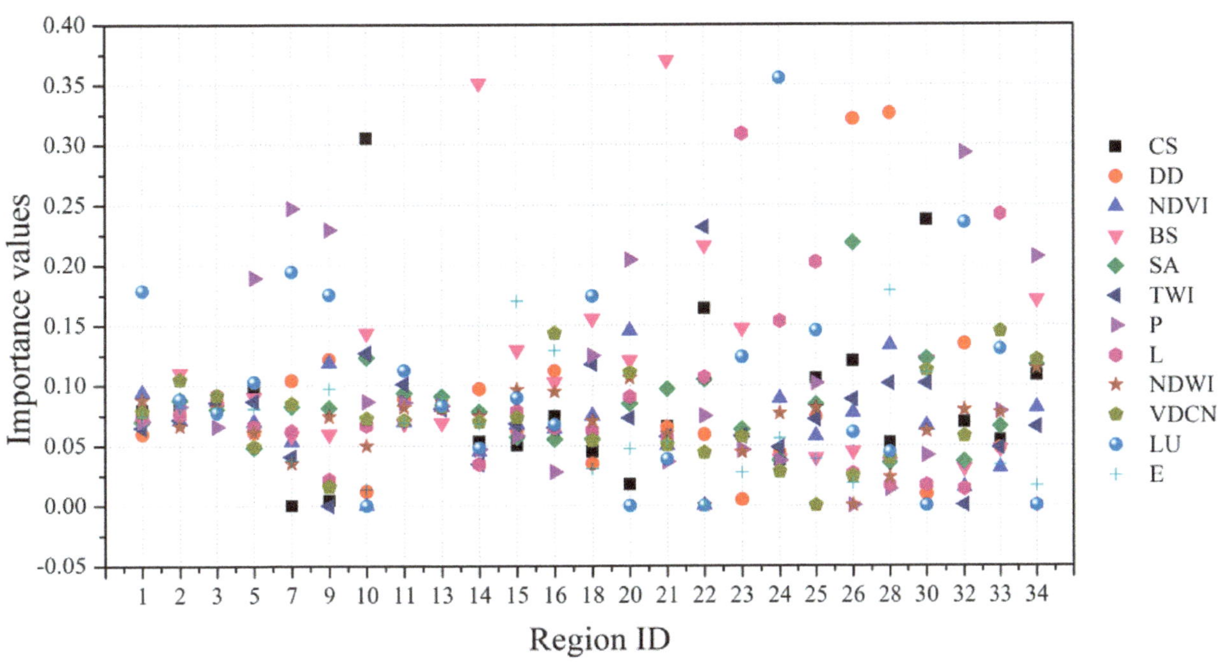

Figure 18: The importance values of the environmental factors in different prediction regions.

Influence of Regions Number

To demonstrate the impact of the performance of segmentation of the study area, the resultant segmentation maps, with respect to different values of N from 2 to 4, are shown in Figure 19. In Figure 19a, the study area is divided into 10 prediction regions when N = 2 which may avoid the problem that the importance rank of each environmental factor is not the same in different prediction regions. However, the impact of each environmental factor in different spatial positions is not taken into account.

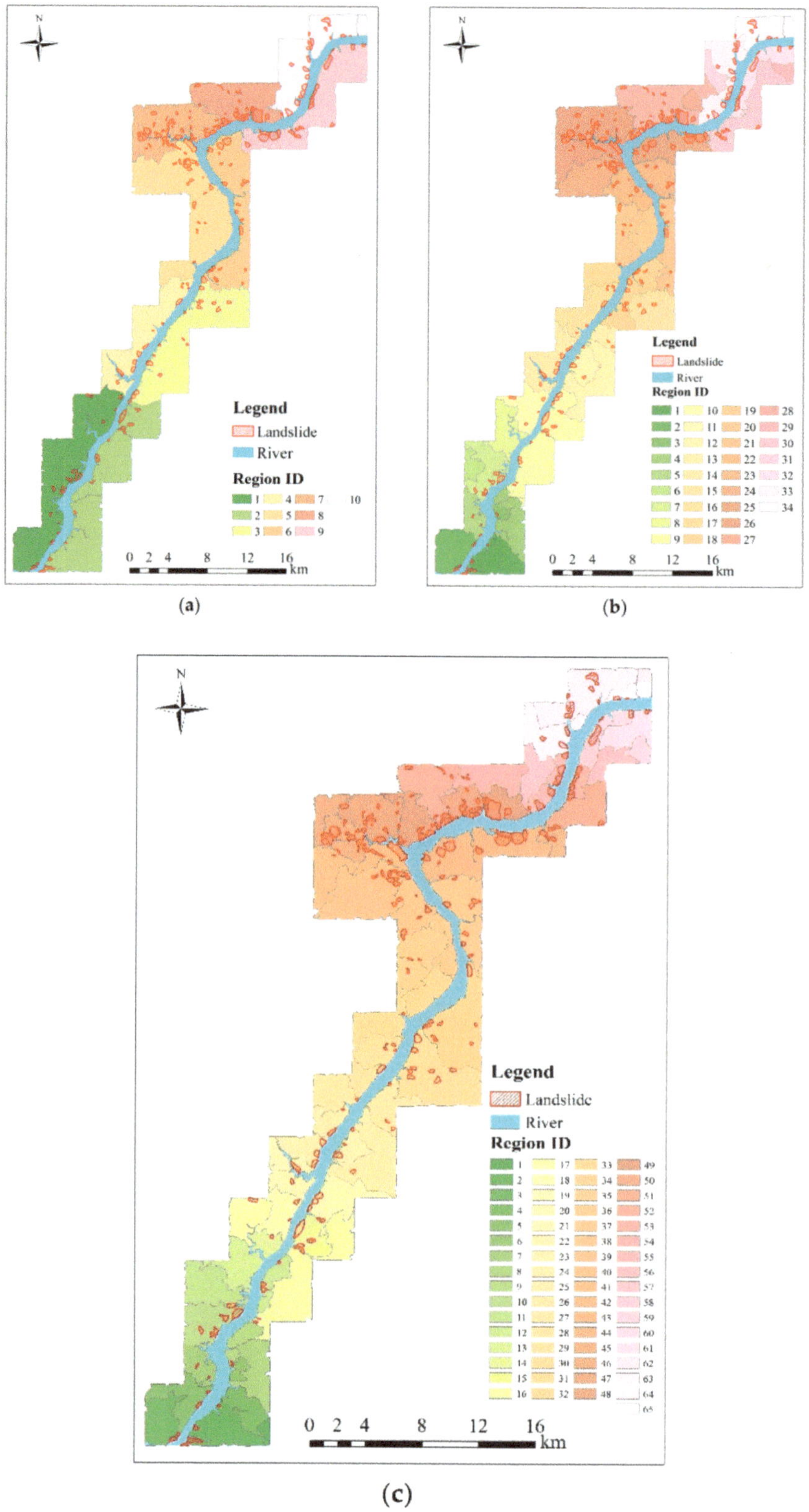

Figure 19: The performance of segmentation with respect to different values ofN. (**a**) N = 2; (**b**) N = 3; (**c**) N = 4.

For instance, all the prediction regions are produced distributing from the Yangtze River to boundaries of the study area, but the importance rank of each environmental factor may greatly change in different parts of each prediction region, which cannot be carefully considered in prediction models if prediction regions are very large. In Figure 19c, the study area is segmented into 65 prediction regions if N = 4. In this way, the slope units may be very few in prediction regions. As a consequence, the landslide and non-landslide slope units in each prediction region are not enough to constitute required samples, which influences landslide prediction accuracies. In contrast, our study area in this work is divided into 34 prediction regions by choosing N = 3 and different impacts of environmental factors in these regions are effectively utilized into prediction models. In addition, the size of each prediction region is appropriate for obtaining the required samples, as shown in Figure 19b.

Model Sensitivity

To evaluate the sensitivity of the proposed model to the number of prediction and verification samples, five prediction regions, which have the most landslide slope-units, are selected to obtain ROC curves of the prediction performance by choosing five different percentages of required sample sets, i.e., 20%, 40%, 60%, 80% and 100%. The corresponding prediction regions in our study area and their ROC curves are depicted in Figure 20. In general, the higher percentage of the required samples we choose, the better the prediction performance, i.e., the prediction accuracy of our model is highest when using all of the required samples, while it is lowest when only 20% of the required samples are used in our model. The prediction results are greatly determined by the selection of samples due to the complexity of landslides in the study area. If training samples are very small, we cannot extract valuable information from environmental factors, which makes it difficult for our model to guarantee accuracies of landslide prediction. In addition, the selection of the required samples in each prediction region results in fewer training samples for prediction. As a result, the prediction accuracy of our model is lower as the training samples are reduced.

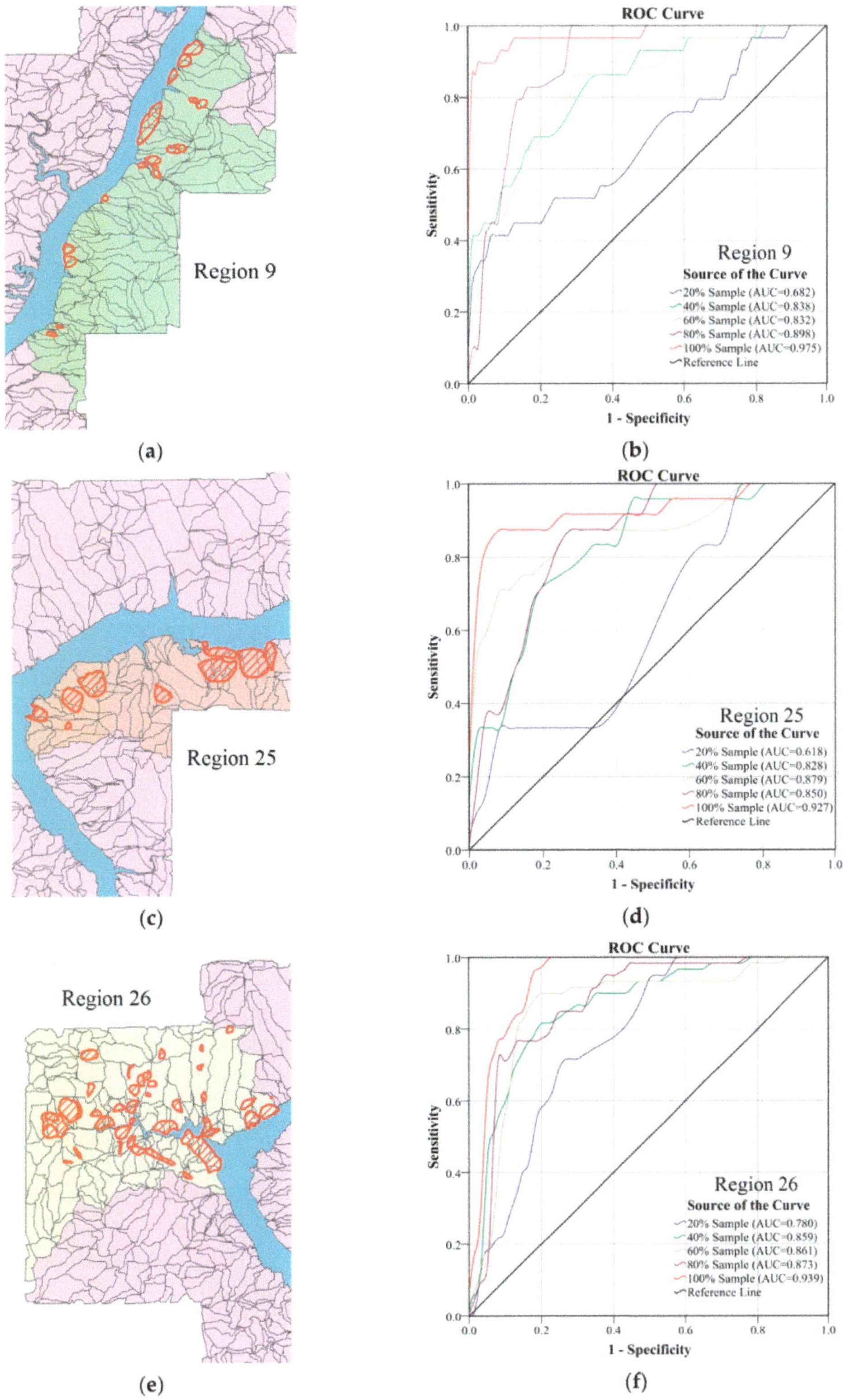

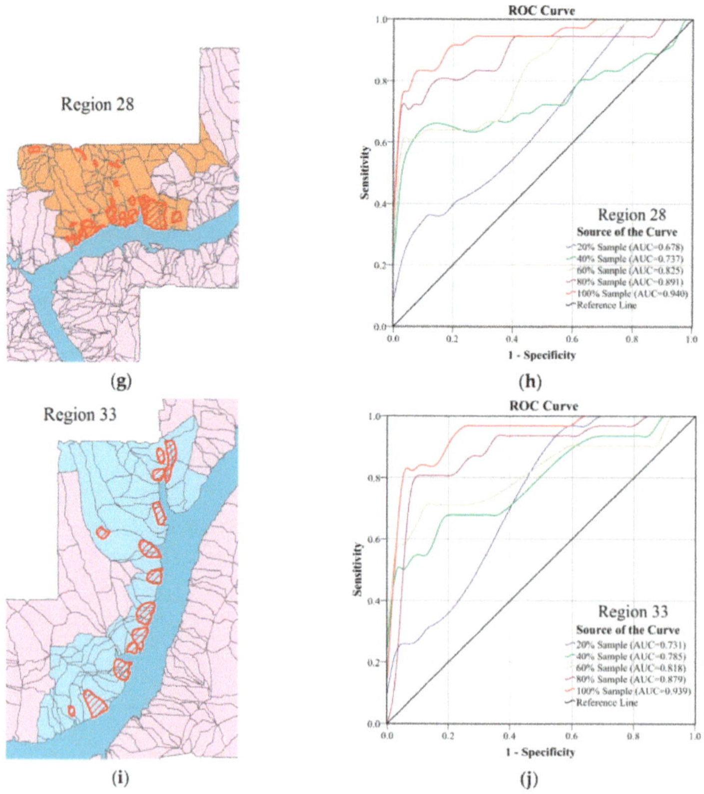

Figure 20: The ROC curves with different sample sets for the five regions. (a) The prediction region No. 9; (b) the ROC curves and AUCs corresponding to (a); (c) the prediction region No. 25; (d) the ROC curves and AUCs corresponding to (c); (e) the prediction region No. 26; (f) the ROC curves and AUCs corresponding to (e); (g) the prediction region No. 28; (h) the ROC curves and AUCs corresponding to (g); (i) the prediction region No. 28; (j) the ROC curves and AUCs corresponding to (i).

CONCLUSIONS

In this paper, an effective PSO-SVM method based on the GWR technique is presented for landslide susceptibility mapping at a local scale by integrating multisource data of the Wanzhou district in the middle of the Three Gorges Reservoir, China. It has been reported that landslide events occurred in the last three years in the main urban area of the Wanzhou district. In this work, a GWR algorithm is used in our model to segment the study area into a series of prediction regions with appropriate sizes by clustering slope units. Then, a PSO-SVM prediction model is applied to each prediction region for landslide susceptibility mapping. This allows the proposed GWR-PSO-SVM model can obtain accurate landslide susceptibility maps at a regional scale. Experimental results demonstrate that coupling different models as in the GWR-PSO-SVM model can achieve better prediction performance, when compared to the traditional SVM-based models. Meanwhile, these landslide prediction models are comprehensively evaluated using three objective measures including the overall prediction accuracy, the landslide susceptibility class-specific accuracies, and the ROC curves and AUC values. We can draw the following conclusions: (1) The GWR-PSO-SVM model can obtain the best overall accuracy of 91.10%; (2) The GWR-PSO-SVM model can achieve the highest class-specific accuracy of 96.27% with respect to the very high-susceptibility zones, which are mainly covered with the previously investigated landslides; (3) The GWR-PSO-SVM model can achieve a more reliable ROC curve and a higher AUC value of 0.971. Therefore, our model can achieve superior prediction performance to the traditional prediction models. In future, a further improvement can be achieved by selecting more reasonable segmentation factors and performing segmentation postprocessing.

ACKNOWLEDGMENTS

This research is supported by the National Natural Science Foundation of China (61271408), the National High-tech R&D Program of China (Grant 2012AA121303) and Key Laboratory of Precise Engineering and Industry Surveying, National Administration of Surveying, Mapping and Geoinformation (Project No.: PF2012-21). We are grateful to the Headquarters of Prevention and Control of Geo-Hazards in Area of the Three Gorges Reservoir for providing data and material. We also thank the editor and anonymous referees for their comments.

REFERENCES

1. Cruden, D.M.; Varnes, D.J. Landslides: Investigation and Mitigation; Chapter 3; Transportation Research Board Special Report: Washington, DC, USA, 1996.
2. Liu, C.; Liu, Y.; Wen, M.; Li, T.; Lian, J.; Qin, S. Geo-hazard initiation and assessment in the three gorges reservoir. In Landslide Disaster Mitigation in Three Gorges Reservoir, China; Springer: Berlin, Germany, 2009; pp. 3–40.
3. Brabb, E.E. The world landslide problem. In Episodes; US International Union of Geological Sciences (IUGS): Bangalore, India, 1991; Volume 14, pp. 52–61.
4. Wan, S.; Chang, S.H. Combined particle swarm optimization and linear discriminant analysis for

landslide image classification: Application to a case study in Taiwan. Environ. Earth Sci. 2014, 72, 1453–1464.

5. Regmi, N.R.; Giardino, J.R.; Vitek, J.D. Assessing susceptibility to landslides: Using models to understand observed changes in slopes. Geomorphology 2010, 122, 25–38.
6. Barredo, J.E.I.; Benavides, A.; Herv, A.S.J.; van Westen, C.J. Comparing heuristic landslide hazard assessment techniques using GIS in the Tirajana basin, Gran Canaria Island, Spain. Int. J. Appl. Earth Obs. Geoinf. 2000, 2, 9–23.
7. Dai, F.; Lee, C. Landslide characteristics and slope instability modeling using GIS, Lantau Island, Hong Kong. Geomorphology 2002, 42, 213–228.
8. Ohlmacher, G.C.; Davis, J.C. Using multiple logistic regression and GIS technology to predict landslide hazard in northeast Kansas, USA. Eng. Geol. 2003, 69, 331–343.
9. Ayalew, L.; Yamagishi, H. The application of gis-based logistic regression for landslide susceptibility mapping in the Kakuda-Yahiko Mountains, Central Japan. Geomorphology 2005, 65, 15–31.
10. Lee, S.; Ryu, J.H.; Min, K.; Won, J.S. Landslide susceptibility analysis using GIS and artificial neural network. Earth Surf. Process. Landforms 2003, 28, 1361–1376.
11. Gomez, H.; Kavzoglu, T. Assessment of shallow landslide susceptibility using artificial neural networks in Jabonosa River Basin, Venezuela. Eng. Geol. 2005, 78, 11–27.
12. Wang, H.; Sassa, K. Rainfall-induced landslide hazard assessment using artificial neural networks. Earth Surf. Process. Landforms 2006, 31, 235–247.
13. Melchiorre, C.; Matteucci, M.; Azzoni, A.; Zanchi, A. Artificial neural networks and cluster analysis in landslide susceptibility zonation. Geomorphology 2008, 94, 379–400.
14. Yao, X.; Tham, L.G.; Dai, F. Landslide susceptibility mapping based on support vector machine: A case study on natural slopes of Hong Kong, China. Geomorphology 2008, 101, 572–582.
15. Marjanović, M.; Kovačević, M.; Bajat, B.; Voženílek, V. Landslide susceptibility assessment using SVM machine learning algorithm. Eng. Geol. 2011, 123, 225–234.
16. Ballabio, C.; Sterlacchini, S. Support vector machines for landslide susceptibility mapping: The Staffora River Basin case study, Italy. Math. Geosci. 2012, 44, 47–70.
17. Xu, C.; Dai, F.; Xu, X.; Lee, Y.H. GIS-based support vector machine modeling of earthquake-triggered landslide susceptibility in the Jianjiang River watershed, China. Geomorphology 2012, 145, 70–80.
18. Erener, A.; Düzgün, H.S.B. Improvement of statistical landslide susceptibility mapping by using spatial and global regression methods in the case of More and Romsdal (Norway). Landslides 2010, 7, 55–68.
19. Sabokbar, H.F.; Roodposhti, M.S.; Tazik, E. Landslide susceptibility mapping using geographically-weighted principal component analysis. Geomorphology 2014, 226, 15–24.
20. San, B.T. An evaluation of svm using polygon-based random sampling in landslide susceptibility mapping: The Candir Catchment area (Western Antalya, Turkey). Int. J. Appl. Earth Obs. 2014, 26, 399–412.
21. Yao, X.; Zhang, Y.; Zhou, N.; Guo, C.; Yu, K.; Li, L.J. Application of two-class SVM applied in landslide susceptibility mapping. In Project Planning and Project Success: The 25% Solution; Taylor & Francis Group: England, UK, 2014; p. 203.
22. Huang, C.-L.; Dun, J.-F. A distributed PSO–SVM hybrid system with feature selection and parameter optimization. Appl. Soft Comput. 2008, 8, 1381–1391.
23. Zhao, H.; Yin, S. Geomechanical parameters identification by particle swarm optimization and support vector machine. Appl. Math. Model. 2009, 33, 3997–4012.
24. Ren, F.; Wu, X.; Zhang, K.; Niu, R. Application of wavelet analysis and a particle swarm-optimized support vector machine to predict the displacement of the shuping landslide in the Three Gorges, China. Environ. Earth Sci. 2015, 73, 4791–4804.
25. Kennedy, J.; Eberhart, R. Particle swarm optimization. In Proceedings of the IEEE International of First Conference on Neural Networks, Perth, Australia, 27 November 1995–1 December 1995; pp. 1942–1948.
26. Holland, J.H. Genetic algorithms and the optimal allocation of trials. SIAM J. Comput. 1973, 2, 88–105.
27. Hassan, R.; Cohanim, B.; De Weck, O.; Venter, G. A comparison of particle swarm optimization and the genetic algorithm. In Proceedings of the 46th AIAA Multidisciplinary Design Optimization Specialist Conference, Austin, Texas, 18–21 April 2005; pp. 18–21.
28. Brunsdon, C.; Fotheringham, A.S.; Charlton, M.E. Geographically weighted regression: A method for exploring spatial nonstationarity. Geogr. Anal. 1996, 28, 281–298.
29. Fotheringham, A.S.; Charlton, M.; Brunsdon, C. The geography of parameter space: An investigation of spatial non-stationarity. Int. J. Geogr. Inf. Syst. 1996, 10, 605–627.
30. Schabenberger, O.; Gotway, C.A. Statistical Methods for Spatial Data Analysis; CRC Press: Boca Raton, FL, USA, 2004.
31. Fischer, M.M.; Getis, A. Handbook of applied spatial analysis: Software tools, methods and applications. Springer Science & Business Media: Berlin, Germany, 2009.
32. Nakaya, T. Local spatial interaction modelling based on the geographically weighted regression approach. In Modelling Geographical Systems; Springer: Dordrecht, The Netherlands, 2002; pp. 45–69.
33. Fotheringham, A.S.; Brunsdon, C.; Charlton, M. Geographically Weighted Regression: The Analysis of Spatially Varying Relationships; John Wiley & Sons: New York, NY, USA, 2003.

34. Feuillet, T.; Coquin, J.; Mercier, D.; Cossart, E.; Decaulne, A.; Jónsson, H.P.; Sæmundsson, B. Focusing on the spatial non-stationarity of landslide predisposing factors in Northern Iceland: Do paraglacial factors vary over space? Prog. Phys. Geogr. 2014, 38, 354–377.
35. Brunsdon, C.; Fotheringham, S.; Charlton, M. Spatial nonstationarity and autoregressive models. Environ. Plann. A 1998, 30, 957–973.
36. Wheeler, D.C. Geographically weighted regression. In Handbook of Regional Science; Springer: Berlin, Germany, 2014; pp. 1435–1459.
37. Fotheringham, A.S.; Charlton, M.; Brunsdon, C. Measuring spatial variations in relationships with geographically weighted regression. In Recent Developments in Spatial Analysis; Springer: Berlin, Germany, 1997; pp. 60–82.
38. Chalkias, C.; Kalogirou, S.; Ferentinou, M. Landslide susceptibility, Peloponnese peninsula in south Greece. J. Maps 2014, 10, 211–222.
39. Fotheringham, A.S.; Charlton, M.; Brunsdon, C. Geographically weighted regression: A natural evolution of the expansion method for spatial data analysis. Environ. Plan. A 1998, 30, 1905–1927.
40. Celik, M.; Kazar, B.M.; Shekhar, S.; Boley, D. Parameter Estimation for the Spatial Autoregression Model: A Rigorous Approach. Available online: http://www-users.cs.umn.edu/~boley/publications/papers/NASA06.pdf (accessed on 23 February 2016).
41. Cleveland, W.S. Robust locally weighted regression and smoothing scatterplots. J. Am. Stat. Assoc. 1979, 368, 829–836.
42. Hirotugu, A. A new look at the statistical model identification. Autom. Control Comput. Sci. 1974, 6, 716–723.
43. Vapnik, V.N. The Nature of Statistical Learning Theory; Springer New York: New York, NY, USA, 1995.
44. Barakat, N.; Bradley, A.P. Rule extraction from support vector machines: A review. Neurocomputing 2010, 74, 178–190.
45. Mountrakis, G.; Im, J.; Ogole, C. Support vector machines in remote sensing: A review. ISPRS J. Photogramm. Remote Sens. 2011, 66, 247–259.
46. Peng, L.; Niu, R.; Huang, B.; Wu, X.; Zhao, Y.; Ye, R. Landslide susceptibility mapping based on rough set theory and support vector machines: A case of the Three Gorges Area, China. Geomorphology 2014, 204, 287–301.
47. Suykens, J.A.; Vandewalle, J. Least squares support vector machine classifiers. Neural Process. Lett. 1999, 9, 293–300.
48. Kennedy, J. Particle swarm optimization. In Encyclopedia of Machine Learning; Springer: Berlin, Germany, 2011; pp. 760–766.
49. Xiao, C.; Hao, K.; Ding, Y. The bi-directional prediction of carbon fiber production using a combination of improved particle swarm optimization and support vector machine. Materials 2014, 8, 117–136.
50. Zhang, L.; Bi, H.; Cheng, P.; Davis, C.J. Modeling spatial variation in tree diameter–height relationships. For. Ecol. Manag. 2004, 189, 317–329.
51. Miller, H.J. Tobler's first law and spatial analysis. Ann. Assoc. Am. Geogr. 2004, 94, 284–289.
52. Jenks, G.F. The data model concept in statistical mapping. Int. Yearb. Cartogr. 1967, 7, 186–190.
53. Li, J.; Xie, S.; Kuang, M. Geomorphic evolution of the yangtze gorges and the time of their formation. Geomorphology 2001, 41, 125–135.
54. Zhou, Q.; Lv, Z.; Ma, Z.; Zhang, Y.; Wang, H. Barrier belt division based on rs and gis in the three gorges reservoir area—A case of Wanzhou district. Procedia. Environ. Sci. 2011, 10, 1257–1263.
55. Hubei Province Geological Survey. Geological Map of Zigui and Badong County (1:50,000); Hubei Province Geological Survey Press: Wuhan, China, 1997.
56. Headquarters of Prevention and Control of Geo-Hazards in Area of Three Gorges Reservoir; 1:10,000 Geological Hazard Mapping Database: Yichang, China, 2011.
57. Guzzetti, F.; Carrara, A.; Cardinali, M.; Reichenbach, P. Landslide hazard evaluation: A review of current techniques and their application in a multi-scale study, central Italy. Geomorphology 1999, 31, 181–216.
58. Liu, J.G.; Mason, P.J.; Clerici, N.; Chen, S.; Davis, A.; Miao, F.; Deng, H.; Liang, L. Landslide hazard assessment in the three gorges area of the Yangtze River using ASTER imagery: Zigui–Badong. Geomorphology 2004, 61, 171–187.
59. Meentemeyer, R.K.; Moody, A. Automated mapping of conformity between topographic and geological surfaces. Comput. Geosci. 2000, 26, 815–829.
60. Kunlong, Y.; Wenxing, J.; Yang, W.; Chunmei, Z.; Changqian, M.; Liling, L.; Lingling, Y.; Yiping, W. The Research of Three Gorges Reservoir in Wanzhou Area of Nearly Horizontal Strata Landslide Formation Mechanism and Control Engineering; China University of Geosciences Press: Wuhan, China, 2007; pp. 39–65.
61. Chen, T.; Niu, R.; Du, B.; Wang, Y. Landslide spatial susceptibility mapping by using GIS and Remote Sensing techniques: A case study in Zigui county, the Three Georges Reservoir, China. Environ. Earth Sci. 2015, 73, 5571–5583.
62. Zweig, M.H.; Campbell, G. Receiver-operating characteristic (ROC) plots: A fundamental evaluation tool in clinical medicine. Clin. Chem. 1993, 39, 561–577.
63. Pawlak, Z. Rough sets. Int. J. Comput. Inf. Sci. 1982, 11, 341–356.
64. Liu, J.; Zeng, Z.; Liu, H.; Wang, H. A rough set approach to analyze factors affecting landslide incidence. Comput. Geosci. 2011, 37, 1311–1317.
65. Komorowski, J.; Pawlak, Z.; Polkowski, L.; Skowron, A. Rough Sets: A Tutorial. Available online: http://eecs.ceas.uc.edu/~mazlack/dbm.w2011/Komorowski.RoughSets.tutor.pdf (accessed on 23 February 2016).
66. Pawlak, Z. Rough Sets: Theoretical Aspects of Reasoning about Data; Springer Science & Business Media: Berlin, Germany, 2012.

67. Pawlak, Z.; Sowinski, R. Rough set approach to multi-attribute decision analysis. Eur. J. Oper. Res. 1994, 72, 443–459.
68. Kryszkiewicz, M. Rough set approach to incomplete information systems. Inf. Sci. 1998, 112, 39–49.
69. Swiniarski, R.W.; Skowron, A. Rough set methods in feature selection and recognition. Pattern. Recognit. Lett. 2003, 24, 833–849.
70. Davvaz, B. Roughness based on fuzzy ideals. Inf. Sci. 2006, 176, 2417–2437.
71. Gorsevski, P.V.; Jankowski, P. Discerning landslide susceptibility using rough sets. Comput. Environ. Urban Syst. 2008, 32, 53–65.
72. Thangavel, K.; Pethalakshmi, A. Dimensionality reduction based on rough set theory: A review. Appl. Soft Comput. 2009, 9, 1–12.
73. Pan, X.; Zhang, S.; Zhang, H.; Na, X.; Li, X. A variable precision rough set approach to the remote sensing land use/cover classification. Comput. Geosci. 2010, 36, 1466–1473.
74. Leung, Y.; Fung, T.; Mi, J.S.; Wu, W.Z. A rough set approach to the discovery of classification rules in spatial data. Int. J. Geogr. Inf. Sci. 2007, 21, 1033–1058.

CHAPTER 3 EVALUATION OF DIFFERENT SOIL SALINITY MAPPING USING REMOTE SENSING TECHNIQUES IN ARID ECOSYSTEMS

Mohamed Elhag

INTRODUCTION

Saudi Arabia has practiced an eminent development in vital sectors over the last three decades. As a result, rapid escalation in terms of water demands in domestic, industrial, and agricultural activities have been countersigned. Agriculture is the keystone water consumption sector; it consumes about 85% of total national water use (Ministry of Water and Electricity) [1]. During the period of 1974–2006, Saudi Arabian government subsidized the agricultural sector to improve living standards in urban areas and to achieve sustainability. These subsidies lead to significant increases in agricultural areas, primarily for rigorous irrigated crops such as wheat, maize, and alfalfa; the irrigated areas increased from less than 400,000 ha in 1971 to about 1.62 million ha in 1992 [2, 3]. The extensive groundwater extrication subsequently leads to higher soil salinity rates in most of the agricultural zones [4–6].

Excessive irrigation of less developed drained soil has led to high floating water tables, waterlogging, and soil salinization [7]. In the absence of well drainage network in shallow soil, water table depth intends to develop remarkably and, eventually, irrigable lands will be abandoned [8]. In the incidence of a higher soil salinity, crop production can be dramatically affected when salts accumulate in the soil surface [9–11]. Conservatively, soil salinity and waterlogging can be regulated through sustaining the salt concentration at the root zone below the crops constraint [12]. This can be realized by maintaining the equilibrium between soil recharge and discharge by draining the surplus irrigational water [13].

Dwivedi [14] used the first four bands of MSS to designate salty soils. Chaturvedi et al. [15] and Singh and Srivastav [16] used brightness and thermal temperature layers to detect soil salinity. Menenti et al. [17] used 3 bands of TM (1, 5, and 7) to detect salt minerals. Moreover, Mulders and Epema [18] used three different bands of TM (3, 4, and 5) to detect calcareous surfaces as an indication of high salt content. Saha et al. [19] used the thermal infrared band of TM to classify inland waterlogged areas. Rao et al. [20] used the false color composition of TM bands (2, 3, and 4) to distinguish between moderate and high soil content soils. Naseri [21] used both unsupervised and supervised classification algorithms to delineate soil salinity on a regional scale using three bands of MSS and four bands of TM. Darvishsefat et al. [22] used principal component analysis to map soil salinity using four bands of ETM+ (3, 4, 5, and 7). Zhang et al. [23] used different bands of OLI-8 to develop a soil salinity detection algorithm based on the Soil Salinity Information Extraction (SSIE).

Implementation of remote sensing data in soil salinity mapping is firm to employ since salinization is not a static process. Therefore, a number of derivatives and alternatives to normalized difference vegetation index have been anticipated to discourse soils salinization monitoring and mapping [24–26]. Different remote sensing indices such as the salinity index (SI), the normalized difference salinity index (NDSI), and the brightness index (BI) in addition to normalized difference vegetation index (NDVI) were implemented to investigate the

Citation: Mohamed Elhag, "Evaluation of Different Soil Salinity Mapping Using Remote Sensing Techniques in Arid Ecosystems," Journal of Sensors, vol. 2016, Article ID 7596175, 8 pages, 2016. doi:10.1155/2016/7596175.

means by which these indices work for soil salinity mapping in the arid environment [27, 28]. The salinity index (SI), which combines the blue and red bands, is sensitive to the surface reflectance of salt-affected land with sparse vegetation cover [27].

The perspective of using remote sensing data and GIS practices has been demonstrated in several scholarly works to be the most resourceful [29–32]. The comprehensive use of satellite remote sensing and GIS has been recognized to be a cost-effective method for monitoring soils salinization in poorly drained basins [33, 34]. However, state-of-the-art outcome specifies that there is no ideal combination of data type and questioning method which can be functional with identical realization across approximately several environmental conditions [25–27].

The aim of current research project is to differentiate the salinized soils from nonsalinized soils implementing different approaches of digital image classification and band math techniques. To reach that objective, several soil salinity indices were applied to the Landsat 8 data verified with ground truth data to validate the percentage of estimations accuracy based on different error's assessment approach.

MATERIALS AND METHODS

Study Area

The study area, Wadi Al Dawasir town, is located in the plateau of Najd at latitude of 44°43′ and longitude of 20°29′, about 300 km south of the capital city Riyadh (Figure 1). This study area is comprised of gravelly tableland disconnected by insignificant sandy oases and isolated mountain bundles. Across the Arabian Peninsula as a whole, the tableland slopes toward the east from an elevation of 1,360 meters in the west to 750 meters at its easternmost limit. Wadi Al Dawasir and Najran regions are the major irrigation water abstraction from Al-Wajid Aquifer. Agriculture in Wadi Al Dawasir area consists of technically highly developed farm enterprises that operate modern pivot irrigation system. The size of center pivot ranges from 30 ha to 60 ha with farms managing hundreds of pivots with the corresponding number of wells. The main crop grown in winter is wheat and occasionally potatoes, tomatoes, or melons. All year fodder consists of alfalfa, which is cut up to 10 times a year for food. Typical summer crops for fodder are sorghum and Rhodes grass, which is perennial but dormant in winter. The shallow alluvial aquifers could not sustain the high groundwater abstraction rates for a long time and groundwater level declined dramatically in most areas. Meteorological features of the area are speckled. Five elements of meteorology are constantly recorded through fixed weather station located within the study area. Temperature varies from 6°C as minimum temperature to 43°C as maximum temperature. Relative humidity is mostly stable at 24%. Solar radiation of average sunrise duration is generally 11 hrs/day. Average wind speed is closer to 13 km/hr and may reach up to 46 km/hr in thunderstorm incidents. Finally, mean annual rainfall is about 37.6 mm [5].

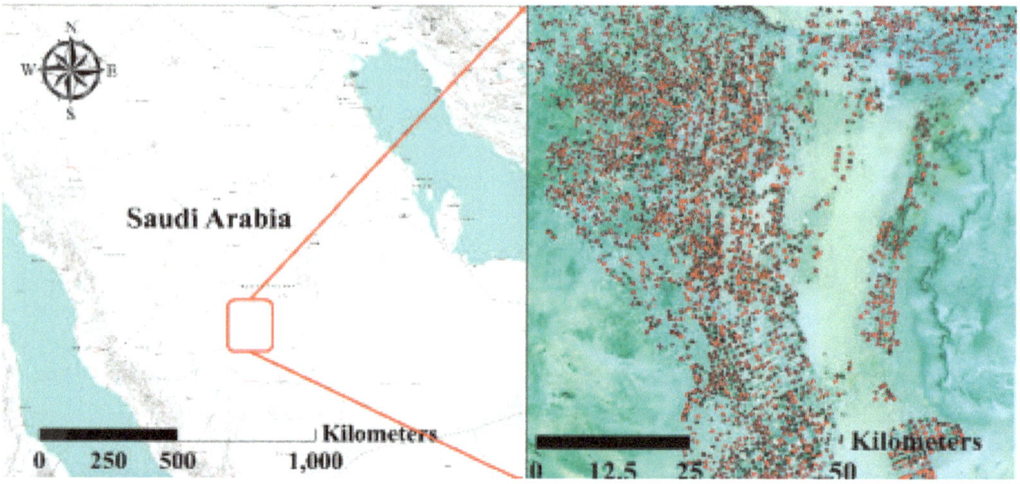

Figure 1: Location of the study area.

Methodological Framework

The current research work is based on assessing a regression correlation between ground truth data and its corresponding salinity index's values conducted from satellite images. Therefore, accurate synchronization of ground truth data collection and satellite bypassing were exercised.

Soil Sampling and Analysis

A total number of 150 random soil samples were collected from Wadi Al Dawasir area with a minimum distance of 100 m between the samples locations to avoid data clumping (Figure 2). The standard procedure of determining soil extract salinity in terms of Electrical Conductivity (EC) was followed according to Shaw [35] under laboratory condition.

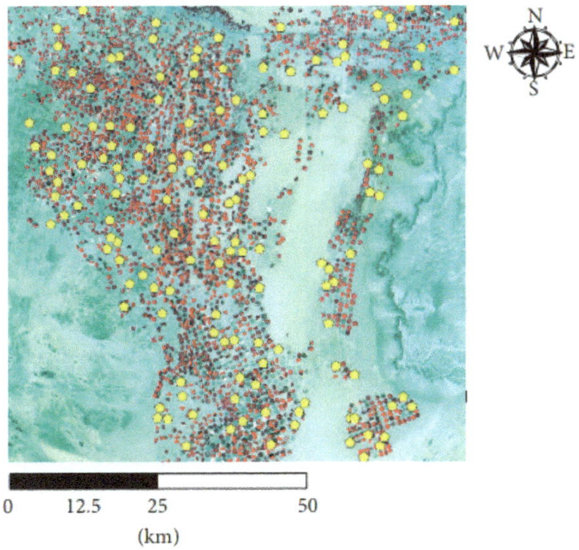

Figure 2: Soil sample location on false-color Landsat 8 image.

Remote Sensing Data Enhancement

Remote sensing data was obtained from Landsat Operational Land Imager (OLI-8) which was acquired on April 17th, 2014. Typical atmospheric and radiometric corrections and spatial resolution enhancement were implemented for each band individually. Furthermore, overcorrection was analyzed for the red and near infrared bands [36]. Atmospheric correction is basically considering blue band offset. Therefore, the largest atmospheric consequence is theoretical to be the most truthful value within the blue band. The calibrated radiance value of this offset allows deciding k value. Therefore, λ^{-k} rule the offset values for the green, red, and near infrared band under investigation [37].

Soil Salinity Indices

Soil salinity detection using remote sensing techniques developed through several scholarly works. Landsat Multispectral Scanning (MSS), TM (Thematic Mapper), and Enhanced Thematic Mapper (ETM+) in addition to Operational Land Imager (OLI-8) images were the most used remote sensing data in soil salinity mapping. Several algorithms of detecting and mapping soil salinity were developed because of the sensors and/or study area variability. There were many attempts to use the remote sensing data and its techniques to detect the salt-affected soil. Soil salinity indices are principally adjusted to detect salt mineral in soils based on the different responses of salty soils to various spectral bands. Table 1 summarizes different indices used to map soil salinity in current study.

Table 1: Soil salinity indices based on different band ratios of Landsat

Salinity index	Band ratios	Reference
Normalized difference salinity index	$\text{NDSI} = \frac{(R - \text{NIR})}{(R + \text{NIR})}$	[43]
Brightness index	$\text{BI} = \sqrt{(R^2 + \text{NIR}^2)}$	[44]
Salinity index-1	$\text{SI} = \sqrt{(B \times R)}$	[44]
Salinity index-2	$\text{SI} = \sqrt{(G \times R)}$	[44]
Salinity index-3	$\text{SI} = \sqrt{(G^2 + R^2 + \text{NIR}^2)}$	[27]
Salinity index-4	$\text{SI} = \sqrt{(G^2 + R^2)}$	[27]
Salinity index-5	$\text{SI} = \frac{B}{R}$	[45]
Salinity index-6	$\text{SI} = \frac{(B - R)}{(B + R)}$	[45]
Salinity index-7	$\text{SI} = \frac{(G \times R)}{B}$	[45]
Salinity index-8	$\text{SI} = \frac{(B \times R)}{G}$	[46]
Salinity index-9	$\text{SI} = \frac{(\text{NIR} \times R)}{G}$	[46]

B, blue band; *G*, green band; *R*, red band; NIR, near infrared band of Landsat 8 image.

Validation

Linear regression model was used to validate the correlation established between the ground truths data measured from the 150 soil samples collected and the corresponding salinity index's values estimated using the aforementioned equations. Root square, root square adjusted, root mean square error, and mean of response are used to signify the inequality of variance and correlation of the linear regression model [38].

RESULTS AND DISCUSSION

The use of different soil salinity equations gives the results the differential preferences to achieve an adequate soil salinity estimation on a large scale using remote sensing data. The evaluation of different soil salinity indices was based on the regression model fit, effect significance, and model parameters. Table 2 summarizes different regression significances.

Table 2: Regression significances

	NDSI	BI	SI-1	SI-2	SI-3	SI-4	SI-5	SI-6	SI-7	SI-8	SI-9
R square	0.62	0.94	0.89	0.90	0.95	0.90	0.44	0.44	0.91	0.86	**0.98**
R square adjusted	0.62	0.94	0.89	0.90	0.95	0.90	0.44	0.44	0.91	0.86	**0.98**
Root mean square error	1.51	760.08	1000.38	1139.67	773.06	1638.87	0.29	0.29	1780.57	1266.35	603.14
Mean of response	0.01	30262	14749	15716	33634	22431	0.18	0.01	19408	14958	**28103**

Based on different "errors" estimated by analysis of variance, SI-9 represents the most straightforward linear regression model. Figure 3 shows the correlation between the estimated soil salinity values using soil salinity index-9 and the collected ground truth values.

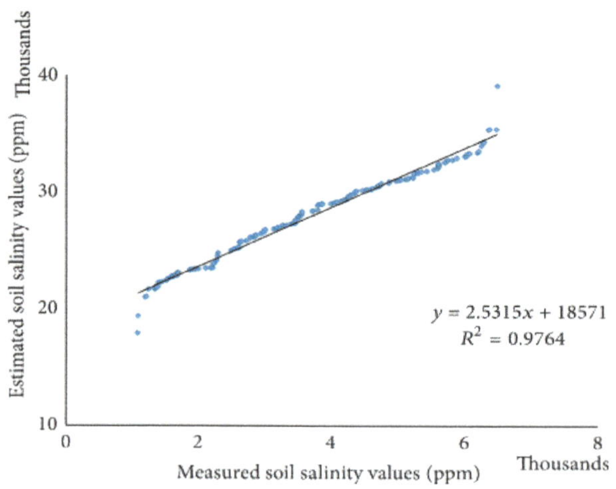

Figure 3: Linear regression between measured and estimated soil salinity values.

Due to the fact that the soil salinity indices used in the current research are varied (11 in total), categorization of different indices using principal component analysis will help decide which is similar/different to which. Figure 4 represents the grouping to different indices according to principal component analysis on covariances.

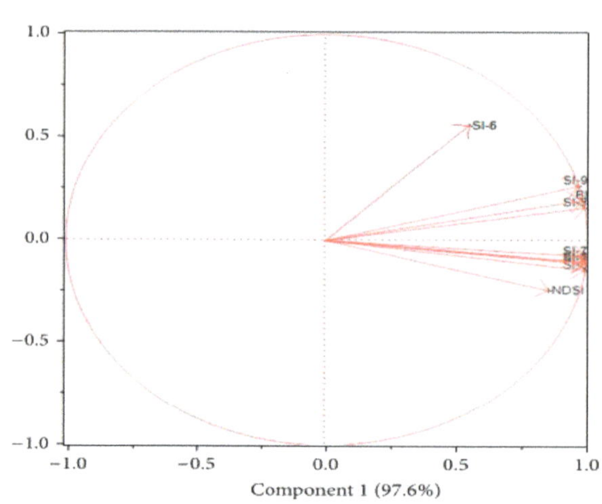

Figure 4: Variance based principal component analysis.

Generally, different soil salinity indices fell into two groups. SI-6, SI-9, SI-3, and BI are in one group and the rest are in the other group. SI-6 and NDSI failed to join their corresponding group and remain as group outlier [39]. Lack of correlation is the main reason of the insignificance of SI-6 and NDSI. High correlation significance between SI-3, SI-9, and BI is the best to represent soil salinity indices according to Table 2.

The results proved the acceleration in the rate of vegetation death and soil salinization based on the high correlation between higher soil salinity values and negative normalized difference vegetation index with values of 0.92 [40]. Moreover, higher soil salinity was related to the improper and intense drainage systems set up in addition to the lack of an operative water resource management plan [41].

The dynamicity of the soil salinization process added further complications to designating salt-affected soils in a systematic uniform perspective. The use of different algorithms based on implementing different combinations and/or ratios of

OLI-8 bands in the form of soil salinity indices evidenced to be more efficient to overcome soil dynamicity problems [23, 42].

The selection of the sensitive bands adequate for proper soil salinity mapping is not systematically generalized [42, 47]. Spatial variability and land cover variations are the most limiting factors of band selection of a sensor [23, 48]. Therefore, some of the used equations may have better results in specific areas implementing specific band ratios than others taking into consideration the use of OLI-8 as a source of remote sensing data [23, 49].

Estimated values of soil salinity index-9 were then used to map and classify soil salinity in the whole study area. Regular interpolation technique with six different smoothing Kernel functions is implemented and evaluated according to Krivoruchko et al. [50] as shown in Table 3 and Figure 5.

Table 3: Evaluation of different interpolation kernels function

	Kernel function					
	Exponential	Gaussian	Quartic	Epanechnikov	Polynomial-5	Constant
Regression function	$0.006X + 29339$	$-0.001X + 29659$	$0.003X + 29477$	$-0.005X + 29778$	$0.005 + 29368$	$-0.001X + 29679$
Mean	1383	1421	1378	1412	**1355**	1428
Root mean square	4251	4257	4250	4257	4253	4266
Mean standardized	0.0065	0.0067	0.0065	0.0066	**0.0064**	0.0067
Root mean square standardized	0.0201	0.0201	0.0202	0.0201	0.0202	0.0200
Average standard error	211114	211797	210334	211526	**209641**	212674

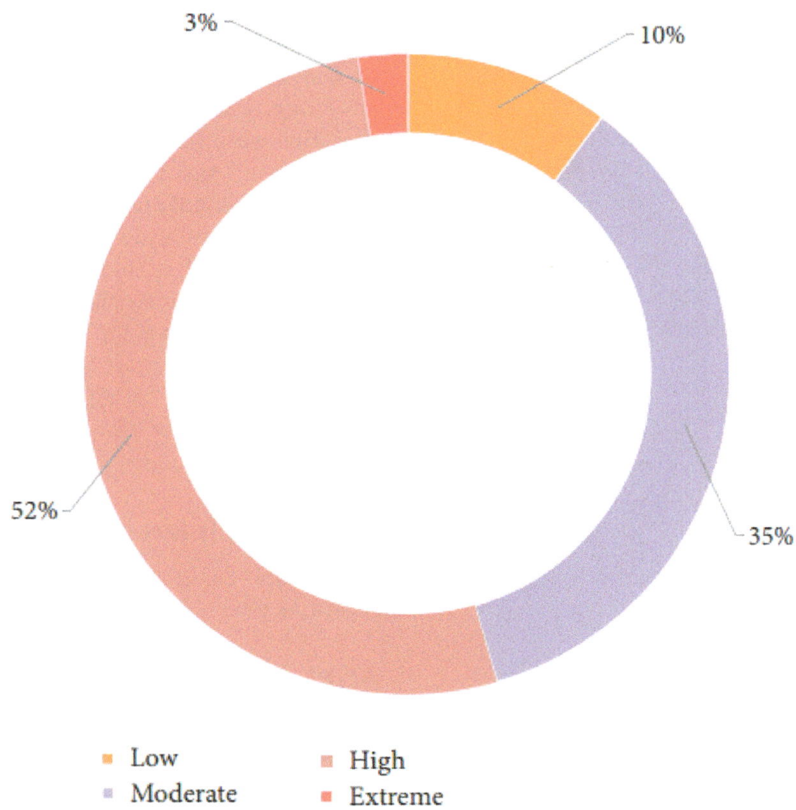

Figure 5: Soil Salinity classes in Wadi Al Dawasir area.

In Figure 5, only 3% of Wadi Al Dawasir area is categorized to be with low soil salinity followed by 10% of moderate soil salinity class. Meanwhile, more than half of the study area is subjected to a higher concentration of soil salinity (52%). In addition, there are soils in Wadi Al Dawasir which cannot be used for the current agricultural practice due to extreme condition of soil salinity (35%).

Kernel function of polynomial-5 was used to produce a thematic soil salinity map in the study area (Figure 6). Based on Soil Survey approach in conjunction with natural breaks classification method was used to reclassify the prediction risk maps into four classes according to their toxicity levels into Low, Moderate, High, and Extreme [51, 52]. The role of the classifier is based on squared error minimization from a class's means by creating internally homogenous groups but maintaining heterogeneity between classes.

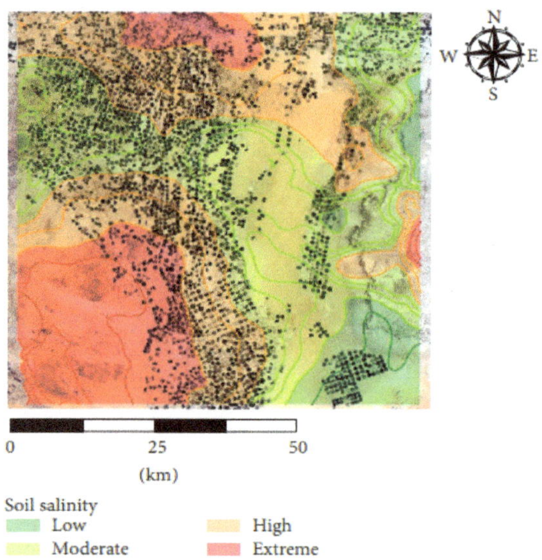

Figure 6: Soil salinity map in Wadi Al Dawasir.

Figure 7 shows the percentages of Wadi Al Dawasir agricultural area which were fallen into the four soil salinity classes based on NDVI estimation. The total agricultural area in Wadi Al Dawasir expressed low soil salinity contribution (6.5%). Those areas are the best to preserve the natural resources and the agricultural practice in Wadi Al Dawasir. About 36% of the total agricultural area constantly needs efforts to sustain the current agriculture practice. Both of the aforementioned classes are less than the half of the cropped area of Wadi Al Dawasir. Closer to 58% of the cultivated area is on improper soils due to its high salinity content.

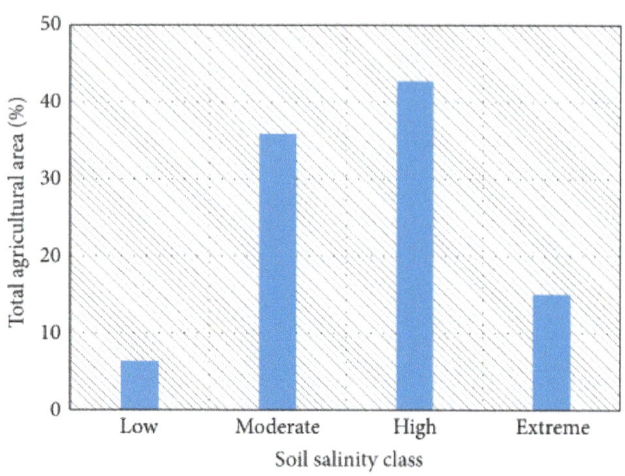

Figure 7: Contribution of different soil salinity classes to the total agricultural area.

Soil salinity index-9, as a remotely sensed indicator, typically depends on the vegetation growing conditions which are helpful for mapping spatially the distribution of soil salinity [53]. The response of certain features of crop stress, dehydration, and senescence using vegetation indices have been examined to determine their response in different spectra [54, 55]. According to Carter [54], cell structure damage and chlorophyll reduction could be reliably measured using reduced near infrared reflectance (NIR) and increased visible reflectance (VIS) among several plant species in response to stress. Tilley et al. [53] also investigated salt stress responses in both of VIS and NIR. Normalized difference vegetation index (NDVI) as one of the most implemented vegetation indices has been inspected in abundant scholarly work by using the different vegetation reflectance to estimate soil salinity [40].

CONCLUSIONS

Crop production in Wadi Al Dawasir is completely dependent on underground water resources represented in artesian wells. The use of the underground water resources is annually overexploited due to the horizontal expansion of the agricultural land. The driving forces of the agricultural land expansion and the use of the underground water resources are controlled only by the Ministry of Water and Electricity but not by the local inhabitants. Such uncontrolled conditions put the water resources and the soils of Wadi Al Dawasir in jeopardy. The use of excess irrigational water to cope with the extremely arid conditions in the study area continuously creates more drainage problems as the agricultural areas in Wadi Al Dawasir have no drainage system by any means. Water logging and soil salinization are common problems in agricultural areas described by no drainage system. Consequently, the hydrological sceneries and the geomorphological patterns of arid lands are very important aspects to be considered in any rational development plan. Soil salinity indices are site/condition specific and soil salinity index-9 is the best to represent soil salinity values in the current study area. The current agricultural practices in Wadi Al Dawasir are obviously not the optimum water and/or soil resources management. More than half of the study area is characterized by either high or extreme soil salinity classes. Crop reallocation and rational water use may serve the goal of sustainability to keep the current agricultural practice in Wadi Al Dawasir. Monitoring and temporal soil salinity investigation are the keystone exercise required for adequate natural resources management plans in the near future.

Conflict of Interests

The author declares that there is no conflict of interests regarding the publication of this paper.

ACKNOWLEDGMENT

This project was funded by the Deanship of Scientific Research (DSR), King Abdulaziz University, Jeddah, under Grant no. 137-155-1436-G. The author, therefore, acknowledges and thanks DSR for technical and financial support.

REFERENCES

1. Ministry of Water and Electricity (MOWE), Annual Report, Ministry of Water and Electricity (MOWE), Riyadh, Saudi Arabia, 2008.
2. A. W. Abderrahman, "Water demand management in Saudi Arabia," in Water Management in Islam, N. I. Faruqui, A. K. Biswas, and M. J. Bino, Eds., IDRC, 2001.
3. World Bank, "A water sector assessment report on the countries of the co-operation Council of the Arab States of the Gulf," Tech. Rep. 32539-MNA, World Bank, 2005.

4. K. H. Al-Zahrani, "Water demand management in the Kingdom of Saudi Arabia," Conference of the International Journal of Arts & Sciences, vol. 2, no. 3, pp. 68–76, 2010.

5. K. H. Al-Zahrani and M. B. Baig, "Water in the kingdom of Saudi Arabia: sustainable management options," The Journal of Animal and Plant Sciences, vol. 21, no. 3, pp. 601–604, 2011.

6. M. Elhag and J. A. Bahrawi, "Conservational use of remote sensing techniques for a novel rainwater harvesting in arid environment," Environmental Earth Sciences, vol. 72, no. 12, pp. 4995–5005, 2014.

7. J. M. Bradd, W. A. Milne-Home, and G. Gates, "Overview of factors leading to dryland salinity and its potential hazard in New South Wales, Australia," Hydrogeology Journal, vol. 5, no. 1, pp. 51–67, 1997.

8. D. R. Williamson, "Land degradation processes and water quality effects, waterlogging and salinity," inFarming Action Catchment Reaction, The Effect of Dry-Land Farming on the Natural Environment, J. Williams, R. A. Hook, and H. L. Gascoigne, Eds., pp. 162–190, CSIRO Publishing, Collingwood, Va, USA, 1998.

9. E. Houk, M. Frasier, and E. Schuck, "The agricultural impacts of irrigation induced waterlogging and soil salinity in the Arkansas Basin," Agricultural Water Management, vol. 85, no. 1-2, pp. 175–183, 2006.

10. R. Jones and G. Marshall, "Land salinization, waterlogging and the agricultural benefits of a surface drainage scheme in Benerembah irrigation district," Review of Marketing and Agricultural Economics, vol. 60, pp. 173–189, 1992.

11. R. George, D. McFarlane, and B. Nulsen, "Salinity threatens the viability of agriculture and ecosystems in Western Australia," Hydrogeology Journal, vol. 5, no. 1, pp. 6–21, 1997.

12. F. Konukcu, J. W. Gowing, and D. A. Rose, "Dry drainage: a sustainable solution to waterlogging and salinity problems in irrigation areas?" Agricultural Water Management, vol. 83, no. 1-2, pp. 1–12, 2006.

13. S. Abdel-Dayem, S. Abdel-Gawad, and H. Fahmy, "Drainage in Egypt: a story of determination, continuity, and success," Irrigation and Drainage, vol. 56, no. 1, pp. 101–111, 2007.

14. R. S. Dwivedi, "Monitoring of salt-affected soils of the Indo-Gangetic alluvial plains using principal component analysis," International Journal of Remote Sensing, vol. 17, no. 10, pp. 1907–1914, 1996.

15. L. Chaturvedi, K. R. Carver, J. C. Harlan, G. D. Hancock, F. V. Small, and K. J. Dalstead, "Multispectral remote sensing of saline seeps," IEEE Transactions on Geoscience and Remote Sensing, vol. 21, no. 3, pp. 239–251, 1982.

16. R. P. Singh and S. K. Srivastav, "Mapping of waterlogged and salt-affected soils using microwave radiometers," International Journal of Remote Sensing, vol. 11, no. 10, pp. 1879–1887, 1990.

17. M. Menenti, A. Lorkeers, and M. Vissers, "An application of thematic mapper data in Tunisia,"International Journal of Applied Earth Observation, vol. 1, pp. 35–42, 1986.

18. M. A. Mulders and G. F. Epema, "The thematic mapper: a new tool for soil mapping in arid areas,"International Journal of Applied Earth Observation, vol. 1, pp. 24–29, 1986.

19. S. K. Saha, M. Kudrat, and S. K. Bhan, "Digital processing of Landsat TM data for wasteland mapping in parts of Aligarh District (Uttar Pradesh), India," International Journal of Remote Sensing, vol. 11, no. 3, pp. 485–492, 1990.

20. B. R. Rao, R. S. Dwivedi, L. Venkataratnam et al., "Mapping the magnitude of sodicity in part of Indo-Gangetic plains of Uttar Pradesh, Northern India using Landsat data," International Journal of Remote Sensing, vol. 12, no. 3, pp. 1419–1425, 1991.

21. M. Y. Naseri, Characterization of salt-affected soils for modeling sustainable land management in the semi-arid environment; a case study in the Gorgan region, Northeast Iran [Ph.D. thesis], Ghent University, Ghent, Belgium, 1998.

22. A. A. Darvishsefat, M. H. Damavandi, M. Jafari, and G. R. Zehtabiyan, "Assessing of Landsat TM images for using in soil salinity classification," Journal of Desert, vol. 5, no. 2, 2000.

23. T. Zhang, G. Zhao, C. Chang et al., "Information extraction method of soil salinity in typical areas of the yellow river delta based on landsat imagery," Agricultural Sciences, vol. 06, no. 01, pp. 71–77, 2015.

24. L. M. Montandon and E. E. Small, "The impact of soil reflectance on the quantification of the green vegetation fraction from NDVI," Remote Sensing of Environment, vol. 112, no. 4, pp. 1835–1845, 2008.

25. B. D. Wardlow and S. L. Egbert, "Large-area crop mapping using time series MODIS 250 m NDVI data: an assessment for the US Central Great Plains," Remote Sensing of Environment, vol. 112, no. 3, pp. 1096–1116, 2008.

26. Z. P. Yang, J. X. Gao, C. P. Zhou et al., "Spatio-temporal changes of NDVI and its relation with climatic variables in the source regions of the Yangtze and Yellow rivers," Journal of Geographical Sciences, vol. 21, no. 6, pp. 979–993, 2011.

27. A. E. K. Douaoui, H. Nicolas, and C. Walter, "Detecting salinity hazards within a semiarid context by means of combining soil and remote-sensing data," Geoderma, vol. 134, no. 1-2, pp. 217–230, 2006.

28. G. Jiapaer, X. Chen, and A. M. Bao, "A comparison of methods for estimating fractional vegetation cover in arid regions," Agricultural and Forest Meteorology, vol. 151, no. 12, pp. 1698–1710, 2011.

29. B. R. M. Rao, R. S. Dwivedi, L. Venkataratnam et al., "Mapping the magnitude of sodicity in part of the Indo-Gangetic plains of Uttar Pradesh, northern India using Landsat-TM data," International Journal of Remote Sensing, vol. 12, no. 3, pp. 419–425, 1991.

30. A. Srivastava, N. K. Tripathi, and K. V. G. K. Gokhale, "Mapping groundwater salinity using IRS-1B LISS II data and GIS techniques," International Journal of Remote Sensing, vol. 18, no. 13, pp. 2853–2862, 1997.

31. R. S. Dwivedi and K. Sreenivas, "Delineation of salt-affected soils and waterlogged areas in the Indo-Gangetic plains using IRS-1C LISS-III data," International Journal of Remote Sensing, vol. 19, no. 14, pp. 2739–2751, 1998.

32. M. Elhag and J. Bahrawi, "Cloud coverage disruption for groundwater recharge improvement using remote sensing techniques in Asir region, Saudi Arabia," Life Science Journal, vol. 11, no. 1, pp. 192–200, 2014.

33. R. Goossens, M. De Dapper, A. Gad, and Th. Ghabour, "A model for monitoring and prediction of soil salinity and waterlogging in the Ismailia area (Egypt) based on remote sensing and GIS," in Proceedings of the International Symposium on Operationalization of Remote Sensing, vol. 6, pp. 97–107, ITC, Enschede, The Netherlands, April 1993.

34. S. Casas, "Salinity assessment based on combined use of remote sensing and GIS," in Use of Remote Sensing Techniques in Irrigation and Drainage Unknown, pp. 141–150, FAO, Rome, Italy, 1995.

35. R. Shaw, "Estimation of the electrical conductivity of saturation extracts from the electrical conductivity of 1:5 soil: water suspensions and various soil properties," Project Report Series QO94025, Department of Primary Industries, Queensland, Australia, 1994.

36. P. S. Chavez Jr., "Image-based atmospheric corrections—revisited and improved," Photogrammetric Engineering and Remote Sensing, vol. 62, no. 9, pp. 1025–1036, 1996.

37. U. Beisl, J. Telaar, and M. Schönermark, "Atmospheric correction, reflectance calibration and BRDF correction for ADS40 image data," in Proceedings of the International Archives of the Photogrammetry, Remote Sensing and Spatial Information Sciences, vol. 87, part B7, pp. 7–12, Beijing, China, 2008.

38. G. E. P. Box, "Some theorems on quadratic forms applied in the study of analysis of variance problems, II. Effects of inequality of variance and of correlation between errors in the two-way classification,"Annals of Mathematical Statistics, vol. 25, no. 3, pp. 484–498, 1954. View at MathSciNet

39. A. Dehni and M. Lounis, "Remote sensing techniques for salt affected soil mapping: application to the Oran Region of Algeria," Procedia Engineering, vol. 33, pp. 188–198, 2012.

40. A. Allbed and L. Kumar, "Soil salinity mapping and monitoring in arid and semi-arid regions using remote sensing technology: a review," Advances in Remote Sensing, vol. 02, no. 04, pp. 373–385, 2013.

41. A. K. Koshal, "Spectral characteristics of soil salinity areas in parts of South-West Punjab through remote sensing and GIS," International Journal of Remote Sensing and GIS, vol. 1, no. 2, pp. 84–89, 2012.

42. L. Lei, T. Tiyip, J.-L. Ding, H.-N. Jiang, and A. Kelimu, "Study on the soil salinization monitoring based on measured hyperspectral and HSI data," Spectroscopy and Spectral Analysis, vol. 34, no. 7, pp. 1948–1953, 2014.

43. D. J. Major, F. Baret, and G. Guyot, "A ratio vegetation index adjusted for soil brightness," International Journal of Remote Sensing, vol. 11, no. 5, pp. 727–740, 1990. ·

44. N. M. Khan, V. V. Rastoskuev, Y. Sato, and S. Shiozawa, "Assessment of hydrosaline land degradation by using a simple approach of remote sensing indicators," Agricultural Water Management, vol. 77, no. 1–3, pp. 96–109, 2005.

45. A. Bannari, A. M. Guedon, A. El-Harti, F. Z. Cherkaoui, and A. El-Ghmari, "Characterization of slightly and moderately saline and sodic soils in irrigated agricultural land using simulated data of advanced land imaging (EO-1) sensor," Communications in Soil Science and Plant Analysis, vol. 39, no. 19-20, pp. 2795–2811, 2008.

46. A. Abbas and S. Khan, "Using remote sensing techniques for appraisal of irrigated soil salinity," inProceedings of the International Congress on Modelling and Simulation (MODSIM ‹07), L. Oxley and D. Kulasiri, Eds., pp. 2632–2638, Modelling and Simulation Society of Australia and New Zealand, Brighton, UK, December 2007.

47. C. W. Zhang, J. K. Tang, X. J. Yu, C. L. Wang, and S. J. Mi, "Quantitative retrieval of soil salt content based on remote sensing in the Yellow River delta," Journal of Graduate University of Chinese Academy of Sciences, vol. 30, pp. 220–227, 2013.

48. J.-L. Ding, M.-C. Wu, H.-X. Liu, and Z.-G. Li, "Study on the soil salinization monitoring based on synthetical hyperspectral index," Spectroscopy and Spectral Analysis, vol. 32, no. 7, pp. 1918–1922, 2012.

49. D. Abdellatif and L. Mourad, "Remote sensing techniques for salt affected soil mapping: application to the Oran Region of Algeria," Procedia Engineering, vol. 33, pp. 188–198, 2012.

50. K. Krivoruchko, A. Gribov, and E. Krause, "Multivariate areal interpolation for continuous and count data," Procedia Environmental Sciences, vol. 3, pp. 14–19, 2011.

51. Soil Survey Staff, Keys to Soil Taxonomy, United States Department of Agriculture, Natural Resources Conservation Service, 10th edition, 2006.

52. B. Jiang, "Head/tail breaks: a new classification scheme for data with a heavy-tailed distribution," The Professional Geographer, vol. 65, no. 3, pp. 482–494, 2013. ·

53. D. R. Tilley, M. Ahmed, J. H. Son, and H. Badrinarayanan, "Hyperspectral reflectance response of freshwater macrophytes to salinity in a brackish subtropical marsh," Journal of Environmental Quality, vol. 36, no. 3, pp. 780–789, 2007.

54. G. A. Carter, "Responses of leaf spectral reflectance to plant stress," American Journal of Botany, vol. 80, no. 3, pp. 239–243, 1993.

55. A. Psilovikos and M. Elhag, "Forecasting of remotely sensed daily evapotranspiration data over Nile Delta region, Egypt," Water Resources Management, vol. 27, no. 12, pp. 4115–4130, 2013.

CHAPTER 4

SOIL CLAY CONTENT MAPPING USING A TIME SERIES OF LANDSAT TM DATA IN SEMI-ARID LANDS

Marouen Shabou, Bernard Mougenot, Zohra Lili Chabaane, Christian Walter, Gilles Boulet, Nadhira Ben Aissa and Mehrez Zribi

Keywords: Topsoil clay content, Landsat TM, MID infrared index, soil maps, TerraSAR-X radar data

INTRODUCTION

Integrated water resource management requires an improved understanding of soil at fine scales. Soil texture is a key parameter allowing the establishment of pedotransfer functions to derive hydrodynamics soil parameters such as wilting point, field capacity or saturated hydraulic conductivity. Clay content (soil fraction <2 μm) is considered the most important parameter to be estimated. It is related to the cation exchange capacity of soils [1] and allow a tight control of soil hydraulic properties like water storage and availability to crop plants, field capacity and wilting point. It is the first statistical factor to be taken into account when building pedotransfer functions [2,3].

Soil texture maps are difficult to derive. Existing soil maps are neither exhaustive nor precise enough, and often describe soil types rather than texture. Point measurements through soil analysis can be performed to complete those maps but are expensive and dense sampling is required to characterize adequately the spatial variability, making broad-scale quantitative evaluation difficult [4]. Moreover, laboratory analysis performed by chemical treatments are expensive and time-consuming [5,6]. New and quick methods to measure soil properties and variability are required for the development of soil interpretations [6,7].

Citation: Shabou, M.; Mougenot, B.; Chabaane, Z.L.; Walter, C.; Boulet, G.; Aissa, N.B.; Zribi, M. Soil Clay Content Mapping Using a Time Series of Landsat TM Data in Semi-Arid Lands. Remote Sens. 2015, 7, 6059-6078. doi:10.3390/rs70506059.

In recent decades, the soil science community has proposed several methodologies to better estimate topsoil texture using remote sensing [8,9,10,11,12]. Remote sensing data are an important component of predictive soil mapping. It provides a spatially contiguous, quantitative measure of surface reflectance, which is related to some soil properties [13]. This tool facilitates mapping inaccessible areas by reducing costly field surveys [14]. It also helps to produce more accurate and updated soil properties' maps through large scale studies by relating field-measured variables such as clay content to surface reflectance through linear regression models to supply continuous estimates for environmental variables [15]. Mapping of continuous variables from high resolution imagery such as Landsat Thematic Mapper (TM) has largely depended on modeling empirical relationships [15]. Landsat TM sensor has been providing near-continuous multispectral coverage, with wavelengths covering the visible (0.45–0.69 μm), near-infrared (0.76–0.9 μm) and shortwave infrared (1.55–2.35 μm) domains, supplying adequate data set for studies of improved management of natural resources [16].

Some studies have shown that physical factors (particle size and surface roughness) and components (surface mineralogy, organic matter content and moisture) control soil spectral reflectance [17]. Mineral composite maps, including ferrous minerals, iron oxide and clay minerals, could be produced using remotely sensed data such as Landsat-5 Thematic Mapper (TM), Landsat-7 Enhanced Thematic Mapper Plus (ETM+), Landsat-8 Operational Land Imager (OLI), Advanced Spaceborne Thermal Emission and Reflection Radiometer (ASTER) and Moderate Resolution Imaging Spectroradiometer (MODIS) through composite indices [18,19,20,21]. Other researchers [8,22] have used hyper spectral data for clay content prediction, calcium carbonate ($CaCO3$), iron and cation-exchange capacity (CEC). Models for soil attribute prediction from spectral data, also named spectral transfer functions, have been developed recently [23,24,25,26]. Spectral transfer functions relate soil properties estimated from samples collected in the field and spectral reflectances measured above those samples, usually in the laboratory. Surface mineralogy can be derived by wavelength-specific charge transfer and crystal field absorptions associated with the presence of iron and iron-oxides (Fe^{2+} and Fe^{3+}), and vibrational absorptions associated with hydroxyl bonds in clay, absorbed water and the carbonate ion [17,27]. The middle infrared index (MID-infrared), which is a normalized difference index between TM5 and TM7 shortwave infrared bands, allows discrimination between sandy and clayey soils [20,21]. These bands are near the two main water absorption bands (1.4 μm–1.9 μm). The presence of clay in the soil composition enhances the absorption in TM7 band (2.2 μm). The use of such bands requires dry bare soil, which makes this method more suited to mapping semi-arid lands. All these studies indicate that the electromagnetic energy interacts with some soil properties in specific wavelengths. An analysis of the behavior of these wavelengths can be utilized in the modeling process to determine and map some soil attributes such as clay content. In our case, we used the MID-infrared index to produce clay content maps over semi-arid lands using time series of Landsat TM data. A similar index as the STI (Soil Tillage Index) was used to detect and quantify crop residue out of the soil baseline [28,29,30].

The aim of this research was to map topsoil clay content at fine spatial resolution over the Kairouan plain. This study used time series of Landsat Thematic Mapper (TM5) images and a dataset of 100 soil samples collected over our study area. To produce clay content map, we proceeded in three steps. The first step consisted in studying relationships between the MID-infrared index and observed clay content from calibration data. The second step was to combine multi-temporal Landsat TM data through cokriging to predict topsoil clay content within the Kairouan plain. The final step was to validate the produced clay content map by independent soil samples and by comparison with existing soil map and with topsoil clay content map produced from TerraSAR-X radar data.

STUDY AREA AND DATA

Study Area: The Kairouan Plain

The Kairouan plain [31] is located in central Tunisia (9°30′E–10°15′E, 35′N, 35°45′N) (Figure 1). The climate in this region is semi-arid, with an average annual rainfall of approximately 300 mm/year, characterized by a rainy season lasting from October to May, with the two rainiest months being October and March. As is generally the case in semi-arid areas, the rainfall patterns in this area are highly variable in time and space. The landscape is mainly flat, and the vegetation is dominated by agricultural production (cereals, olive groves, fruit trees and winter and summer market gardening). Soil units reflect high spatial and vertical variability of alluvial deposits, except along wadi. The last flood of 1969 changed some soil horizons with sandy deposits of up to 100 cm and existing soil maps may become outdated. The dominant major soil group is classified as Fluvisols and saline soils [32]. Soil texture presents a large spatial variability from the south to the north of our region [33]. The clay content is ranging from 15.3 g/kg to 684.58 g/kg. Mainly four soil units exist in the study area. The first unit includes Vertic saline soils. The second unit is Fluvisols characterized by a loamy soil texture located near the wadi bed of Merguellil. Another soil unit includes Fluvisols with coarse soil texture on the downstream part of the Merguellil River. The last soil unit, located in the north of Kairouan, includes clayey soil. Vertic soil is an indicator of the presence of montmorillonite clay. The organic matter rate is very low, near 0.5% from the surface to 10 cm, which corresponds to an isohumic diagnostic property of all soil units.

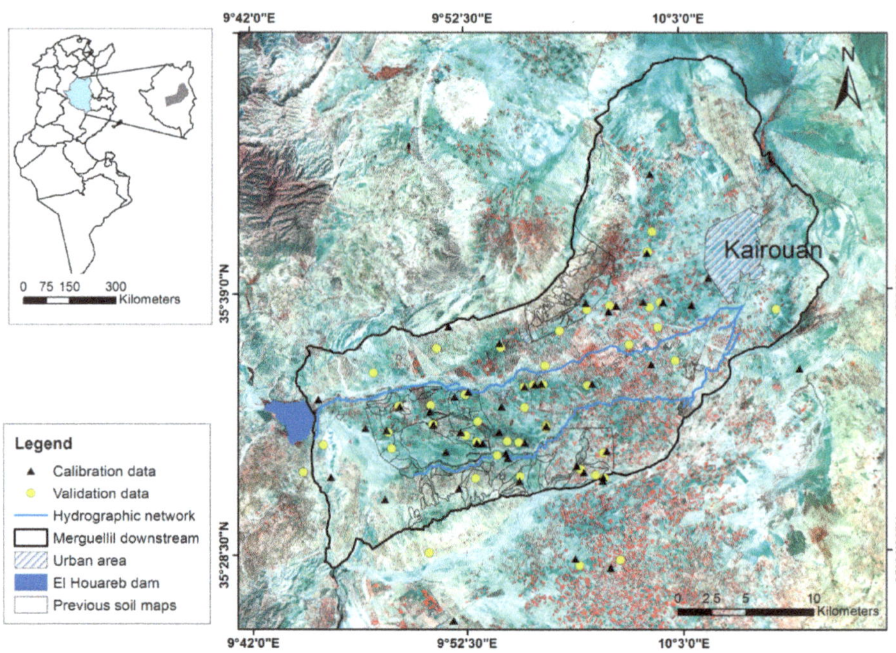

Figure 1: Study area and location of calibration and validation samples.

Data Collection and Processing

Ground Measurements

The dataset used in this study includes 100 samples. A first dataset of 30 samples was collected in March 2010 [12] over a limited area within the plain of Kairouan. An additional campaign was conducted during the summer season of 2013 after cereal harvesting over a larger part of the plain within a 670 km^2 area (Figure 1). During this campaign, 70 samples with a large spatial range of soil textures were selected along main and agricultural roads of the study area. Sample points were selected based on visual assessment of topsoil textural changes. These samples were located in bare soil areas and cereals fields [12] corresponding to homogeneous areas extracted from MID-Infrared index maps, and over some olive fields with a vegetation fraction coverage lower than 10%. These plots are of sufficiently representative size (3 × 3 pixels) and easily accessible. The samples were collected from the surface to a depth of 5 cm. About 20 g were extracted for topsoil texture analysis. The initial samples were air-dried and sieved up to 2 mm. Topsoil texture measurements were carried out over the selected fields to measure clay content (soil fraction <2 μm), fine silt content (between 2 and 20 μm), coarse silt content (between 20 and 50 μm) and sand content (between 0.05 and 2 mm) by means of a classical laboratory analysis (Robinson pipette). To divide the dataset into calibration set and validation set, we kept one sample out of two. We considered 50 samples for calibration and 50 samples to validate the clay content map produced in this study. For clay content, the mean and standard deviation were 325.74 g/kg and 179.04 g/kg, respectively, and the measured values ranged between 15.3 and 684.6 g/kg. The coefficient of variation was high, about 55%. For sand content and silt content, the coefficients of variation were, respectively, about 85% and 60%. It confirms the high variability in the distribution of sand content over the Kairouan plain. Analyzed fields are plotted into texture triangle according to the USDA classification as illustrated by Figure 2. We can distinguish three dominant soil classes: clayey soil, silty soil and sandy soil as a continuous distribution between sandy and silty clay classes.

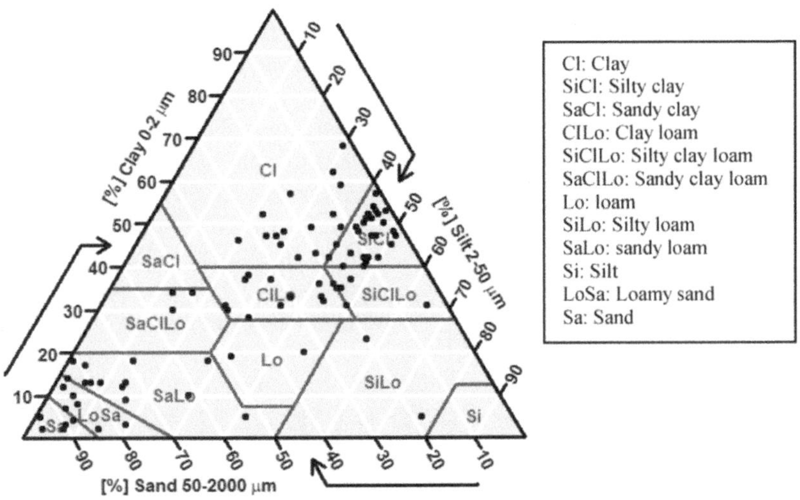

Figure 2: Distribution of 100 analyzed soil samples over the Kairouan plain across the USDA soil textural triangle.

Remote Sensing Data

To produce the topsoil texture map, time series of Landsat TM5 (2009) archival images were used. The Landsat TM5 was launched on 1 March 1984 and decommissioned on 5 June 2013. These images were downloaded fromhttp://earthexplorer.usgs.gov/. Landsat TM5 data are acquired at 30 m spatial resolution. These images were geometrically and atmospherically corrected. All images are projected in the Universal Transverse Mercator system (UTM/Zone 32/North). The root mean square error of ortho-rectification is less than 1/2 pixel for each image. Visible, near- and mid-infrared bands calibration were carried out as developed in [34]. The method is based on a multi-temporal algorithm for detecting clouds, cloud shadows, free water and for estimating aerosol optical thickness (AOT). Subsequent preprocessing included conversion of digital numbers to radiance, radiance to top-of-atmosphere reflectance and top of atmosphere reflectance to surface reflectance. The detailed method for correcting these images is illustrated in [35,36]. From the time series of Landsat TM data, we selected images during the seasons that minimize the influence of green vegetation (cereals mainly) and soil moisture due to rain. The latter is observed at the meteorological station. Calibration data were located on Landsat TM data. A polygon with 3 × 3 pixels was digitized to extract statistical data from Landsat TM images around each measurement. Polygons were represented within the sample plots in order to avoid selecting pixels at the edge.

Legacy Data

The 1:50,000 previous soil map used in our study was produced in 1975 by the Ministry of Agriculture, whose extent is shown in Figure 1. This map covers an area of about 13 km^2. It includes four soil type units: Fluvisols, Vertisols or Vertic soils, Calcisols and Isohumic soils.

We also used an existing topsoil clay content map derived from two TerraSAR-X radar images acquired over the Kairouan plain with a ground pixel spacing of 1 m [12]. The first TerraSAR-X radar image was acquired on 8 March 2010, just after a period of a heavy rain. The second one was acquired on 30 March 2010, when the soil was dry. The method used to produce clay content map from radar data was based on analyzing the difference in soil moisture content between the two images. The largest decrease in topsoil moisture is statistically related to the coarsest soil material based on a soil sample dataset of 15 samples [12,37].

METHODS

Bare Soil Detection

Estimation of soil properties such as clay content is hampered if the pixels have a vegetation cover over 20% and also if the pixels are covered by crop residues [38]. We need rapid, accurate, and objective methods to detect residue cover and green vegetation in individual fields. Crop residue and soil are often spectrally similar and differ only by the magnitude of the reflectance in visible and near infrared wavelengths [39,40,41,42]. In fact, to extract bare soil extent over the Kairouan plain, we have to select Landsat TM data, which minimize the influence of soil moisture due to rain or irrigation. We also need to avoid crop residue after crop harvesting. To do so, we selected, in a first step, nine Landsat TM images with little influence of soil moisture due to rain on the basis of rainfall observations at meteorological stations. High soil moisture content decreases soil reflectance. In the second step, we retained only four images (17 July 2009, 1 August 2009, 2 September 2009 and 18 September 2009) in the summer season to avoid soil moisture due to local rain and irrigation. These images contain three types of pixels: bare soil pixels, plowed bare soil pixels and bare soil with crop residue. To extract only bare soil and to identify the agricultural practices, we analyzed simultaneously the TM3 profile, the NDVI profile and MID-infrared profile over all test fields as illustrated in Figure 3. In fact, using TM3 band, we can detect, approximately, the date of some agricultural practices such as crop harvesting and plowing. The plowing date can be identified by the decrease in red band reflectance values.

For each selected image, we masked urban areas, using a GIS shape file. The Normalized Difference Vegetation Index (NDVI) was used to mask green vegetation from each image by applying a sill of 0.2. The image acquired on 18 September 2009 is contaminated by clouds. In this case, we used the provided cloud mask, generated during the image pre-processing, to mask them. The next step consists in surface changes detection (crop residues). We selected two MID-infrared images. One acquired at the beginning of the summer season, just after cereals harvesting (26 June 2009). This image contains a maximum of pixels affected by crop residue. The second image is acquired in 18 September 2009, when farmers are starting to plow their lands. The image of difference between these two MID-infrared dates was computed and analyzed. We observe that the highest values on the difference image correspond to fields with laying straw (crop residues) and not yet plowed. A value ranging from −0.08 to −0.04 extracted from the difference image was selected as a criterion for masking crop residues areas. Once we have no more crop residue, the MID-Infrared index becomes more stable. The image acquired on 2 September 2009 allows the best prediction of topsoil clay content. However, for the illustrated example in Figure 3, on this date the field was still covered by crop residues. Based on the MID-Infrared index profile extracted within the studied field, it becomes stable from the date of 18 September 2009. This date can be considered the best date for topsoil clay prediction within this field.

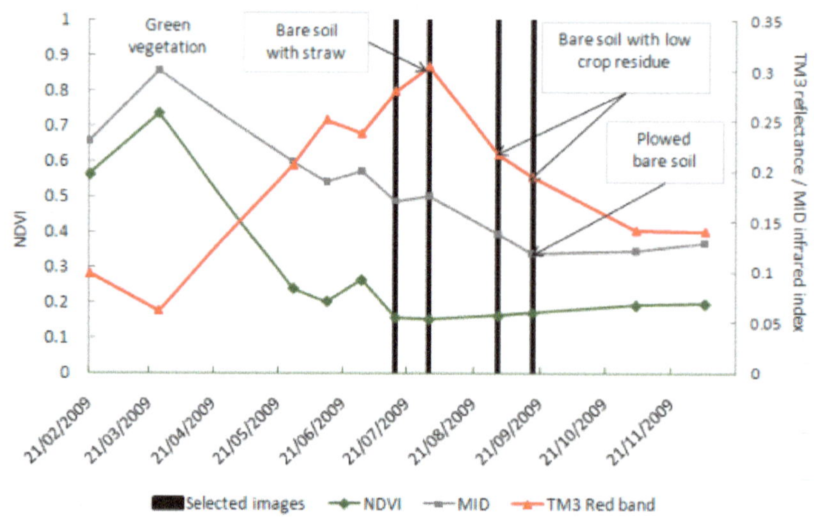

Figure 3: Example of Normalized Difference Vegetation index (NDVI), Landsat Red band (TM3) and MID-Infrared index profiles extracted from selected Landsat TM images, over a cereal test plot of 8 hectares (clayey soil texture) during the period ranging from 21 February 2009 until 21 November 2009.

Prediction of Clay Content

Calibrating MID Infrared Index and Topsoil Clay Content

The MID-infrared index is defined in Equation (1) and represents a normalized index between TM band 5 and TM band 7. Applying this index over the driest bare soils allows separating sandy soils from clayey ones.

This index was applied over bare soil for each selected image in order to produce mid-infrared index maps at 30 m spatial resolution.

$$MID - Infrared\ index = \frac{TM5 - TM7}{TM5 + TM7} \quad (1)$$

One out of two samples from the 100 samples was selected in order to build the calibration dataset. We applied green vegetation and crop residue mask for each selected image: 20 samples corresponding to fields with either green vegetation or crop residue were thus masked and not available for calibration. The remaining 30 samples were used for calibration with topsoil clay content values ranging from 50 g/kg to 684 g/kg with a mean value of about 323 g/kg. These values span the whole range of observed topsoil clay content with a satisfying distribution amongst dominant texture classes. The use of these 30 samples seems therefore sufficient to represent topsoil clay content variability within the study area. They were distributed in the different textural classes in our study area. A mean value of MID-infrared index was extracted from each test field and used with clay content. This step allowed establishing a statistical relationship (one for each selected image) between the mid-infrared index and the soil clay content using a simple linear regression model. These relationships were used to produce clay content maps at 30 m spatial resolution over our study area and for each selected image. The resulting maps contain two types of pixels: pixels with clay content estimates and others with no data over the masked areas that differ from one image to another, except for orchards. To get an exhaustive coverage of clay content at 100 m spatial resolution over the masked area, we combined the four clay content images using ordinary cokriging (OCK).

Ordinary Cokriging

Ordinary cokriging is a multivariate variant of ordinary kriging and computes the best linear unbiased estimator with minimum error variance [43,44]. Ordinary cokriging takes advantages of the correlation between variables and allows for local variability of the means by restricting the stationarity of both primary and secondary variables to a local neighborhood *W(u)* centered on the location *u* being estimated. Ordinary cokriging examines the spatial relations among two or more variables by determining their coregionalization. If coregionalization exists, then it is feasible to use the spatial information of co-variables to improve the predictions of the target variable through cokriging. For each variable a variogram describing spatial dependence of a spatial random field or a stochastic process can be estimated by Equation (2):

$$\gamma(u) = \frac{1}{2\,n_i(u)} \sum_{\alpha_i=1}^{n_i(u)} \left[Z_i(u_{\alpha_i}) - Z_i(u_{\alpha_i}+u)\right]^2 \quad (2)$$

where u is the spatial lag distance between two locations; $n_i(u)$ is the number of observed data pairs with the lag u; and Z i (uαi) and Zi(uαi+ u) are two measured values at locations, uαi and uαi+ u, respectively.

The spatial interdependence between any two variables can be examined by the cross variogram $_{ij}(u)$, which is calculated according to Equation (3):

$$\gamma_{ij}(u) = \frac{1}{2\,n_i(u)} \sum_{\alpha_i=1}^{n_i(u)} \left[Z_i(u_{\alpha_i}) - Z_i(u_{\alpha_i}+u)\right]\left[Z_i(u_{\alpha_i}) - Z_j(u_{\alpha_i}+u)\right] \quad (3)$$

In order to describe the experimental variograms and cross variograms for use in cokriging, a model of coregionalization has to be fitted to these variograms, which consider the spatial dependence of two or more sets of variables and their interdependence simultaneously. By incorporating related secondary information, cokriging takes into consideration spatial cross correlation between primary and secondary variables to improve interpolation accuracy.

For the case of two or more secondary variables, the ordinary cokriging estimator is given by Equation (4):

$$Z_{OCK}^{(1)*}(u) = \sum_{i=1}^{N_v} \sum_{\alpha_i=1}^{n_i(u)} \lambda_{\alpha_i}^{OCK}(u)\, Z_i(u_{\alpha_i}) \quad (4)$$

where N_v is the number of variables Z_i and $\lambda_{\alpha i}^{OCK}$ represents data weights:

$$\sum_{\alpha_1=1}^{n_1(u)} \lambda_{\alpha_i}^{OCK}(u) = 1 \quad (5)$$

$$\sum_{\alpha_i=1}^{n_i(u)} \lambda_{\alpha_i}^{OCK}(u) = 0 \qquad i = 2,\ldots, N_v \quad (6)$$

The cokriging variance can be estimated from the covariance C_{ll} and the cross-covariances C_{il} through Equation (7):

$$\sigma_{OCK}^2(u) = C_{11}(0) - \mu_1^{OCK}(u) - \sum_{i=1}^{N_v} \sum_{\alpha_i=1}^{n_i(u)} \lambda_{\alpha_i}^{OCK}(u)\, C_{i1}(u_{\alpha_i} - u) \quad (7)$$

Where μ_1^{OCK} is the Lagrange multiplier of the cokriging system and $(u_i - u)$ denotes the distance between u_i and u.

From each clay content image, a point vector layer at 100 m with clay content estimates was extracted and used for interpolating clay content over masked areas. For each date, an experimental semi-varioram was calculated. The Geostatistical operations were performed with ArcMap Software (version 10.1) using the Geostatistical wizard.

Validation Method

Clay content predicted by cokriging at each point of the validation data was extracted and compared to the measured clay content at the same location. The performance of the prediction models was evaluated with the following statistical criteria: the coefficient of determination (R^2) (Equation (8)), and the Root Mean Square Error (RMSE) (Equation (9)). The coefficient R^2 measures the effectiveness of a variable to predict another variable, whereas RMSE measures the average error of prediction. The indices were calculated according to [45]. To perform the calibration model estimation, we used only the R^2 and the RMSE. In addition, to independent soil samples, validation was performed using existing soil maps. The validation consists on comparing topsoil clay content and the soil typology extracted from existing soil maps.

$$R^2 = \frac{\sum_{i=1}^{n}\left(z^*(x_i) - \bar{z}\right)^2}{\sum_{i=1}^{n}\left(z(x_i) - \bar{z}\right)^2} \qquad (8)$$

$$RMSE = \sqrt{\frac{1}{n} \times \sum_{i=1}^{n}\left(z(x_i) - z^*(x_i)\right)^2} \qquad (9)$$

where point observations were denoted by n, measured and predicted clay content at ith location as $z(x_i)$ and $z^*(x_i)$, the horizontal bar represents the mean of the observed values.

RESULTS

Relationships between MID-Infrared Index and Soil Clay Content

To predict clay content over our study area, a linear regression model was built to establish the relationship between mid-infrared index and clay content corresponding to the 30 selected sites for each of the four selected dates: 16 July 2009, 1 August 2009, 2 September 2009 and 18 September 2009. Resulting relationships indicated that MID infrared index and clay content are significantly correlated (Figure 4). The coefficients of determination R^2 were, respectively, 0.48, 0.53, 0.73 and 0.63, while RMSE was, respectively, estimated at 13%, 12%, 10% and 11% of the mean clay content, which corresponds to about 130 g/kg.

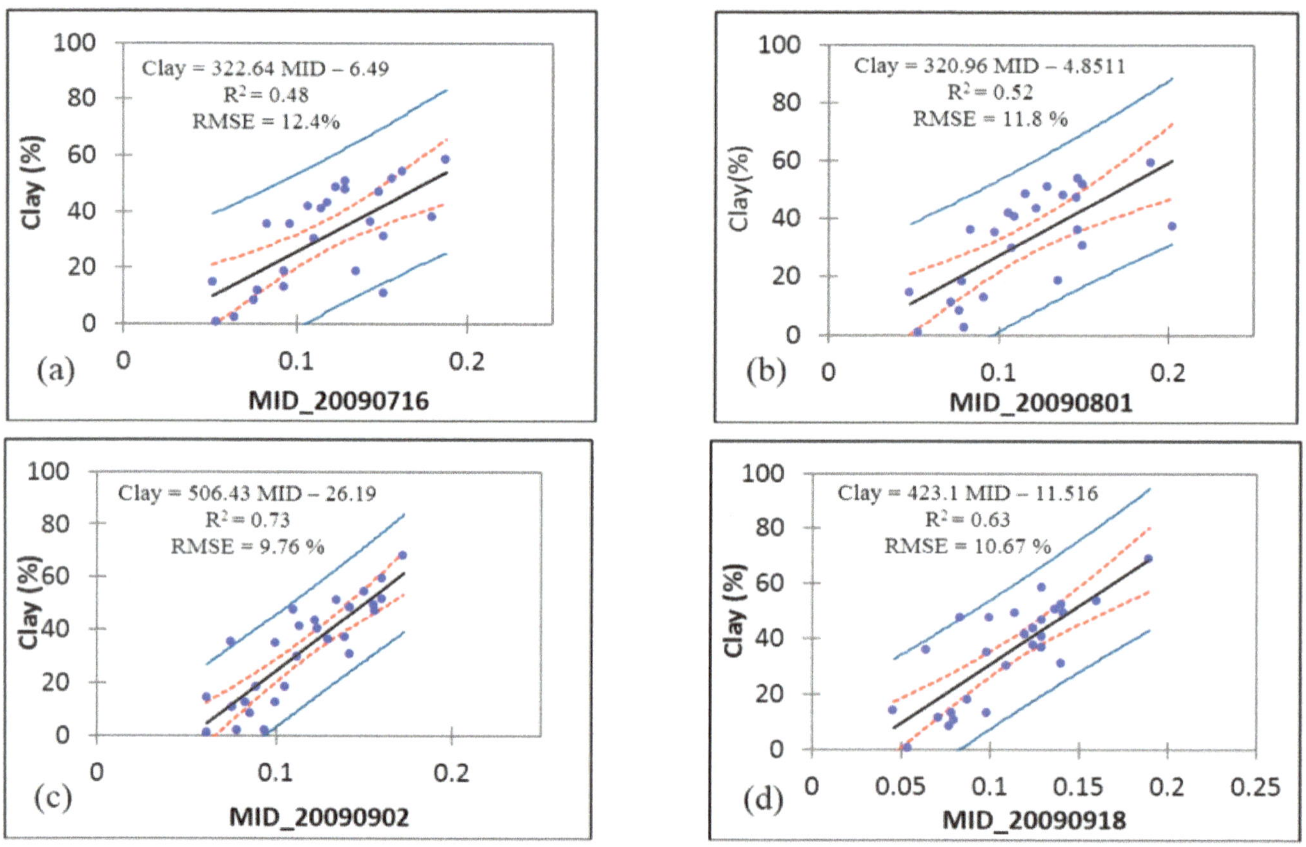

Figure 4: Linear relationship between MID-infrared index and observed clay content: (**a**) 16 July 2009; (**b**) 1 August 2009; (**c**) 2 September 2009; and (**d**) 18 September 2009. (Blue dots) observed clay content values. The red dotted line is the confidence interval of the mean. The blue line is the confidence interval for clay content observation. The continuous black line is the y-axis data *versus* x-axis data regression line.

The clay content maps at 30 m spatial resolution generated by the relationships between the mid-infrared index and the clay content are illustrated in Figure 5. We can distinguish between two types of pixels: white ones correspond to the masked areas and grey ones correspond to clay content values. The predicted clay content is ranging from 15.3 to 684 g/kg. The vegetation and crop residues masks differ from one image to another and the coverage of the remaining area is incomplete. The percentage of information contained in each image is 49%, 47%, 52% and 49%, respectively, for 16 July 2009 image, 1 August 2009, 2 September 2009 and 18 September 2009. The decrease in the percentage of information that the image contains for the 18 September 2009 image is due to cloud mask applied to mask cloud coverage.

From these maps, we extracted the clay content value of each 100 m grid and regular lattices at 100 m spatial resolution were prepared in order to interpolate the clay content over masked areas.

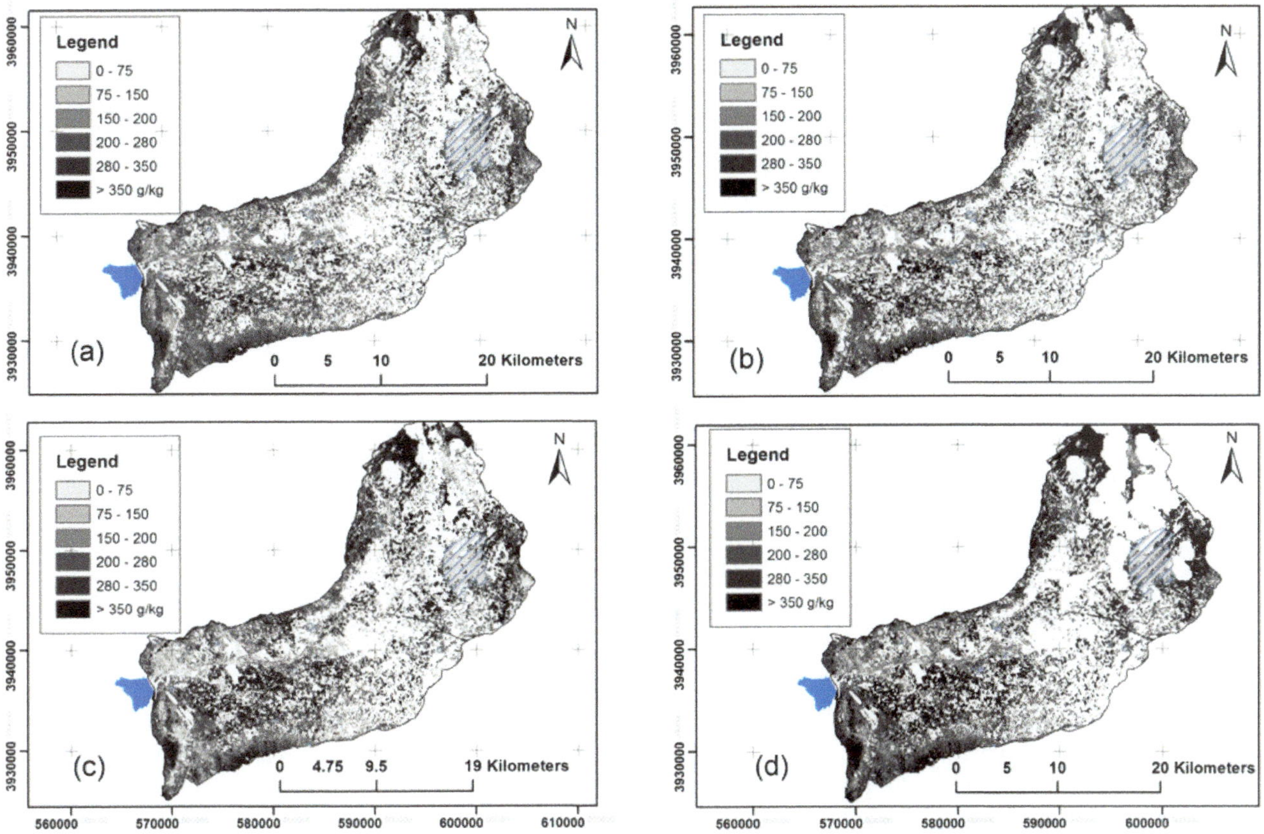

Figure 5: Clay content maps at 30 m spatial resolution obtained using relationships between observed clay content and mid-infrared index for each selected date: (**a**) 16 July 2009; (**b**) 1 August 2009; (**c**) 2 September 2009; and (**d**) 18 September 2009.

Ordinary Cokriging Parameters

The image acquired on 2 September 2009 was used as the primary variable for cokriging. Images acquired on 16 July 2009 and 1 August 2009 and 18 September 2009 were used as secondary variables. To fit the experimental variogram and the cross variogram of the four selected images, we used a stable model. Parameters of established variograms are listed in Table 1. The same range and lag size values were used in the fitting procedure, which are 3681 m and 102 m, respectively. The resulting variograms show a nugget effect. It represents random variance often caused by measurement error or micro variability of topsoil clay content, which cannot be detected at the scale of sampling.

Table 1: Stable model variogram parameters for the image acquired on 16 July 2009, 1 August 2009, 2 September 2009 and 18 September 2009

Variables	C_0 (Nugget)	C (Partial Sill)	Total sill (C_0 + C)	Range (m)	$C_0/(C_0+C)$ (%)
16 July 2009	23.84	68.01	91.85	3681	25.9
1 August 2009	29.62	69.65	99.27	3681	29.8
2 September 2009	55.38	86.17	141.55	3681	39.1
18 September 2009	54.57	122.81	177.38	3681	30.7

The resulting variogam for topsoil clay content extracted from the image acquired on 2 September 2009 and cross-variograms between the primary variable and the secondary variable are illustrated in Figure 6.

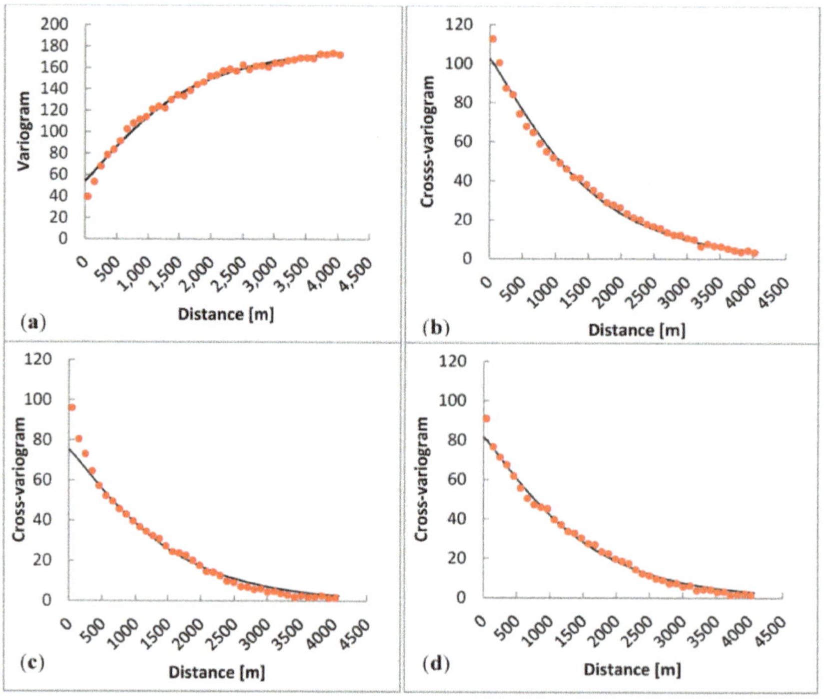

Figure 6: (**a**) Experimental semi-variograms (red dots) and model fits (black line) for topsoil clay content extracted from Landsat TM5 image acquired on 2 September 2009; (**b**) Cross-variogram between clay content from 2 September 2009 and 18 September 2009; (**c**) Cross-variogram between clay content from 2 September 2009 and 1 August 2009; (**d**) Cross-variogram between clay content from 2 September 2009 and 16 July 2009.

Topsoil Clay Content Map Produced by Ordinary Cokriging

Cokriging was used to combine the 4 images and to produce a clay content map at 100 m spatial resolution, which is illustrated in Figure 7. The obtained clay content map is classified into six classes of clay content according to the USDA texture. Defined textural classes boundaries allow mapping homogeneous areas. The two first classes include areas with clay content values ranging from 10 to 150 g/kg. These areas correspond to the Merguellil wadi bed represented by a linear structure in the north of the western half of the plain. A medium texture class includes clay content values between 200 and 280 g/kg. Within the red areas, the clay content values are higher than 280 g/kg. These areas are cultivated essentially by irrigated cereals and winter and summer vegetables. In the north of Kairouan city, we obtained a fine topsoil texture with clay content ranging from 280 g/kg to more than 350 g/kg. It corresponds to the last soil unit with clayey salt-affected soils described in Section 2.1.

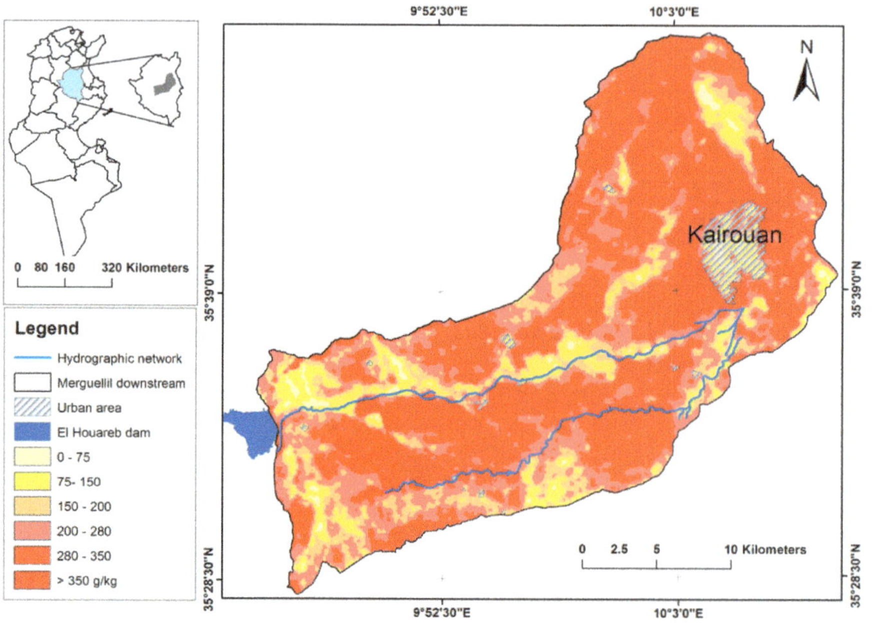

Figure 7: Clay content map at 100 m spatial resolution produced by combining four Landsat TM images: 16 July 2009, 1 August 2009, 2 September 2009 and 18 September 2009. (Blue line) corresponds to hydrographic network.

Validation of Clay Content Map from Landsat TM Data Using Independent Soil Samples and Existing Soil Maps

The resulting clay content map was validated by comparison with 50 samples over the study area (Figure 1). RMSE is about 10%, corresponding to 100 g/kg of clay. The R^2 between predicted and observed values is close to 0.65. Comparison between validation data set and clay content values predicted by cokriging is illustrated in Figure 8. The model used to predict topsoil clay content overestimates low observed clay content.

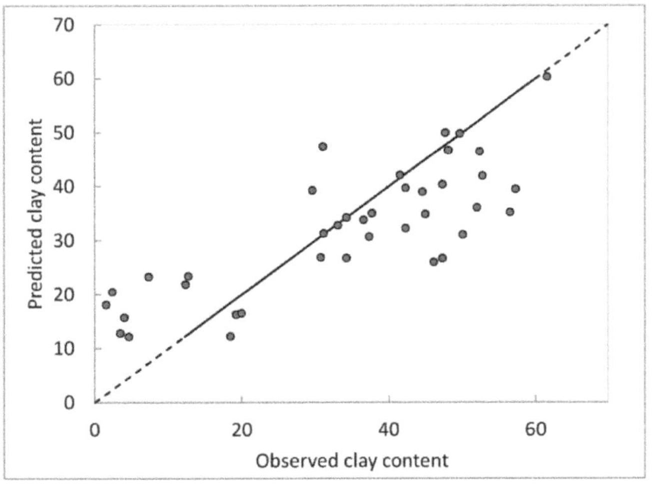

Figure 8: Comparison of observed topsoil clay content and cokriging predicted topsoil clay content at 50 independent sampling points. The dashed line is 1:1 line.

Table 2: Area percentage of each combination between soil units and produced clay content map from Landsat TM data over an area of 13 km^2

			Clay Content Classes					
			1	2	3	4	5	6
			0–75	75–150	150–200	200–280	280–350	>350 g/kg
Soil units from existing soil map	1	Fluvisols	0.19	3.95	9.18	15.44	9.00	8.02
	2	Vertisols	0.09	0.73	0.99	4.14	10.44	16.45
	3	Isohumic	0	0.97	4.31	6.94	2.79	1.28
	4	Calcisols	0	0.33	1.20	1.52	1.04	1.00

Another validation of the clay content map was achieved by comparison with existing soil maps, produced in 1975 by the Ministry of Agriculture at 1:50,000 scale. To validate our map, we compared those existing soil type maps and the soil clay map derived from Landsat TM data. The percentage of clay content classes is calculated for each soil map unit in Table 2. The first soil unit (Fluvisols) is marked by a high variability in clay content values in agreement with known characteristics of these soils. The dominant class represents 15% of the area and the clay content value is ranging from 200 to 280 g/kg. It corresponds to a medium topsoil texture class. The distribution of clay content classes higher than 350 g/kg is representative of Vertisols (16% of the area) and in agreement with clay content values observed by [46] (clay content value ranging from 350 to 520 g/kg). The isohumic soil is present in the Kairouan plain with different clay content values as identified with Landsat TM data. In addition, the Merguellil wadi bed is well identified as a linear structure in the map obtained from Landsat TM. The results show that the distribution of clay content produced by Landsat TM data is consistent with the soils units. Thus, Landsat TM type data can be used to update existing soil maps, to improve complex soil units according to clay criterion and to produce soil texture information where not available.

DISCUSSION

The model developed for clay content prediction was built from time series of Landsat TM data. Its main advantage is to provide an acceptable clay content estimation over fields with a limited number of samples (only 30 samples were used for calibration), which can be applied over large regions.

Calibration Models

For each selected image, results showed that the MID infrared index is correlated with topsoil clay content. Results were approximately in the range of error estimations derived from other remote sensing techniques to estimate clay content. With hyper-spectral Hymap images, authors obtained an R^2 of 0.73 [47]. Using TerraSAR-X radar data, authors obtained an R^2 of 0.6 [12]. The relative scattering of the plots can be due to sites with a heterogeneous surface: incomplete plowing, mixed pixels and the related MID infrared sill used for bare soil mapping. The effect of the atmosphere on the signal attenuation decreases in the MID-infrared domain.

Comparison between Clay Content Maps Produced from Landsat TM and TerraSAR-X Radar Data

The accuracy of our clay content prediction model ($R^2 = 0.64$ and a RMSE of about 10% corresponding to 100 g/kg) is in agreement with an existing study [12] who obtained an $R^2 = 0.6$ and an RMSE about 12% (120g/kg). A comparison was therefore carried out with this study based on TerraSAR-X radar data on a smaller area (Figure 9a,b) where neither calibration nor validation points were present. Before comparing these two maps, the same mask of green vegetation is applied over the clay content map extracted from Landsat TM data. The clay map produced from TerraSAR-X radar data displays sharp contrasts over short distances compared to the smoother fields of Landsat TM. This is due to the effect of residual speckle. Cokriging and the rough resolution of Landsat TM lead to more continuous fields and therefore a smoother variation from one spatial element of similar texture to the other. Both maps show similar features such as the wadi bed of Merguellil with low clay content values. The other regions show similarities and differences in clay content values. Additional field observations were made without analysis to differentiate coarse and fine textural classes over some areas.

Within region (I), we obtained high clay content value of about 60% when using Landsat TM data and low clay content value of about 15% using TerraSAR-X radar data. Field observation shows that the fields were not cultivated and this situation corresponds to bare soil areas with fine topsoil texture. Landsat TM, therefore, estimates clay content better than TerraSAR-X radar data within this area.

Within region (II), the use of TerraSAR-X data overestimates the clay content compared to Landsat TM data. It can be due to the presence of olive trees, which cover 5% of the soil surface. Concerning the third region, we predicted high clay content when using Landsat TM data. This is in agreement with our *in-situ* observations, but only in partial agreement according to radar data. In these areas, the dominant crops are cereals, beans, chilies and tomatoes; all of them being irrigated and over fine topsoil texture. For region (IV), more consistent outputs were provided when using Landsat TM data than when using TerraSAR-X radar data. We predicted fine topsoil texture using TM data with clay content ranging from 35% to 45%. The difference between both approaches can be due to soil moisture variation and to green vegetation at the date of image acquisition and to the erroneously non-masked crop residue areas.

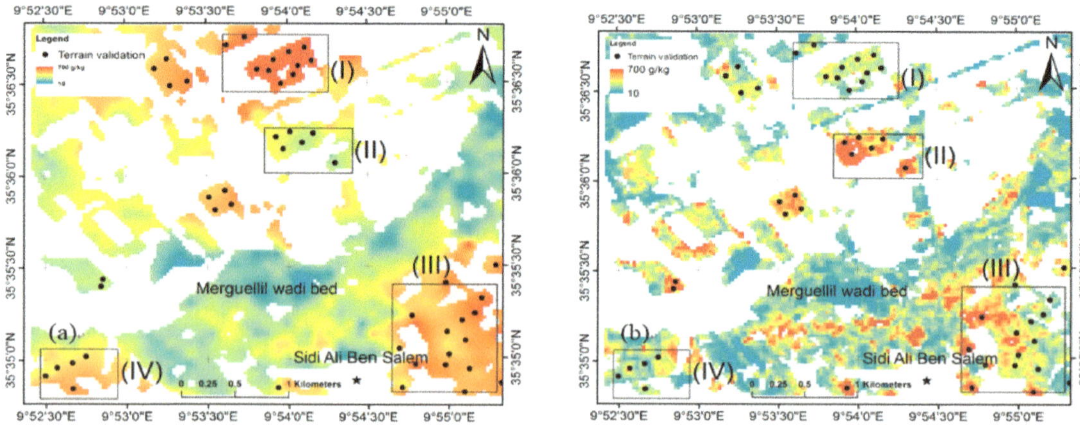

Figure 9: (**a**) Extract of the clay content map produced by Landsat TM data (2009). (**b**) Clay content map produced by TerraSAR-X data over only bare soil areas (2010). Black dots are the field observation in January 2015. Black frames are the compared sub sites.

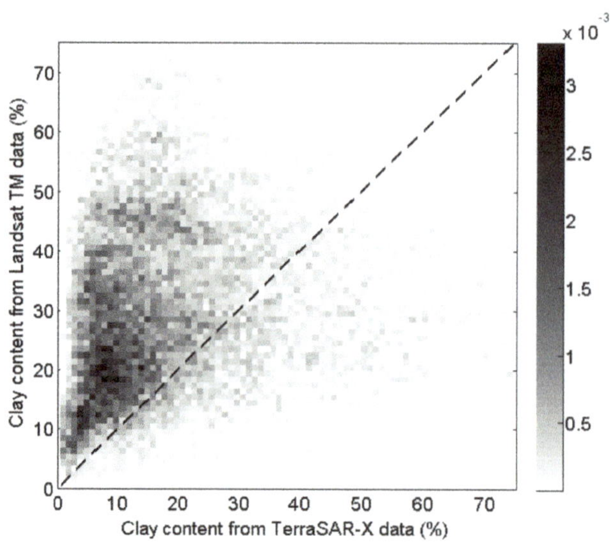

Figure 10: Density plot of clay content obtained from Landsat TM data and clay content value from TerraSAR-X radar data. The dashed line is the 1:1 line.

The difference between both approaches is validated by the density plot of clay content values obtained from Landsat TM data and TerraSAR-X radar data (Figure 10). In this area, the bias is about 5% for radar data. The central part of the scatter plot is occupied by clay content values ranging from 10% to 25% for Landsat TM data and from 5% to 10% for TerraSAR-X radar data. The average clay content value from TerraSAR-X radar data is significantly lower than the average clay content estimated using Landsat TM data. Disagreement between Landsat TM and TerraSAR-X products could be particularly explained by the presence of irrigation over some agricultural fields during experimental campaigns, which induce large errors for radar methodology.

CONCLUSIONS

The purpose of this paper was to develop a multi-temporal approach using Landsat TM data in order to map topsoil clay content at a fine spatial resolution. To predict topsoil clay content spatial interpolation can be performed with ordinary cokriging using more than one variable. The results show that remote sensing is an efficient and low-priced tool that enables generating a topsoil clay content map. We can predict clay content for each pixel of the image, thus avoiding sampling a large number of fields. The resulting topsoil clay content map is at 100 m spatial resolution and is reasonably accurate with an R^2 about 0.64 and a RMSE of 10% (100 g/kg). Cokriging made it possible to fill in the information gaps created by the vegetation mask and to validate model predictions. Improvement of relationships between MID-infrared index and clay content need more sample sites. The method needs to be tested with other satellites e.g., the new Landsat-8 and the future Sentinel2 (ESA, launch scheduled in April 2015). The ground resolution of the latter one is more accurate (20 m) to detect homogenous surface. The revisit frequency of this sensor is less than five days. With this type of data, we can improve mapping of bare soils in dry conditions. Comparisons with radar show differences. Moreover, these two approaches are complementary if we consider that visible and middle infrared reflectances are limited to surface characteristics, and radar data are limited to the drying speed of the first centimeters. Such topsoil clay content maps will be used to calculate soil hydrodynamic properties with pedotransfer functions. The surfaces, detected in the visible and MID-infrared bands, represent the agronomic horizon. This horizon is regularly homogenized by plowing. This makes it difficult to estimate clay content of horizons beyond 30 cm. To do so, we need to combine remote sensing data with some auxiliary data, such as digital elevation model, geological and geomorphologic maps.

ACKNOWLEDGMENTS

This study was funded by: the ANR-Groundwater Arena (CEP S/11-09), the PHC Maghreb 14 MAG 22 (Estimation spatialisée de l'utilisation de l'eau par l'agriculture pluviale et irriguée au Maghreb) and the ANR AMETHYST—ANR-12-TMED-0006-01. We would like to thank the SICMED projects. The authors extend their thanks to USGS for Landsat TM data, CESBIO/CNES for applying kindly the processing of Landsat TM images used in this study (framework of VenμS and Sentinel-2 preparation). Agrocampus-Ouest Rennes, INRA, UMR 1069 SAS for their collaboration. Thanks To the Institut National des Grandes Cultures (INGC, Chébika station), the Institut National Météorologique, the Soil Science Department (DG/ACTA) for providing us with soil studies, The Commissariat Régional de Développement Agricole (CRDA) of Kairouan for their assistance during the measurements on test fields. We also wish to thank the technical team from IRD and the National Institute of Agronomy of Tunisia (INAT) for their collaboration to acquire field data.

REFERENCES

1. Russel, E.J. Soil Conditions and Plant Growth, 10th ed.; Longman: London, UK, 1973.
2. Bresler, E.; Dagan, G.; Wagenet, R.J.; Laufer, A. Statistical analysis of salinity and texture effects on spatial variability of soil hydraulic conductivity1. Soil Sci. Soc. Am. J. 1984, 48, 16–25.
3. Frenkel, H.; Goertzen, J.O.; Rhoades, J.D. Effects of clay type and content, exchangeable sodium percentage, and electrolyte concentration on clay dispersion and soil hydraulic conductivity. Soil Sci. Soc. Am. J. 1978, 42, 32–39.
4. Dent, D.; Young, A. Soil Survey and Land Evaluation; HarperCollins Publishers Ltd.: London, UK, 1981.
5. Demattê, J.A.M.; Galdos, M.V.; Guimarães, R.V.; Genú, A.M.; Nanni, M.R.; Zullo, J. Quantification of tropical soil attributes from etm+/landsat-7 data. Int. J. Remote Sens. 2007, 28, 3813–3829.
6. Shepherd, K.D.; Walsh, M.G. Development of reflectance spectral libraries for characterization of soil properties. Soil Sci. Soc. Am. J. 2002, 988–998.
7. McKenzie, N.J.; Cresswell, H.P.; Ryan, P.J.; Grundy, M. Contemporary land resource survey requires improvements in direct soil measurement. Commun. Soil Sci. Plant Anal. 2000, 31, 1553–1569.
8. Gomez, C.; Lagacherie, P.; Coulouma, G. Regional predictions of eight common soil properties and their spatial structures from hyperspectral VIS-NIR data. Geoderma 2012, 189–190, 176–185.
9. Mulder, V.L.; de Bruin, S.; Schaepman, M.E.; Mayr, T.R. The use of remote sensing in soil and terrain mapping—A review. Geoderma 2011, 162, 1–19.
10. Odeh, I.O.A.; McBratney, A.B. Using AVHRR images for spatial prediction of clay content in the lower namoi valley of Eastern Australia. Geoderma 2000, 97, 237–254.
11. Sullivan, D.G.; Shaw, J.N.; Rickman, D. IKONOS imagery to estimate surface soil property variability in two alabama physiographies. Soil Sci. Soc. Am. J. 2005, 69, 1789–1798.
12. Zribi, M.; Kotti, F.; Lili-Chabaane, Z.; Baghdadi, N.; Ben Issa, N.; Amri, R.; Duchemin, B.; Chehbouni, A. Soil texture estimation over a semiarid area using TerraSAR-X radar data. IEEE Geosci. Remote Sens. Lett. 2012, 9, 353–357.
13. Scull, P.; Franklin, J.; Chadwick, O.A.; McArthur, D. Predictive soil mapping: A review. Progr. Phys. Geogr. 2003, 27, 171–197.

14. Dogan, H.M.; Kilic, O.M. Modelling and mapping some soil surface properties of central Kelkit basin in Turkey by using Landsat-7 ETM+ images. Int. J. Remote Sens. 2013, 34, 5623–5640.
15. Cohen, W.B.; Maiersperger, T.K.; Gower, S.T.; Turner, D.P. An improved strategy for regression of biophysical variables and landsat ETM+ data. Remote Sens. Environ. 2003, 84, 561–571.
16. Moran, M.S.; Bryant, R.; Thome, K.; Ni, W.; Nouvellon, Y.; Gonzalez-Dugo, M.P.; Qi, J.; Clarke, T.R. A refined empirical line approach for reflectance factor retieval from Landsat-5 TM and Landsat-7 ETM +. Remote Sens. Environ. 2001, 78, 71–82.
17. Irons, J.R.; Weismiller, R.A.; Pterson, G.W. Soil reflectance. In Theory and Applications of Optical Remote Sensing; Asar, G., Ed.; John Wiley: New York, NY, USA, 1989; pp. 66–106.
18. Dogan, H.M. Applications of remote sensing and geographic information systems to assess ferrous minerals and iron oxide of Tokat Province in Turkey. Int. J. Remote Sens. 2008, 29, 221–233.
19. Dogan, H.M. Mineral composite assessment of kelkit river basin in Turkey by means of remote sensing. J. Earth Syst. Sci. 2009, 118, 701–710.
20. Madeira, N.J. Etude Quantitative des Relations Constituants Minéralogiques-Réflectance Diffuse des Latosols Brésiliens: Application à l'utilisation Pédologique des Données Satellitaires TM (Région de Brasilia); Universite Pierre et Marie Curie: Paris, France, 1993.
21. Scull, P.; Franklin, J.; Chadwick, O.A.; McArthur, D. The application of classification tree analysis to soil type prediction in a desert landscape. Ecol. Modell. 2005, 181, 1–15.
22. Ouerghemmi, W.; Gomez, C.; Naceur, S.; Lagacherie, P. Applying blind source separation on hyperspectral data for clay content estimation over partially vegetated surfaces. Geoderma 2011, 163, 227–237.
23. Ben-Dor, E.; Taylor, R.G.; Hill, J.; Demattê, J.A.M.; Whiting, M.L.; Chabrillat, S.; Sommer, S. Immaging spectrometry for soil applications. In Advances in Agronomy; Donald, L.S., Ed.; Academic Press: Waltham, MA, USA, 2008; Volume 97, pp. 321–392.
24. Chang, C.W.; Laird, D.A.; Musbach, M.J.; Helder, D.L. Near-infrared reflectance spectroscopy–principal components regression analyses of soil properties. Soil Sci. Soc. Am. J. 2001, 65, 480–490.
25. COLEMAN, T.L.; MONTGOMERY, O.L. Soil moisture, organic matter and iron content effects on thespectral characteristics of selected vertisols and Alfisols of Alabama. Photogramm. Eng. Remote Sens. 1987, 53, 1695–1663.
26. Dunn, B.W.; Batten, G.D.; Beecher, H.G.; Ciavarella, S. The potential of near-infrared reflectance spectroscopy for soil analysis—A case study from the riverine plain of South-Eastern Australia. Austr. J. Exp. Agric. 2003, 42, 607–614.
27. Madeira, J.N. Spectral reflectance properties of soils. Photointerpretation 1996, 34, 59–70.
28. Daughtry, C.S.T.; Doraiswamy, P.C.; Hunt, E.R., Jr.; Stern, A.J.; McMurtrey, J.E., III; Prueger, J.H. Remote sensing of crop residue cover and soil tillage intensity. Soil Tillage Res. 2006, 91, 101–108.
29. Jacques, D.C.; Kergoat, L.; Hiernaux, P.; Mougin, E.; Defourny, P. Monitoring dry vegetation masses in semi-Arid areas with MODIS SWIR bands. Remote Sens. Environ. 2014, 153, 40–49.
30. van Deventer, A.P.; Ward, A.D.; Gowda, P.H.; Lyon, J.G. Using thematic mapper data to identify contrasting soil plains and tillage practices. Photogramm. Eng. Remote Sens. 1997, 63, 97–93.
31. Zribi, M.; Chahbi, A.; Shabou, M.; Lili-Chabaane, Z.; Duchemin, B.; Baghdadi, N.; Amri, R.; Chehbouni, A. Soil surface moisture estimation over a semi-arid region using ENVISAT ASAR radar data for soil evaporation evaluation. Hydrol. Earth Syst. Sci. 2011, 15, 345–358.
32. Batjes, N.H. Soil Property Estimates for Tunisia Derived from Soter and Wise (Sotwis-Tunisia, Ver. 1.0); ISRIC-World Soil Information: Wageningen, The Netherlands, 2010; p. 41.
33. Barbery, J.; Mohdi, M. Carte des Ressources en Sols de la Tunisie, Feuille de Kairouan; Direction des ressources en eau et en sol: Division des sols, Tunisie, 1987; p. 53.
34. Chander, G.; Markham, B.L.; Helder, D.L. Summary of current radiometric calibration coefficients for Landsat MSS, TM, ETM+, and EO-1 ALI sensors. Remote Sens. Environ. 2009, 113, 893–903.
35. Hagolle, O.; Dedieu, G.; Mougenot, B.; Debaecker, V.; Duchemin, B.; Meygret, A. Correction of aerosol effects on multi-temporal images acquired with constant viewing angles: Application to Formosat-2 images. Remote Sens. Environ. 2008, 112, 1689–1701.
36. Hagolle, O.; Huc, M.; Pascual, D.V.; Dedieu, G. A multi-temporal method for cloud detection, applied to Formosat-2, Venμs, Landsat and Sentinel-2 images. Remote Sens. Environ. 2010, 114, 1747–1755.
37. Paris Anguela, T.; Zribi, M.; Baghdadi, N.; Loumagne, C. Analysis of local variation of soil surface parameters with TerraSAR-x radar data over bare agricultural fields. IEEE Trans. Geosci. Remote Sens. 2010, 48, 874–881.
38. Bartholomeus, H.; Epema, G.; Schaepman, M. Determining iron content in Mediterranean soils in partly vegetated areas, using spectral reflectance and imaging spectroscopy. Int. J. Appl. Earth. Obs. Geoinform. 2007, 9, 194–203.
39. Aase, J.; Tanaka, D. Reflectances from four wheat residue cover densities as influenced by three soil backgrounds. Agron. J. 1991, 83, 753–757.
40. Baird, F.; Baret, F. Crop residue estimation using multiband reflectance. Remote Sens. Environ. 1997, 59, 530–536.
41. Nalger, P.; Daughtry, C.S.T.; Goward, S. Plant litter and soil reflectances. Remote Sens. Environ. 2001, 71, 207–215.

42. Streck, N.A.; Rundquist, D.; Connot, J. Estimating residual wheat dry matter from remote sensing measurements. Photogramm. Eng. Remote Sens. 2002, 68, 1193–1201.

43. Goovaerts, P. Geostatistics for Natural Resources Evaluation; Oxford University Press: New York, NY, USA, 1997; p. 483.

44. Wackernagel, H. Multivariate Geostatistics; Spinger-Verlag: London, UK, 1995.

45. Williams, P.C.; Norris, K. Variables affecting near-infrared reflectance spectroscopy analysis. In Near-Infrared Technology in the Agricultural and Food Industries; American Association of Cereal Chemists Inc.: Saint Paul, MN, USA, 1987; pp. 143–166.

46. Belkhodja, K. Etude Pédologique des Zones de Draa Affane nord Merguellil et Chebika Mellouche (Kairouan); Direction des ressources en eau et en sol: Division des sols, Tunisie, 1975; p. 16.

47. Lagacherie, P.; Baret, F.; Feret, J.-B.; Madeira Netto, J.; Robbez-Masson, J.M. Estimation of soil clay and calcium carbonate using laboratory, field and airborne hyperspectral measurements. Remote Sens. Environ. 2008, 112, 825–835.

CHAPTER 5 MAPPING AND MODELLING SPATIAL VARIATION IN SOIL SALINITY BASED ON REMOTE SENSING INDICATORS AND REGRESSION TECHNIQUES

Amal Allbed, Lalit Kumar and Priyakant Sinha

Keywords: Soil salinity, electrical conductivity, remote sensing, Salinity Index, regression analysis

INTRODUCTION

Soil salinity refers to surface or near-surface accumulation of salts [1]. It is a worldwide environmental problem that mainly occurs in arid and semiarid regions and causes soil degradation [2]. The spatial variability of soil salinity over the landscape is highly sensitive and controlled by a variety factors. These factors include soil factors (parent material, permeability, water table depth, groundwater quality and topography), management factors (irrigation and drainage) and climatic factors (rainfall and humidity) [3]. The characterization of soil salinity is generally done measuring the electric conductivity (EC) in a saturated soil paste or in aqueous extracts with different soil/water ratios and using a spectrometer [4,5]. To elaborate detailed maps, density of soil samples using the previous technique is required, through an extensive design, which makes mapping time consuming and expensive.

In recent decades, there has been a widespread application of remote sensing data to map soil salinity, either directly from bare soil or indirectly from vegetation in a real-time and cost-effective manner at various scales [6]. Beside, assessing soil salinity spatial modelling, which is the utilization of numerical equations to simulate and predict real phenomena and processes, has followed several approaches. The approaches used range from artificial neural network [7–10], to classification and regression tree [11,12], to fuzzy logic [13], to generalized Bayesian analysis [14], to geostatistics (e.g., Kriging, CoKriging and regression kriging) [15–18] and statistical analysis (e.g., regression, ordinary least squares) [10,19–21]. An overview of these techniques and how they provide optimal results under certain circumstances is given in the review papers of McBratney *et al.* [22] and Scull *et al.* [23].

An integrated approach using RS in addition to various statistical methods has great potential for developing soil prediction models. In the case of soil salinity, statistical analysis, in particular linear regression, has created a tremendous potential among other techniques for improvement in the way that soil salinity is modelled, because of its rapid, practical and cost-effective manner [22,24]. A variety of statistical models based on remote sensing data has been developed and has revealed reasonable predictors of soil salinity in the literature [16,18,20,25–35]. In Thailand, Shrestha [27] developed several salinity prediction models containing spectral variables, including Normalized Difference Vegetation Index (NDVI), Normalized Difference Salinity Index (NDSI), the eight original bands of Landsat Enhanced Thematic Mapper plus (Landsat ETM+) and soil properties. The results indicated that mid-infrared (band 7) and near-infrared (band 4) had the

Citation: Allred A., Kumar L., Sinha, P. Mapping and Modelling Spatial Variation in Soil Salinity Based on Remote Sensing Indicators and Regression Techniques. Remote Sens. 2014, 6, 1137-1157. doi:10.3390/rs6021137.

highest association with the measured EC. Combining these variables yielded salinity prediction models to infer soil salinity over a large area. In contrast, Mehrjardi *et al.* [30] found that among the Landsat ETM+ bands 1–5 and 7, band 3 (red band) had the highest correlation with EC, and based on that result, a regression model fitted to relate EC to band 3 and the exponential relation was found to be the best type of model.

A regression model based on image enhancement techniques (spectral indices, Principal Components Analysis (PCA) and Tasseled Cap Transformation (TCT)) have also been extensively used to predict soil salinity and to improve the characterised variability of salinity. For example, Tajgardan *et al.* [36] combined Principal Components Analysis (PCA) techniques and regression analysis to predict and map soil salinity from data collected by the Advanced Spaceborne Thermal Emission and Reflection Radiometer (ASTER) at the north of the Aq-Qala Region in northern Iran. From this study, a suitable regression model was developed with electrical conductivity (EC) to predict and map soil salinity. Similarly, Afework [33] built a reliable model to predict soil salinity in the Metehara sugarcane farms in Ethiopia by relating EC to the Normalized Difference Salinity Index (NDSI) using linear regression.

Other researchers found that incorporating satellite images spectral bands with enhanced images has great promise for soil salinity modelling and mapping. Bouaziz *et al.* [37] conducted a study to detect soil salinity based on the Moderate Resolution Imaging Spectroradiometer (MODIS) and a multiple linear regression. They found that incorporating Salinity Index SI2 with near-infrared (NIR) (band 3) into a statistical model allowed researchers to gain great insight into the spatial detection of the spread of soil salinity. Recently, Judkins and Myint [20] found that Landsat band 7, Transformed Normalized Vegetation Index (TNDVI) and Tasselled Cap 3 and 5, derived from TCT, provided high correlation to the variation in soil salinity. Combining these spectral variables into a multiple linear regression model enabled them to predict and map soil salinity surface variation levels efficiently.

Most of the reviewed studies and others found in the literature modelled soil salinity using statistical analysis and multispectral images with moderate spatial resolution (e.g., Landsat, MODIS, *etc.*), while only in limited studies multispectral high spatial resolution images such as IKONOS, were used [19]. Moreover, several studies have been undertaken for mapping and modelling soil salinity over vegetation species other than date palms, and so far, a limited study has been undertaken to map soil salinity in a primarily date palm region. One such region is Al Hassa oasis in the eastern province of Saudi Arabia, which is the most productive date palm (*Phoenix dactylifera*) farming regions in Saudi Arabia and is seriously threatened by soil salinity. Although the date palm is highly tolerant of soil salinity, the growth and productivity of date palms in this oasis are being negatively impacted by an increasing soil salinity problem [38]. Thus, predicting the variability of soil salinity and mapping its spatial distribution are becoming increasingly important in order to implement or support effective soil reclamation programs that minimize or prevent future increases in soil salinity.

The overall aim of this study was to develop effective combined spectral-based statistical regression models using IKONOS high-resolution images to predict and map spatial variation in soil salinity in the Al Hassa oasis, a region dominated by date palms.

MATERIALS AND METHODS

Study Area

The Al Hassa oasis is situated approximately 70 km inland of the gulf coast between a latitude of 25°05′ and 25°40′N and a longitude of 49°10′ and 49°55′E (Figure 1). This oasis covers an area of approximately 20,000 ha and is at an altitude of approximately 130 to 160 m above sea level [39,40]. The Al Hassa oasis is L-shaped and is actually composed of two separate oases [41]. The main water sources include the Neogene groundwater aquifer and some free flowing springs that are distributed across the area [42]. The oasis groundwater is primarily used for domestic, irrigation and industrial purposes. The oasis is characterized by an arid climate with a high potential evaporation rate that goes above the annual average precipitation of approximately 488 mm. The absolute ambient temperature exceeds 45 °C during the summer season (from June to August). During the winter (December to February), the temperature is between 2 and 22 °C. The study area covers six different soil types, which are Torripsamments, Torriorthents, Calciorthids, Salorthids, Gypsiorthids and Haplaquepts [43,44]. The particle size distribution reveals that soils are sandy loam in texture.

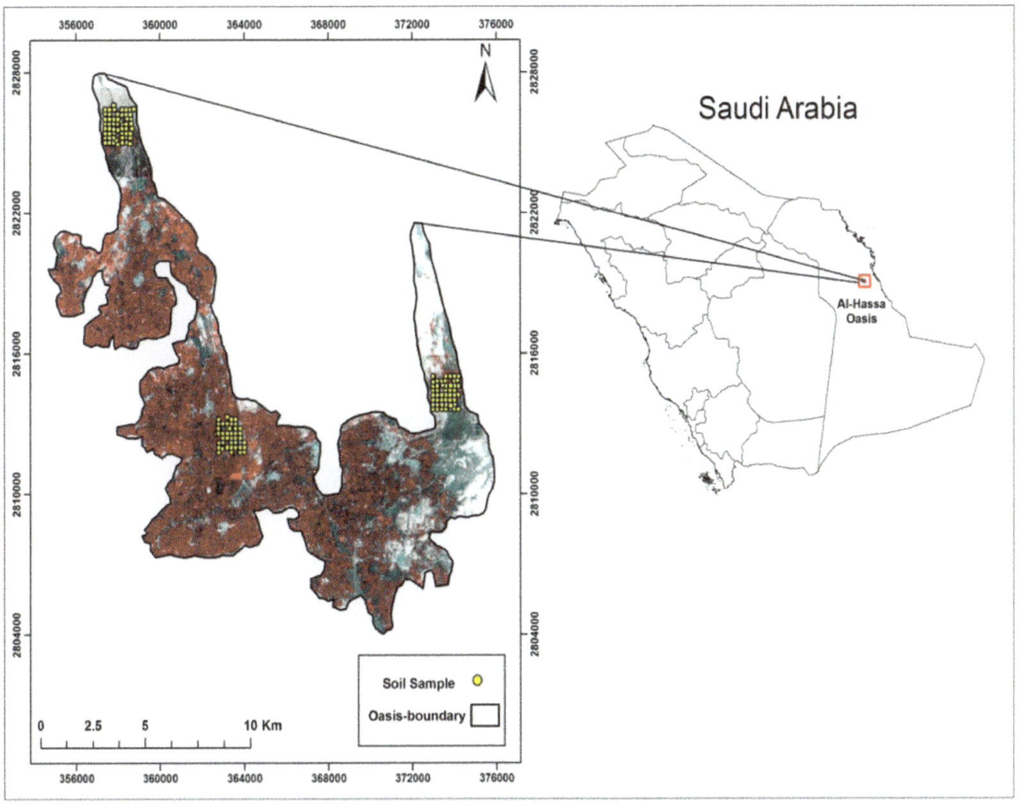

Figure 1: Study area with the location of the study sites.

Field Sampling

Three sampling sites were selected based on the division of the oasis and different amounts of vegetation. The first site was located in the northern part of the oasis at Al-Uyoun city, which is characterised by low vegetation cover. The second site, in the middle of the oasis at Al-Bataliah village, had high vegetation cover. The last set of samples was collected under medium vegetation cover in the eastern oasis, which is located in the town of Al-Umran (Figure 1).

Composite soil sampling was performed during January and February (the dry season) of 2012 following the sampling procedures of Bouaziz *et al.* [45]. The exact coordinates of each composite sample were registered using a global positioning system (GPS) with an accuracy of ±5 m. Each composite soil sample was comprised of four core sub-samples that were collected at a distance of 20 m north, south, east and west of the centre sampling point. The sub-samples were collected from the surface horizon (0–20 cm) with a hand auger (10 cm diameter) and were crushed and mixed together to form one sample. A total of 149 composite soil samples were collected from the three defined sites. Soil salinity can be measured directly by measuring the EC in the field and remotely, including the lab measurement. However, since the aim of this study is to establish a relationship between EC and satellite spectral band and extrapolate point information to generate a soil salinity map of study area, soil salinity direct measurement was performed by measuring the EC in the soil saturation extracts in the laboratory, as described by Richards [46].

Satellite Data Acquisition and Processing

High spatial resolution cloud-free IKONOS satellite images were used in this study and were acquired near the actual soil sampling date on 20 April 2012. The IKONOS images include multispectral bands (blue, 0.40–0.52 μm; green, 0.52–0.60 μm; red, 0.63–0.69 μm; near-infrared (NIR), 0.76–0.90 μm) and record the reflected or emitted radiation from the Earth's surface [47]. The images were geo-rectified to a Universal Transverse Mercator (UTM) coordinate system using World Geodetic System (WGS) 1984 datum assigned to north UTM zone 39. Atmospheric correction was performed using the Dark-Object Subtraction (DOS) technique [48]. All the remote sensing data processing was performed using the Environment for Visualizing Images (ENVI) version 4.8 software.

Data Analysis, Model Generation and Selection

Initially; the EC data was tested to establish whether it conformed to a normal distribution. The normality test exhibited that EC data had positive-skewed frequency distributions; thus, Box-Cox transformation was carried out to improve sample symmetry and to stabilise the spread. As part of the model generation process, various spectral soil salinity indices were tested for assessing and enhancing the variations in surface soil salinity. Out of all indices tested, the Salinity Index (SI) (Equation (1)), which has been proposed by Tripathi *et al.* [49], was used to create enhanced images for soil salinity in this study, due to its very highly significant correlation with EC. To ascertain the spatial location of the soil samples; a convolution low pass filter with a kernel size of 5 × 5 was applied to the enhanced images, then digital values were extracted at the location of sample points over those enhanced images.

$$\text{Salinity Index (SI)} = \frac{R}{NIR} \times 100 \quad (1)$$

where R is the red band and NIR is the near-infrared band of the IKONOS image.

Subsequently, Pearson Correlation analysis between the four bands (blue, green, red and near-infrared) and SI with EC were conducted to reveal the relationship between these variables and assess their efficiency in predicting soil salinity. The explanatory variables chosen were those showing the highest significant correlations with EC.

To build the regression model, samples were randomly split into two subsets. One subset was used for training (n = 98), the other for testing purposes (n = 51). Deciding which explanatory variables to include in the regression model is not always easy, and increasing the number of variables in a model may lead to an over-fit and provide poor prediction when used with a different data set [50]. To overcome these issues, stepwise regression was used to determine the variables that best explained most of the variability of the dependent variable, which was EC. Once all the developed regression models were tested, models with (i) a high R^2, signifying a strongly linear relationship, (ii) low standard errors of the model's variables and (iii) few variables with a p-value of <0.05 were selected for evaluation using the testing data. Consequently, the best performed regression model that met all the model selection and validation criteria was chosen and used to predict and map the spatial variation in soil salinity. All statistical analyses were undertaken in JMP®10 (JMP statistical discovery software from SAS), and significance levels set to $p < 0.05$.

Model Validation

The performance of the developed regression models that met the model selection criteria was quantified using the testing subset to ensure that they not only worked on one particular data set but also yielded an accurate result on different data sets. Two quantitative criteria between measured and predicted values were calculated (Table 1). R^2 values indicate the strength of the statistical linear relationship between measured and predicted soil salinity values, and Root Mean Square Error (RMSE) indicates absolute estimation errors [51]. In addition to these criteria, histograms, normal probability plots and Shapiro–Wilk tests (W) were employed to assess whether or not the residuals present a normal distribution. If the W test is significant ($p < 0.05$) or highly significant ($p < 0.001$) then the distribution is non-normal [52].

Table 1: Statistical criteria for evaluating the regression model

Function Name	Equation	Equation Number
Coefficient of determination	$R^2 = \left(\frac{\sum_{i=1}^{n}(\chi_i - \overline{\chi})(\gamma_i - \overline{\gamma})}{\sqrt{\sum_{i=1}^{n}(\chi_i - \overline{\chi})^2 + \sum_{i=1}^{n}(\gamma_i - \overline{\gamma})^2}}\right)^2$	(2)
Root mean square error	$RMSE = \sqrt{\frac{\sum_{i=1}^{n}(\chi_i - \gamma_i)^2}{n}}$	(3)

* χ_i and γ_i are measured and predicted values, respectively; $\overline{\chi}$ and $\overline{\gamma}$ represent the means of the measured and predicted values, respectively; n is the number of samples.

RESULTS

Data Analysis

The main statistical parameters for EC data are given in Table 2. According to the soil salinity classification of the Food and Agriculture Organization (FAO), EC values of the study area vary from very strongly saline (>16 dS/m) to non-saline (0–2 dS/m). The high Co-efficient of Variation (CV) of 85.39% confirms the variations of the EC values over the study area. About 73% of the total samples were classified as very strongly saline soil, signifying that this is the dominant soil salinity class. Correlation analysis showed a significant positive correlation ($p < 0.001$) between EC and remotely sensed data of the blue (B1), green (B2), red (B3) and SI, respectively, but not with near-infrared (B4) (Table 3).

Table 2: Descriptive statistics of electrical conductivity (EC)

	Mean	Max	Min	SD	CV (%)
EC	73.37	202	1.43	62.65	85.39

Max: maximum; Min: minimum; SD: standard deviation; CV: coefficient of variation.

Table 3: Correlation coefficient between EC and remotely sensed data

Variables	B1	B2	B3	B4	SI
EC	0.41 ***	0.42 ***	0.45 ***	0.06 ns	0.70 ***

Significant: * $p < 0.05$; ** $p < 0.01$; *** $p < 0.001$; ns = not.

Models Development and Valuations

Remotely sensed data with a significant correlation to EC were considered for developing the regression models. The developed regression models are shown in Figure 2 and their statistical results are summarized in Table 4, showing how well spatial variation in soil salinity can be predicted by applying the different developed regression models. All the developed regression models were highly significant; however, models 1, 2, 3, 4 and 9 were best able to predict soil salinity spatial variation, as they met all the model selection criteria. Among these models, model 4, which combines SI with B3, provided the best fit overall. It had the highest R^2, signifying a strongly linear relationship between estimated and predicted EC and indicated that 65% of the variance in the EC values could be explained by this model with relatively low standard errors for its variables at 29.99, 0.52 and 0.26, respectively. Each of these variables had significant *p*-values, indicating a strong correlation with EC.

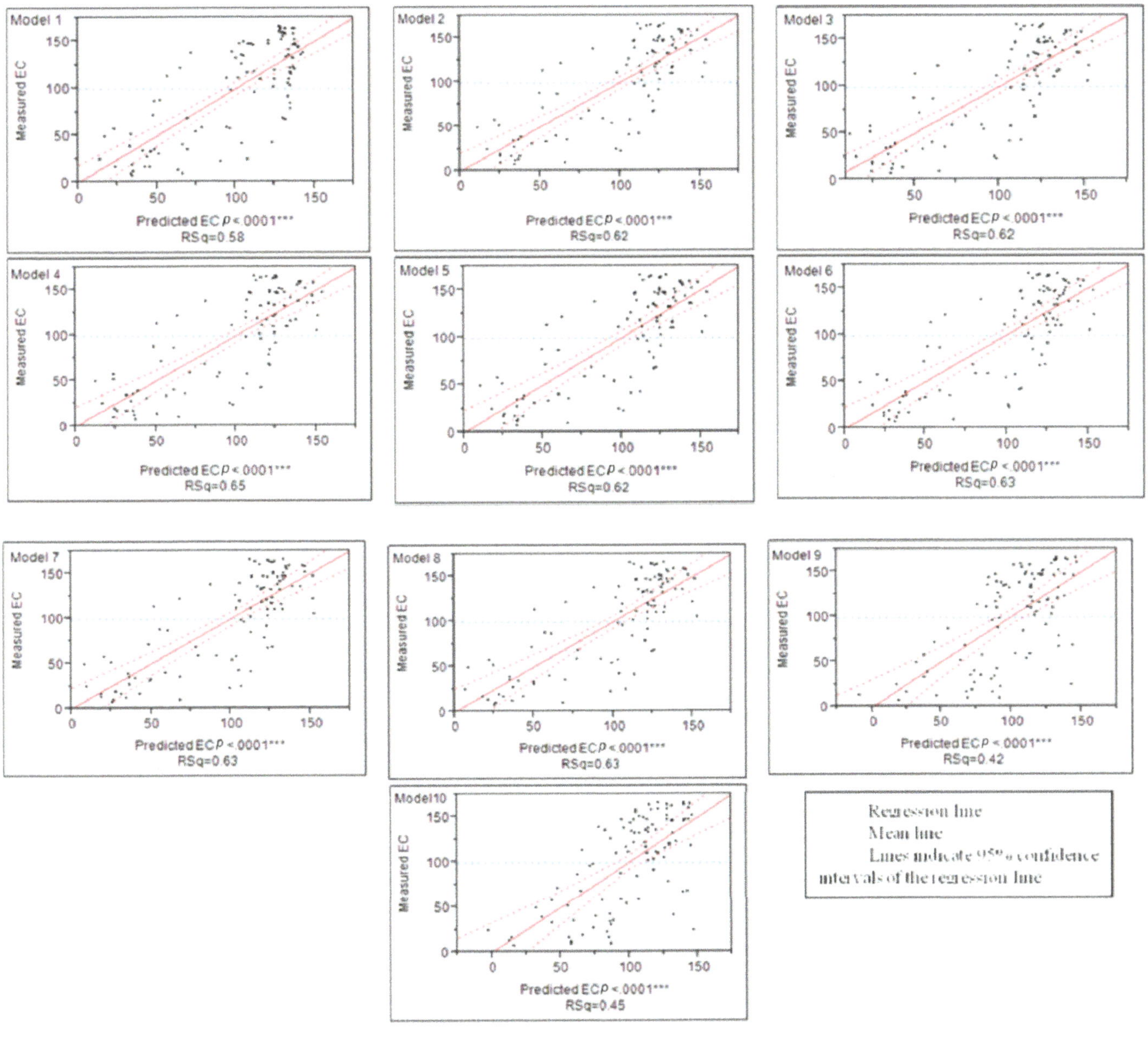

Figure 2: Scatter plots of predicted *vs.* measured EC using the developed regression models.

Table 4: Developed regression models to predict EC based on remotely sensed data. SI, Salinity Index

Model	Variable	Regression Coefficient	Standard Error	*p*-Value
1	Intercept	−239.49	29.82	<0.0001 ***
	SI	3.52	0.31	<0.0001 ***
		$R = 0.76^{***}$, $R^2 = 0.58^{***}$		
2	Intercept	−205.25	29.95	<0.0001 ***
	SI	−1.77	0.511	0.0008 ***
	B1	4.83	0.48	<0.0001 ***
		$R = 0.79^{***}$, $R^2 = 0.62^{***}$		
3	Intercept	−239.60	28.31	<0.0001 ***
	SI	4.83	0.48	<0.0001 ***
	B2	−1.011254	0.30	0.001 **
		$R = 0.79^{***}$, $R^2 = 0.62^{***}$		
4	Intercept	−269.13	29.99	<0.0001 ***
	SI	4.87	0.52	<0.0001 ***
	B3	−0.83	0.26	0.002 **
		$R = 0.81^{***}$, $R^2 = 0.65^{***}$		
5	Intercept	−193.18	83.48	0.02 *
	SI	4.81	0.49	<0.0001 ***
	B1	−2.39	4.04	0.5559
	B2	0.36	2.35	0.8771
		$R = 0.79^{***}$, $R^2 = 0.62^{***}$		
6	Intercept	−136.12	93.24	0.1476
	SI	4.64	0.54	<0.0001 ***
	B1	−3.59	2.38	0.1356
	B3	0.95	1.21	0.4355
		$R = 0.79^{***}$, $R^2 = 0.63^{***}$		

The validation results for the best regression models (1, 2, 3, 4 and 9) are shown in Figure 3. The results show that model 4 was most accurate, whereas model 2 was the worst. Model 4 outperformed the other regression models with regard to the normality test of the residuals. Furthermore, the W test for model 4 upgraded to 0.98 with a non-significant *p*-value ($p < 0.05$), and the bell-shaped histogram indicates the normal distribution of the residuals. Furthermore, values of R^2 equalling 0.34 and RMSE of 39 dS/m indicate that this regression model had the best fit compared to the others. Values of R^2 of 0.28 and RMSE values of 42 dS/m for regression model 9 indicate that this model would not predict soil salinity with high accuracy using remotely sensed data. Thus, these statistical results reveal that regression model 4 met both the model selection and model evaluation criteria.

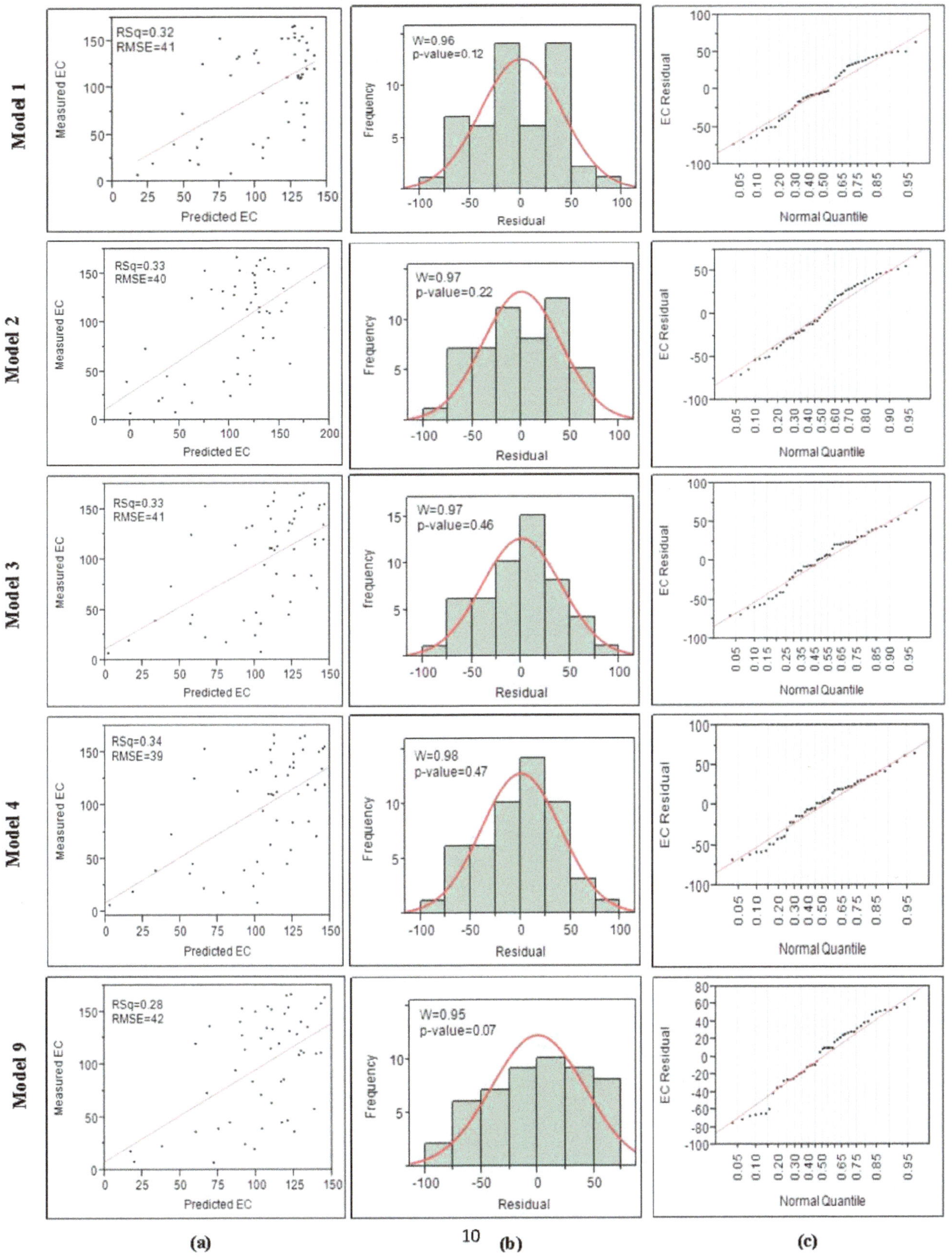

Figure 3: Validation of the developed regression models: (**a**) scatterplots of the predicted *vs.* measured EC; (**b**) the histogram of the residuals; (**c**) the normal plot of the residuals.

Spatial Variation in Soil Salinity Maps

Maps of the study areas generated using the selected model (model 4) are presented in Figure 4. In general, these maps show that most of the areas with very strongly saline soil (>16 dS/m) are non-vegetated, and areas with vegetation have soils with lower salinity levels, although still in the >16 dS/m class.

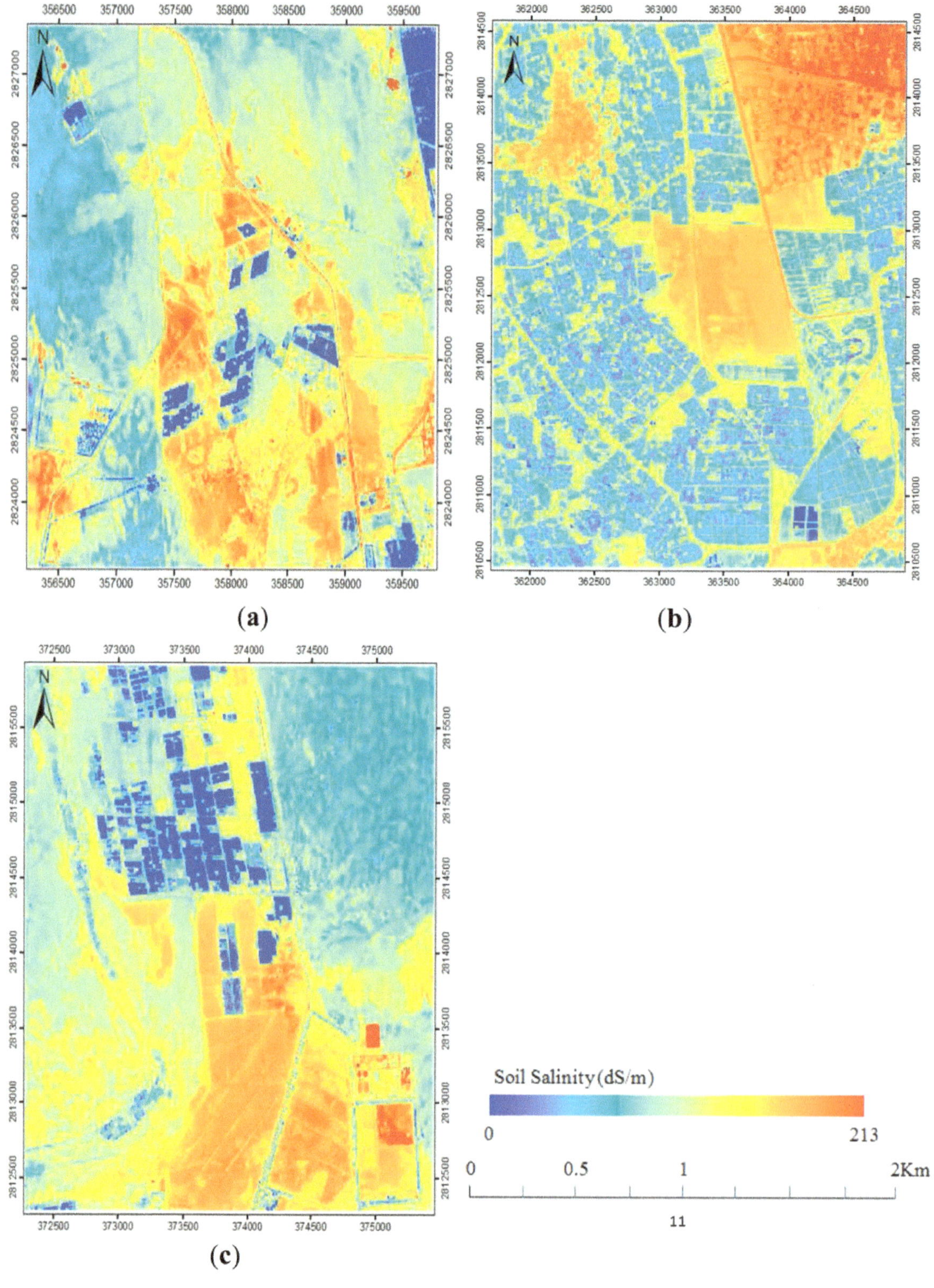

Figure 4: Soil salinity maps for different sites in the study region: (**a**) site one; (**b**) site two; and (**c**) site three.

DISCUSSION

The Developed Regressions Models

The efficiency of the selected regression model to predict and map the spatial variation in soil salinity is shown by the good relationship ($R^2 = 0.65$) at the 99% probability level, RMSE of 39 dS/m and the normality of the residuals. This is in part due to the high spatial resolution of the IKONOS images. The selected model in this study showed superiority in the prediction power ($R^2 = 0.65$) of soil salinity over those reported by Shrestha [27] ($R^2 = 0.23$) and recently by Shamsi *et al.* [35] ($R^2 = 0.39$), which have been developed using different moderate spatial resolution satellite images. Moderate spatial resolution images, such as the Advanced Spaceborne Thermal Emission and Reflection Radiometer (ASTER), Moderate Resolution Imaging Spectroradiometer (MODIS) and Landsat, are economically priced or free, more accessible and typically offer broader spatial coverage than more expensive high spatial resolution imagery. Nonetheless, differences in spatial resolution can have a high impact on predicting soil salinity. Our finding that the prediction of soil salinity based on IKONOS images yields better results than those based on moderate resolution images is in agreement with Eldeiry and Garcia [19]. Given this concern, it is important to take into account spatial resolution as one of the key factors to consider when using satellite imagery to infer soil salinity.

Moreover, the good performance of the selected model in this study is due to the enhanced images efficacy in highlighting information from soil salinity and suppressing the other details. Image enhancement is data processing that aims to increase the overall visual quality of an image or to enhance the visibility and interpretability of certain features of interest in it [53]. Several studies have shown that image enhancement techniques consisting of spectral indices (e.g., NDVI, SI, NDSI, TNDVI) have a great potential in enhancing and delineating soil salinity detail in an image [29,31,54–59]. For example, Tripathi *et al.*[49] found and emphasized that identifying salt-affected soils based on the image enhancement method, represented by the salinity index, yields better results than individual bands, due to its ability to enhance the saline patches by suppressing the vegetation. Recently, Shamsi *et al.* [35] conducted a study to characterise soil salinity in the south-east of Fars Province, Iran, using remote sensing and statistical analysis, and found that using an image enhancement method (Salinity Index (SI)) reduced estimation errors and increased the model's efficiency.

Beside this, the superiority of the visible red band over the other bands in retrieving soil salinity has contributed to improving the regression model. This result is supported by those of Arasteh [60] and Mariappan [61] who found that the visible red band performs best among the Landsat ETM+ bands at characterizing the pattern and features of soil salinity due to its high correlation with EC ground measurements. Soil salinity spectral reflectance is affected by the physical-chemical properties of soil: quality and mineralogy of salt, together with soil moisture, colour and surface roughness [62]. Salts influenced surface features are crusts without or with only a little evidence of salt; thick salt crusts and puffy structures. Salt causes variations in the surface roughness which induces variation in the soil spectral reflectance [4,5,63]. Most salt-affected soils can be identified by a white salt crust that will form on the soil surface; thus, these soils tend to increase spectral reflectance [64–66]. Crusted soil, which affects soil structure and reduces the soil infiltration rate [67] is characterized with significant spectral changes due to the structural crust formation and colour [66]. Salt crust at its inception (high infiltration rate) presents low spectral reflectance, whereas in intense salt crust soil, the spectral reflectance will be significantly higher [66]. Besides, smooth crust surfaces have higher spectral reflectance than rougher crust surfaces [66,68–70].

According to the investigated samples in this study, saline soils with a smooth and light salty crust surface show high spectral reflectance in the red band, in contrast saline soils characterized by coarse dark puffy surface crust exhibit a decrease in spectral reflectance (Figure 5). These findings are in agreement with the those of Metternicht and Zinck [71], Schmid *et al.* [72] and Shamsi *et al.* [35], and confirm the fact that saline soil reflectance results from spectral properties such as the presence of salt crust, soil colour and moisture content, which have a combined effect on the amount of reflectance.

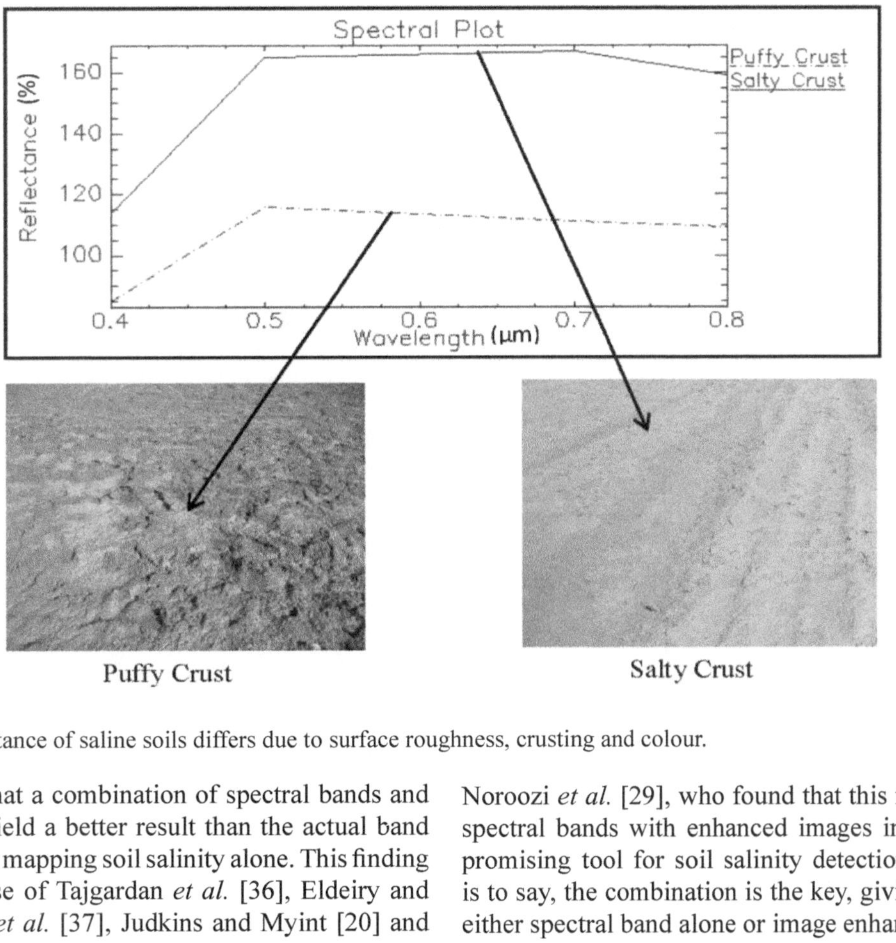

Figure 5: Spectral reflectance of saline soils differs due to surface roughness, crusting and colour.

Thus, it is clear that a combination of spectral bands and image enhancement yield a better result than the actual band used for modelling and mapping soil salinity alone. This finding is consistent with those of Tajgardan *et al.* [36], Eldeiry and Garcia [19], Bouaziz *et al.* [37], Judkins and Myint [20] and Noroozi *et al.* [29], who found that this method of combining spectral bands with enhanced images in a single model is a promising tool for soil salinity detection and mapping. That is to say, the combination is the key, giving better results than either spectral band alone or image enhancement alone.

Mapping Spatial Variation in Soil Salinity

Factors causing soil salinity include inappropriate and excessive irrigation without an adequate drainage system, irrigation water quality, a rising water table, climate, rainfall history, local topography, soil composition and farming practices [73–77]. Therefore, increasing soil salinity at the surface is most likely to vary according to the distribution of these different factors across the landscape. For example, Bilgili [73] found that the spatial distribution of saline soils in the Harran Plain, southeast Turkey, is likely due to inappropriate irrigation coupled with high evaporation and topographical factors.

In this study, soil salinity maps that were generated using the selected model showed large surface areas with very strongly saline soil (>16 dS/m). The spatial distribution in this soil salinity class was variable over the investigated areas. Patches with strongly saline soil along the three sites were most pronounced in non-vegetated wet and dry areas. Wet areas, shown in red and orange colours in the maps (Figure 4), form due to a rising water table and are characterized by a moisture-filled soil [74]. The rising water table brings salt from deep in the soil up to the surface, causing salt accumulation [75]. On the other hand dry lands, which are shown in a colour graduation from yellow to light blue, occurs when a saline water table comes close to the ground level and a high evaporation rate leaves salts at the soil surface [76]. Thus, these findings suggested that a rising water table and salt accumulation at the surface combined with a high evaporation rate are one of the most likely factors that have resulted in the spatial variation in soil salinity over these lands. Similar results in an arid region, Sultanate Oman, were found by [77].

On the other hand, vegetated areas occupy strongly saline soil but within lower salinity levels compared to the non-vegetated areas. The lower observed salinity levels on the vegetated areas may occur because vegetated areas are subjected to a leaching process which reduces salinity levels. In spite of this, there were pronounced salinity differences between the three sites over the vegetated areas. These differences were potentially caused by variation in topography, soil type and structure, poor drainage and irrigation water quality. All of these parameters are known to affect soil salinity distribution across the landscape [77–81]. However, while irrigation water quality can be problematic, this can be overcome by proper irrigation management. Therefore, the observed salinity differences in the salinity levels on vegetated areas could be caused by different irrigation management practices, including irrigation scheduling. For example, instruments that measure and monitor soil moisture were not used for irrigation scheduling at any of the three study sites as farmers were not aware of them and/or lacked the required skills for use. In addition, many farmers cannot afford these instruments, and no extension services are available to encourage their use. Consequently, excessive water applications and poor timing result in various levels of salt build-up in these soils which adversely affect date palm. Several studies have found that soil salinity can cause significant effects on date palm growth and productivity, even though date palm is a high salt-tolerant crop [82–85]. Recently, Al-Abdoulhadi *et al.* [38] conducted a study to describe the effects of soil salinity on date palms. The results of their study revealed that salinity depressed plant growth and the biomass of date palms, and as the salinity increased the leaf length of the fronds was significantly reduced.

This study shows how regression analysis, coupled with high spatial resolution remote sensing images, could successfully predict and map spatial variation in soil salinity over an area vegetated mainly with date palms. Thus, the information presented here can help agricultural workers, scientists and engineers to manage soil salinity problems affecting the ecosystem. Additionally, the simplicity of this approach, with its satisfactory accuracy, can contribute greatly to soil salinity prediction and mapping, at lower costs than conventional approaches.

While, this study focuses on mapping and modelling soil salinity on a spatial variation basis at one point in time, further research requires investigating the temporal variation of soil salinity in this oasis in order to assess the pattern of soil salinity change over time as soil salinity is a space-time variation phenomena. This timely detection of soil salinity, prediction and mapping of its severity and extent will enable decision makers to decide what necessary actions should be taken, especially in areas of strongly saline soils, to protect the date palm outputs, sustain agricultural lands and natural ecosystems.

Besides, although this study contains promising results for modelling and mapping soil salinity based on regression analysis and IKONOS high spatial resolution images, the absence of the thermal band, which has been found a useful tool in several soil salinity studies [71,86,87], and the poor spectral resolution of the images, most likely limit the model capability. Thus, this study can be extended in the future by using hyperspectral images and investigating how this can increase the accuracy of spatial variation in similar modelling and mapping environments. Different studies have reported that hyperspectral images have a promising potential in the assessment and mapping of soil salinity [7,25,88–92]. For example, Weng *et al.* [88] found that soil salinity can be predicted and mapped successfully based on Partial Least Squares Regression (PLSR) techniques with Hyperion hypespectral data in a large area. More recently, in the Jezre'el Valley, northern Israel, Goldshleger *et al.* [5], based on PLSR techniques, assessed the relationships between salinity in tomato plants and soil spectral reflectance obtained using a hyperspectral radiometer and found that the results promising. They concluded that a hyperspectral radiometer is useful for characterizing salinity in growing vegetation and assessing its salt quality. To the best of our knowledge, hyperspectral remote sensing data have never been used to model and map soil salinity in communities vegetated mainly with date palms in the remote sensing domain. Therefore, further research is needed to investigate the capability of hyperspectral remote sensing data in mapping and modelling soil salinity under such conditions.

CONCLUSIONS

The present study demonstrates that combining the IKONOS red band and the salinity index into a regression model offers a potentially quick and inexpensive method to map and model the spatial variation in soil salinity of communities vegetated mostly with date palm. The combination of these remotely sensed variables into one model were able to explain 65% of the spatial variation in the soil salinity of the study area.

The great capacity of this combined model over the other developed models is attributed to the enhanced images and the red band efficacy in highlighting information from soil salinity. The developed model's simplicity and acceptable degree of accuracy makes it a promising tool for continued use in soil salinity prediction. Thus, this model can be used by the decision makers in Al Hassa oasis municipality and similar regions to implement or support effective soil reclamation programs that minimize or prevent future increases in soil salinity.

Although this study demonstrates that soil salinity mapping and modelling can be undertaken with good accuracy based on high spatial resolution multispectral images, further research is needed to focus on investigating the possibility of hyperspectral data in mapping and modelling soil salinity over areas dominated by date palm and investigating whether it can increase the accuracy of modelling and the mapping process.

ACKNOWLEDGMENTS

The authors would like to thank the Space Research Institute at the King Abdul-Aziz City for Sciences and Technology for providing the IKONOS satellite images.

REFERENCES

1. Tanji, K. Salinity in the Soil Environment. In Salinity: Environment-Plants-Molecules; Lauchli, A., Luttge, U., Eds.; Kluwer Academic Publisher: Dordrecht, The Netherlands, 2004; pp. 21–51.
2. Jordán, M.; Navarro-Pedreño, J.; García-Sánchez, E.; Mateu, J.; Juan, P. Spatial dynamics of soil salinity under arid and semi-arid conditions: Geological and environmental implications. Environ. Geol 2004, 45, 448–456.
3. Douaik, A.; Meirvenne, M.; Toth, T. Stochastic Approaches for Space–Time Modeling and Interpolation of Soil Salinity. In Remote Sensing of Soil Salinization: Impact on Land Management; Metternicht, G., Zinck, J.A., Eds.; CRC Press: Boca Raton, FL, USA, 2008; pp. 273–289.
4. Ben-Dor, E.; Goldshleger, N.; Eshel Mor, V. Combined Active and Passive Remote Sensing Methods for Assessing Soil Salinity: A Case Study from Jezre'el Valley, Northern Israel. In Remote Sensing of Soil Salinization: Impact on Land Management; Metternicht, G., Zinck, J.A., Eds.; CRC Press: Boca Raton, FL, USA, 2008; pp. 236–253.
5. Goldshleger, N.; Chudnovsky, A.; Ben-Binyamin, R. Predicting salinity in tomato using soil reflectance spectra. Int. J. Remote Sens 2013, 34, 6079–6093.
6. Metternicht, G.; Zinck, A. Remote Sensing of Soil Salinization: Impact on Land Management; CRC Press: Boca Raton, FL, USA, 2008; p. 377.
7. Farifteh, J.; van der Meer, F.; Atzberger, C.; Carranza, E. Quantitative analysis of salt-affected soil reflectance spectra: A comparison of two adaptive methods (PLSR and ANN). Remote Sens. Environ 2007, 110, 59–78.
8. Akramkhanov, A.; Vlek, P. The assessment of spatial distribution of soil salinity risk using neural network. Environ. Monit. Assess 2012, 184, 2475–2485.
9. Patel, R.; Prasher, S.; God, P.; Bassi, R. Soil salinity prediction using artificial neural networks. J. Am. Water Resour. Assoc 2002, 38, 91–100.
10. Fethi, B.; Magnus, P.; Ronny, B.; Akissa, B. Estimating soil salinity over a shallow saline water table in semiarid Tunisia. Open Hydrol. J 2010, 4, 91–101.
11. Taghizadeh-Mehrjardi, R.; Minasny, B.; Sarmadian, F.; Malone, B. Digital mapping of soil salinity in Ardakan region, central Iran. Geoderma 2014, 213, 15–28.
12. Tóth, T.; Pasztor, L.; Kabos, S.; Kuti, L. Statistical prediction of the presence of salt-affected soils by using digitalized hydrogeological maps. Arid Land Res. Manag 2002, 16, 55–68.
13. Malins, D.; Metternicht, G. Assessing the spatial extent of dryland salinity through fuzzy modeling. Ecol. Model 2006, 193, 387–411.
14. Douaik, A.; van Meirvenne, M.; Toth, T.; Serre, M. Space-time mapping of soil salinity using probabilistic bayesian maximum entropy. Stoch. Environ. Res. Risk Assess 2004, 18, 219–227.
15. Triantafilis, J.; Odeh, I.; McBratney, A. Five geostatistical models to predict soil salinity from electromagnetic induction data across irrigated cotton. Soil Sci. Soc. Am. J 2001, 65, 869–878.
16. Tajgardan, T.; Ayoubi, S.; Shataee, S.; Sahrawat, K.L.; Gorgan, I. Soil surface salinity prediction using ASTER data: Comparing statistical and geostatistical models. Aust. J. Basic Appl. Sci 2010, 4, 457–467.
17. Eldeiry, A.A.; García, L.A. Using Deterministic and Geostatistical Techniques to Estimate Soil Salinity at the Sub-Basin Scale and the Field Scale. Proceedings of the 31th Annual Hydrology Days, Fort Collins, CO, USA, 21–23 March 2011.
18. Douaoui, A.E.K.; Nicolas, H.; Walter, C. Detecting salinity hazards within a semiarid context by means of combining soil and remote-sensing data. Geoderma 2006, 134, 217–230.
19. Eldeiry, A.A.; Garcia, L.A. Detecting soil salinity in alfalfa fields using spatial modeling and remote sensing. Soil Sci. Soc. Am. J 2008, 72, 201–211.
20. Judkins, G.; Myint, S. Spatial variation of soil salinity in the Mexicali valley, Mexico: Application of a practical method for agricultural monitoring. Environ. Manag 2012, 50, 478–489.
21. Fan, X.; Pedroli, B.; Liu, G.; Liu, Q.; Liu, H.; Shu, L. Soil salinity development in the yellow river delta in relation to groundwater dynamics. Land Degrad. Dev 2012, 23, 175–189.
22. McBratney, A.; Santos, M.d.L.M.; Minasny, B. On digital soil mapping. Geoderma 2003, 117, 3–52.
23. Scull, P.; Franklin, J.; Chadwick, O.; McArthur, D. Predictive soil mapping: A review. Progr. Phys. Geogr 2003, 27, 171–197.
24. Lesch, S.M.; Strauss, D.J.; Rhoades, J.D. Spatial prediction of soil salinity using electromagnetic induction techniques: 1. Statistical prediction models: A comparison of multiple linear regression and cokriging. Water Resour. Res 1995, 31, 373–386.

25. Yonghua, Q.; Siongb, J.; Xudonga, L. A Partial Least Square Regression Method to Quantitatively Retrieve Soil Salinity using Hyper-Spectral Reflectance Data. Proceedings of the SPIE 7147, Geoinformatics 2008 and Joint Conference on GIS and Built Environment: Classification of Remote Sensing Images, Guangzhou, China, 31 October 2008.
26. Wang, H.; Wang, J.; Liu, G. Spatial Regression Analysis on the Variation of Soil Salinity in the Yellow River Delta. Proceedings of the SPIE 6753, Geoinformatics 2007: Geospatial Information Science, Nanjing, China, 10 June 2007.
27. Shrestha, R. Relating soil electrical conductivity to remote sensing and other soil properties for assessing soil salinity in northeast Thailand. Land Degrad. Dev 2006, 17, 677–689.
28. Pakparvar, M.; Gabriels, D.; Aarabi, K.; Edraki, M.; Raes, D.; Cornelis, W. Incorporating legacy soil data to minimize errors in salinity change detection: A case study of Darab Plain, Iran. Int. J. Remote Sens 2012, 33, 6215–6238.
29. Noroozi, A.A.; Homaee, M.; Farshad, A. Integrated application of remote sensing and spatial statistical models to the identification of soil salinity: A case study from Garmsar Plain, Iran. Environ. Sci 2012, 9, 59–74.
30. Mehrjardi, R.T.; Mahmoodi, S.H.; Taze, M.; Sahebjalal, E. Accuracy assessment of soil salinity map in Yazd-Ardakan Plain, Central Iran, based on Landsat ETM+ imagery. Am.-Eurasian J. Agric. Environ. Sci 2008, 3, 708–712.
31. Fernandez-Buces, N.; Siebe, C.; Cram, S.; Palacio, J. Mapping soil salinity using a combined spectral response index for bare soil and vegetation: A case study in the former lake Texcoco, Mexico. J. Arid Environ 2006, 65, 644–667.
32. Eldeiry, A.A.; Garcia, L.A. Spatial modeling using remote sensing, GIS, and field data to assess crop yield and soil salinity. Hydrol. Days 2004, 7, 55–66.
33. Afework, M. Analysis and Mapping of Soil Salinity Levels in Metehara Sugarcane Estate Irrigation Farm Using Different Models. Ms.C. Thesis,. Addis Ababa University, Ethiopia, Addis Ababa, 2009.
34. Weng, Y.; Gong, P.; Zhu, Z. Reflectance spectroscopy for the assessment of soil salt content in soils of the Yellow River Delta of China. Int. J. Remote Sens 2008, 29, 5511–5531.
35. Shamsi, F.R.S.; Sanaz, Z.; Abtahi, A.S. Soil salinity characteristics using moderate resolution imaging spectroradiometer (MODIS) images and statistical analysis. Arch. Agron. Soil Sci 2013, 59, 471–489.
36. Tajgardan, T.; Shataee, S.; Ayoubi, S. In Spatial Prediction of Soil Salinity in the Arid Zones Using ASTER Data, Case study: North of Ag Ghala, Golestan Province, Iran. Proceedings of Asian Conference on Remote Sensing (ACRS), Kuala Lumpur, Malaysia, 12–16 November 2007.
37. Bouaziz, M.; Matschullat, J.; Gloaguen, R. Improved remote sensing detection of soil salinity from a semi-arid climate in Northeast Brazil. Comptes Rendus Geosci 2011, 343, 795–803.
38. Al-Abdoulhadi, I.A.; Dinar, H.A.; Ebert, G.; Bttner, C. Effect of salinity on leaf growth, leaf injury and biomass production in date palm (Phoenix dactylifera L.) cultivars. Indian J. Sci. Technol 2011, 4, 1542–1546.
39. Al-Dakheel, Y.; Massoud, M. Towards Sustainable Development for Groundwater in Al-Hassa and the Role of Geographic Information Systems. Proceeidngs of the The 2nd International Conference on Water Resources & Arid Environment, Riyadh, Saudi Arabia, 26–29 November 2006.
40. Hussain, G.; Al-Zarah, A.I.; Latif, M.S. Influence of groundwater irrigation on chemical properties of soils in the vicinity of wastewater drainage canals in Al-Alisa Oasis. Res. J. Environ. Toxicol. (RJET) 2012, 7, 1–17.
41. Al-Naeem, A.A. Evaluation of groundwater of al-hassa oasis, eastern region Saudi Arabia. Res. J. Environ. Sci 2011, 5, 624–642.
42. Al Tokhais, A.S.; Rausch, R. In The Hydrogeology of Al Hassa Springs. Proceedings the 3rd International Conference on Water Resources and Arid Environments and the 1 Arab Water Forum, st Riyadh, Saudi Arabia, 16–19 November 2008; pp. 16–19.
43. AI-Barrak, S.A. Characteristics of some soils under date palm in AI-Hassa eastern oasis, Saudi Arabia. J. King Saud Univ. Agric. Sci 1990, 2, 115–130.
44. Elprince, A.M. Model for the soil solution composition of an oasis. Soil Sci. Soc. Am. J 1985, 49, 1121–1128.
45. Bouaziz, M.; Gloaguena, R.; Samirb, B. Remote mapping of susceptible areas to soil salinity, based on hyperspectral data and geochemical, in the Southern Part of Tunisia. Proc. SPIE 2011, 8174.
46. Richards, L. Determination of the Properties of Saline and Alkali Soils. In Diagnosis and Improvement of Saline and Alkali Soils, Agriculture Handbook No. 60; US Regional Salinity Laboratory: Riverside, CA, USA, 1954; pp. 7–33.
47. Dial, G.; Bowen, H.; Gerlach, F.; Grodecki, J.; Oleszczuk, R. IKONOS satellite, imagery, and products. Remote Sens. Environ 2003, 88, 23–36.
48. Chavez, P.S., Jr. Image-based atmospheric corrections-revisited and improved. Photogramm. Eng. Remote Sens 1996, 62, 1025–1035.
49. Tripathi, N.K.; Rai, B.K.; Dwivedi, P. Spatial modelling of soil alkalinity in GIS environment using IRS data. Proceedings of the 18th Asian Conference in Remote Sensing, ACRS Kuala Lumpur, Malaysia, 20–25 October 1997; pp. 81–86.
50. Royston, P.; Sauerbrei, W. Multivariable Model-Building: A Pragmatic Approach to Regression Anaylsis Based on Fractional Polynomials for Modelling Continuous Variables; John Wiley & Sons: Chichester, UK, 2008; p. 322.
51. Moriasi, D.; Arnold, J.; van Liew, M.; Bingner, R.; Harmel, R.; Veith, T. Model evaluation guidelines for systematic quantification of accuracy in watershed simulations. Trans. ASABE 2007, 50, 885–900.

52. Field, A.; Miles, J.; Field, Z. Discovering Statistics Using R; SAGE Publications: London, UK, 2012; p. 992.
53. Gao, J. Digital Analysis of Remotely Sensed Imagery, 1st ed; McGraw Hill Professional: New York, NY, USA, 2008; p. 674.
54. Odeh, I.O.; Onus, A. Spatial analysis of soil salinity and soil structural stability in a semiarid region of New South Wales, Australia. Environ. Manag 2008, 42, 265–278.
55. Iqbal, F. Detection of salt affected soil in rice-wheat area using satellite image. Afr. J. Agric. Res 2011, 6, 4973–4982.
56. Khan, N.M.; Rastoskuev, V.V.; Sato, Y.; Shiozawa, S. Assessment of hydrosaline land degradation by using a simple approach of remote sensing indicators. Agric. Water Manag 2005, 77, 96–109.
57. Lobell, D.; Lesch, S.; Corwin, D.; Ulmer, M.; Anderson, K.; Potts, D.; Doolittle, J.; Matos, M.; Baltes, M. Regional-scale assessment of soil salinity in the Red River Valley using multi-year MODIS EVI and NDVI. J. Environ. Qual 2010, 39, 35–41.
58. Hamzeh, S.; Naseri, A.A.; AlaviPanah, S.K.; Mojaradi, B.; Bartholomeus, H.M.; Clevers, J.; Behzad, M. Estimating salinity stress in sugarcane fields with spaceborne hyperspectral vegetation indices. Int. J. Appl. Earth Obs. Geoinformation 2012, 21, 282–290.
59. Naumann, J.C.; Young, D.R.; Anderson, J.E. Spatial variations in salinity stress across a coastal landscape using vegetation indices derived from hyperspectral imagery. Plant Ecol 2009, 202, 285–297.
60. Arasteh, P.D. Soil Salinity Change Detection in Irrigated Area Under Gazvin Plain Irrigation Network Using Satellite Imagery. Proceedings of the 9th International Drainage Symposium, Québec City, QC, Canada, 13–16 June 2010; pp. 1–9.
61. Mariappan, V.E.N. Soil salinity assessment using geospatial technology, perspectives, approaches and sctrategies. Indian Cartogr 2010, 30, 25–30.
62. Metternicht, G.; Zinck, J.A. Spectral Behavior of Salt Types. In Remote Sensing of Soil Salinization: Impact on Land Management; Metternicht, G., Zinck, J.A., Eds.; CRC Press: Boca Raton, FL, USA, 2008; pp. 21–36.
63. Goldshleger, N.; Livne, I.; Chudnovsky, A.; Ben-Dor, E. New results in integrating passive and active remote sensing methods to assess soil salinity: A case study from Jezre'el Valley, Israel. Soil Sci 2012, 177, 392–401.
64. Panah, S.K.A.; Goossens, R.; Matinfar, H.R.; Mohamadi, H.; Ghadiri, M.; Irannegad, H.; Asl, M.A. The efficiency of Landsat TM and ETM+ thermal data for extracting soil information in arid regions. J. Agric. Sci. Technol 2008, 10, 439–460.
65. De Jong, S.M.; Addink, E.A.; van Beek, L.P.H.; Duijsings, D. Physical characterization, spectral response and remotely sensed mapping of Mediterranean soil surface crusts. Catena 2011, 86, 24–35.
66. Ben-Dor, E.; Goldlshleger, N.; Benyamini, Y.; Agassi, M.R.; Blumberg, D.G. The spectral reflectance properties of soil structural crusts in the 1.2-to 2.5-μm spectral region. Soil Sci. Soc. Am. J 2003, 67, 289–299.
67. Agassi, M.; Morin, J.; Shainberg, I. Effect of raindrop impact energy and water salinity on infiltration rates of sodic soils. Soil Sci. Soc. Am. J 1985, 49, 186–190.
68. Metternicht, G.I.; Zinck, J.A. Remote sensing of soil salinity: Potentials and constraints. Remote Sens. Environ 2003, 85, 1–20.
69. Elnaggar, A.A.; Noller, J.S. Application of remote-sensing data and decision-tree analysis to mapping salt-affected soils over large areas. Remote Sens 2010, 2, 151–165.
70. Goldshleger, N.; Ben-Dor, E.; Benyamini, Y.; Agassi, M. Soil reflectance as a tool for assessing physical crust arrangement of four typical soils in Israel. Soil Sci 2004, 169, 677–687.
71. Metternicht, G.; Zinck, J. Spatial discrimination of salt- and sodium-affected soil surfaces. Int. J. Remote Sens 1997, 18, 2571–2586.
72. Schmid, T.; Koch, M.; Gumuzzio, J. Applications of Hyper-Spectral Imagery to Soil Salinity Mapping. In Remote Sensing of Soil Salinization: Impact on Land Management; Metternicht, G., Zinck, J., Eds.; CRC Press: Boca Raton, FL, USA, 2009; pp. 113–140.
73. Bilgili, A.V. Spatial assessment of soil salinity in the Harran Plain using multiple kriging techniques. Environ. Monit. Assess 2013, 185, 777–795.
74. Keddy, P.A. Wetland Ecology: Principles and Conservation; Cambridge University Press: Cambridge, UK, 2010; p. 497.
75. Ashraf, M.; Öztürk, M.A.; Athar, H.R. Salinity and Water Stress: Improving Crop Efficiency; Springer: Dordrecht, The Netherlands, 2009; p. 260.
76. Jardine, A.; Speldewinde, P.; Carver, S.; Weinstein, P. Dryland salinity and ecosystem distress syndrome: Human health implications. EcoHealth 2007, 4, 10–17.
77. Hussain, N.; Al-Rawahy, A.S.; Rabee, J.; Al-Amri, M. Causes, origin, genesis and extent of soil salinity in the Sultanate of Oman. Pak. J. Agric. Sci 2006, 43, 1–2.
78. Thiruchelvam, S.; Pathmarajah, S. An Economic Analysis of Salinity Problems in the Mahaweli River System H Irrigation Scheme in Sri Lanka; Economy and Environment Program for Southeast Asia (EEPSEA): Ottawa, Canada, 1999; p. 39.
79. Cetin, M.; Kirda, C. Spatial and temporal changes of soil salinity in a cotton field irrigated with low-quality water. J. Hydrol 2003, 272, 238–249.
80. Zheng, Z.; Zhang, F.; Ma, F.; Chai, X.; Zhu, Z.; Shi, J.; Zhang, S. Spatiotemporal changes in soil salinity in a drip-irrigated field. Geoderma 2009, 149, 243–248.
81. Sakadevan, K.; Nguyen, M.-L. Extent, impact, and response to soil and water salinity in arid and semiarid regions. Adv. Agron 2010, 109, 55–74.
82. Alhammadi, M.; Glenn, E. Detecting date palm trees health and vegetation greenness change on the eastern coast of the United Arab Emirates using SAVI. Int. J. Remote Sens 2008, 29, 1745–1765.
83. Ramoliya, P.; Pandey, A. Soil salinity and water status affect growth of Phoenix dactylifera seedlings. N. Z. J. Crop Hortic. Sci 2003, 31, 345–353.

84. Alhammadi, M.S.; Edward, G.P. Effect of salinity on growth of twelve cultivars of the United Arab Emirates date palm. Commun. Soil Sci. Plant Anal 2009, 40, 2372–2388.

85. Hussein, A.; El-Desouki, M.; El-Kased, F.; Nour, G.; Abd-El-Hamid, N. Effect of salinity on date palm seeds germination and early seedling growth. J. Agric. Sci. Mansoura Univ 1993, 18, 1306–1314.

86. Verma, K.; Saxena, R.; Barthwal, A.; Deshmukh, S. Remote sensing technique for mapping salt affected soils. Int. J. Remote Sens 1994, 15, 1901–1914.

87. Alavipanah, S.K.; Goossens, R. Relationship between the Landsat TM, MSS data and soil salinity. J. Agric. Sci. Technol 2001, 3, 101–111.

88. Weng, Y.; Gong, P.; Zhu, Z. Soil salt content estimation in the Yellow River delta with satellite hyperspectral data. Can. J. Remote Sens 2008, 34, 259–270.

89. Lu, N.; Zhang, Z.; Gao, Y. Recognition and Mapping of Soil Salinization in Arid Environment with Hyperspectral Data. Proceedings of the IGARSS'05. IEEE International Geoscience and Remote Sensing Symposium, Seoul, Korea, 25–29 July 2005; pp. 4520–4523.

90. Ben-Dor, E.; Patkin, K.; Banin, A.; Karnieli, A. Mapping of several soil properties using DAIS-7915 hyperspectral scanner data-a case study over clayey soils in Israel. Int. J. Remote Sens 2002, 23, 1043–1062.

91. Zhang, T.T.; Zeng, S.L.; Gao, Y.; Ouyang, Z.T.; Li, B.; Fang, C.M.; Zhao, B. Using hyperspectral vegetation indices as a proxy to monitor soil salinity. Ecol. Indic 2011, 11, 1552–1562.

92. Hamzeh, S.; Naseria, A.A.; Panah, S.K.A.; Mojaradic, B.; Bartholomeus, H.M.; Herold, M. Mapping salinity stress in sugarcane fields with hyperspecteral satellite imagery. Proc. SPIE 2012.

CHAPTER 6

OPERATIONAL MAPPING OF SOIL MOISTURE USING SYNTHETIC APERTURE RADAR DATA

Nicolas Baghdadi, Maelle Aubert, Olivier Cerdan, Laurent Franchistéguy, Christian Viel, Eric Martin, Mehrez Zribi and Jean François Desprats

Keywords: SAR sensors, soil moisture, ISBA model

INTRODUCTION

Soil moisture plays a crucial role in the continental water cycle, more specifically in the distribution of precipitation between surface runoff and infiltration, which is the main driver behind most hydrological and geomorphologic processes [1-3]. The retrieval of soil moisture data is of considerable importance in various applications, such as hydrology, risk prediction, agriculture, and meteorology. It is a key indicator for constraining the initial conditions of infiltration/runoff rates when modeling overland flow. The knowledge of the initial water content at the beginning of the rainfall event is a prerequisite because all infiltration models integrate this information. In the context of implementing the Water Directive and the forthcoming Soil Framework Directive in Europe, there is a need for operational tools to evaluate land management scenarios and to provide sound references for targeting land use planning and the protection of natural resources. More specifically, operational services in charge of flood prevention and forecasting, such as the SCHAPI (National Flood Forecasting Agency) in France are investing in the continuous distributed modeling of soil surface moisture.

Only spatial remote sensing allows monitoring of environmental problems over large areas at regular intervals. Moreover, radar sensors allow mapping whatever the meteorological conditions (clouds, fog, etc.), both day and night. This is not the case with optical sensors, which are difficult to operate if there is cloud cover, a frequent situation in periods where flood events occur. The Soil Surface Characteristics (SSC) can be estimated from microwave remote sensing sensors due to the sensitivity of radar signal to soil characteristics such as the soil's roughness and dielectric constant [4-6]. In addition, the radar signal depends on various radar parameters such as the polarization, incidence angle and frequency.

The SAR sensors currently operational are PALSAR/ALOS, ASAR/ENVISAT, RADARSAT-1 and ERS-2. ERS-2 and RADARSAT-1 provide data based on a single polarization (VV for ERS-2, and HH for RADARSAT-1) whereas ASAR allows the scientific community to acquire images in dual-polarization mode (HH/HV, HH/VV, VV/VH). The radar incidence angle is 23° for ERS-2, between 20° and 50° for RADARSAT-1, and ranges from 14° to 45° for ASAR. The nominal swath width for ASAR and ERS-2 is greater or equal to 100 km, with a spatial resolution of, at best, 25 m (12.5 m pixel size). For RADARSAT-1, the nominal swath width is greater than or equal to 50 km, with a spatial resolution of, at best, 10 m (6.25 m pixel size). These three radar sensors operating in C-band (~5.3 GHz), have been widely used for retrieving soil moisture [e.g. 4, 7-15]. The SAR data are more and more used to derive soil parameters. Although many improvements have been achieved, this field is still not fully operational.

When using ERS-1/2, RADARSAT-1 and ASAR, a better estimate of soil moisture is obtained with a radar configuration that minimizes the effects of other soil surface characteristics

Citation: Baghdadi, N.; Aubert, M.; Cerdan, O.; Franchistéguy, L.; Viel, C.; Eric, M.; Zribi, M.; Desprats, J.F. Operational Mapping of Soil Moisture Using Synthetic Aperture Radar Data. Sensors 2007, 7, 2458-2483. doi:10.3390/s7102458.

(mainly surface roughness). Several studies have shown that the best C-band estimates of soil moisture when using only one radar channel are obtained with SAR images acquired at low and medium incidence angles [13, 15, 16]. These optimal configurations of incidence angle (≤37°) can be attained with current SAR sensors (ERS-2, RADARSAT-1, ASAR, and ALOS) about 10 times per month on a study site in Europe. The arrival of new sensors (TerraSAR-X, COSMO-SkyMed, and RADARSAT-2) should improve this temporal frequency.

The use of two SAR images acquired at both low and high incidence angles markedly improves the precision of the moisture estimate because the backscattering coefficients ratio eliminates the effects of roughness and thus permits linking of the radar backscattering coefficients to only the moisture [13, 15, 16]. This solution is not possible with current SAR sensors and will not be possible either with the sensors planned for the near future (ALOS, RADARSAT-2 and Terra SAR-X). Indeed, the time separating two SAR images acquired by the same sensor at two selected incidence angles (20° and 40°) is of the order of a few days (about 7 days with ASAR over the Touch basin), which limits the use of this possibility because the soil surface characteristics could change rapidly between the two acquisitions. Only the single incidence case is thus possible. The use of SAR images with two polarizations (case of ASAR sensor) does not provide a significant improvement in estimating soil moisture (improvement <1%, [15, 17]) when compared with the results obtained with a single polarization.

Soil moisture can be measured (in situ or using remote sensors) and also modeled. In situ soil moisture measurements are performed using two techniques: TDR (Time Domain Reflectometry) probes and gravimetry. TDR measurements are quicker to carry out, but less precise than those stemming from the gravimetric method. A drift in TDR measurements is often observed, and it is for this reason that the calibration of probes by several gravimetric measurements is often recommended. In practice, the volumetric water content on a field scale is assumed to be equal to the mean value estimated from several representative samples collected from the top layer of soil.

Estimation of soil moisture by inversion of SAR data can be performed in using physical or semi-empirical approaches. The physical approach uses backscattering models capable of reproducing the radar backscattering coefficient from the sensor configuration (wavelength, polarization, and incidence angle) and soil parameters (soil moisture and surface roughness for bare soils). However, several studies have reported a poor agreement between measured radar signals and those predicted by the models [18-21]. Differences between simulations and measurements may reach several decibels, rendering the inversion results inaccurate. The second approach consists of establishing experimental calibration relationships linking the signal radar to soil parameters (surface roughness and soil moisture of bare soils) and to sensor parameters (frequency, polarization, and incidence). This approach requires an important experimental database, acquired from several study sites and representative of possible physical soil conditions. The validity of these relationships for use in operational mode depends on the quality of the database and its representativeness. With current SAR sensors, the cartography in operational mode of the surface moisture of bare soils is carried out on images with one channel (one incidence and one polarization), which leads to the elaboration of experimental relationships between the signal radar and the soil moisture whatever the surface roughness [11, 15, 16]. Using this technology the precision on the prediction of soil surface moisture varies between 3 and 10 % of moisture depending on the incidence angle. Uncertainty analyses carried out with the Green and Ampt infiltration/runoff model showed that this degree of accuracy is adapted to the use of this kind of hydrological models [22].

Another approach for the estimation of soil moisture is to use a Soil-Vegetation-Atmosphere-Transfer model (SVAT). In this case, the SVAT is forced by analyzed or observed meteorological data to estimate mesoscale soil moisture over large areas. Such an approach is used within the SAFRAN-ISBA-MODCOU chain in France, which is composed of a meteorological analysis system (SAFRAN), a SVAT (ISBA) and a hydrological model (MODCOU, [23]). This system is able to simulate the water and energy budget at the surface, the soil moisture and the discharge of the main French rivers. It is run on an operational mode to evaluate the soil moisture at the scale of France. The typical scale of application of the SAFRAN-ISBA-MODCOU model is 8 km × 8 km over all of France. A comparison with satellite data is of great interest to validate the variables simulated by the system.

The objective of the present paper is to test the applicability of experimental relationships established between the C-band radar backscattering coefficient (ERS-2, RADARSAT-1, and ASAR sensors) and the soil moisture from a great experimental database collected between 1995 and 2005 at several study sites. Next, in the soil moisture mapping process, these relationships will be inverted and applied to ERS-2 and ASAR images, acquired in 2006 and 2007 over the Touch catchment basin (located near Toulouse in France) to provide additional key indicators to improve flood forecasting over the Touch catchment basin. The correlation between textural units and soil moisture will also be studied. Moreover, the soil moisture maps estimated from SAR images will be compared with simulations obtained by the ISBA soil-vegetation-atmosphere transfer model. This model, developed by Météo-France models the main surface processes and is currently coupled to atmospheric prediction model. As the soil moisture is prognostic variable of ISBA, it is thus necessary to quantify its quality. Soil moisture in situ data, measured by gravimetry, will be used to validate obtained results. Given that the presence of dense and high vegetation cover prevents C-band radar signal (wavelength ~ 6 cm) from reaching the ground, the soil moisture mapping from ERS and ASAR sensors is focused on bare soils or zones with little vegetation cover.

DESCRIPTION OF DATABASES

Database 1995-2005

An extensive experimental database collected at seven study sites was used for building relationships between radar backscattering coefficient and soil moisture. This database is composed of SAR images acquired between 1995 and 2005, andin situ soil moisture measurements (bare soils only) carried out jointly with radar acquisitions:

- Touch site, near Toulouse, France (long. 01°02′ E, lat. 43°27′ N). Soil composition is about 55% silt, 21% clay, and 24% sand. Data were acquired in 2004 and 2005.

- Villamblain site, near Orléans, France (long. 01°34′ E, lat. 48°00′ N). Soils composed of about 60% silt, 30% clay, and 10% sand. Data were acquired in 2003, 2004, and 2005.
- Châteauguay site, near Montreal, Canada (long. 73°46′ W, lat. 45°19′ N). Soil composition at both sites is about 42% silt, 36% clay, and 20% sand. Fieldwork was carried out in 1999.
- Brochets site, near Montreal, Canada (long. 72°54′ W, lat. 45°08′ N). Soil composition at both sites is about 42% silt, 36% clay, and 20% sand. Fieldwork was carried out in 1999.
- Alpilles site, Rhône valley, France (long. 4°45′ E, lat. 43°47′ N). Soil composition is 54% silt, 40% clay, and 6% sand. Fieldwork was carried out in 1997.
- Orgeval site, near Paris, France (long. 3°07′ E, lat. 48°51′ N). Soil composition is about 78% silt, 17% clay, and 5% sand. Fieldwork was carried out in 1995 to measure soil moisture and surface roughness.
- Pays de Caux site, Normandy, France (long. 0°50′ W, lat. 49°47′ N). Soil composition is approximately 67% silt, 13% clay, and 17% sand. Fieldwork was carried out in 1994, 1998, and 1999 to measure soil moisture and surface roughness.

These sites are characterized by agricultural fields intended for growing wheat and corn. The measurements of volumetric soil moisture have been carried out on the first top 10 cm using two techniques: a TDR probe and gravimetry. The soil moisture ranges from 5.4% to 47.3%. SAR images used were acquired by ERS-2, RADARSAT-1 and ASAR, with a great range of incidence angles (20° to 48°), and at HH/HV/VV polarizations.

The database available for this study consists of about 400 doublets of backscattering coefficients and soil moisture. Each point of this database corresponds to the mean radar signal over a reference plot and the corresponding in situ soil moisture. The reference plots were chosen with a low local topography (generally flat).

Soil roughness measurements were also carried out, using a 1 or 2 m long profilometer with a 1 or 2 cm sampling interval (pinmeter or laser). For radar applications, the surface roughness of a given bare soil is defined statistically by two variables, determined from the surface height profiles: the standard deviation of surface height (rms), which specifies the vertical scale of the roughness, and the correlation length (L), representing the horizontal scale. The rms values vary in our database between 0.25 cm and 5.29 cm; the lowest ones correspond mainly to sown fields and the highest ones to ploughed fields. The correlation length ranges from 2 to 20 cm for 95% of measurements. Recent investigations have indicated that roughness variables estimated on the basis of field measurement data and simulations are very sensitive to profile length [e.g. 24-26]. The limited length of conventional profilers (one or two metres long) leads to large uncertainties in the estimated roughness variables. In a theoretical study based on simulated profiles, Oh and Kay [24] demonstrated that in order to measure rmssurface height and correlation length with a precision of ±5% of their mean values, the length of the roughness profiles had to be at least 100L and 270L respectively. Measurement precision can be improved by averaging multiple profiles. For a correlation length that ranges from 2 to 20 cm, and for 10 averaged profiles, we deduce that the 2 m profiles provide a precision better than ±5% for rms surface height and between ±5% and ±15% for correlation length. At 1 m profile length, the precision can reach 10% for rms height and 20% for correlation length.

Database 2006-2007

Study Site

Our study site is the Touch catchment basin, located at the south-west of Toulouse in France (latitude: 43°17′ N to 43°37′ N, longitude: 0°48′ E to 01°24′ E; Figure 1). It covers a total area of about 500 km^2. The site is composed mainly of agricultural fields intended for growing wheat and corn. It is flat or slightly undulated with altitudes varying between 135 and 377 m. The Touch basin was selected as it is a site representative of the region where several catchments suffer from excessive runoff due to the conjunction of fragile soils and intensive agricultural practices, responsible for many damages. It is now selected as reference basin for several official authorities of flood forecasting and prevention, such as DIREN (Regional Direction of the Environment) and SCHAPI (National Flood Forecasting Agency).

Figure 1: Location of Touch basin.

Radar and Optical Images

Images used in this study were acquired by the ERS-2 and ASAR SARs between 08 March 2006 and 12 March 2007. Fifteen ASAR images and three ERS-2 images were obtained with incidence angles between 18° and 44°, and at HH/VV/VH polarizations (Table 1). Optical images obtained from SPOT and ASTER sensors were also used to allow a reliable mapping of bare soils. Before analyzing the SAR images, the data were radiometrically calibrated, which enables extraction of the backscattering coefficient ($\sigma°$) from the signal intensity of each pixel (BestW software, [27]). The radiometric calibration of SAR images allows converting the raw digital numbers (DN) to backscattering coefficients ($\sigma°$):

Table 1: Main characteristics of SAR images used for soil moisture mapping. Images are acquired over the Touch basin in 2006-2007. For each SAR date, there are the optical images used for mapping bare soils. 33 gravimetric measurements were performed on March 12, 2007

Date (dd/mm/yy)	SAR sensor	Acquisition Time (TU)	Polarization	Incidence angle	Optical image Date	Percentage of bare soil
12/03/2007	ASAR	10 :13	VV , VH	IS2 : 23°	SPOT 15/02/2007	10.9
17/01/2007	ASAR	10 :10	VV , VH	IS3 : 28°	SPOT 15/02/2007	19.9
07/01/2007	ASAR	21 :43	VV , VH	IS3 : 28°	SPOT 15/02/2007	19.9
04/01/2007	ASAR	10 :13	VV , VH	IS2 : 23°	SPOT 15/02/2007	19.9
01/01/2007	ASAR	10 :13	VV , VH	IS2 : 23°	SPOT 15/02/2007	19.9
14/08/2006	ERS-2	10 :42	VV	23°	ASTER 01/07/2006	20.9
10/07/2006	ERS-2	10 :42	VV	23°	ASTER 01/07/2006	20.9
15/04/2006	ASAR	21 :34	VV , HH	IS1 : 18°	SPOT 07/04/2006	16.1
15/04/2006	ASAR	10 :16	VV , HH	IS1 : 18°	SPOT 07/04/2006	16.1
12/04/2006	ASAR	10 :10	VV , HH	IS3 : 28°	SPOT 07/04/2006	16.1
02/04/2006	ASAR	21 :43	VV , HH	IS3 : 28°	SPOT 07/04/2006	16.1
30/03/2006	ASAR	10 :19	VV , HH	IS1 : 18°	SPOT 07/04/2006	8.7
27/03/2006	ERS-2	10 :42	VV	23°	SPOT 07/04/2006	16.0
18/03/2006	ASAR	09 :56	VV , HH	IS7 : 44°	SPOT 15/03/2006	24.1
17/03/2006	ASAR	21 :45	VV , HH	IS4 : 33°	SPOT 15/03/2006	24.1
11/03/2006	ASAR	21 :34	VV , HH	IS1 : 18°	SPOT 15/03/2006	24.1
11/03/2006	ASAR	10 :16	VV , HH	IS1 : 18°	SPOT 15/03/2006	24.1
08/03/2006	ASAR	10 :10	VV , HH	IS3 : 28°	SPOT 15/03/2006	24.1

$$\sigma° = \frac{DN^2 \sin\theta}{K} \quad (1)$$

where θ is the radar incidence angle, and K the calibration constant given in the image header. The backscattering coefficients are then expressed in decibels (dB):

$$\sigma°_{dB} = 10\log_{10}(\sigma°) \quad (2)$$

This calibration makes it possible to carry out multi-temporal analysis of different images (with same incidence and polarization). All SAR and optical images were then georeferenced using topographic maps with a root mean square error of the control points of about 30 m. The registration error of SAR images is taken into account by selecting areas of interest (AOI) within each training site and the border pixels were removed (defined by GPS control points).

Speckle noise, due to the coherent interference of waves reflected from many elementary scatterers, is present on SAR images and makes the pixel-by-pixel interpretation of SAR images extremely difficult. This explains why the estimation of soil characteristics is generally carried out on homogeneous sectors with several pixels or at field scale (which helps reducing speckle). In practice, the mean backscattering coefficients are calculated from calibrated SAR images by averaging the linear $\sigma°$ values of all pixels within the field (or sub-field). A reduction in speckle and an improvement in the quality of our estimations are highly dependent on the size of the fields [e.g. 28, 29]. Consequently, parcels with a homogeneous surface of 200 pixels or more are often used to study the behaviour of the signal as a function of SSC. In the case of ERS and ASAR sensors, this corresponds to fields of around two hectares or more.

ISBA Data

The soil moistures simulated by the ISBA (Interactions between Soil Biosphere and Atmosphere) SVAT [30, 31] were also used. The simulations were compared to the soil moistures estimated from SAR images and measured by gravimetry on March 12, 2007. ISBA is a SVAT designed to run in stand alone mode or in atmospheric models. The version used in the SAFRAN-ISBA-MODCOU is the force restore version of ISBA. The temperature evolution is described using two temperatures (surface temperature and deep soil). Three layers are used to describe the hydrological processes: the surface layer (one centimeter), the root zone layer (~1.50 m, depending of the vegetation type in the grid), the sub-root zone (usually down to 2 m below the surface). The hydrological processes which influence the liquid water content of the surface layer are precipitation (except the part intercepted by the vegetation) and evaporation from bare ground. In addition, it is relaxed to an equilibrium value which account for the balance between gravity and capillary forces (this value depends on the soil texture). The second layer accounts for the root depth zone. It is in relation with the surface layer via the relaxation term and with the sub-root zone via the gravitational drainage and the diffusion term (capillary versus gravitational drainage). The surface runoff occurs on the saturated part of the ISBA grid, which depends on the root zone wetness. It is calculated using the Variable infiltration capacity scheme [23]. Since the purpose of this study was to depict the state of surface moisture, only the surface level soil moisture was used in this study. For temperature, a force restore approach is used. The soil and vegetation parameters are derived from the ECOCLIMAP database [32] and from textural information (clay and sand fractions).

The surface water and energy budget were calculated by ISBA using the SAFRAN [33, 34] meteorological data as input. SAFRAN used meteorological model outputs and meteorological observations to derive the variables used to force ISBA (near surface air temperature, wind and humidity, precipitation and incoming long wave and short wave radiation). The analysis is produced for climatologically homogeneous regions (approximately 30 km × 30 km), then interpolated on the 8 km × 8 km grid used to run ISBA. The data used in this paper are the data produced on an operational mode by Météo-France and used for soil moisture and drought monitoring.

It is important to underline that in this study the radar soil moisture estimates are obtained exclusively for bare soils or for soils with little vegetation, whereas ISBA simulations are carried out whatever the landuse (bare soils, forest, cultures …). A preliminary study has shown the absence of correlation between the ISBA moisture values and the percentage of bare soil in ISBA grids. Comparisons between radar moisture estimated on only bare soil and ISBA moisture calculated on 8 km × 8 km grids regardless of land use can therefore be carried out.

In situ measurements

Simultaneously with the acquisition of the SAR image of 12 March 2007, true ground measurements of soil moisture were performed on several test fields. Gravimetric moisture measurements on 33 test fields of bare soil well distributed on the whole of the catchment basin within ± 3 h of the SAR overpass were carried out. Gravimetric measurements were carried out on the upper 0-10cm soil layer because the radar signal penetration depth is only of few centimetres at C-band [e.g. 5, 35]. The soil moisture on a field scale is assumed to be equal to the mean value estimated from 5 samples collected from the top layer of soil. The gravimetric soil moisture content is then transformed into volumetric moisture content by multiplying it by the bulk density of the soil. The volumetric soil moisture measurements range from 19.4% to 32.6% (standard deviation on the field scale about ±3%), and the soil bulk density ranges from 0.9 g/cm^3 to 1.6 g/cm^3 (standard deviation on the field scale about ±0.15 g/cm^3). During the ground campaign of 12 March 2007, the surface roughness of test fields varied from smooth (wheat and corn sown fields) to rough (ploughed fields). No roughness measurement was taken on March 12, 2007.

The lack of in situ moisture data encourages the analysis of radar and ISBA moistures using climatic and pedologic factors. For climatic factors, daily precipitation and temperature data recorded on six meteorological stations installed on the basin (Toulouse Blagnac, Toulouse Francazal, Fabas, La Bastidette, Lherm, and Rieumes) were used. Concerning pedologic factors, soil texture was used as its influence on soil moisture has already been demonstrated in numerous studies [36]. Other pedological factors are usually used to determine moisture, as organic carbon content, density, and depth of root zone [37, 38]. These factors were not considered in this study because they show little variation within the basin.

Textural Map

The textural map of the Touch basin, extracted from the Europe map at 1:1000000, has been thus used (Soil Geographical Database of Europe, SGBDE, [39]). This database contains information on soil texture (Figure 2) and it was possible to define four textural units on the Touch basin. The South-western part of the basin is mainly composed of clayey material. In the center of the basin, the majority formation is the clayey-sandy soils. Silty soils are primarily present in the North-East of the catchment.

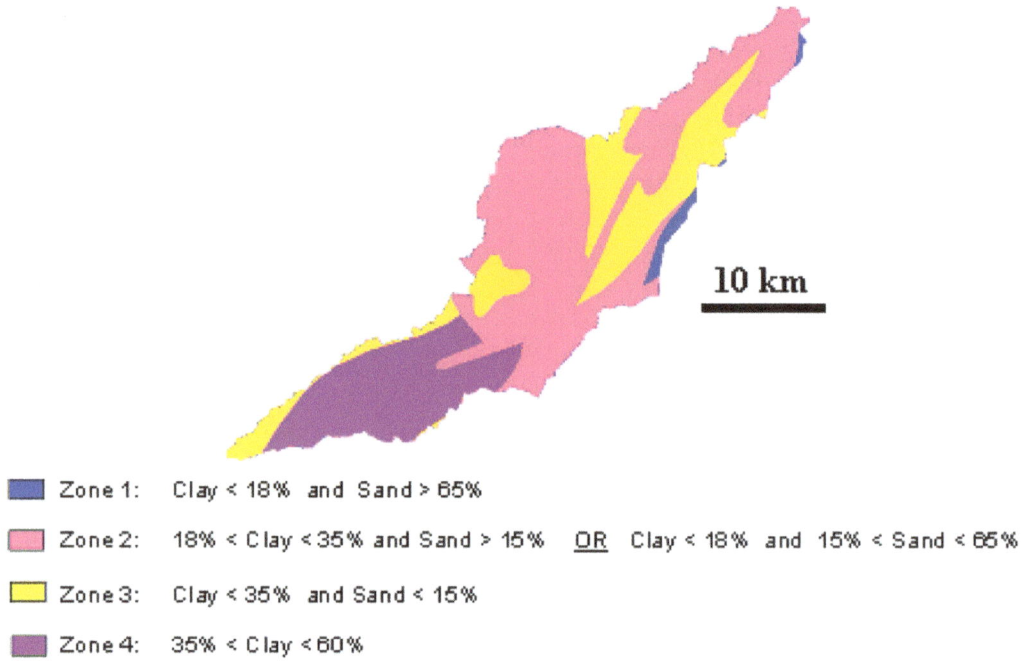

Figure 2: Soil textural zones of the Touch catchment defined according to the 1:1000000 Soil Geographical database.

SOIL MOISTURE MAPPING

Relationship between radar signal and soil moisture

For bare soils, the radar backscattering coefficient in decibels can be written as the sum of two functions : the first one, f(linear), describes the dependence of radar signal on volumetric surface soil moisture, while the second, g (exponential), illustrates the dependence of $\sigma°$ on surface roughness [13, 15]:

$$\sigma^0_{dB} = a m_v + b e^{-krms} + c \tag{3}$$

where k is the wave number ($\approx$1.11 cm^{-1} for C-band), and rms is the root mean square surface height (characteristic parameter of surface roughness). For a given radar wavelength, the coefficients a, b, and c are observed to be dependent on both radar incidence angle and polarization [15, 16, 17]. In this equation, the correlation length is not taken into account.

To retrieve soil moisture (m_v) from a single radar configuration, it is necessary to establish a relationship between the radar backscattering coefficient ($\sigma°$) and m_v alone, without having any knowledge of the rms surface height. As a first approximation, the radar backscattering coefficient (in dB) may be expressed as follows [15, 40]:

$$\sigma^0_{dB} = a m_v + d \tag{4}$$

This simplified relationship ignores the surface roughness. The coefficient a is dependent on both incidence angle and polarization. The coefficient d is primarily controlled by incidence angle, polarization and surface roughness. However, the coefficients describing the linear relationship between radar signal and soil moisture can be different from one catchment to another and also from one year to the next, and sometimes need to be calibrated. This difference is mainly due to the effects of roughness variation [41].

The coefficients a and d of the equation (4) were calculated using the database of 1995-2005 for 7 radar configurations defined by one polarization and a range of incidence angle (Figure 3): (VV 21°-25°), (HH 20°-28°), (HH 34°-37°), (HH 39°-45°), (HV 20°-24°), (HV 34°-37°), and (HV 40°-44°). These coefficients were established for radar configurations in C-band, as a function of polarization and incidence angle, regardless of soil roughness. They are valid only for the same ranges of soil texture and surface roughness. The large database acquired over the last twelve years from study sites in France and Canada shows that the sensitivity of the radar signal to soil moisture is important for low and medium incidence angles ($\leq$37°). The sensitivity of radar signal to soil moisture is slightly different with polarization (for low incidences, 0.23 dB/% in VV compared to 0.19dB/% in HH and 0.15 dB/% in HV). It is of the same order of magnitude for incidences between 20° and 37° (between 0.19 dB/% for HH20°-28° and 0.23 dB/% for HH 34°-37°) and seems to decrease for incidences greater than 39° (0.11 dB/%). These results therefore show that moisture mapping is optimal at low and medium incidences ($\leq$37°).

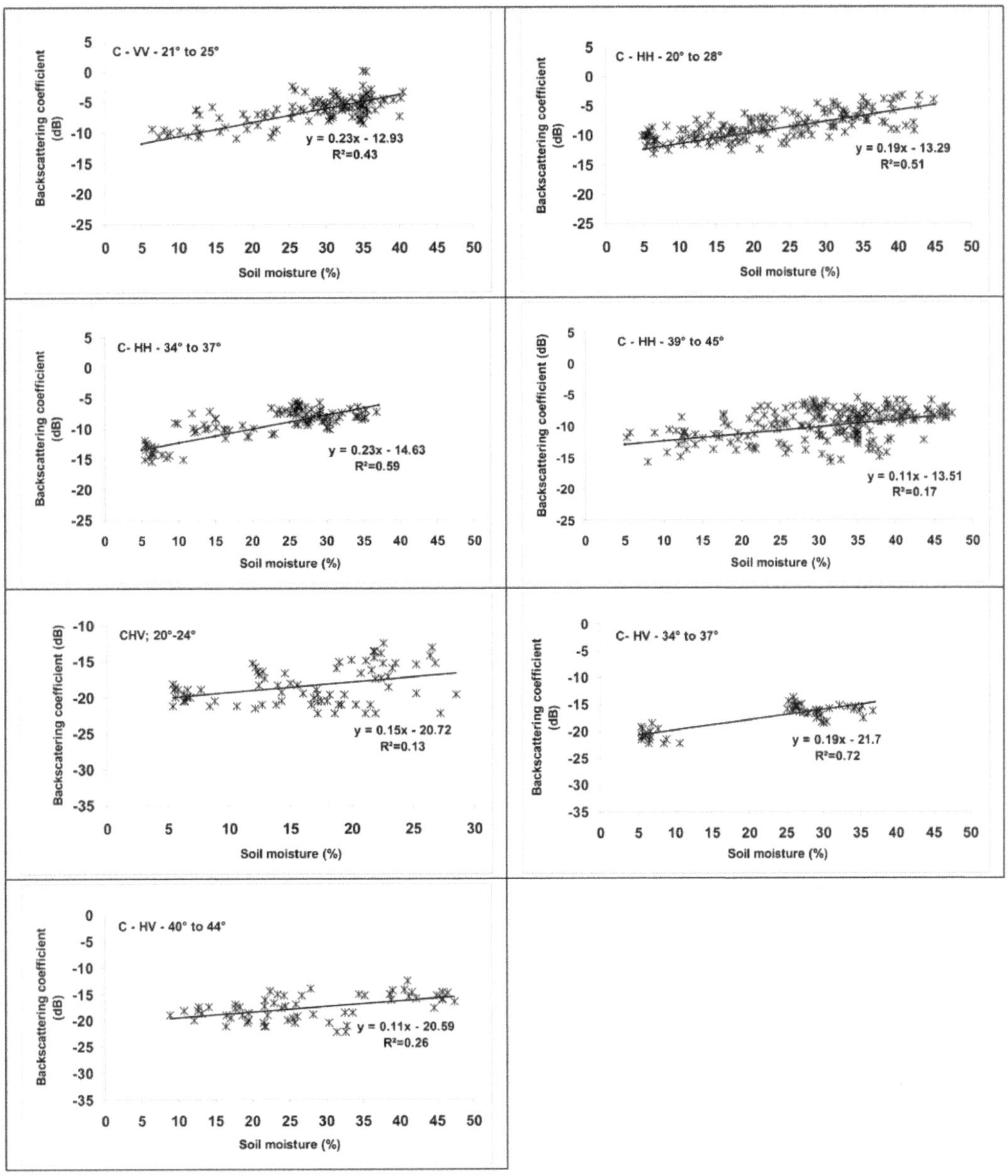

Figure 3: Sensitivity of radar signal in C-band (~5.6 GHz) to surface soil moisture as a function of incidence angle and polarization. Results were obtained from ERS, RADARSAT-1 and ASAR data for clay loam sites (silt: 42 to 78%, clay: 13 to 40%, sand: 5 to 24%). Each point corresponds to the average backscattering coefficient in decibels for one test field.

These results are comparable with the observed sensitivities by various studies [11, 16, 17, 42]. Next, the inversion procedure was applied to the database of 2006-2007 for the estimation of soil moisture. The validity of this procedure was verified by comparing the inversion output with experimental data of soil moisture.

Methodology

Based on single SAR images (one polarization and one incidence), a simple procedure was used for mapping the surface soil moisture over bare soils [equation (4)]. The mapping of soil moisture was carried out on the Touch catchment using SAR data acquired in 2006 and 2007. For each SAR image, the relationship $\sigma^\circ_{dB}(m_v)$ corresponding to the same radar configuration (incidence and polarization) was used (equation (4), Figure 3). All images were used, even those which correspond to low sensitivities of radar signal to soil moisture (18 March 2006 – IS7), except the VV-33° and VV-44° configurations because the relationships between σ° and m_v could not be established for lack of data. Three soil moisture mapping scales estimates were carried out: an estimate for each ISBA cell (8 km × 8 km), an estimate for each textural unit, and an average estimate on the basin. For each SAR image, the bare soils map used corresponds to the one obtaining from the nearest optical image (SPOT and ASTER). The surface of bare soil covered between 8.7% (30 March 2006) and 24.1% (08 to 18 March 2006) of the basin›s total area (cf. Table 1).

For mapping soil moisture on the ISBA grid (8 km × 8 km), the basin was divided into contiguous windows before calculating the mean backscattering coefficient on the whole of bare soil pixels of each cell. For the mapping according to textural units, the mean radar signal is calculated for each textural unit (Figure 2). The estimation of mean soil moisture on the scale of catchment basin uses the mean backscattering coefficient of all bare soils pixels present on the basin. Finally, soil moisture estimates are obtained by inverting the relationship between radar signal and the soil moisture [equation (4)]. The gravimetric soil moisture on ISBA grid scale is assumed to be equal to the mean value of all gravimetric measurements carried out in each grid. On the scale of catchment basin, the gravimetric soil moisture is obtained in averaging all in situ measurements.

The inversion performance of radar signal for retrieving soil moisture is evaluated to show the robustness of the presented technique. The ground truth measurements of soil moisture carried out the March 12, 2007 allow comparing effectively estimated soil moisture by SAR image and observed soil moisture. As suggested by Willmott [43], three statistical indexes were used:

Mean Absolute Error:

$$MAE = \frac{1}{N}\sum_{i=1}^{N}|Pi - Oi| \quad (5)$$

Root Mean Square Error:

$$RMSE = \sqrt{\frac{1}{N}\sum_{i=1}^{N}(Pi - Oi)^2} \quad (6)$$

Index of agreement:

$$d = 1 - \left[\frac{\sum_{i=1}^{N}(Pi - Oi)^2}{\sum_{i=1}^{N}\left(\left|Pi - \frac{1}{N}\sum_{i=1}^{N}Oi\right| + \left|Oi - \frac{1}{N}\sum_{i=1}^{N}Oi\right|\right)^2}\right], \quad 0 \le d \le 1 \quad (7)$$

where P is the field or simulated variable, O the estimated variable, and N the data number. MAE and RMSE are expressed in %, while d is dimensionless.

Results and Discussion

Comparison between Radar Estimates and Gravimetric Data

The comparison between soil moistures estimated from SAR images of March 12, 2007 and gravimetric moistures shows a good agreement between estimated and measured soil moistures, with a bias about 2% and a RMSE of the order of 3% (Figure 4, Table 2). VV and VH polarizations give similar performances. According to the radiometric accuracy of ASAR and ERS sensors, typically of ±1dB, the accuracy on the soil moisture estimation will be about ±5% (for a sensitivity of radar signal to soil moisture of 0.20 dB/%). This value is consistent with results presented in Table 2. The relationships between radar signal and soil moisture are then robust and can be used in the mapping process of soil moisture.

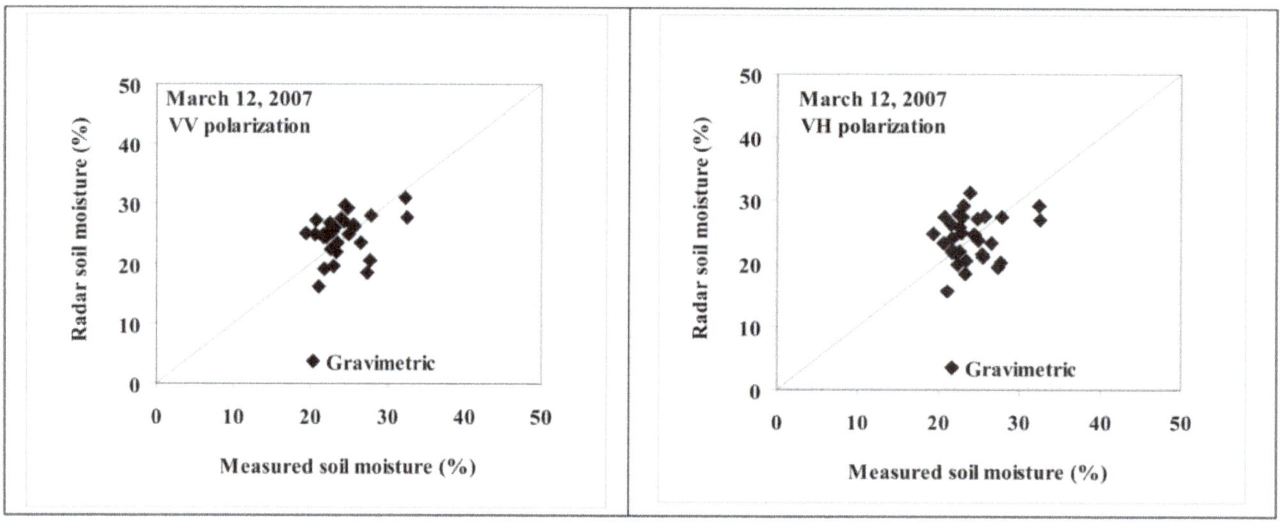

Figure 4: Comparison between soil moisture estimated by ASAR image acquired the March 12, 2007 and soil moisture measured by gravimetry (at the field scale). Bias "Radar-Gravimetry"=0.34%; standard deviation=3.78%; MAE=3.06%; RMSE=3.74%; d=0.58.

Table 2: Comparison between soil moistures measured by gravimetry, simulated by ISBA model, and estimated from ASAR images of March 12, 2007 (scale of ISBA grids)."Rad[Mean]" corresponds to radar soil moisture obtained by the mean estimate of VV and VH polarizations.

Figure 5 shows the procedure applied for soil moisture mapping on the SAR image acquired the 12 March 2007 (VV polarization; incidence angle ~ 23°). The map of bare soils was recognized by SPOT 5 image acquired the 15 February 2007. The soil moisture estimated was next compared to in situ soil moisture measurements.

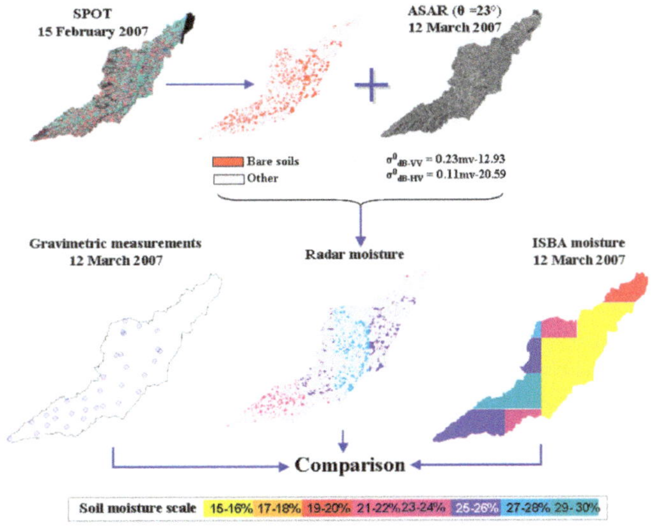

Figure 5: Procedure for mapping soil moisture from SAR images. Optical images are used for bare soils mapping. Radar soil moisture is estimated only over bare soils. The class denominated "other" includes forests, soils covered with vegetation, etc. For the date of March 12, 2007 a comparison between radar soil moisture, in situ soil moisture measurements (gravimetric measurements), and ISBA simulations is realized.

Analysis of ISBA Simulations

ISBA moisture values were analyzed spatially according to the ISBA grid (Figures 6a and 6b). For very wet soil conditions (approximately 35%), the soil moistures simulated by ISBA show a maximum variation between ISBA cells of about 8%. For moistures about 25%, the ISBA moistures vary strongly between grids to reach 15%. For low moistures (approximately 15%), the maximum variation between grids is about 5% except for January 04, 2007 where a strong difference between some grids is observed (about 23%). Indeed, for January 04, 2007, ISBA simulates curiously low moistures on grids 20419, 20420 and 20421 (moistures 5% approximately whereas the other grids are about 20%). These comparisons are made at radar acquisition time (T_radar). Only ISBA moistures obtained by the average of ISBA simulations over all the day of January 04 showed moisture values of the same order for all grids. The three grids having an atypical behavior (20419 to 20421) are located in the SW part of the catchment basin.

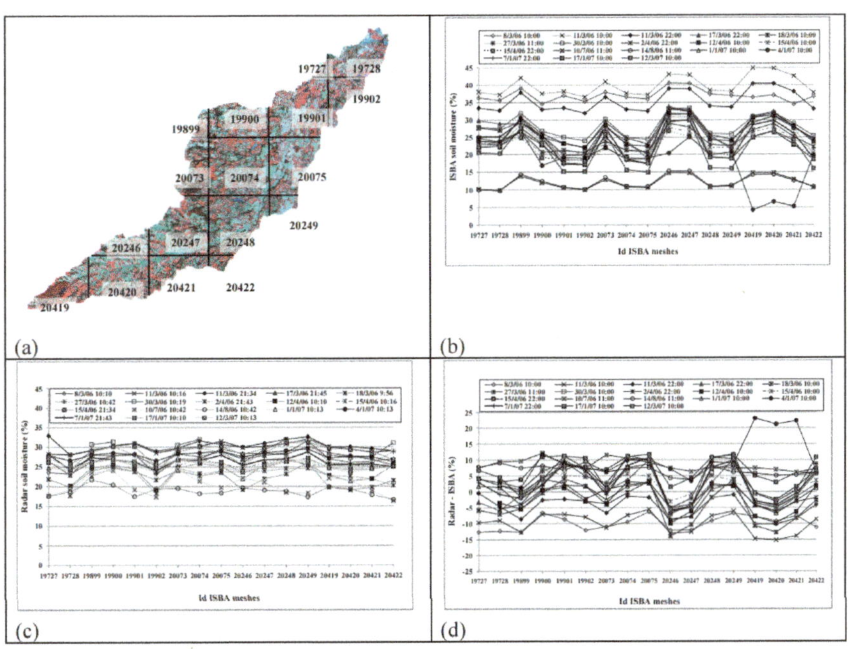

Figure 6: Comparison between radar estimations and ISBA simulations for each ISBA grid (grids size 8km × 8km). (a) Id ISBA grids, (b) ISBA simulations, (c) Radar estimations, and (d) Difference (Radar – ISBA). The standard deviation of radar moisture values estimated on ISBA grid scale varies between 3% and 6% except for March 18, 2006 where the standard deviation varies between 5% and 9%.

The precipitations measured in the basin by the six meteorological stations show that it rained as much in December on the various stations. Indeed, the total amount of rainfall over December varies from 26 to 37 mm according to station. The nearest station to the three grids (20419 to 20421), located at the SW of the basin (Fabas station), had recorded a total amount of rainfall over December about 35 mm, while the station located at the centre of the basin (Rieumes station) indicated approximately 37 mm. The analysis of precipitations over the period January 01-04, 2007 shows a total rainfall which varied between 04 and 11 mm, according to stations, with a total rainfall of 7 mm on the Fabas station (situated to the SW of basin). Moreover, according to meteorological stations and Safran previsions, there no was rain on January 04, 2007 in the basin before the acquisition time of the radar image (10 h). In conclusion, the precipitations data measured on the stations do not explain the atypical behavior of ISBA model on January 04, 2007 on the 3 grids (20419 to 20421). The analysis of ISBA input parameters for January 04, 2007 at 10 h (T_radar) shows that the soil was frozen. Freezing was much more marked on the three grids 20419 to 20421 than on the other grids because they are in another meteorological zone. By adding the frozen bulk soil water content, the total soil water content in the superficial reservoir becomes about 35% on the three grids 20419 to 20421. With the use of the total soil water content instead of the liquid soil water content only, the difference between radar and ISBA moistures becomes lower than 5%.

Figure 6b shows also that moistures simulated by ISBA are distributed according to two groups. Indeed, cells located at the North-East of the basin correspond to ISBA moistures lower than those of other cells. Moreover, moistures of each group are similar for a given date (variation inter group inferior to 5%). The comparison between ISBA simulations and textural map (Figure 2) shows that ISBA moistures are correlated with the soils texture, due to the fact that infiltration and runoff processes are driven in ISBA by the texture characteristics of the soil.

Comparison between ISBA and radar moistures

The radar moisture estimation was aggregated to produce moisture estimation for each ISBA cell. It shows that the maximum variations between grids are of 7% (18 March 2006, Figure 6c) for the radar, while they are around 15% with the ISBA model. For radar acquisitions dates (March 2006 to March 2007), the moisture estimated by radar on ISBA cells shows values ranging from 17% to 33%, whereas ISBA simulations varied between 10 and 45%. The comparison carried out between radar and ISBA soil moistures in using the same ISBA cells show that the difference between radar and ISBA moistures vary from -15% to +12%, except for January 04, 2007 where the difference reaches 23% (Figure 6d). Results show also that ISBA simulations averaged on the basin are well comparables to radar estimates with a mean difference lower than 5% for 12 dates among the 18 examined dates. For the six remaining dates, differences between 5% and 10% are observed (Figure 7).

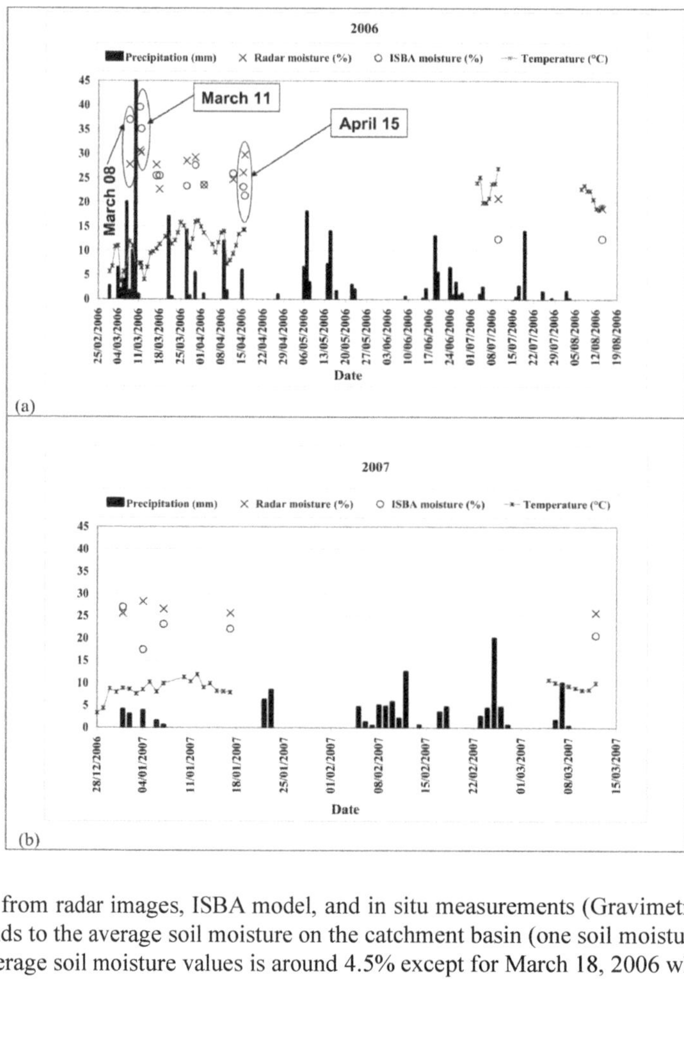

Figure 7: Soil moistures obtained from radar images, ISBA model, and in situ measurements (Gravimetry) according to daily precipitation. Each soil moisture value corresponds to the average soil moisture on the catchment basin (one soil moisture value for the whole of the basin). The standard deviation of basin average soil moisture values is around 4.5% except for March 18, 2006 where the standard deviation is 9%.

Correlation Analysis between Precipitations and Soil Moistures (ISBA And Radar)

The lack of in situ moisture data on the basin at the time of radar acquisitions (except for March 12, 2007) encouraged the analysis of radar moistures according to daily precipitations. Precipitations were given by six stations installed in the basin. Figure 7 shows the temporal variation of soil moisture estimated by SAR sensors between March 08, 2006 and March 12, 2007. This variation correlated with precipitations, which confirms that radar estimates are qualitatively coherent. For example, after important precipitations on March 10, 2006 (45 mm) the radar soil moisture increased from 28% on March 08, 2006 to 31% on March 11, 2006. From March 11, 2006 the radar soil moisture starts to decrease to reach 23% on March 18, 2006. This tendency is completely coherent because there no were precipitations between these two dates and the soil has thus dried. A good correlation between radar moisture and precipitations is also observed on the database of 2007. The radar moisture increased from 26% on January 01 to 28% on January 04, and decreases then to 27% on January 07, 2007. This behavior is coherent with precipitations about 13 mm between 01 and 04 January, and 2 mm between 04 and 07 January. The mean temperature during the first fortnight of January was of approximately 10°C.

ISBA simulations follow well the distribution of precipitations. Figure 7 shows ISBA moistures varying between 35 and 40% for the period of 08 to 11 March 2006. These high moisture values correspond to important precipitations between 07 and 10 March 2006 (20mm for 07 March and 45mm for 10 March 2006). Moistures simulated by ISBA decrease after the 11 March to reach soil moisture of 25%. This value is coherent with the absence of precipitations between the 10 and 18 March 2006.

The strong values simulated by ISBA on March 08, 2006 at 10h and on March 11, 2006 at 10 h (time of radar acquisitions) are higher than radar moistures from approximately 9%. This strong over-estimate of moisture by ISBA for these two dates does not seem to correspond to meteorological conditions. Indeed, simulations ISBA from March 11, 2006 at 10 h is about 40%, against 35% at 21 h. It would seem that the strong ISBA value on March 11, 2006 at 10 h is relates well to strong precipitations of March 10, 2006 (45 mm) whereas that of March 08 at 10 h seems to over-estimate the soil moisture. The decrease of ISBA soil moisture between 10 h and 21 h on March 11, 2006 is well justified, since there no was rain between 10 h and 21 h.

From 17 March to 15 April 2006, ISBA moistures fluctuated between 28% and 21%, with values close to radar moistures (difference inferior to 5%). Radar and ISBA moistures of April 15, 2006 are very close for the 10 h radar acquisition time (difference about 3%), but they are very different for the 21h radar acquisition (difference about 9%). According to meteorological data, it rained 0.6 mm between 0 h and 10 h and 4.8 mm between 10 h and 21 h on April 15, 2006. The precipitation of April 15 explains the increase of the soil moisture observed on the radar image acquired at 21 h (increase of 4% between 10 h and 21 h), and shows that the decrease of moisture in the ISBA simulations (3%) is not coherent. This wrong simulation of soil moisture by the ISBA model on April 15 at 21 h can be explained by the wrong precipitation distribution used in the input of the ISBA model (Safran model). For some meteorological situations, the hourly interpolation of precipitations carried out by Safran gives sometimes values which are wrongly distributed by the hourly resolution. Indeed, in the ISBA model, it was predicted and then used precipitations of 0.8 mm between 7 h and 10 h and 0 mm between 10 h and 21 h. Radar estimates are superior to ISBA simulations of about 9% on July 10 and 7% on August 14, 2006. However, it is difficult to know if radar or the ISBA model better estimates the soil moisture for these two dates.

From 01 to 04 January 2007, it rained approximately 11 mm (Figure 7b). The soil moistures estimated by radar increases of 3% between 01 and 04 January, whereas ISBA moistures decrease approximately of 10% (from 27 to 17%). This decrease of ISBA moisture is not correlated with meteorological conditions of this period. From 04 to 17 January 2007, the behavior of radar moisture is coherent with the weak precipitations recorded between these dates. For March 12, 2007 the soil moisture is of 25.7% with radar, 20.6% with ISBA, and 23.4% with gravimetric measurements.

Soil Moisture Mapping According to Soil Surface Texture of the SGBDE

Spatial variations of soil moisture obtained by gravimetry, radar images, and ISBA simulations were studied at textural zones scale. The textural zones (Figure 2) were defined according to the textural information contained in the 1:1000000 Soil Geographical database. The gravimetric data, measured on 33 test fields, were classified according to their textural zones. Concerning radar data, the mean soil moisture was calculated for each date and each textural zone in using the mean radar signal of all bare soil pixels presents inside each zone (Figure 8a). For ISBA moistures, moisture values corresponding to bare soils were averaged on each textural zone (Figure 8b).

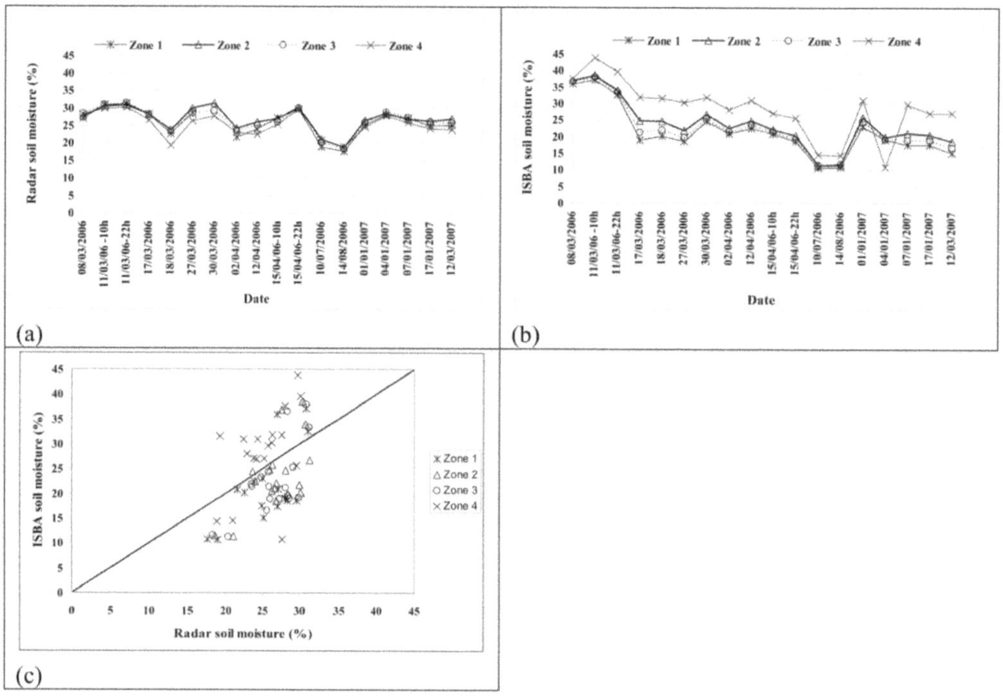

Figure 8: Correlation between textural zones and surface soil moisture simulated from ISBA model, and estimated from ASAR images. Each symbol corresponds to the mean soil moisture on each textural zone. Zone 1 is mainly sandy and zone 2 is mainly composed of clay (cf. Figure 2).

The behavior of the soil moisture calculated for textural zones is different for ISBA, radar, and gravimetric data. Figure 8ashows that there is no dependence between soil moistures estimated by radar images and the textural zones. However, an important dependence is obtained between ISBA soil moisture and textural zones (Figure 8b). Indeed, the ISBA soil moisture for zone 4, mainly composed of clayey material, is stronger than that for the other textural zones. The mean difference between radar estimates and ISBA simulations at textural class scale varies from -3.27% for zone 4 to +4.58 for zone 1 (bias=3.27% for zone 2 and 4.16 for zone 3). The standard deviation varies from 5.57% to 5.98% except for zone 4 where the value is 7.45% (Figure 8c).

Moreover, spatial distribution of moisture heterogeneities was displayed at textural zone scale, using box plots (Figure 9). For gravimetric database, 33 data points were used. For both radar and ISBA databases, 72 moisture values were used (18 dates × 4 textural zones). Results obtained between gravimetric or radar moistures do not show dependence with soil texture. On the contrary, ISBA data are well correlated to textural units.

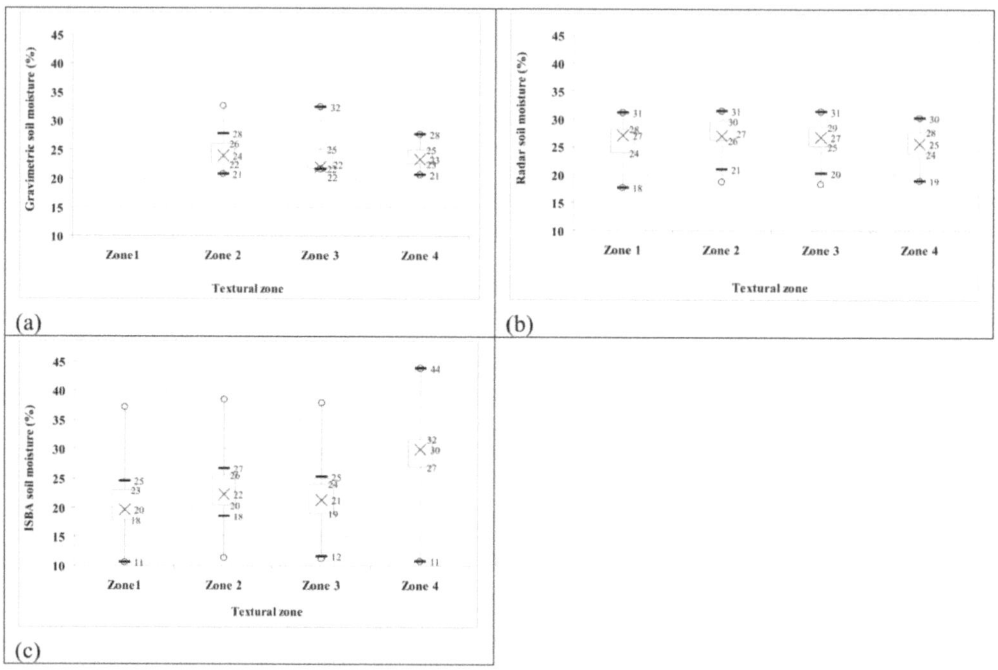

Figure 9: Box plot comparing the results obtained for soil moistures according to textural zones. (a) Gravimetric, (b) Radar, (c) ISBA.

Concerning gravimetry and radar data, the differences between the main quartiles (median, first and third quartiles) of 4 textural zones do not exceed 2% (Figures 9a and 9b). Indeed, moisture would have had to be higher for zone 4 (mainly clay) and to decrease while going towards zone 1 (rather sandy).

As the ISBA model uses the pedological characteristics among input parameters, it would be evident that the soil moisture is dependent of soil texture. In fact, ISBA database shows that zone 4, mainly composed of clay, is well differentiable of other zones (1, 2, and 3). Indeed, the difference between the quartiles of zone 4 and other zones varies between 6% and 10% (Figure 9c). Results show also that pedological zones 1, 2 and 3 differ only of about 2%.

In order to consolidate these results, a nonparametric Kruskal Wallis test was carried out with a confidence threshold of 5%. This test confirms that for radar and gravimetric data, the difference between soil moistures of textural zones is not significant. For ISBA moistures, a significant dependence was observed between the textural zones and soil moisture.

Radar estimations and gravimetric measurements of March 12, 2007 have shown that the differences of moisture observed between textural zones are not significant. The low spatial resolution of the textural map (1:1000000 scale) is not adapted to illustrate the moisture variations inside the basin. Indeed, in situ gravimetric data of 12 March 2007 show a moisture variation between the 33 test fields around 12% (moisture values between 21% and 33%), whereas at the textural zones scale the moisture variation is very weak (2%). It is then probable that the textural zones defined by the 1:1000000 map are not homogeneous and thus contain a mixture of textures. However, following the comparison between gravimetric moistures and the soil texture at the SGBDE textural zones scale, it was concluded that this scale is not adapted to explain the moisture variations on the basin. It would be thus more judicious to use in ISBA model a finer textural grid, at least on the scale of ISBA grids. Nevertheless, many studies concerning pedotransfer functions showed the importance of texture on the moisture estimation [36]. The hypothesis of dependence between moisture and soil texture is not excluded, but with the low spatial resolution of SGBDE map, this is not possible to quantify. In fact, textural spatial heterogeneities determining moisture values not appear at this scale. Thus, ISBA simulations based on textural criteria is legitimate but texture variations must be more precisely determined for obtain good moisture simulation.

CONCLUSIONS

This study examined the potential of ASAR and ERS-2 data for estimating volumetric soil moisture (m_v) over bare soils. Soil moisture estimates obtained from SAR acquisitions were compared to those obtained by the operational Météo-France ISBA model and by in situ measurements. These results appear promising for the development of simplified algorithms for retrieving soil moisture from SAR data, and for monitoring multi-temporal moisture changes. The mapping algorithm of soil moisture from SAR data performed very well. The results on the soil moisture estimates may be summarized as follows:

- The comparison between radar soil moisture and in situ measurements shows a RMSE for the soil moisture estimate of the order of 3% (in comparison with gravimetric moistures of March 12, 2007). Other SAR images with a wide range of soil moisture (<20% and >35%) are required to better validate the linear regressions established between radar signal and soil moisture.
- The mean difference between ISBA simulations and radar estimates is lower than 5% for 12 dates among the 18 examined dates. The 6 remaining dates show a mean difference between 5% and 10%. The comparison between ISBA and gravimetric measurements of the 12 March 2007 shows a RMSE of about 6%. Generally, these results are very encouraging.
- An important difference between ISBA and radar moistures is observed for some grids when the soil is frozen. The use of both liquid and frozen bulk soil water contents improves the results of the comparison between radar and ISBA model.
- No dependence was observed between gravimetric or radar moistures and textural zones according to 1:1000000 textural map. The soil moisture mapping from SAR images at the textural scale shown a weak variation between textural units. The correlation observed between texture map and ISBA moisture is induced by the use of the texture map as an input parameter in the ISBA model. Even if this parameter is very important for soil moisture estimation, radar and gravimetric results shown that the scale at 1:1000000 of the used textural map is not appropriate to differentiate moistures zones. Finally, additional studies on the sensibility of soil moisture to each pedological parameter (texture, density, ...) would be necessary in using higher resolution texture data.

Moreover, radar measurements are acquired with a given periodicity (some days) and cannot describe the continuous variations of moisture, whereas ISBA model provides moisture simulations every hour. Thus, it would be important to associate the radar remote sensing in the hydrological modelling. Assimilation of soil moisture data estimated from radar images could be used as calibration data to improve the forecast models (ISBA model for example).

The recent radar sensor (PALSAR/ALOS) and those that will come into operation early (TerraSAR-X, COSMO-SkyMed, RADARSAT-2) should provide us with polarimetric data (all polarizations), with the possibility of better characterising the soil surface. Polarimetry plays an important role as it allows a separation of soil moisture and surface roughness effects. Several inversion models are based on the use of fully polarimetric SAR images [14]. Polarimetric parameters such as, for example, the entropy, the α angle and the anisotropy should allow us to map two soil surface characteristics simultaneously [44]. Finally, the very high spatial resolution (metric) of future SAR sensors offers great potential in terms of improving the quality of SSC mapping. These new SAR sensors will provide a diagnosis suited to catchment areas where the parcels are of small size.

Currently it is difficult to map the moisture of soils covered by vegetation using C-band data. The arrival of L-band (ALOS) will enable to extend the soil moisture mapping from SAR data of bare soils only to soils with vegetation (depending on the density and height of the vegetation). Indeed, the penetration depth of radar wave in vegetable cover is more important at L-band than at C-band. Moreover, the use of L-band enables to map soils with vegetation without using optical images, which is not currently the case with C-band because an optical image is always used for this purpose. Dubois et al. [9] showed that the ratio of two polarizations HV/VV in L-band is strongly correlated with the NDVI (Normalized Difference Vegetation Index). If this information is confirmed, the cartography of soil moisture could be carried out directly from radar images without the use of optical data.

ACKNOWLEDGMENTS

This work was carried out within the framework of a project supported by MEDD (French Environmental Ministry) and BRGM (French Geological Survey) in the frame of SCHAPI (National Flood Forecasting Agency) activities. The authors wish to thank ESA (European Space Agency) for kindly providing the SAR images under Dragon project n° 2551 and ENVISAT/ASAR project n° 351. The SPOT images were obtained at special rates through the CNES (French spatial agency) ISIS program. Authors are grateful for R. Malosti and H. Yesou, who very kindly helped for SAR data ordering and INRA-Orléans for providing the soil texture data. The authors thank M. Saplairoles, J.L. Izac, and N. Baran for their participation in ground measurements during the field campaigns. We would also like to acknowledge Ms Florence Habets which helped to clarify us the ISBA model.

REFERENCES

1. Beven, K.J.; Fisher, J. Remote sensing and scaling in hydrology. In Scaling up hydrology using remote sensing; Stewart, J.B., Engman, E.T., Fedes, R.A., Kerr, Y., Eds.; Wiley Press: Chichester, UK, 1996; pp. 1–18.
2. Le Bissonnais, Y.; Jamagne, M.; Lambert, J.J.; Le Bas, C.; Daroussin, J.; King, D.; Cerdan, O.; Léonard, J.; Bresson, L.M.; Jones, R. Pan-European soil crusting and erodibility assessment from the European Soil Geographical Database using pedotransfer rules. Adv. Environ. Monitor. Model. 2005, 2, 1–15.
3. Valentin, C.; Poesen, J.; Li, Y. Gully erosion: Impacts, factors and control. Catena 2005, 63, 132–153.
4. Dobson, M.C.; Ulaby, F.T. Active microwave soil moisture research. IEEE Trans. Geosci. Remote Sens. 1986, 24, 23–36.
5. Ulaby, F.T.; Moore, R.K.; Fung, A.K. Microwave Remote Sensing, Active and Passive, From Theory to Applications; Artech House Inc.: Norwood, MA, USA, 1986; Vol. 3.
6. Fung, A.K. Microwave scattering and emission models and their applications; Artech House, Inc.: Boston, London, 1994.
7. Ulaby, F.T.; Batlivala, P.P.; Dobson, M.C. Microwave backscatter dependence on surface roughness, soil moisture, and soil texture: Part I – Bare soil. IEEE Trans. Geosci. Remote Sens. 1978, GE-16, 286–295.
8. Fung, A.K.; Chen, K.S. Dependence of the surface backscattering coefficients on roughness, frequency and polarization states. Int. J. Remote Sens. 1992, 13, 1663–1680.
9. Dubois, P.; Van Zyl, J.J.; Engman, T. Measuring of soil moisture with imaging radars. IEEE Trans. Geosci. Remote Sens.1995, 33, 915–926.
10. Shi, J.; Wang, J.; Hsu, A.Y.; O'Neill, P.E.; Engman, E.T. Estimation of bare surface soil moisture and surface roughness parameter using L-band SAR image data. IEEE Trans. Geosci. Remote Sens. 1997, 35, 1254–1266.
11. Le Hégarat, S.; Zribi, M.; Alem, F.; Weisse, A.; Loumagne, C. Soil moisture estimation from ERS/SAR data: Toward an operational methodology. IEEE Trans. Geosci. Remote Sens. 2002, 40, 2647–2658.
12. Baghdadi, N.; Gaultier, S.; King, C. Retrieving surface roughness and soil moisture from synthetic aperture radar (SAR) data using neural networks. Can. J. Remote Sens. 2002, 5, 701–711.
13. Zribi, M.; Dechambre, M. A new empirical model to retrieve soil moisture and roughness from C-band radar data.Remote Sens. Environ. 2002, 84, 42–52.
14. Oh, Y. Quantitative retrieval of soil moisture content and surface roughness from multipolarized radar observations of bare soil surfaces. IEEE Trans. Geosci. Remote Sens. 2004, 42, 596–601.
15. Baghdadi, N.; Holah, N.; Zribi, M. Soil moisture estimation using multi-incidence and multi-polarization ASAR SAR data. Int. J. Remote Sens. 2006, 27, 1907–1920.
16. Srivastava, H.S.; Patel, P.; Manchanda, M.L.; Adiga, S. Use of multi-incidence angle RADARSAT-1 SAR data to incorporate the effect of surface roughness in soil moisture estimation. IEEE Trans. Geosci. Remote Sens. 2003, 41, 1638–1640.
17. Holah, H.; Baghdadi, N.; Zribi, M.; Bruand, A.; King, C. Potential of ASAR/ENVISAT for the characterisation of soil surface parameters over bare agricultural fields. Remote Sens. Environ. 2005, 96, 78–86.
18. Mattia, F.; Le Toan, T.; Souyris, J.C.; De Carolis, G.; Floury, N.; Posa, F.; Pasquariello, G. The effect of surface roughness on multifrequency polarimetric SAR data. IEEE Trans. Geosci. Remote Sens. 1997, 33, 915–926.
19. Zribi, M.; Taconet, O.; Le Hegarat-Mascle, S.; Vidal-Madjar, D.; Emblanch, C.; Loumagne, C.; Normand, M. Backscattering behavior and simulation: Comparison over bare soils using SIR-C/X-SAR and ERASME 1994 data over Orgeval. Remote Sens. Environ. 1997, 59, 308–320.
20. Baghdadi, N.; Holah, N.; Zribi, M. Calibration of the Integral Equation Model for SAR data in C-band and HH and VV polarizations. Int. J. Remote Sens. 2006, 27, 805–816.
21. Baghdadi, N.; Zribi, M. Evaluation of Radar Backscatter Models IEM, OH and Dubois Using Experimental Observations. Int. J. Remote Sens. 2006, 27, 3831–3852.
22. Cerdan, O.; Baghdadi, N.; Denux, J.P.; Desprats, J.F.; Gay, M.; Albergel, C.; Dubus, I.; Dupros, F.; Holah, N.; El Hajj, M. APIBAR: Appui à la Prévision des Inondations. Cas des Bassins Rapides du Sud de la France Final Report BRGM/RP-54218-FR. 2006.

23. Habets, F.; Boone, A.; Champeaux, J.L.; Etchevers, P.; Franchistéguy, L.; Leblois, E.; Ledoux, E.; Le Moigne, P.; Martin, E.; Morel, S.; Noilhan, J.; Quintana Segui, P.; Rousset-Regimbeau, F.; Viennot, P. The SAFRAN-ISBA-MODCOU hydrometeorological model applied over France. J. Geophys. Res. 2007. submitted.

24. Oh, Y.; Kay, Y. Condition for precise measurement of soil surface roughness. IEEE Trans. Geosci. Remote Sens. 1998, 36, 691–695.

25. Baghdadi, N.; Paillou, P.; Davidson, M.; Grandjean, G.; Dubois, P. Relationship between profile length and roughness parameters for natural surfaces. Int. J. Remote Sens. 2000, 21, 3375–3381.

26. Davidson, M.; Le Toan, T.; Marria, F.; Satalino, G.; Maninnen, T.; Borgeaud, M. On the characterisation of agricultural soil roughness for radar remote sensing studies. IEEE Trans. Geosci. Remote Sens. 2000, 38, 630–640.

27. BEST - Basic Envisat SAR Toolbox; Version 4.0.3. European Space Agency (ESA), July 2005. http://envisat.esa.int/.

28. Joughin, I.; Percival, D.; Winebrenner, D. Maximum Likelihood Estimation of K Distribution Parameters for SAR Data.IEEE Trans. Geosci. Remote Sens. 1993, 31, 989–999.

29. Lee, J.S.; Jurkevich, I.; Dewaele, P.; Wambacq, P.; Oosterlinck, A. Speckle filtering of synthetic aperture radar images: A Review. Remote Sens. Rev. 1994, 8, 313–340.

30. Noilhan, J.; Mahfouf, J.F. The ISBA land surface parameterization scheme. Global Planet. Change 1996, 13, 145–159.

31. Noilhan, J.; Planton, S. A simple parameterization of land surface processes for meteorological models. Month. Weath. Rev. 1989, 117, 536–549.

32. Masson, V.; Champeaux, J.L.; Chauvin, F.; Méringuet, C.; Lacaze, R. A global database of land surface parameters at 1 km resolution for use in meteorological and climate models. J. Clim. 2003, 16, 1261–1282.

33. Durand, Y.; Brun, E.; Mérindol, L.; Guyomarc'h, G.; Lesaffre, B.; Martin, E. A meteorological estimation of relevant parameters for snow schemes used with atmospheric models. Ann. Glaciol. 1993, 18, 65–71.

34. Quintana Seguí, P.; Le Moigne, P.; Durand, Y.; Martin, E.; Habets, F.; Baillon, M.; Franchisteguy, L.; Morel, S.; Noilhan, J. Analysis of near surface atmospheric variables: validation of the SAFRAN analysis over France. J. Appl. Meteorol. Climatol. 2007. in press.

35. Bruckler, L.; Wittono, H.; Stengel, P. Near surface moisture estimation from microwave measurements. Remote Sens. Environ. 1988, 26, 101–121.

36. Bruand, A.; Pérez-Fernandez, P.; Duval, O.; Quétin, P.; Nicoullaud, B.; Gaillard, H.; Razison, L.; Pessaud, J.F.; Prudhomme, L. Estimation des propriétés de rétention en eau des sols : Utilisation de classes de pédotransfert après stratification texturale et texturo-structurale. Etude Gest. Sols 2002, 9, 105–125.

37. Gupta, S.C.; Larson, W.E. Estimating soil water retention characteristics from particle size distribution, organic matter percent and bulk density. Water Resour. Res. 1979, 15, 1633–1635.

38. Puckett, W.E; Dane, J.H; Hajeh, B.F. Physical and mineralogical data to determine soil hydraulic properties. Soil Sci. Soc. Am. J. 1985, 49, 831–836.

39. Soil Geographical Database of Europe at scale 1:1,000,000 (SGDBE), Version 2; http://eusoils.jrc.it/ESDB%5FArchive/ESDBv2/.

40. Weimann, A.; Von Schönermark, M.; Schumann, A.; Jörn, P.; Günther, R. Soil moisture estimation with ERS-1 SAR data in the East-German loess soil area. Int. J. Remote Sens. 1998, 19, 237–243.

41. Zribi, M.; Baghdadi, N.; Holah, N.; Fafin, O. New methodology for soil surface moisture estimation and its application to ENVISAT-ASAR multi-incidence data inversion. Remote Sens. Environ. 2005, 96, 485–496.

42. Beaudoin, A.; Gwyn, Q.H.J.; Le Toan, T. SAR observation and modelling of the C-band backscatter variability due to multi-scale geometry and soil moisture. IEEE Trans. Geosci. Remote Sens. 1990, 28, 886–894.

43. Willmott, C.J. Some comments on the evaluation of model performance. Bull. Am. Meteorol. Soc. 1982, 63, 1309–1313.

44. Hajnsek, I.; Pottier, E.; Cloude, S. Inversion of surface parameters from polarimetric SAR. IEEE Trans. Geosci. Remote Sens. 2003, 41, 727–744.

CHAPTER 7

LAND SURFACE WATER MAPPING USING MULTI-SCALE LEVEL SETS AND A VISUAL SALIENCY MODEL FROM SAR IMAGES

Chuan Xu, Haigang Sui and Feng Xu

Keywords: Water mapping, level sets, visual saliency model, SAR image

INTRODUCTION

Water resources are an irreplaceable strategic resource for human survival. Land surface water (LSW) is an important part of the water cycle. LSW mapping, using remote sensing techniques, plays an important role in wetland monitoring, flood monitoring, flood disaster assessment, surface water area estimation, and water resources management. Although multi-spectral optical images have been found to be effective when the sky is clear, there are some limitations in overcast weather conditions. However, due to their capabilities of large-area coverage, cloud penetration and all-weather acquisition, the SAR image data could compensate for the weakness in optical images in such conditions for LSW mapping [1]. This paper proposes a fast and reliable land surface water mapping method of large-area and high-resolution SAR images. Only calm water was considered including calm rivers and lakes, not including puddles and paddy fields.

Water classification is a particular case of SAR image classification, with only two classes to assign; it is widely used in applications such as wetland monitoring, flood monitoring, surface water area estimation and so on. However, due to the presence of coherent speckle, which can be modeled as strong, multiplicative noise, segmentation of SAR images is generally acknowledged as a difficult problem [2]. Up to now, many segmentation algorithms have been proposed ,such as the threshold method [3], edge detection method [4,5], region-based method [6], and fuzzy-based method [7]. In addition, in the recent research, Markov random fields (MRFs) and active contour models (ACMs) are two of the most useful methods for SAR image segmentation. As concerns the MRF segmentation method, Deng *et al.* [8] proposed a novel Markov random field (MRF) model to generate accurate unsupervised sea and ice segmentation results. Stéphane and Grégoire [9] developed a vector hidden Markov chain (HMC) model that was adapted to a multi-scale description to segment oil slick from sea SAR images. Gan *et al.* [10] and Wang *et al.* [11] proposed an unsupervised SAR image segmentation method based on Triplet Markov Fields, respectively. After that, A Reversible Jump Markov Chain Monte Carlo scheme is designed to simulate the SAR image segmentation model [12]. As concerns the active contour models method, Huang *et al.* [13] used a partial differential equation (PDE)-based level set method for oil slick segmentation in SAR images. Silveira and Heleno [14] adopted a mixture of lognormal densities for SAR image segmentation between water and land, and results demonstrated the good performance of their proposed method. Moreover, a novel level set segmentation method that uses a heterogeneous clutter model is proposed to overcome the problem of strong speckle and texture in fine spatial resolution SAR images [15]. Yin and Yang [16] investigates the application of a level set method for the automated multiphase segmentation of multiband and polarimetric synthetic aperture radar (SAR) images.

However, these segmentation methods cannot be used directly for water segmentation in SAR images because of the following reasons: (1) Most of these segmentation methods are

Citation: Xu, C.; Sui, H.; Xu, F. Land Surface Water Mapping Using Multi-Scale Level Sets and a Visual Saliency Model from SAR Images. ISPRS Int. J. Geo-Inf. 2016, 5, 58.

not designed for water segmentation in SAR images, as water features in SAR images are not considered; moreover, there are also some problems in the existing segmentation methods. For example, the edge detection technique is plagued by difficulties that include discontinuous boundary points and a pseudo-boundary in the homogeneous region, which cause significant difficulties in subsequent processing. The region-growing approach is user-dependent during the growing and merging of neighboring small regions, while segmentation methods based on the thresholding of grey levels are often inappropriate for SAR images because of the presence of speckle noise. The MRF method has a large computational cost and hence is not widely used in practice, especially for large-area images; (2) The geometric ACMs, known as level sets [17], have the advantage of being independent from parameterization and are able to automatically change topology. In the level set method, a curve is embedded as a zero level set of a higher dimensional surface. Next, the entire surface is evolved to minimize a metric defined by the curvature and image gradient,*i.e.*, the speed terms of the level set will reduce to zero when reaching the object boundary. Although level set methods present a good performance in SAR image segmentation, there are still some problems. The curve in level sets is usually evolved on the whole SAR image, so objects other than the water body are also extracted. In particular, the segmentation results for large-area images at a high spatial resolution often yield a wide variety of very small-scale image objects (*i.e.*, "salt and pepper" effects); (3) Objects with a low radar cross section similar to calm water, such as roads, airstrips, radar shadows, or paddy fields, may affect the results of water extraction.

Water regions are salient objects in SAR images because these regions often occupy a large area and appear as the darkest grey value; therefore, human eyes usually focus their initial attention on water regions. Selective visual attention is a mechanism by which we can rapidly direct our gaze towards objects of interest in our visual environment, and ignore some information that is not important [18]. This mechanism reduces the calculation time during information processing. Thus, the introduction of this mechanism into the processing of remote sensing images could significantly improve its efficiency. Recently, Itti *et al.* [19] proposed a classical model of selective visual attention, and this model can be used to search for salient objects in complex scenes. Since then, many objects detection methods have been developed using this model. For example, Itti *et al.* [20] applied this model to a wide range of target detection tasks, using synthetic and natural stimuli. Walther *et al.* [21] proposed a method for detecting multiple objects in cluttered scenes based on a bottom-up visual attention model. Li *et al.* [22] proposed a wavelet transform saliency model for generating a saliency map to detect natural objects. Gao *et al.* [23] presented a novel detection model based on a visual attention mechanism to detect ships in remote sensing images. Li and Itti [24] integrated saliency features acquired by a visual salience model, and with gist features, to detect objects, such as ships, buildings and airplanes in remote sensing images. However, these models are aimed at analyzing images of nature or at extracting ships, airplanes, *etc.* from optical satellite images, and they are not designed for SAR images which make it difficult to apply them directly.

The main purpose of this paper is to propose a fast and accurate land surface water mapping method, and most important is that it can solve some complex situation, such as shadow, farmland confusion, *etc.* The remainder of this paper is structured as follows. In Section 2, our method is formulated and applied to SAR image water extraction. Firstly, a TW-Itti model is proposed. Secondly, we describe the adaptive multi-scale level set method based on Gamma model. Thirdly, a post-processing method is described. Some experimental results are shown in Section 3, before we conclude the paper in Section 4.

PROPOSED METHOD

Given the above problems, a novel water segmentation method was proposed based on the integration of multi-scale level sets and a visual salience model, as follows, (1) To improve the accuracy and efficiency of water extraction, the classic Itti model is first introduced to detect the SWR. After that, the level set method can be applied only to these regions, and at the same time, the visual salience map can be used for initializing the zero level set function. Thus, we can detect water regions rapidly using Itti model and we can accurately segment water from those regions using level sets; (2) The classical Itti model is not designed for SAR water images, thus, an improved visual saliency attention model was presented. First, we replace the color images in the Itti model with textured images, which are generated using the gray-level co-occurrence matrix (GLCM) algorithm. Based on the grey values of the water body, a weight coefficient is introduced to calculate the center-surround differences; (3) In order to better avoid "salt and pepper" effects, multi-scale technology is introduced in the level sets method. A multi-scale level set method was proposed by Sui *et al.* [25], the difference with our approach being that we use OTSU algorithms [26] to adaptively define the initial level set function; (4) To detect objects that may be confused with calm water, a post-processing method was proposed based on object-oriented geometrical features.

In this section, the framework of our rapid water extraction method for SAR imagery was presented, as shown in Figure 1.

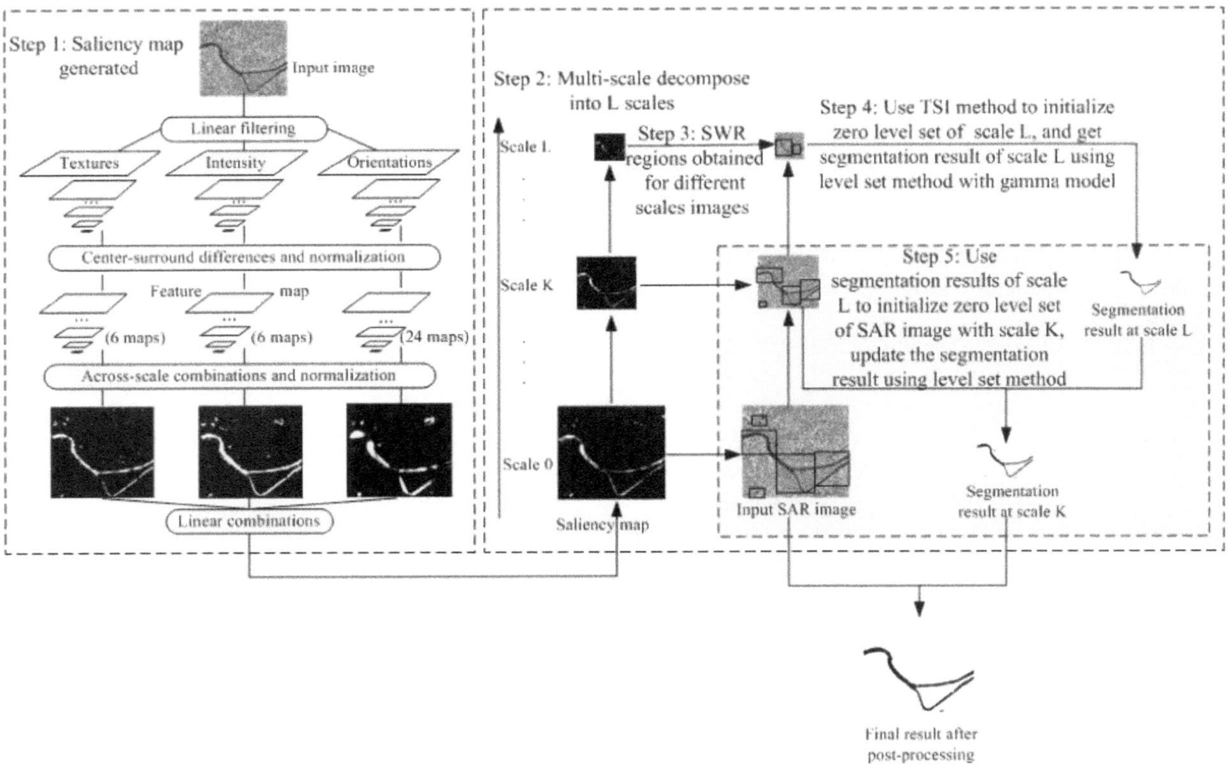

Figure 1: The basic principle of the proposed method. (Step1, saliency map generated from original image; Step 2, Decomposed the saliency map into L scales; Step3, Obtained SWR regions in different scales; Step4, Segmented the water in SWR regions; Step5, refine the segmented result; Step6, post-processing to acquire accurate water body.)

The principal steps of our method are as follows.

Step 1: Generate a visual salience map using the improved Itti model.

Step 2: Decompose the original image and visual saliency map into L scales by the bilinear interpolation algorithm. Let K = L.

Step 3: SWR regions are obtained: segment the saliency map into binary images at each scale using the OTSU method, and then objects in these binary images can be marked using the connecting area algorithm. Next, SWR regions (black rectangles in the SAR image shown in Figure 1) can be obtained by calculating the maximum external rectangle of each marked object.

Step 4: Use the OTSU algorithms to initialize the level set function at scale L. Obtain the scale K segmentation results using the level set method with the Gamma model.

Step 5: K = K − 1. If K ≥ 0, return to Step 4, until the final segmentation result is obtained.

Step 6: Post-processing: the weighting of the smoothness within the shape is applied to the final binary image to distinguish paddy fields from rivers or lakes (see Section 2.3).

Improved TW-Itti Model

The model proposed in this paper draws its inspiration from the bottom-up attention model proposed by Itti *et al.* [20] (see the left part of Figure 1). Our contribution is focused mainly on the extraction of early visual features. Instead of using color information, the textures generated by the GLCM are applied to obtain the feature maps. Since we used single-polarized and single-band SAR images, no colors can be acquired; thus, we fused the texture feature to generate color images. The benefit of doing this is to enhance the visibility of the water body and the background. Moreover, in view of the dark grey value of water, a new calculation method for center-surround differences is presented. Center-surround is implemented in the model as the difference between fine and coarse scales, The center is a pixel at scale $c\in\{2,3,4\}$, and the surround is the corresponding pixel at scale $s=c+\delta$, with $\delta\in\{3,4\}$. The across-scale difference between two maps, denoted "$\ominus$" below, is obtained by interpolation to the finer scale and point-by-point subtraction. The input is provided as a static gray image, and nine spatial scales are generated using dyadic Gaussian pyramids, which progressively low-pass filter and subsample the input image. I is used to create a Gaussian pyramid *I()*, where $\sigma\in(0...8)$ is the scale. The center-surround differences ($\ominus$ defined previously) between a"center" fine scale *c* and a "surround" coarser scale *s* yield the feature maps.

The first set of feature maps is concerned with intensity contrast, which, in mammals, is detected by neurons sensitive either to dark centers on bright surrounds or to bright centers on dark surrounds. Here, both types of sensitivities are simultaneously computed (using a rectification) in a set of six maps I(c,s).

$$I(c,s)=|I(c)\ominus I(s)|,\ c\in\{2,3,4\},s=c+\sigma,\sigma\in\{3,4\} \qquad (1)$$

Considering the dark grey value of water, a weight coefficient is introduced to calculate the center-surround differences, mainly to maintain the low reflective value of water. Thus, Equation (1) can be rewritten as below:

$$I(c,s)=|\omega\cdot I(c)\ominus I(s)| \qquad (2)$$

where ω is the weight coefficient and $\omega\in[0,1]$.

A second set of maps is similarly constructed for the texture channels. A texture image T is generated using GLCM and principal components analysis (PCA) algorithm. Here, T is comprised of five conventional textures, *i.e.*, energy, correlation, inertia, entropy and inverse difference moment). The window size is 7 × 7, the distance value is 1 pixel and there are 16 grey levels. After acquiring these textures, the PCA method is used to obtain the first principal component that represents texture image T. The texture feature maps $T(c,s)$ can be represented as follows:

$$T(c,s)=|T(c)\ominus T(s)| \quad (3)$$

The GLCM-based features, proposed by Haralick *et al.* [27], have been used widely in texture segmentation. They are defined for the neighborhood of an input image and represented by using the second order joint condition probability density p_{ij}, which is the relative frequency at which two image pixels are related. The construction of the GLCM depends on three parameters, *i.e.*, the relative distance (d), the window size (w), and the orientation (θ) between pixels in a pair. In this paper, the distanced is set to one pixel, the window size is set to 7 × 7 pixels, while the orientation is equal to the average value of four directions (0°, 45°, 90° and 135°). Next, we extract five features from the GLCM, as follows:

Energy:

$$ASM = \sum_i \sum_j P(i,j)^2 \quad (4)$$

Correlation:

$$CON = \left[\sum_i \sum_j ((ij)p(i,j) - u_x u_y) \right] / \sigma_x \sigma_y \quad (5)$$

Inertia:

$$INT = \sum_i \sum_j (i-j)^2 p(i,j) \quad (6)$$

Entropy:

$$ENT = -\sum_i \sum_j p(i,j) \log p(i,j) \quad (7)$$

Inverse Difference Moment

$$IDM = \sum_i \sum_j \frac{1}{1+(i-j)^2} p(i,j) \quad (8)$$

where $u_{x,\ uy,\ \sigma x,\ \sigma y}$ are the means and standard deviations of the marginal distribution associated with P(i,j)/R, and R is a normalizing constant.

After acquiring these five texture features, they are fused using PCA where the principal components are then used for segmentation purposes. Finally, the fuzzy-C means algorithm is performed with the first three principal components during image segmentation.

The third set of maps is the local orientation O, here O is obtained from I using oriented Gabor pyramids O(σ,θ), where σ∈(0…8) represents the scale and θ∈{0°,45°,90°,135°} is the preferred orientation. The orientation feature maps, O(c,s,θ), are encoded as a group, with a local orientation contrast between the center and surround scales, as follows:

$$O(c,s,\theta)=|O(c,\theta)\ominus O(s,\theta)| \quad (9)$$

In total, 36 feature maps were computed: six for intensity, six for texture, and 24 for orientation. After processing of across-scale combinations, normalization and linear combinations with six for intensity, six for texture, and 24 for orientation. As mentioned in [20], the improved salience map is generated.

After obtaining the saliency map, suspected water regions can be found by using the Otsu method. The Otsu algorithm is a nonparametric and unsupervised method of automatic threshold selection for image segmentation, and it is a classical method for optical image segmentation. However, due to the strong speckle noise of SAR images, especially for fine spatial resolution SAR images, Otsu method could be invalidated. Thus, it was applied to the visual saliency map, which has no speckle noise, and it is used to segment the visual saliency map and obtain the suspected water regions. At the same time, we can also use this segmentation result of the visual saliency map to initialize the zero level set function, supposing the segmentation result of the visual saliency map is t(x).The details for using the visual saliency map to initialize the zero level set function can be seen in the last part of Section 2.2.

Adaptive Multi-Scale Level Set Method Based on the Gamma Model

In order to remove the influence of speckle and preserve important structural information, a multi-scale level set algorithm is introduced. We acquire multi-scale images at several scales by decomposing the SAR image using the block averaging algorithm. Because the Gamma model can well represent the distribution of SAR images if the speckle is fully developed and spatially uncorrelated, and the radar backscatter of the region is constant, it is employed to define the energy functional of the level set function for the original SAR image. Even if there is correlation between neighboring pixels, or the radar backscatter features texture, the Gamma distribution can still be used as an approximation by adjusting the number of looks L. For scale images decomposed by the block averaging algorithm, the probability density function (PDF) of the pixel intensity can be also given by a Gamma distribution. For a given image $u(x)$, $\tilde{u}(x)$ is the decomposed image generated by block averaging algorithm. If u(x) is modeled by a Gamma distribution as u(x)~Γ(μ,L), then $\tilde{u}(x) \sim \Gamma(\mu, 4L)$. After using the threshold level set initialization method at the coarsest scale, the initial segmentation result, which is marked by a label, is generated. This label is used not only for the level set initialization at finer scales but also as an additional constraint to guide contour evolution. Consequently, the finest scale image is extracted as the final result.

Chan and Vese [17] proposed a model that implemented the Mumford–Shah functional via the level set function for the purpose of bimodal segmentation. The segmentation is performed using an active contour model without boundaries. Let Ω be a bounded open subset of R^2, with ∂Ω being its boundary. Let $u_0(x,y) : \overline{\Omega} \rightarrow R$ be a given image, and C be a curve in the image domain Ω. Segmentation is achieved via the evolution of curve C, which is the basic idea of the active contour model. In the level set method, $C\subset\Omega$ is represented by the zero level set of a Lipschitz function $\phi:\Omega\rightarrow R$, and we replace the unknown variable C with the unknown variable ϕ.

In this paper, assume that the image is partitioned into two classes Ω_1 and Ω_2, which are separated by a curve C, and that classes Ω_1 and Ω_2 are modeled using the probability density functions (pdfs) P_1 and P_2, respectively.

The partition is obtained by minimizing the following energy function:

$$F(C,p_1,p_2) = \mu Length(C) - \lambda_1\int_\Omega (\log p_1 dx) - \lambda_2\int_\Omega (\log p_2 dx) \quad (10)$$

A gamma model was used for high resolution SAR image segmentation. Suppose uSAR(x,y) is a SAR image, we model the image in each region R_i by a Gamma distribution of mean intensity u_i and number of looks L:

$$P(u_{SAR}(x,y)) = \frac{L^L}{u_i\Gamma(L)}\left(\frac{u_{SAR}(x,y)}{u_i}\right)^{L-1} e^{-\frac{Lu_{SAR}(x,y)}{u_i}} \quad (11)$$

Using the Heaviside function H, and the one-dimensional Dirac function δ_0, which are defined, respectively, by

$$H(z) = \begin{cases} 1 & \text{if } z \geqslant 0 \\ 0 & \text{if } z < 0 \end{cases} \text{ and } \delta(z) = \frac{dH(z)}{dz} \quad (12)$$

The segmentation is performed via the evolution of ϕ by minimizing the following energy function:

$$F(\varnothing,p_1,p_2) = \mu\int \nabla H(\varnothing)\,dx - \lambda_1\int H(\varnothing)\log p_1 dx - \lambda_2\int (1-H(\varnothing))\log p_2 dx \quad (13)$$

where μ, λ_1, λ_2 are non-negative weighted parameters. Function $\varnothing$ represents class Ω_1 for $\varnothing>0$, and Ω_2 for $\varnothing<0$.

The evolution of ϕ is governed by the following partial differential motion equation:

$$\frac{\partial\phi}{\partial t} = \delta_\varepsilon(\phi)\left[\mu\,\mathrm{div}\left(\frac{\nabla\phi}{|\nabla\phi|}\right) + v + \lambda_1\log p_1\left(y|\hat{\theta}^1\right) + \lambda_2\log p_2\left(y|\hat{\theta}^2\right)\right] \quad (14)$$

where $\delta\varepsilon(\phi)$ is a regularized version of the Dirac function.

Next, we aim to estimate the Gamma parameters $\theta = \{\mu_i\}$. Using maximum likelihood estimation $\theta^* = \arg\max_\theta \log p(y|\theta)$. Assuming that the samples $y_j, j=1,\ldots,N$ in each region are independent and identically distributed, the log likelihood is $\log p(y|\theta) = \log\prod_{j=1}^N p(y_j|\theta) = \sum_{j=1}^N \log p(y_j|\theta)$. Taking the derivative of $\log p(y|\theta)$ with respect to θ and setting them equal to zero, $\mu_i = \sum_{j=1}^{N_i} y_j/N_i$ was obtained, where N_i is the pixel number in Ω_i.

As we know that the segmentation result of the visual saliency map is t(x), then we simply initialize the level set functionϕ as follows:

$$\phi=t(x) \quad (15)$$

The main step of the level sets method is as follows (also see the details for Step 4 of our method):

- Initialize the level set function using Equation (15);
- Evolve the level set function ϕ according to Equations (13) and (14);
- Check whether the evolution is stationary. If not, return to Step 2.

Post-Processing: Using Object-Oriented Geometrical Feature

Water objects in SAR images are continuous regions with low light, and some objects with a low radar cross-section similar to calm water, such as urban roads, airstrips, building shadows or paddy fields, may influence the results of water extraction.

Compared with a large water area, urban roads, airstrips and building shadows are not significant, thus, these areas are ignored after we acquire the SWR regions using the visual salience model. However, some paddy fields remain as salient objects because of their occupation of large areas. In this study, geometrical features are used for post-processing to distinguish paddy fields from water bodies. First, a fast pixel labeling method is applied to search for connected areas in binary images (more details can be found in [28]). Then, the attributes of each object can be acquired, such as the area, perimeter and number of objects. In SAR images, the geometrical features of paddy fields usually appear as regular rectangles, whereas, water bodies such as rivers and lakes, mostly appear as irregular geometrical shape. Thus, inspired by the user guide of eCognition software (Definiens AG, Munich, Germany) [29], the weighting of smoothness (WS) within the shape was introduced to distinguish paddy fields from water bodies, as shown in Equation (16).

$$WS = \frac{\text{Area of Minimum Enclosing Rectangle}}{\text{Area of the object}} \quad (16)$$

We obtained good segmentation results using the proposed method, and most paddy fields had geometrical shapes such are as shown in Figure 2a. However, after the evolution of the level sets, some paddy fields were connected each other, such as the shape shown in Figure 2b. Figure 2c shows the object shape of a river and Figure 2d shows the object shape of a lake. The red dashed rectangles are the minimum enclosing rectangles for each object. During image segmentation, the ratio of the minimum enclosing rectangle's area and the object's area is used to make the segmentation object's shape more regular, and the value of the ratio is very close to 1, so the object's shape is very close to a rectangle. The ratio values for Figure 2a–d are 1.344, 2.143, 6.524 and 2.5869, respectively. Thus, based on the experience value (3000 objects with 15 images were tested), we assume that an object is "water class", if the value of smoothness weighting in an object is greater than 2, and all others water classes can be removed during post-processing.

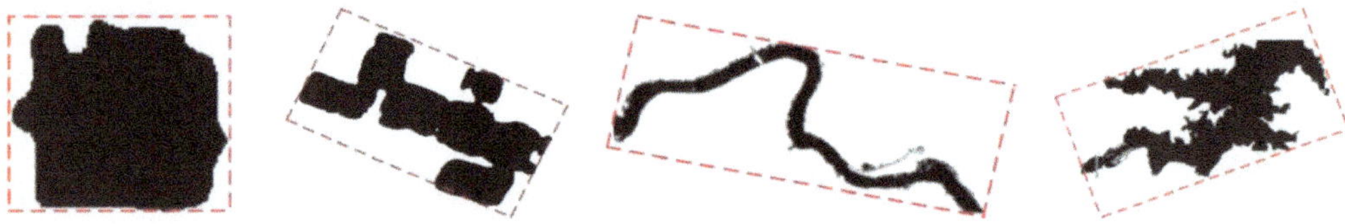

(a) (b) (c) (d)

Figure 2: The demonstration of the weighting of smoothness within the objects. (a) Single paddy field; (b) connected paddy fields; (c) river; (d) lake.

EXPERIMENTAL DATA SET

SAR amplitude images (Radarsat2 and TerraSAR-X) of river basin regions in China were used in the experiment, detailed descriptions of the experiment data can be seen in Table 1.

Table 1: Experimental data of different water bodies

Name of water bodies	Huai River	Hanjiang River	Changjiang River
Sensor	Radarsat 2 (VV polarization)	TerraSAR-X (VV polarization)	TerraSAR-X (VV polarization)
Orbit	descending	ascending	ascending
Mode	Ulra-Fine	Spotlight	Spotlight
Date	7 December 2009	9 October 2008	9 October 2008
Resolution	3 m	1 m	1 m
Image Size (pixels)	3024 × 2263	7153 × 1948	15540 × 14550

1. *Study Area of Huai River*: Huai River catchment around eastern China is taken as one of the study areas. The Radarsat 2 images (VV polarization) at a spatial resolution of 3 m were acquired on 7 December 2009. A Google map image of 0.5m resolution with the same region and same season is used as the true water class image, then manual water extraction in this google image is used for reference to check the accuracy of the water extraction process.
2. *Study Area of Hanjiang and Changjiang River*: Hanjiang River is the largest branch of Changjiang River, and they are located in Wuhan City, Hubei Province, China. The TerraSAR-X images (VV polarization) at a spatial resolution of 1 m were acquired on 9 October 2008. A vector map of the same region supported by Map Institute of Hubei Province is used for reference to check the accuracy of the water extraction process.

RESULTS AND DISCUSSION

The following three methods were compared to evaluate the performance of the proposed method.

ALG1: A MRF segmentation method proposed by Picco and Placio [30].

ALG2: A method for the separation of land and water in SAR images which was proposed by Silveira and Heleno [14].

ALG3: A multi-scale level set method for SAR image segmentation which was proposed by Sui *et al.* [25].

Note that all of our images were tested without a filter when using the proposed method and the scale level was 2. The value of the smooth index WS = 2, *i.e.*, if the value of the object smoothing index was less than 2, this object could be removed. The weight coefficient of the center-surround differences is equal to 0.1.

Experiment on Radarsat-2 Image

Figure 3 compares the salience maps generated using the Itti model and our model. This shows clearly that our model readily distinguished areas of water. Figure 3 a shows the Huai River in eastern China, which was acquired by Radarsat-2 in 2009 in descending VV polarization mode, with a resolution of 3 m and a size of 3024 × 2263 pixels. Figure 3b,c shows the saliency map generated using Itti model and our TW-Itti model, respectively. The experiment validates our model performance much better than the Itti model. In Figure 3b, except for water body, other features are also incorrectly considered to be salient, which made the water body discontinuous. Figure 3d shows the Otsu segmentation result after the post-processing of Figure 3c. It can be seen from Figure 3d that segmentation using the saliency map can acquire main outlines, however, missing edge details. It can be seen that the edges are jagged and irregular. As such, we need to apply the level set method to refine the edge and make it smooth.

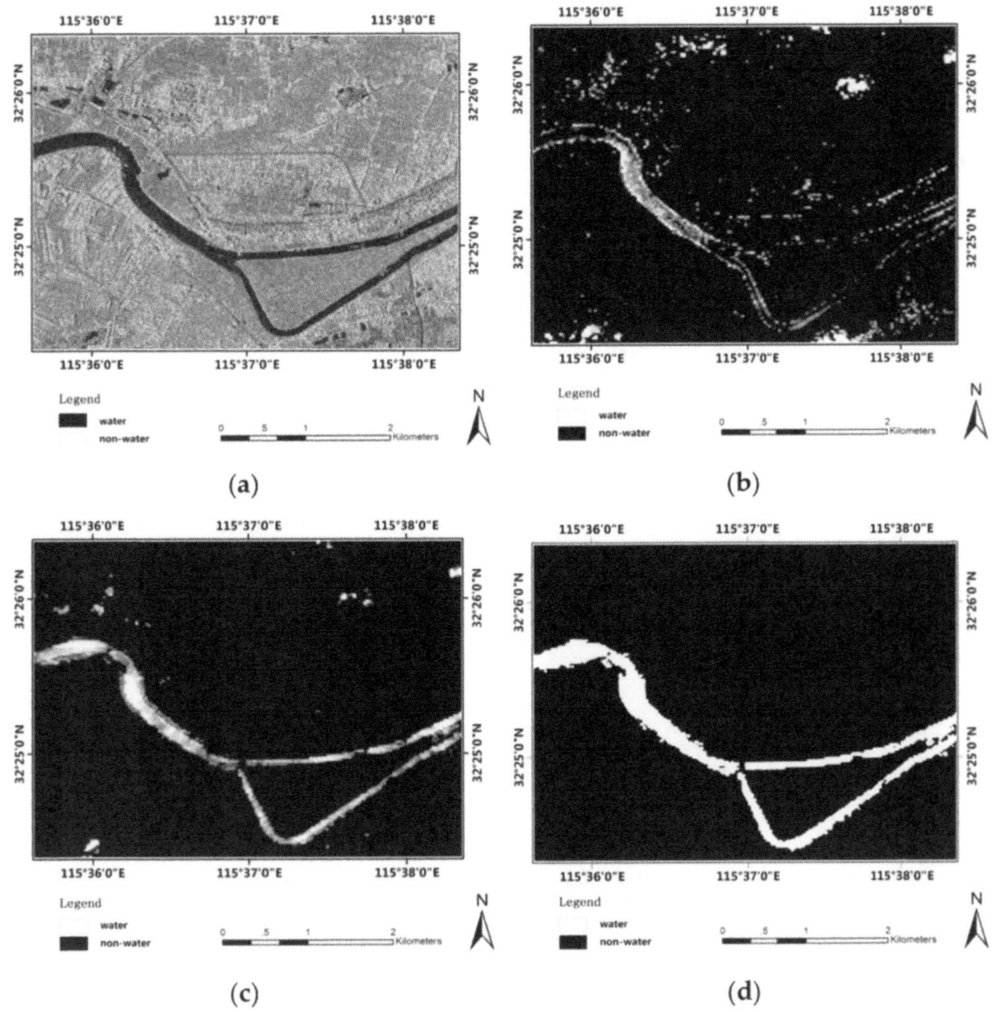

Figure 3: Saliency map comparison (Huai River). (**a**) Original imagery; (**b**) Salience map generated by Itti model; (**c**) Salience map generated by our TW-Itti model; (**d**) Otsu segmentation result after post-processing of image (c).

Figure 4 shows the water extraction results for the Huai River. Figure 4a is the SWR generated by the saliency map (seeFigure 3d) after post-processing. In general, the actual positions of boundaries within a SAR scene are unknown. Thus, the segmentation quality measures were modified to allow a comparison of the automatic segmentation approaches and manual segmentation, as shown in Figure 4b. Figure 4c–f shows segmentation results by ALG1, ALG2, ALG3 and our method, respectively. Figure 4g shows the post-processing results of the proposed method. The experiment validates that our method (see Figure 4g) performs best in distinguishing water and preventing speckle. Detailed comparisons of accuracy and efficiency are given in Table 1.

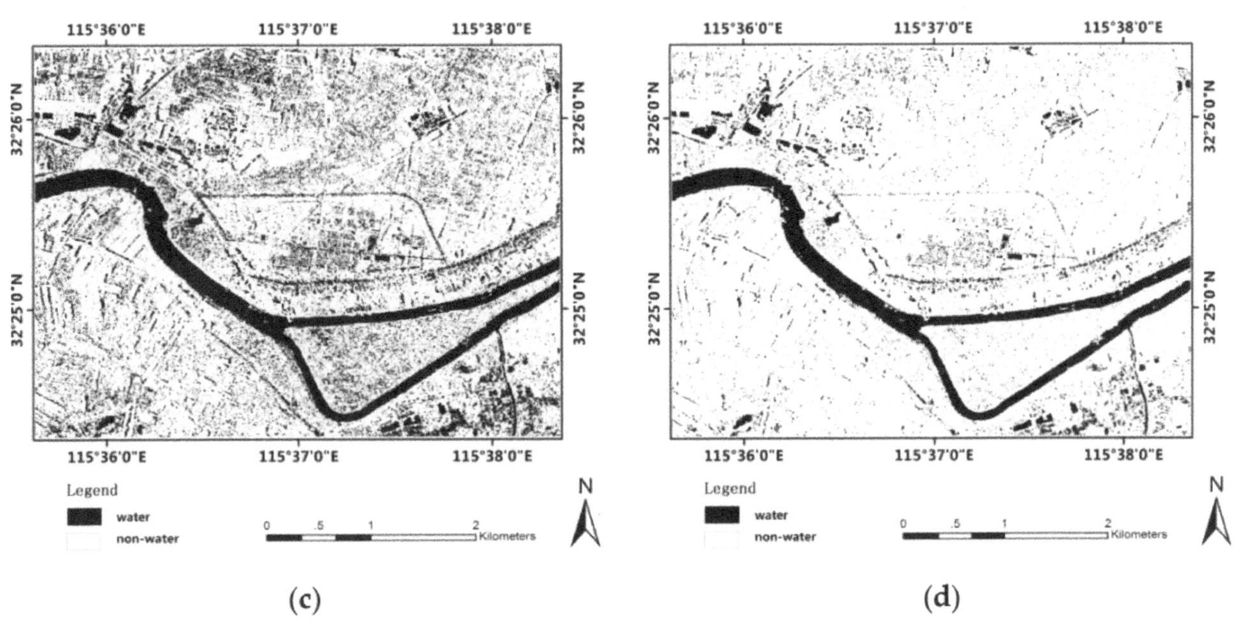

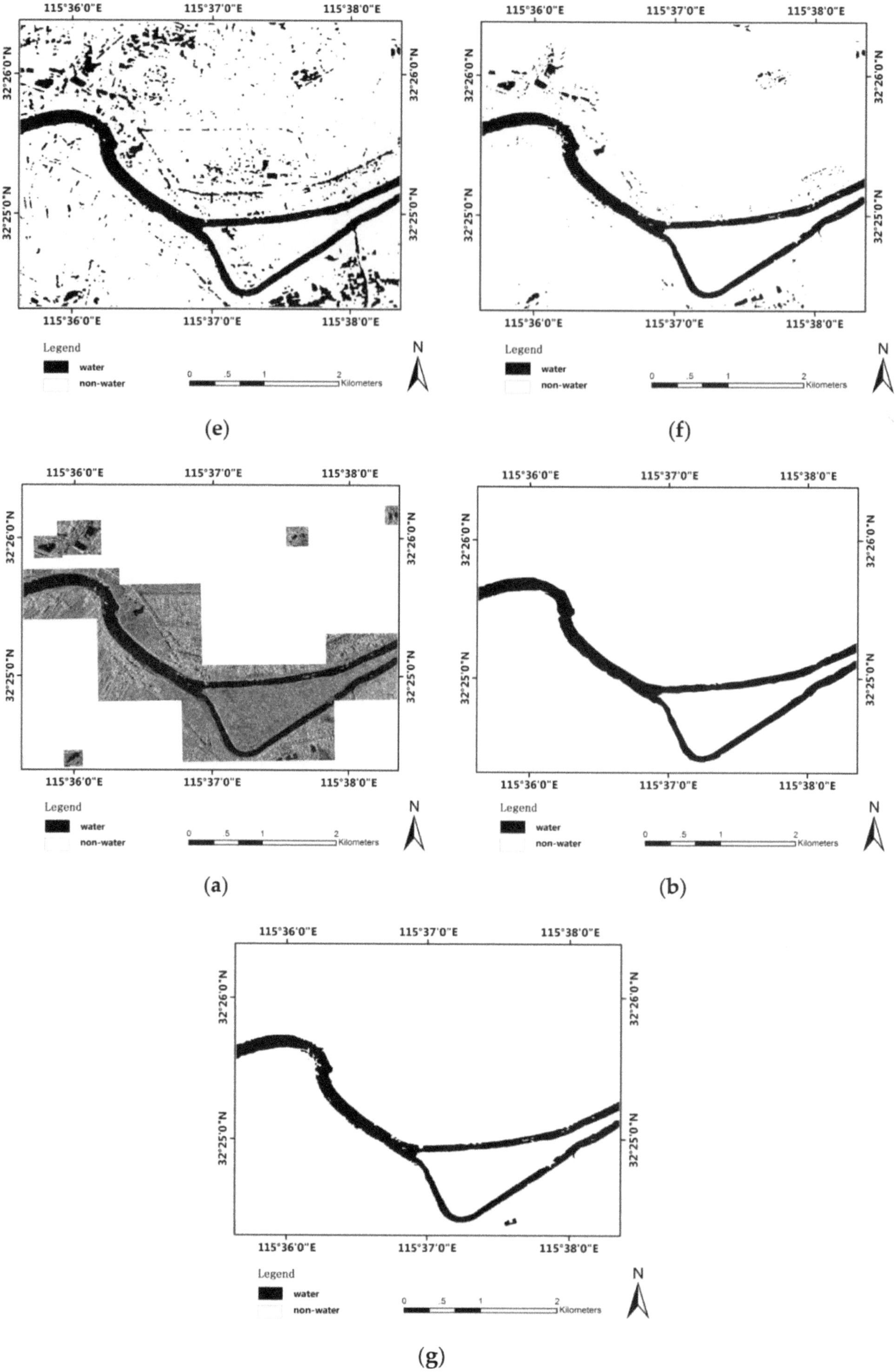

Figure 4: Segmentation results (Huai River). (**a**) SWR region generated by Figure 3d; (**b**) Manual segmentation; (**c–e**) LSW map obtained using ALG1, ALG2 and ALG3,respectively; (**f**) LSW map obtainedusing proposed method; (**g**) LSW map obtained after post-processing.

Experiment on TerraSAR-X Imagery

Figure 5 shows a comparative saliency map using a TerraSAR-X image. Figure 5a shows the Hanjiang River in central China, which was acquired by TSX-1 in 2008 in a descending VV polarization mode, with a resolution of 1 m and a size of 7153 × 1948 pixels. After comparing Figure 5b,c, the same conclusions about the model improvements in Radarsat-2 were applicable.

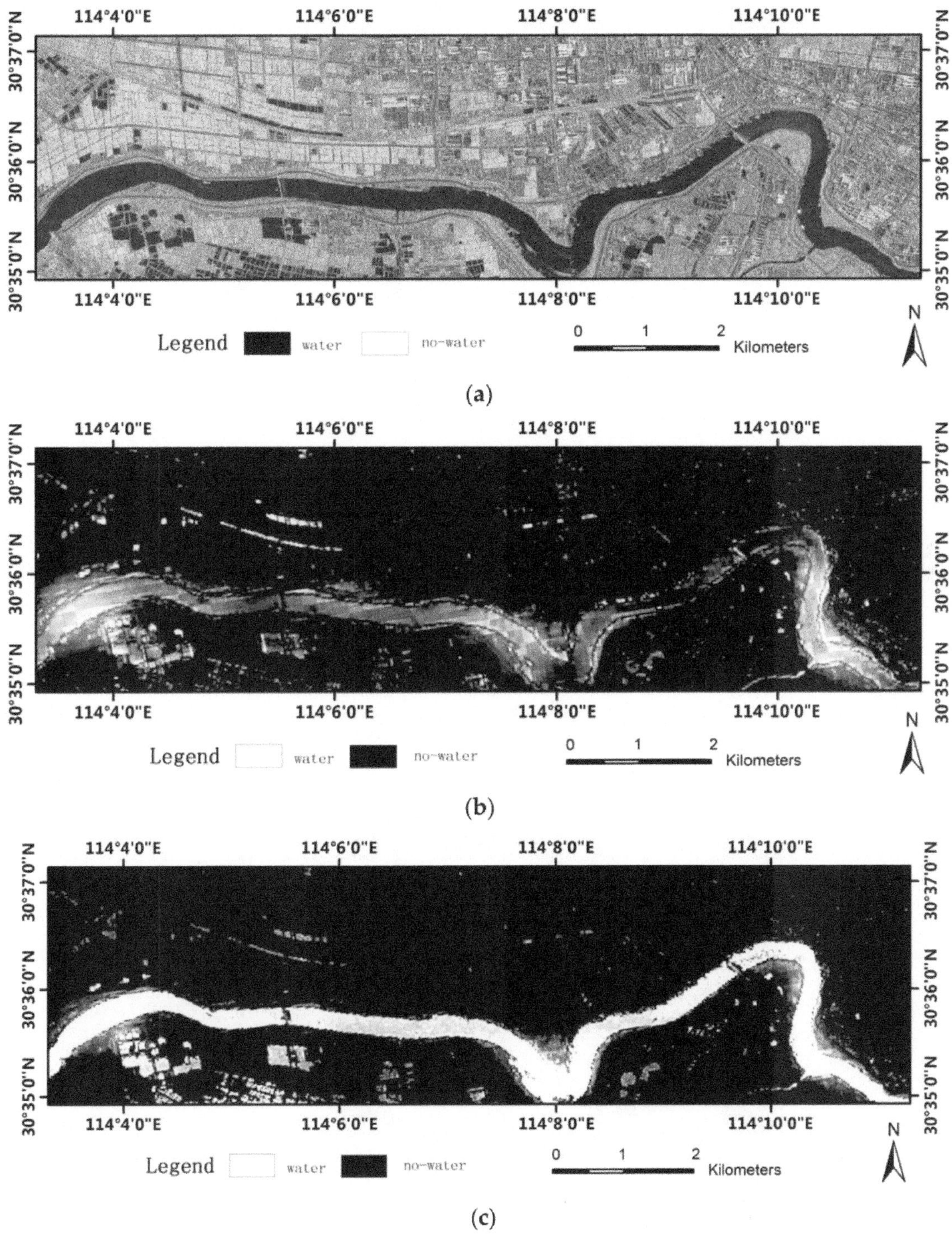

Figure 5. Saliency map comparison (Hanjiang River). (**a**) Original imagery; (**b**) Salience map generated by Itti model; (**c**) Salience map generated by our model.

Figure 6 shows the water extraction results for the Hanjiang River. Figure 6a shows the SWR generated using the saliency map (see Figure 5c). Figure 6b is the manual segmentation result. Figure 6c–f show segmentation result by ALG1, ALG2, ALG3 and our method, respectively. Figure 6g shows the post-processing result of the proposed method. The experiment also validates that our method can achieve the best segment result. The speckle noise is very strong in this image, and we can see that there are many very small-scale objects (*i.e.*, "salt and pepper" effects) in Figure 6c,d, and there are many other features misrepresented as water. Figure 6e is much better than the previous methods, however, compared with Figure 6f, there are still some noise effects. Detailed comparisons of accuracy and efficiency are given in Table 1.

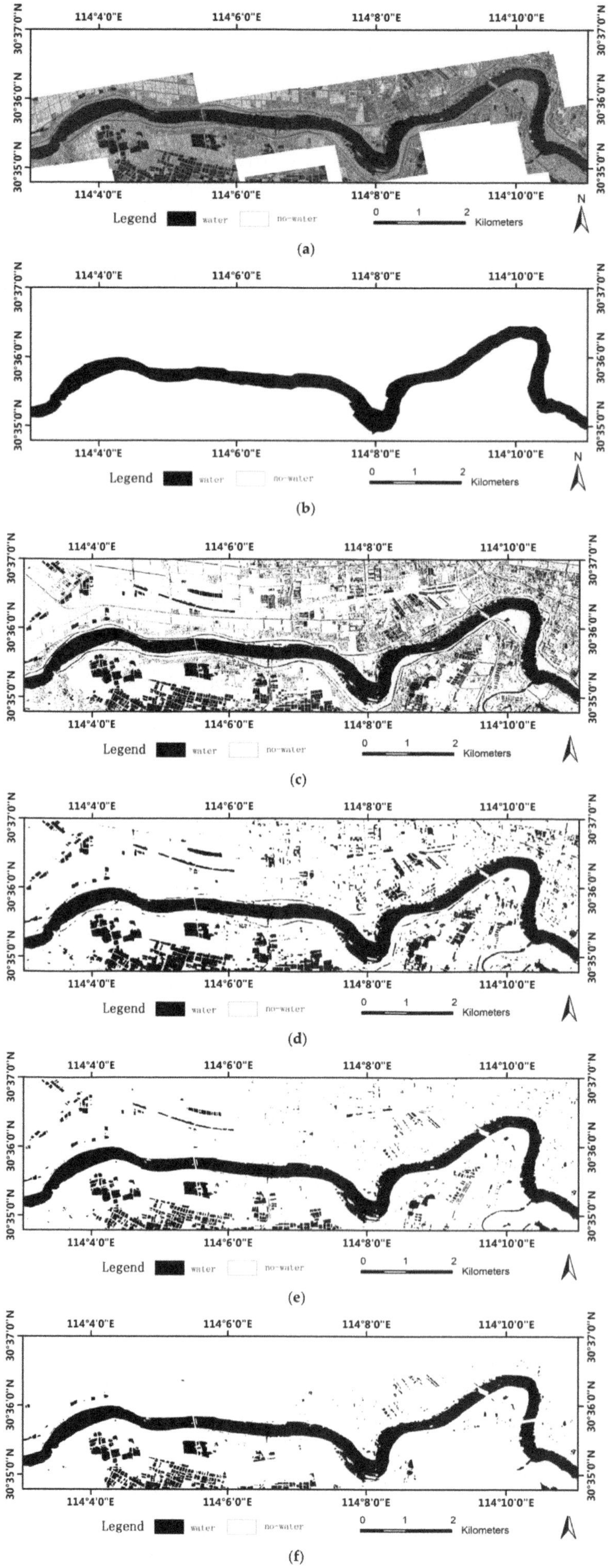
114°4'0"E
114°6'0"E
114°8'0"E
114°10'0"E
30°37'0"N
30°36'0"N
30°35'0"N
Legend
water
no-water
0 1 2
Kilometers
N
(a)
(b)
(c)
(d)
(e)
(f)

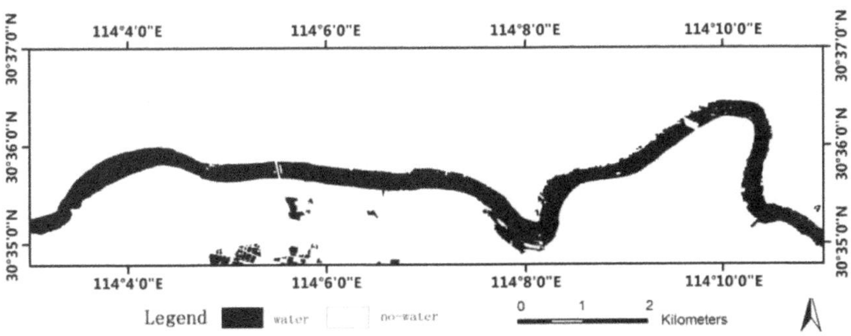

(g)

Figure 6: Segmentation results (Hanjiang River). (**a**) SWR region generated by Figure 4d; (**b**) Manual segmentation; (**c–e**) LSW map obtained using ALG1, ALG2 and ALG3,respectively; (**f**) LSW map obtained using our method; (**g**) LSW map obtained after post-processing.

Applications to Large-Area SAR Images

To demonstrate the speed and reliability of our method, we tested the proposed method using real large-area SAR images. The Changjiang River basin, which is one of the main river basins in China, was selected for this experiment. Figure 7 shows the water extraction results for the Changjiang River. Figure 7a shows the original SAR image with an image size of 15540 ×14550 pixels. Figure 7b shows the saliency map generated by Itti model. Figure 7c is the salience map generated using the TW-Itti model and Figure 7d shows the water segmentation result using our method. The time required to generate this image was 1209.463s.

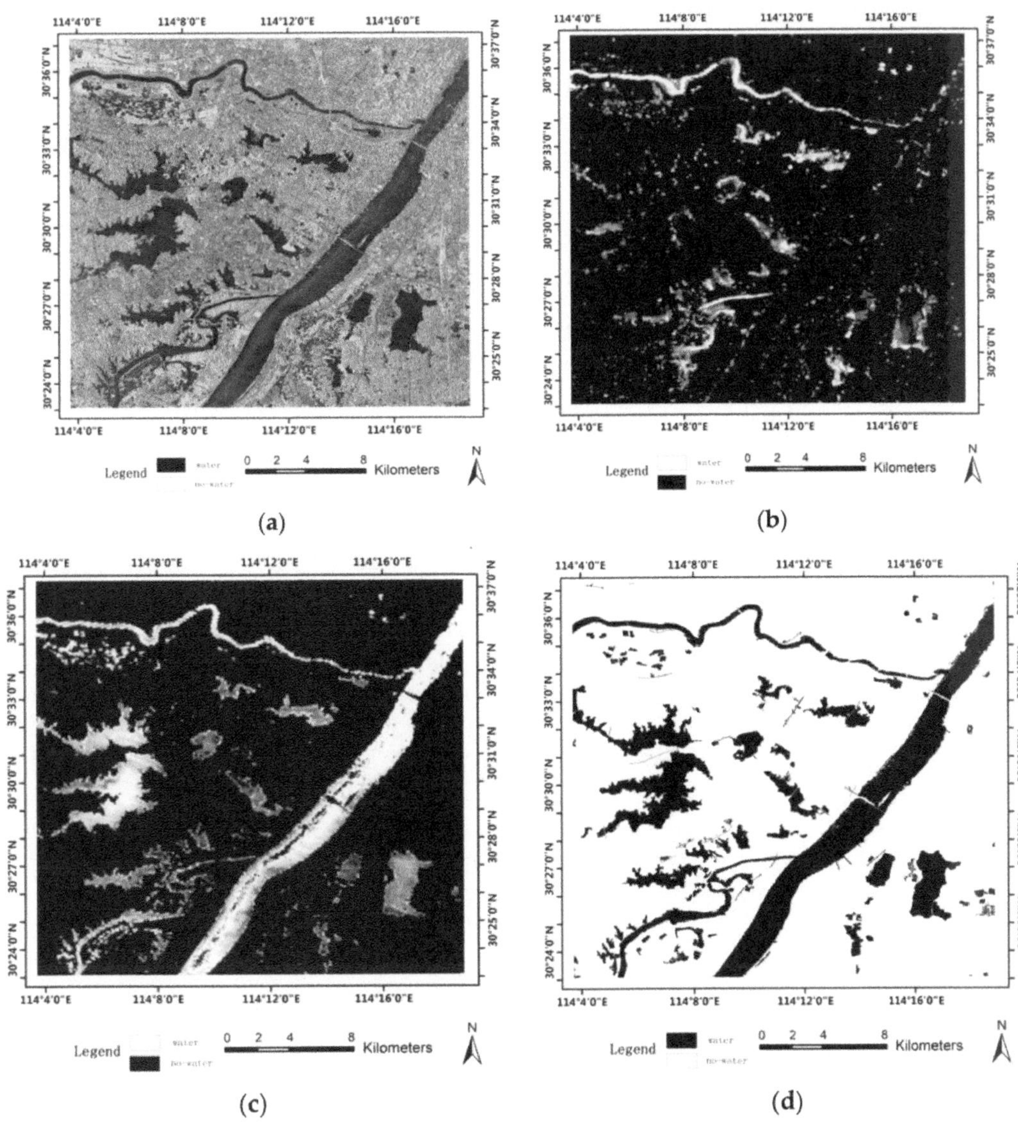

Figure 7: Salience map comparison and segmentation result (Changjiang River). (**a**) original SAR image; (**b**) the saliency map generated by Itti model; (**c**) the saliency map generated by TW-Itti model; (**d**) LSW map obtained using our method.

Accuracy Analysis and Discussion

In order to perform quantitative analysis, overall accuracy (short for OA) and Kappa coefficient are adopted as the accuracy assessment tools in this paper. Assuming that the size of an original SAR image is M×N, we denote the label image as X', whose size is also M×N. Correspondingly, *R* represents the label image from manual extraction (it also indicates an ideal water extraction) for the same original SAR image. The error image is therefore defined as E=X′−R.

$$OA = 1 - \frac{l}{M \times N} \times 100\% \quad (17)$$

where *l* denotes the number of non-zero pixels in *E*.

Kappa coefficient can be described as below:

$$Kappa = \frac{(M \times N)\sum_{i}^{r} x_{ii} - \sum (x_{i+} x_{+i})}{(M \times N)^2 - \sum (x_{i+} x_{+i})} \quad (18)$$

where r is the row numbers of the confusion matrix, xii is the value of principal diagonal, xi+ and x+i are the sum value of row i and column i, respectively.

Table 2 compares the overall accuracy and Kappa coefficient of the above methods.

Table 2: Overall Accuracy and Kappa Coefficient Comparison

Method	Overall Accuracy (%)	Kappa Coefficient
	Experiment on Radarsat-2 imagery	
ALG1	79.22	0.262
ALG2	92.39	0.530
ALG3	91.38	0.511
Proposed Method	**98.48**	**0.856**
	Experiment on TerraSAR-X imagery	
ALG1	81.80	0.434
ALG2	88.71	0.577
ALG3	94.17	0.731
Proposed Method	**98.42**	**0.913**

Based on the qualitative analysis (Experiments 1 and 2) and the quantitative analysis (Table 1), the following conclusions can be reached:

1. Compared with the classical Itti model, the improved visual saliency model provides better enhancement of water information and it suppresses other surface features (See Figure 2c,d for Experiment 1and Figure 4b,c for Experiment 2).
2. Compared with the state-of-the-art methods (ALG1, ALG2 and ALG3), the method of integration using the multi-scale technique and visual saliency model produced better accuracy.
3. In all cases, our method frequently produced superior results and it produced the most accurate results in the shortest time (see Figure 4g for Experiment 1, Figure 6g for Experiment 2 and the bold line in Table 1).

Moreover, in order to better evaluate the performance of the proposed method, the error matrix of the results for ALG1, ALG2, ALG3 and our method were calculated for validation; in this paper, we only take the dataset of Radarsat-2 (Experiment 1) as an example. An error matrix is a very effective way to assess accuracy in that the accuracies of each category are plainly described along with the errors of omission and commission present in the classification. The error matrixes of Experiment 1 are shown in Table 3, Table 4, Table 5 and Table 6.

Table 3: Error matrix of the ALG1 classified image (Dataset 1)

Class	Reference Data		
	Water	Non-Water	Total
Water	351523	1383003	1734526
Non-water	38789	5069997	5108786
Total	390312	6453000	6843312
Overall accuracy = 79.22% Kappa coefficient = 0.262			
Producer accuracy		User accuracy	
Water=351523/390312=90.06%		Water=351523/1734526=20.27%	
Non-water=5069997/6453000=78.57%		Non-water=5069997/5108786=99.24%	

Table 4: Error matrix of the ALG2 classified image (Dataset 1)

Class	Reference Data		
	Water	Non-Water	Total
Water	340153	470767	810920
Non-water	50159	5982233	6032392
Total	390312	6453000	6843312
Overall accuracy = 92.39% Kappa coefficient = 0.53			
Producer accuracy Water=340153/390312=87.15% Non-water=5982233/6453000=92.7%		User accuracy Water=340153/810920=41.95% Non-water=5982233/6032392=99.17%	

Table 5: Error matrix of the ALG3 classified image (Dataset 1)

Class	Reference Data		
	Water	Non-Water	Total
Water	361014	560267	921281
Non-water	29298	5892733	5922031
Total	390312	6453000	6843312
Overall accuracy = 91.38% Kappa coefficient = 0.511			
Producer accuracy Water=361014/390312=92.49% Non-water=5892733/6453000=91.32%		User accuracy Water=361014/921281=39.19% Non-water=5892733/5922031=99.51%	

Table 6: Error matrix of the proposed method classified image (Dataset 1)

Class	Reference Data		
	Water	Non-Water	Total
Water	329671	43053	390312
Non-water	60641	6409947	6453000
Total	390312	6453000	6843312
Overall accuracy = 98.48% Kappa coefficient = 0.856			
Producer accuracy Water=329671/390312=84.46% Non-water=5069997/6453000=99.33%		User accuracy Water=329671/390312=88.45% Non-water=6409947/6453000=99.06%	

Overall accuracies for ALG1, ALG2, ALG3 are 79.22% and 92.39%, 91.38%, and the Kappa coefficients are 0.262, 0.53 and 0.511. Individual class accuracies were highly variable, ranging from 20.27% to 99.51%. For our method, overall accuracy is 98.48% and the Kappa coefficient is 0.856. Moreover, individual class accuracies are all very high, above 80%. This proves that our classifier is accurate and robust.

CONCLUSIONS

The main purpose of this study was to devise a water mapping method that improves accuracy by using a visual attention-based method for multi-scale level sets, particularly in areas with roads, airstrips, radar shadows and paddy fields, which are often major causes of low classification accuracy. Experiments demonstrated that our method significantly improved accuracy in areas where shadow and other dark surfaces were the main sources of classification errors. Our method was aimed at calm water bodies and, therefore, did not consider dynamic water bodies, such as flood water with high wind. In future work, we will focus more on other types of water bodies, looking at, for instance, flood-induced backscatter changes in SAR data [31] and so on.

Although experiments have testified that our method gets satisfactory results, much work remains to be done. The multi-scale analysis framework is a new component in the level set method. In future work, multi-region level sets method can be merged with multi-scale analysis, and, as such, more land features can be considered. Moreover, we will develop our method to be used in full polarized SAR images for acquiring more accurate classification.

ACKNOWLEDGMENTS

This work was supported by National Key Fundamental Research Plan of China (973) (No.2012CB719906), major projects of high resolution earth observation system (No. 03-Y20A10-9001-15/16) and National Natural Fund of China (NSFC) (No.41101414).

REFERENCES

1. Thomas, H.; Sandro, M.; Andre, T.; Achim, R.; Manfred, B. Extraction of water and flood areas from SAR data. In Proceedings of the 7th European Conference on Synthetic Aperture Radar (EUSAR), Friedrichshafen, Germany, 2–5 June 2008; pp. 1–4.
2. Lee, J.S.; Iurkevich, I. Segmentation of SAR Image. IEEE Trans. Geosci. Remote Sens. 1989, 27, 981–990.
3. Sahoo, P.K.; Soltani, S.; Wong, A.K.C. A survey of thresholding techniques. Comput. Vis. Graph. Image Process. 1988, 41, 233–260.
4. Fiortoft, R.; Lopes, A.; Marthon, P. An optimal multi edge detector for SAR image segmentation. IEEE Trans. Geosci. Remote Sens. 1998, 36, 793–802.
5. Oliver, C.; Connell, I.M.; White, R.G. Optimum edge detection in SAR. SPIE Satell. Remote Sens. 1995, 2584, 152–163.
6. Cook, R.; Connell, I.M.; Oliver, C.J. MUM Segmentation for SAR Images. Proc. SPIE 1994, 2316, 92–103.
7. Udupa, J.K.; Samarasekera, S. Fuzzy connectedness and object definition: Theory, algorithms, and applications in image segmentation. Graph. Models Image Process. 1996, 58, 246–261.
8. Deng, H.; Clausi, D. Unsupervised segmentation of synthetic aperture radar sea ice imagery using a novel Markov random field model. IEEE Trans. Geosci. Remote Sens. 2005, 43, 528–538.
9. Stéphane, D.; Grégoire, M. Unsupervised multiscale oil slick segmentation from SAR images using a vector HMC model. Pattern Recognit. 2007, 40, 1135–1147.
10. Gan, L.; Wu, Y.; Wang, F.; Zhang, P.; Zhang, Q. Unsupervised SAR image segmentation based on triplet Markov fields with graph cuts. IEEE Geosci. Remote Sens. Lett. 2014, 11, 853–857.
11. Wang, F.; Wu, Y.; Zhang, Q.; Zhao, W.; Li, M.; Liao, G.S. Unsupervised SAR image segmentation using higher order neighborhood-based triplet Markov fields model. IEEE Trans. Geosci. Remote Sens. 2014, 52, 5193–5205.
12. Wang, Y.; Li, Y.; Zhao, Q.H. Segmentation of high-resolution SAR image with unknown number of classes based on regular tessellation and RJMCMC algorithm. Int. J. Remote Sens. 2015, 36, 1290–1306.
13. Huang, B.; Li, H.; Huang, X. A level set method for oil slick segmentation in SAR images. Int. J. Remote Sens. 2005, 26, 1145–1156.
14. Silveira, M.; Heleno, S. Separation between water and land in SAR images using region-based level sets. IEEE Geosci. Remote Sens. Lett. 2009, 6, 471–475.
15. Zou, P.F.; Li, Z.; Tian, B.S.; Guo, L.J. A level set method for segmentation of high-resolution polarimetric SAR images using a heterogeneous clutter model. Remote Sens. Lett. 2015, 6, 548–557.
16. Yin, J.J.; Yang, J. A modified level set approach for segmentation of multiband polarimetric SAR images. IEEE Trans. Geosci. Remote Sens. 2015, 52, 7222–7232.
17. Chan, T.; Vese, L. Active contours without edges. IEEE Trans. Image Process. 2001, 10, 266–277.
18. Itti, L. Modeling primate visual attention. In Computational Neuroscience: A Comprehensive Approach; Feng, J., Ed.; CRC Press: Boca Raton, FL, USA, 2003; pp. 635–655.
19. Itti, L.; Koch, C.; Niebur, E. Model of saliency-based visual attention for rapid scene analysis. IEEE Trans. Pattern Anal. Mach. Intell. 1998, 20, 1254–1259.
20. Itti, L.; Gold, C.; Koch, C. Visual attention and target detection in cluttered natural scenes. Opt. Eng. 2001, 40, 1784–1793.
21. Walther, D.; Rutishauser, U.; Koch, C. Selective visual attention enables learning and recognition of multiple objects in cluttered scenes. Comput. Vis. Image Underst. 2005, 100, 41–63.
22. Li, Z.Q.; Fang, T.; Huo, H. A saliency model based on wavelet transform and visual attention. SCIENCE CHINA Inf. Sci. 2010, 53, 738–751.
23. Gao, L.N.; Bi, F.K.; Yang, J. Visual attention based model for target detection in large-field images. J. Syst. Eng. Electron. 2011, 22, 150–156.
24. Li, Z.C.; Itti, L. Saliency and gist features for target detection in satellite images. IEEE Trans. Image Process. 2011, 20, 2017–2029.
25. Sui, H.G.; Xu, C.; Liu, J.Y.; Sun, K.M.; Wen, C.F. A novel multi-scale level set method for SAR image segmentation based on a statistical model. Int. J. Remote Sens. 2012, 33, 5600–5614.
26. Otsu, N. A threshold selection method from gray-level histogram. IEEE Trans. Syst. Man Cybern. 1979, 9, 62–66.
27. Haralick, R.M.; Shanmugam, K.; Dinstein, I. Textural features for image classification. IEEE Trans. Syst. Man Cybern. 1973, 3, 610–621.
28. He, L.F.; Chao, Y.Y.; Suzuki, K.J. A run-based two-scan labeling algorithm. IEEE Trans. Image Process. 2008, 17, 749–756.
29. User Guide of eCognition in Chinese. Available online: http://vdisk.weibo.com/s/zt_sYprKIffXK (accessed on 12 October 2015).
30. Picco, M.; Palacio, G. Unsupervised classification of SAR images using markov random fields and g0I Model. IEEE Geosci. Remote Sens. Lett. 2011, 8, 350–353.
31. Martinis, S.; Twele, A.; Voigt, S. Unsupervised extraction of flood-induced backscatter changes in SAR data using Markov image modeling on irregular graphs. IEEE Trans. Geosci. Remote Sens. 2011, 49, 251–263.

CHAPTER 8

AUTOMATED SUBPIXEL SURFACE WATER MAPPING FROM HETEROGENEOUS URBAN ENVIRONMENTS USING LANDSAT 8 OLI IMAGERY

Huan Xie, Xin Luo, Xiong Xu, Haiyan Pan and Xiaohua Tong

Keywords: Water mapping, urban, subpixel classification, Landsat 8 OLI, endmember selection

INTRODUCTION

Remote sensing has played an important role in various water studies, including flood monitoring [1], water resource estimation [2,3], water quality assessment [4,5], and shoreline/coastline extraction [6,7]. All of these applications are closely related to surface water mapping from remote sensing images.

Optical remote sensing of water bodies is based on the difference in the spectral reflectance of land and water. Water absorbs most of the energy in the near-infrared and middle-infrared wavelengths, whereas other surface materials have a higher reflectance in these wavelengths. To date, various water body extraction algorithms for optical imagery have been developed, and they can be categorized into four basic types: (1) hard classification [8,9]; (2) single-band thresholding [10,11]; (3) spectral water indexes [12,13,14,15,16]; and (4) spectral unmixing [17,18,19]. The hard classification method can achieve high precision in the water extraction, however, its algorithm is always complicated and time-consuming. Single-band thresholding is the simplest water extraction method, it mainly depends on the reflectivity difference of water and others in the infrared wave bands, and it can achieve high precision in most cases. The two-band spectral water index method has the dual advantages of simplicity and high precision; it is the most frequently used method for water extraction. Among these methods, the first three groups of methods can only obtain water maps at the pixel level, which may not meet the precise requirements of the practical applications in urban areas.

The spectral mixture analysis (SMA) can be used as a pre-processing step for water mapping at sub-pixel resolution. Reliable SMA is critical for the post-sub-pixel mapping process [20,21]. However, the typical processes of SMA consist of two steps—endmember determination and abundance estimation [22]—and for each step, prior knowledge or manual intervention is usually required [23,24,25]. As all of the land-cover types are usually involved in a SMA analysis, this is too complicated when surface water is the only land-cover type of concern. Several previous studies have simplified the SMA process for subpixel water mapping, e.g., Sun et al. [26], and Ma et al. [27] proposed subpixel water mapping methods based on MODIS 250-m data using a linear mixture model, and Pardo-Pascual et al. [7] used the single-band intensity gradient to determine the shoreline position at a subpixel precision. Sethre et al. [28] applied the IMAGINE Subpixel Classifier [29] to detect subpixel-scale ponds.

Even though several subpixel-level water body extraction methods have been proposed in the literature, the research for urban regions is still few and the existing methods seem ineffective for urban regions. In the highly heterogeneous urban environments, the water bodies, like narrow rivers and small ponds, are distributed sparsely in a subpixel-scale size, and water extraction methods often fail to distinguish the narrow rivers in urban areas because the mixed land-water pixels are defined by the neighborhoods of the pure water pixels. The large amount of low albedo surfaces and building shadows, which are easily confused with water bodies in urban areas, also bring a big challenge to urban subpixel water mapping.

Citation: Xie, H.; Luo, X.; Xu, X.; Pan, H.; Tong, X. Automated Subpixel Surface Water Mapping from Heterogeneous Urban Environments Using Landsat 8 OLI Imagery. Remote Sens. 2016, 8, 584. doi:10.3390/rs8070584.

In this paper, an automatic subpixel water mapping method (ASWM) was introduced for use in urban areas using Landsat 8 OLI data. The objectives of this research are: (1) to develop an automatic mixed land-water pixel extraction technique utilizing a water index; (2) to derive the most representative endmembers of both water and land by utilizing neighboring water pixels and the optimal neighboring land pixels, respectively; and (3) to apply a linear unmixing model for subpixel water fraction estimation.

The organization of this paper is structured as follows: Section 2 introduces the study areas and datasets. Section 3describes the proposed ASWM method in detail, including automated mixed land-water pixel extraction, representative endmember selection, and water abundance estimation by a linear unmixing method. In Section 4, the Landsat OLI data used in the experiments were described, and the surface water mapping results and quantitative accuracy assessment are given. Finally, a discussion and the conclusions are presented in Section 5 and Section 6, respectively.

STUDY AREAS AND DATA SOURCES

Four different cities in China were selected as study areas in order to include diverse water body types and complex ground surface features. The selected cities were Beijing, Shanghai, Hangzhou, and Guangzhou (see Figure 1). Beijing is the capital of China and has the second-largest population of any city in China, just behind Shanghai. The selected scene is located in the center of Beijing. The surface water bodies in this area are mostly clear water present as artificial lakes, ponds, and small rivers. The Shanghai scene is located at Lujiazui, the city center of Shanghai, where the Huangpu River and its small tributaries surround the area. In this area, it is difficult to distinguish the water bodies because of the shadows caused by the high-rise buildings. The selected Hangzhou scene is located in the suburbs of the city, where there are abundant water bodies, including the Qiantang River (which varies from clear to turbid) and its tributaries, and many small ponds. The Guangzhou scene is surrounded by the Pearl River and its tributaries, and this area can be roughly divided into half urban and half suburban areas. The water bodies in this area are of great diversity. The details of the study areas and the corresponding Landsat 8 OLI images are summarized and listed in Table 1.

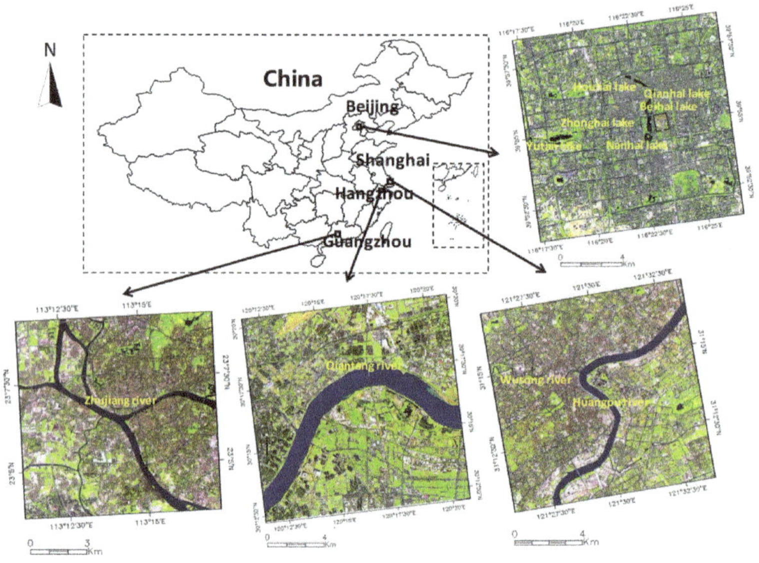

Figure 1: Study areas: Beijing, Shanghai, Hangzhou, and Guangzhou, China, illustrated in false-color composite (RGB: bands 6, 5, 2) Landsat OLI images overlaid by the main water bodies of these areas.

Table 1: Details of the study areas

Study Area	Main Water Bodies (km^2)	Water Characteristics	Acquisition Date of OLI Image	Acquisition Date of Reference Data
Beijing	Nanhai Lake (0.20) Zhonghai Lake (0.25) Beihai Lake (0.35) Qianhai Lake (0.08) Houhai Lake (0.16) Yutan Lake (0.40)	Small and clear lakes Small and clear rivers	16 April 2015	13 April 2015
Shanghai	Huangpu River (7.61) Wusong River (0.38)	Moderately turbid river Small and clear lake	10 April 2014	25 November 2014 22 November 2014
Hangzhou	Qiantang River (28.75)	Clear lake Moderately turbid and turbid river	13 Decemebr 2014	24 December 2014 3 May 2014
Guangzhou	Pearl River (7.47)	Clear and moderately turbid river	15 October 2014	5 October 2014

All of the OLI images used in this study were free of clouds, and atmospheric correction using the Fast Line-of-sight Atmospheric Analysis of Spectral Hypercubes (FLAASH) module was applied to all of the selected scenes. The coastal/aerosol band (band 1 of OLI) and the cirrus band (band 9 of OLI), which include little information under cloudless conditions, were removed and the remaining six bands were used in the experiments. To obtain the reference data (true water fraction of each pixel), fine spatial resolution images (<1 m) were acquired from Google Earth™. The acquisition dates between the Google Earth™ images and the test images could be considered to be closely matched, since the water bodies in urban areas are usually stable over a short time period, so the slight bias of the surface water boundaries between the reference data and test images could be ignored. The "true" boundaries of all the water bodies in the study areas were digitized manually on-screen and used for the accuracy assessment of the water body subpixel mapping. The OLI images were registered to the reference images using a first order polynomial with at least 20 ground control points in each scene, with root-mean-square errors (RMSEs) of 0.23, 0.3, 0.22, and 0.29 pixels, corresponding to the test sites of Beijing, Shanghai, Hangzhou, and Guangzhou, respectively.

METHODOLOGY

The ASWM approach can be divided into three major steps: (1) automated mixed land-water pixel extraction; (2) determination of endmember spectra from neighboring pixels; and (3) linear SMA with the endmember spectra, and the detailed process is shown in the flowchart of the ASWM approach (see Figure 2).

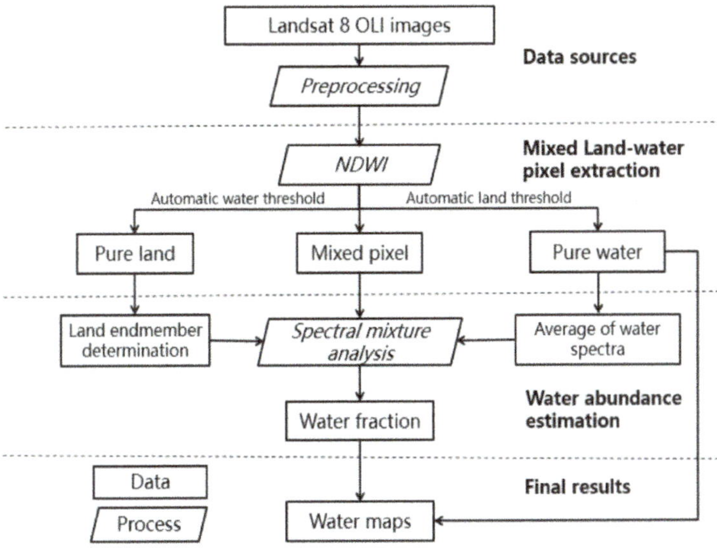

Figure 2: Flowchart of the proposed ASWM method.

Automatic Mixed Land-Water Pixel Extraction

Normalized Difference Water Index (NDWI) Calculation

The NDWI is a quantitative spectral indicator designed for characterizing water bodies [12,13]. It has been widely used for water body extraction through enhancing the water information while suppressing the land information in remote sensing imagery [30]. In this study, the NDWI is used for obtaining the mixed land-water pixels through identifying those pixels which are neither enhanced nor suppressed. A previous study showed that, for Landsat 8 OLI images, the green band and short-wave infrared band 2 (SWIR-2) are the best option to calculate the NDWI [31].

$$NDWI = \frac{\rho_{Green} - \rho_{SWIR2}}{\rho_{Green} + \rho_{SWIR2}}, \tag{1}$$

where ρGreenρGreen and ρSWIR2ρSWIR2 are the reflectance of the respective bands.

Initial Extraction of Water Pixels and Land Pixels

The Otsu algorithm [32] is utilized for the initial extraction of water pixels and land pixels. The Otsu algorithm is used to select an optimal threshold for the separation by calculating the maximum inter-class variance. The fundamental theorem of the Otsu algorithm can be described as follows: the water index image value is assumed to be within the range [a, ..., b], and the water index image can be divided into two parts by a threshold t. The inter-class variance σ^2 can be calculated as follows:

$$\sigma^2 = w_0 (\mu_0 - \mu)^2 + w_1 (\mu_1 - \mu)^2, \tag{2}$$

$$w_0 \cdot \mu_0 + w_1 \cdot \mu_1 = \mu \tag{3}$$

where μ_0, μ_1, and μ are the mean values of the two parts and the whole of the image. w_0 and w_1 are the percentage of each part of the image. From Equations (2) and (3), the variance σ^2 and threshold t can be simply expressed as follows. When σ^2 reaches the maximum value, the corresponding t is regarded as the optimal threshold.

$$t = argmax\left[w_0 \cdot w_1 \cdot (\mu_1 - \mu_0)^2\right] \quad (4)$$

The Ostu method was applied to the histogram of the NDWI image, thereby obtaining the initial water and land pixels.

Automated Mixed Land-Water Pixel Extraction

To automatically extract mixed land-water pixels, the Ostu threshold is used as the start value to search for mixed land-water pixels in the NDWI image histogram between the land threshold and the water threshold (see Figure 3). Locally weighted scatter plot smoothing is first applied to the original histogram curve [33]. The slopes of the histogram curve were then used as the discriminant criteria (see Equation (6)). Considering that mixed pixels are much more easily confused with complex land-cover types than pure water pixels, the slopes of the pure land threshold and pure water threshold are set to 60° and 30°, respectively. The corresponding tangential curvatures for pure land and pure water selection can be calculated as follows:

$$|\tan\alpha|_{land,left} \geqslant 1.732, \ |\tan\alpha|_{water,right} \geqslant 0.5 \quad (5)$$

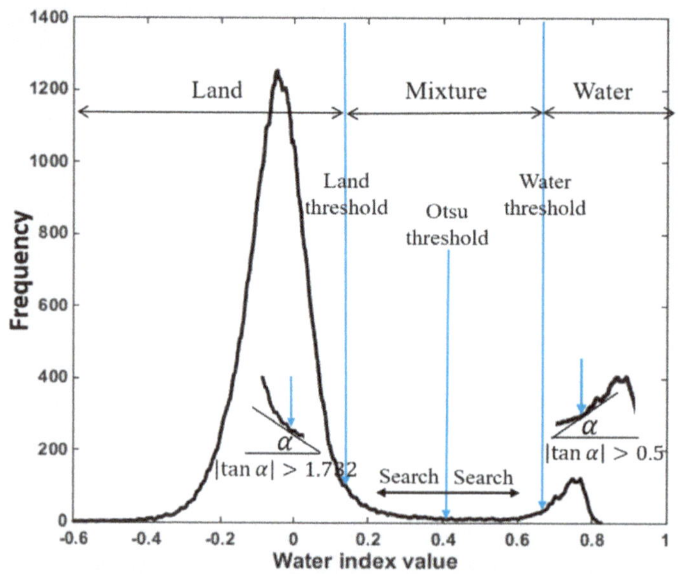

Figure 3: Histogram of the NDWI at the Shanghai test site.

When the curve points of each side first meet the criteria, the corresponding abscissa values are selected as the land and water thresholds.

Misclassification Removal by Simple Post-Processing

Therefore, in this step, two rules are built for misclassification removal as follows:

$$\rho_{blue} - \rho_{green} \leqslant 0, \ \rho_{SWIR1} \leqslant 0.2 \quad (6)$$

where ρ represents the reflectance of the respective bands. The extracted mixed land-water pixels obtained in the previous steps are eliminated if they do not satisfy the above rules. A map of the mixed land-water pixels is then produced (Figure 4).

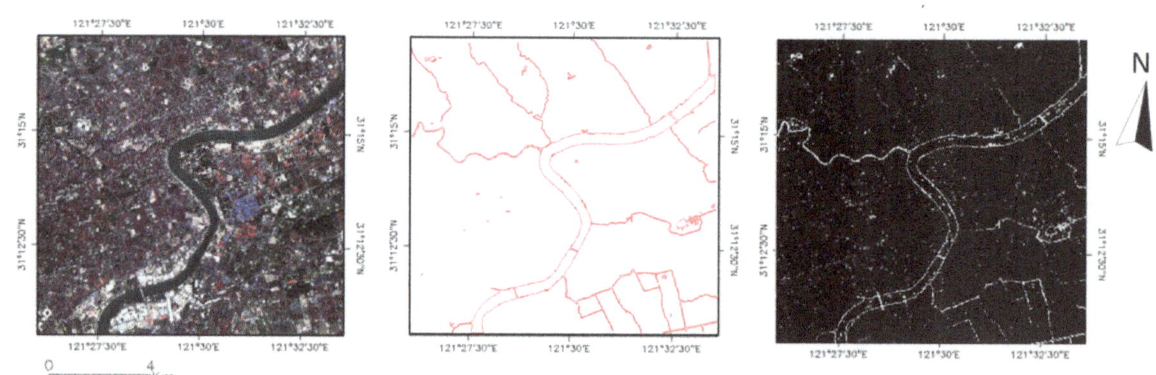

Figure 4. Maps showing the processing of the mixed land-water pixel extraction. (**a**) Landsat OLI true-color composite (RGB: bands 4, 3, 2) image of the Shanghai test site; (**b**) true water boundaries acquired from the high spatial resolution Google Earth™images; and (**c**) mixed land-water pixels derived from the water index.

Endmember Determination by Local Spatial Information

Under the assumption that the endmember signature of a target pixel should be similar to the spatially-adjacent pixels, based on spatial dependence theory, the endmembers of land and water are determined by the neighboring pure land or pure water pixels within a certain distance (see Figure 5). Through experimental comparison, the optimal window size is set to 9 × 9. Based on the previous process described in Section 3.1, the image is divided into three parts: water pixels, land pixels, and mixed land-water pixels. The local window traverses the test image centered at each mixed pixel. The spectrum of the land endmember is then determined by selecting the most representative land pixel by adaptive iteration in the local window, and the spectrum of the water endmember is determined by calculating an average of the water pixels in the local window.

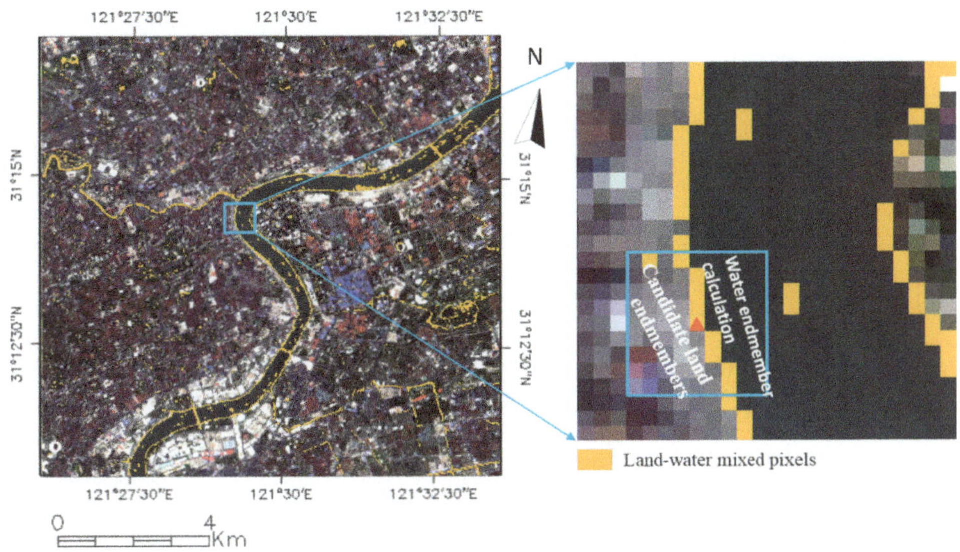

Figure 5: Endmember determination based on adjacent pixels (Landsat OLI true-color composite (RGB: bands 4, 3, 2)).

Linear Unmixing Method

In the linear spectral mixture analysis (LSMA) model, the spectrum of a mixed pixel is assumed to be a linear combination of the endmembers' spectra weighted by the area coverage of each endmember within the pixel. In this study, the relationship between the reflectivity of a mixed pixel and its endmembers is also considered to be linear. The constraints were imposed that the sum of the endmembers' abundances should be equal to one and the range of the endmember abundance should be limited to between 0 and 1. The unmixing equations can then be defined as follows:

$$\rho_m = \rho_l \times f_l + \rho_w \times f_w + \sigma, \tag{7}$$

$$f_l + f_w = 1,\ 0 \leqslant f_l, f_w \leqslant 1 \tag{8}$$

where ρ_l, ρ_w, and ρ_m are the reflectance values of pure land, pure water, and mixed pixels, respectively. f_l and f_w are the respective abundances of the land and water endmembers, and σ is the residual error. According to Equations (7) and (8), the water abundance can be derived as follows:

$$f_w = \frac{\rho_m - \rho_l - \sigma}{\rho_w - \rho_l}, \tag{9}$$

subject to:

$$\left\|\sigma\right\|_1 = \text{min and min} \leqslant E\left(\left\|\sigma\right\|_1\right) + 3 \times \sqrt{D\left(\left\|\sigma\right\|_1\right)}, \tag{10}$$

In Equation (9), ρ_m is the known spectrum of the mixed land-water pixel; ρ_w is the average spectrum of the water pixels in the local window; and ρ_l is determined by selecting the optimal candidate land pixel in the local window, with the constraint of the minimal L1 norm $||\sigma||_1$. Meanwhile, the minimal L1 norm $||\sigma||_1$ should meet the requirement of being less than a threshold, which means that the selected optimal endmembers are representative of the true fractions in the mixed pixel. In this study, the threshold is automatically set according to the statistics of the $||\sigma||_1$ sets, and $E(||\sigma||_1)$ and $\sqrt{D\left(\left\|\sigma\right\|_1\right)}$ in Equation (10) represent the average value and mean-square error of the $||\sigma||_1$ sets, respectively.

Accuracy Assessment

A quantitative accuracy evaluation is often difficult to implement for the subpixel mapping of surface features because of the difficulty of obtaining a reference map and the lack of quantitative accuracy evaluation indexes. In this study, the water boundaries were extracted from high-precision Google Earth™ images and applied for the reference water fraction map generation, using a 10 × 10 grids for each pixel for the test images (see Figure 5). As for reducing the impact of the geometric error between the Landsat OLI images and the reference images, a 90 × 90 m² sampling size (3 × 3 pixels size for the OLI image) was adopted for both the reference water maps and the estimated water maps [34,35,36,37].

Spectral unmixing is considered to be a soft classification method, and its output result is the percentage of ground features in each pixel. According to previous research, three types of error measurements are widely used in subpixel classification [34,35,38,39], i.e., RMSE, mean absolute error (MAE), and systematic error (SE). The RMSE and MAE are both used to quantify the relative error of the estimated surface abundance at the pixel level, and the SE indicates the overall trend of the over- or under-estimation. However, all these accuracy evaluation indexes have their limits and cannot comprehensively reflect subpixel classification accuracy. Therefore, the producer's accuracy, user's accuracy, and Kappa coefficient expressions were modified to fit with subpixel classification evaluation. The five accuracy measurements can be calculated as follows:

$$Pro.acc = \sum_{i=1}^{N} Min\{f_i(ref), f_i(esti)\} / \sum_{i=1}^{N} f_i(ref) \tag{11}$$

$$Use.acc = \sum_{i=1}^{M} Min\{f_i(ref), f_i(esti)\} / \sum_{i=1}^{M} f_i(esti) \tag{12}$$

$$\begin{cases} Kappa = (Po - Pc)/(1 - Pc) \\ Pc = \sum_{i=1}^{nm} f_i(esti) \times \sum_{i=1}^{nm} f_i(ref) + \sum_{i=1}^{nm} (1 - f_i(esti)) \times \sum_{i=1}^{nm} (1 - f_i(ref)) / (nm \times nm) \\ Po = \sum_{i=1}^{nm} (Min\{f_i(ref), f_i(esti)\} + 1 - Max\{f_i(ref), f_i(esti)\}) / nm \end{cases}, \tag{13}$$

$$RMSE = \sqrt{\frac{1}{nm} \sum_{i=1}^{nm} (f_i(esti) - f_i(ref))^2} \tag{14}$$

$$SE = \frac{1}{nm} \sum_{i=1}^{nm} (f_i(esti) - f_i(ref)) \tag{15}$$

where Pro.acc and Ues.acc represent the producer's accuracy and user's accuracy, respectively. $f_i(ref)$ and $f_i(esti)$ are the water abundance of the pixels of the reference image and estimated image, respectively, and i is the location of each pixel. N and M are the number of mixed land-water pixels in the reference image and estimated image, respectively, and nm is the number of pixels in the whole image.

RESULTS

Water Maps from the Different Methods

To compare the accuracy of ASWM with other methods, both the MNDWI water index method (selecting the optimal water thresholds manually according to the reference image) and a target detection method of constrained energy minimization (CEM) were applied for water extraction at the pixel level [13]. Furthermore, two SMA methods were also adopted for water extraction at the subpixel level in comparison with the proposed method. The first SMA method was the traditional fully-constrained least squares (FCLS) unmixing method [40]. When implementing this method, a pixel purity index (PPI) was first applied to the test images for pure ground feature spectra collection by referencing the high-resolution Google Earth™ images. The endmember spectra (including vegetation, impervious, soil, and water) were then acquired by averaging the spectra of each ground feature, and finally applied to FCLS for water fraction estimation. The second subpixel method was multiple endmember spectral mixture analysis (MESMA) [41,42]. In this study, with the PPI image computation and Google Earth™ reference images, four endmember libraries of vegetation, impervious, soil, and water were built. Specifically, as the area of soil in urban areas is not usually large, and the spectral characteristics of soil and impervious surfaces are similar in some cases [34], the spectral libraries of soil and impervious were combined into one endmember library, and the corresponding endmember was called the impervious (soil) endmember. Meanwhile, considering that a considerable amount of shade exists in urban areas, a photometric shade endmember with a uniform reflectance of zero in all bands was introduced to account for the variation in illumination [43]. Then, with the endmember libraries of water, vegetation, impervious (soil), and photometric shade, hundreds of four-endmember models for each test image were built for the MESMA application.

In the first column of Figure 6, the reference maps (Figure 6a,d,g,j) were shown. In the second column of Figure 6, the water maps by the best-performing contrast method are displayed, i.e., the water mapping results of MNDWI for study area 1—Beijing (Figure 6b), MESMA for study area 2—Shanghai (Figure 6e), FCLS for study area 3—Hangzhou (Figure 6h), and CEM for study area 4—Guangzhou (Figure 6k). In the last column of Figure 6, the water mapping results of the ASWM method for the different study areas (Figure 6c,f,i,l) were shown. From a visual inspection, it can be seen that MNDWI and CEM (Figure 6b,k) omit the small water bodies in most cases, while FCLS and MESMA (Figure 6e,h) often misclassify land surface as water. The visual inspection of the surface water maps derived from ASWM suggests that ASWM is able to extract urban water maps with a higher degree of accuracy.

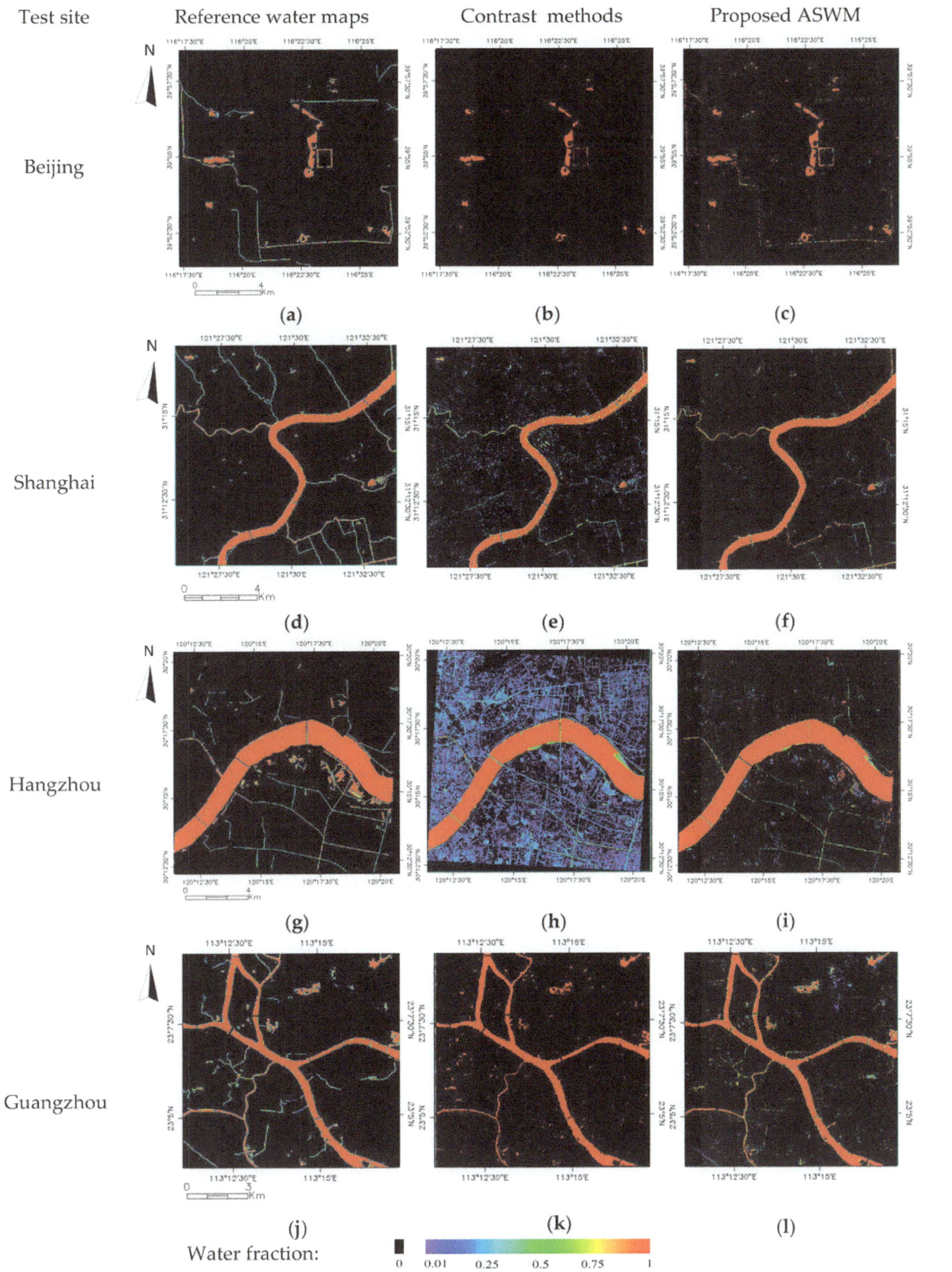

Figure 6: Comparison of the water mapping results. First row: reference map, water map from the contrast method of MNDWI, and result of the proposed method of Beijing corresponding to (**a–c**); Second row, reference map, water map from the contrast method of MESMA, and result of the proposed method of Shanghai corresponding to (**d–f**); Third row, reference map, water map from the contrast method of FCLS, and result of the proposed method of Hangzhou corresponding to (**g–i**); Fourth row, reference map, water map from the contrast method of CEM, and result of the proposed method of Guangzhou corresponding to (**j–l**).

Accuracy Assessment Results

A scatterplot was first used to examine the correlation between the reference water fractions and the estimated water fractions obtained by the proposed ASWM. As shown in Figure 7, the regression models all have a near 1:1 relationship, supported by their slopes of 0.97, 1.02, 1.04, and 0.98, R-squared values of 0.9454, 0.9486, 0.9665, and 0.9607, and intercepts of −0.01, −0.06, −0.07, and −0.01, for Beijing, Shanghai, Hangzhou, and Guangzhou, respectively.

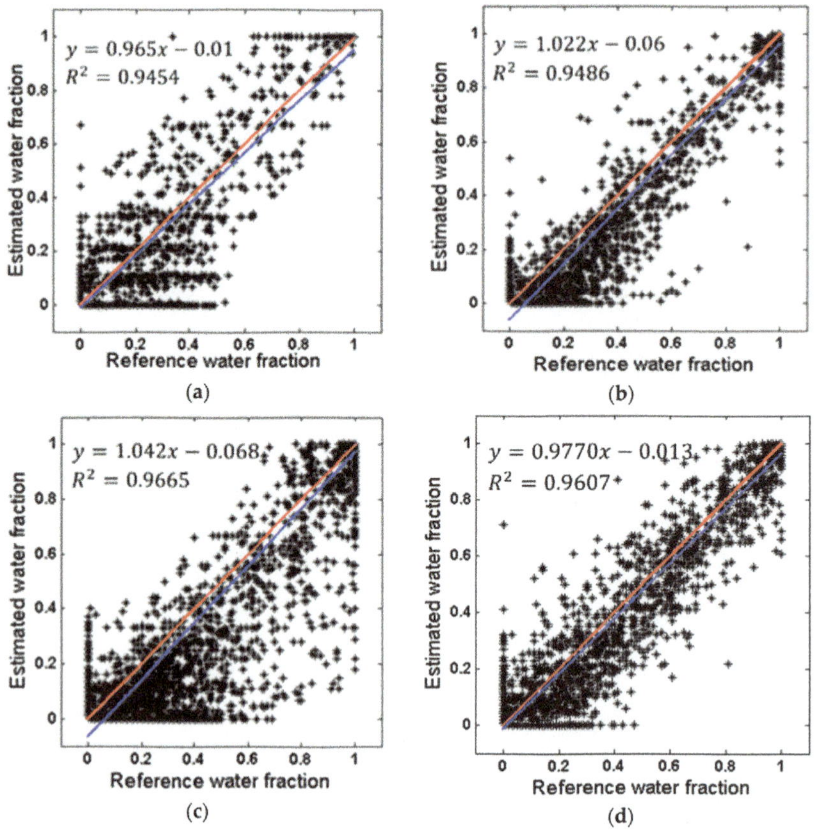

Figure 7: Comparison between the reference and estimated water fractions obtained by ASWM, where the red line is the 1:1 reference line, and the blue line is the regression line. **(a–d)** represent the results for the study areas of Beijing, Shanghai, Hangzhou, and Guangzhou, respectively.

Five different accuracy assessment approaches (see Section 3.4) are used to compare the water mapping accuracies: user's accuracy, producer's accuracy, kappa coefficient, RMSE, and SE. The quantitative subpixel accuracy assessment results are summarized in Table 2, where it is clear that the proposed ASWM method significantly outperforms the other four related methods. A close look at the conditional kappa coefficient for each test site reveals that the water maps derived from the proposed method have the highest consistency with the reference water map. The water maps of the four methods are the least accurate in the Beijing test site. For the other test sites, only ASWM has the ability to achieve producer's accuracy and user's accuracy values of higher than 84%. The accuracy assessment indicates that ASWM is able to provide an adequate estimation of the surface water fraction, with an average RMSE of 0.055 and an average SE of −0.007. Negative SE values are observed for all the study areas, suggesting that the overall water abundance is under-estimated.

Table 2: Summary of the classification accuracy of the four classifiers by study area

Study Area	Kappa					Producer's Accuracy				
	ASWM	MNDWI	FCLS	MESMA	CEM	ASWM	MNDWI	FCLS	MESMA	CEM
Beijing	0.739	0.610	0.279	0.088	0.211	70.2%	45.8%	87.3%	88.2%	45.22%
Shanghai	0.899	0.842	0.548	0.837	0.442	84.8%	74.3%	88.3%	77.4%	788.0%
Hangzhou	0.900	0.890	0.754	0.855	0.881	88.7%	82.7%	89.1%	49.8%	84.3%
Guangzhou	0.910	0.817	0.729	0.739	0.797	87.5%	71.4%	91.4%	85.9%	79.0%
	User's accuracy					RMSE				
	ASWM	MNDWI	FCLS	MESMA	CEM	ASWM	MNDWI	FCLS	MESMA	CEM
Beijing	79.2%	93.7%	18.4%	0.7%	16.0%	0.046	0.059	0.110	0.209	0.157
Shanghai	97.2%	99.6%	54.8%	93.6%	36.1%	0.045	0.067	0.085	0.058	0.078
Hangzhou	93.3%	99.5%	71.2%	97.5%	94.4%	0.050	0.098	0.110	0.189	0.065
Guangzhou	97.3%	99.2%	64.6%	69.0%	85.9%	0.080	0.085	0.093	0.084	0.151
	SE									
	ASWM	MNDWI	FCLS	MESMA	CEM					
Beijing	−0.002	−0.011	0.079	0.186	0.039					
Shanghai	−0.009	−0.017	0.041	−0.012	0.078					
Hangzhou	−0.004	−0.023	0.034	−0.100	−0.010					
Guangzhou	−0.014	−0.025	0.037	0.022	−0.011					

DISCUSSION

Mixed Land-Water Pixel Extraction

This study has explored the extraction of mixed land-water pixels by the use of a water index. Compared to the traditional FCLS and MESMA approaches, the proposed ASWM unmixing algorithm applied only to mixed pixels achieved an improvement both in accuracy and efficiency in the abundance estimation in the experiments.

According to the histogram of the water index image, the mixed land-water pixels always locate between the pure land and pure water thresholds. Therefore, the histogram slopes were used to find these mixed pixels. In order to ensure that the mixed pixels are correctly extracted from the remote sensing images, the pure land and pure water thresholds should locate at the furthest left and right of the bottom of the water index histogram. Accounting for the fact that mixed pixels are mainly confused with land pixels, in the experiments, the thresholds of the slopes of pure water and pure land were set to 30° and 60°, respectively (Figure 3), and the corresponding thresholds of the absolute value of the tangential curvature were 0.5 and 1.732. The following unmixing process also has the effect of removing the wrongly extracted pixels by estimating the water abundances which approach 0 for the wrongly-extracted land pixels, and 100% for the wrongly-extracted water pixels. Several experiments were implemented using images from urban areas of China, and the corresponding kappa coefficient range from 0.8825–0.8994 showed that the final water abundance map result only varied a little when the slope thresholds were set to 30° and 60° for both pure water and pure land extraction, respectively.

In the mixed land-water pixel extraction process, some post-processing is implemented to improve the accuracy of the extraction. In this study, the difference between the green and blue bands was used to eliminate dark building pixels (Equation (6)). This approach works for two main reasons: (1) as a result of eutrophication and human disturbance in urban areas, surface water always contains higher amounts of chlorophyll and suspended sediment. In addition, with the strong spatial dependence between water and vegetation, the vegetation may be the key component of the land-water pixels. Taken together, these factors mean that the green band reflectance of mixed land-water pixels is higher than the blue band [44,45,46]; and (2) most of the land covers that show higher reflectance in the green band than the blue band possess similar characteristics to water bodies; however, dark buildings present opposite characteristics. Above all, the green band and the blue band can be used for separation of mixed land-water pixels and dark buildings. However, it is noticeable that this approach ignores the situation of the mixed land-water pixels being made up of water and dark buildings. Therefore, to retain all of the mixed land-water pixels in the elimination of dark building pixels, although the threshold of Equation (6) was set to zero in this study, the threshold can also be set a little higher than zero to prevent the mixed land-water pixels being removed. In summary, the post-processing aims to remove those land pixels that are clearly distinct from the mixed land-water pixels.

Endmember Selection

The key to successful SMA is appropriate endmember selection [24,47]. Accounting for spectral variability when selecting endmembers has been pointed out to be an important factor that can severely affect the accuracy of subpixel land-cover fractions [22,35,36,48]. Traditional SMA uses an invariant set of endmembers and it is quite limited when mapping heterogeneous urban surfaces [49,50,51]. MESMA takes the spectral heterogeneity in urban environments into account, and this method divides the image into several classes, where the classes are subdivided into sub-classes as much as possible. The experimental results showed that MESMA achieved better results than the traditional SMA, but consumed more time.

Above all, two issues should be stressed here. Firstly, the endmember spectrum acquisition of the two methods depends on the collection of pure pixels. However, especially for MESMA, it is impossible to obtain all the sub-classes of endmembers through pure pixels in the image. Secondly, to solve the endmember variability problem, researchers need to acquire all of the endmembers of materials in the whole image and consider the compositions by different endmembers of material as much as possible for every pixel; this is hard work for most cases. Actually, in some cases, the "purest" endmembers are not necessarily representative of the fractions within the pixel, and the representative spectra may not necessarily be selected as "pure" endmembers [51]. Therefore, how to select the representative endmembers becomes the most important issue. In ASWM, the whole image is divided into two classes (water and land), and then the most representative endmembers are selected by adaptive iteration in the neighboring pixels. The endmember spectra of water and land are then determined by the adjacent pixels of each mixed pixel. Frankly speaking, this technique for the selection of the most representative endmembers may not be precise enough, and error between the selected land endmember and the true land endmember can exist and can seriously affect the final result. Even so, in this study, the proposed ASWM method still achieved the highest subpixel water extraction accuracy with the lowest RMSE and SE when compared with the other methods.

Error Analysis

ASWM achieved the highest water mapping accuracy when compared with the other methods; however, errors did exist in the final water maps. In the experiments, the Beijing test area resulted in the lowest water mapping accuracy, with a kappa coefficient of 0.7392, followed by the Shanghai test area, with a kappa coefficient of 0.8994. Through a comprehensive analysis of the error statistics, we believe that the error source comes from three aspects: (1) all four test images were acquired in urban areas with complex ground surface conditions, especially for the Beijing and Shanghai test areas, which include many dark buildings and building shadows, which are easily confused with water bodies due to the high similarity in the spectral features [15]; (2) the technique of the selection of the most representative endmembers, accounting for land surface similarity in an adjacent space, is proposed. However, this may not agree with the spectra of the fractions in the mixed pixels, and could result in large residuals in the unmixing process; and (3) the linear unmixing method was applied for the water abundance estimation in the study;

however, the measured signal of the sensor always results from the interactions of electromagnetic radiation with the multiple constituents within each pixel [52]. The multiple scattering between at least two materials in the field of view can make the spectra of a mixed pixel and its fractions not simply linear, so errors will inevitably occur with the application of a linear method.

As stated above, to verify the generalization ability of the new proposed method, four urban areas in the diverse complex urban environments were selected for testing. All of the experiments show the best performance of the proposed method, however, as the proposed method utilized some empirical knowledge for the mixed land-water pixel extraction, further tests are still necessary. Future studies will need to improve the details based on the general processing framework proposed in this study, and the detailed improvements should be able to achieve an even higher accuracy in water extraction.

CONCLUSIONS

For urban and environmental applications at regional/national scales, urban surface water information should be extracted accurately and not be limited to the pixel level. In this paper, a novel method was presented for urban water mapping with a subpixel precision. In the proposed method, mixed land-water pixels was first exacted from the remote sensing images based on a water index. A linear unmixing method is then applied to each mixed pixel, and the most representative endmembers are selected using the adjacent spatial information. As all the fixed parameters in the process are set in advance, the new ASWM method can also be regarded as a fully automatic approach. Analysis of the results allows a number of conclusions to be made. Firstly, the mixed land-water pixel extraction before unmixing can help to both improve the unmixing accuracy and the computational efficiency. Secondly, with the adjacent spatial similarity, the endmember signatures derived from the adjacent ground surfaces provide a new means of approximating the most representative endmember signatures. Finally, the results of both the water maps and quantitative accuracy assessment indicate that the proposed ASWM method performs reasonably well in the mapping of surface water, with a relatively high subpixel precision.

The future research directions will include reference water map acquisition and accurate registration between reference and test image to better assess the accuracy of the estimated water abundance. In addition, as the confusion between water and other urban materials is an unsolved problem in water extraction, the distinction between surface water and other surface features requires further study. Finally, the method developed in this study has extended the unmixing technique into water mapping in a complex urban environment. Its core idea could also be extended to the extraction of other land covers at the subpixel level. However, there is certainly room for improvement. For instance, the selection of the most representative endmembers by integrating spectral and spatial information, and the consideration of nonlinear factors that influence water abundance estimation should be able to contribute to the achievement of an even higher precision.

ACKNOWLEDGMENTS

The work described in the paper was substantially supported by the Shanghai Rising-Star Program (Project No. 15QA1403700), the National Natural Science Foundation of China (Project Nos. 41201426, 41325005, 41571407 and 41611130113), the National Basic Research Program of China—973 Program (Project No. 2012CB957701), and the Fundamental Research Funds for the Central Universities.

REFERENCES

1. Proud, S.R.; Fensholt, R.; Rasmussen, L.V.; Sandholt, I. Rapid response flood detection using the MSG geostationary satellite. IEEE J. Sel. Top. Appl. Earth Obs. Remote Sens. 2011, 13, 536–544.
2. Feng, L.; Hu, C.M.; Chen, X.L.; Li, R.F.; Tian, L.Q.; Murch, B. MODIS observations of the bottom topography and its inter-annual variability of Poyang Lake. Remote Sens. Environ. 2011, 115, 2729–2741.
3. Prigent, C.; Papa, F.; Aires, F.; Jiménez, C.; Rossow, W.B.; Matthews, E. Changes in land surface water dynamics since the 1990s and relation to population pressure. Geophys. Res. Lett. 2012, 39, 85–93.
4. Güttler, F.N.; Niculescu, S.; Gohin, F. Turbidity retrieval and monitoring of Danube Delta waters using multi-sensor optical remote sensing data: An integrated view from the delta plain lakes to the western–northwestern Black Sea coastal zone. Remote Sens. Environ. 2013, 132, 86–101.
5. Tong, X.; Xie, H.; Qiu, Y.; Zhang, H.; Song, L.; Zhang, Y.; Zhao, J. Quantitative monitoring of inland water using remote sensing: An example of upper reaches of Huangpu River, China. Int. J. Remote Sens. 2010, 31, 2471–2492.
6. Ekercin, S. Coastline change assessment at the Aegean Sea Coasts in Turkey using multitemporal Landsat imagery. J. Coast. Res. 2009, 23, 691–698.
7. Pardo-Pascual, J.E.; Almonacid-Caballer, J.; Ruiz, L.A.; Palomar-Vazquez, J. Automatic extraction of shorelines from Landsat TM and ETM multi-temporal images with subpixel precision. Remote Sens. Environ. 2012, 123, 1–11.
8. Hung, M.C.; Wu, Y.H. Mapping and visualizing the Great Salt Lake landscape dynamics using multi-temporal satellite images, 1972–1996. Int. J. Remote Sens. 2005, 26, 1815–1834.
9. Tulbure, M.G.; Broich, M. Spatiotemporal dynamic of surface water bodies using Landsat time-series data from 1999 to 2011. ISPRS J. Photogramm. 2013, 79, 44–52.
10. Bryant, R.G.; Rainey, M.P. Investigation of flood inundation on playas within the Zone of Chotts, using a time-series of AVHRR. Remote Sens. Environ. 2002, 82, 360–375.
11. Jain, S.K.; Singh, R.D.; Jain, M.K.; Lohani, A.K. Delineation of flood-prone areas using remote sensing technique. Water Resour. Manag. 2005, 19, 337–347.
12. McFeeters, S.K. The use of normalized difference water index (NDWI) in the delineation of open water features. Int. J. Remote Sens. 1996, 17, 1425–1432.

13. Xu, H. Modification of normalised difference water index (NDWI) to enhance open water features in remotely sensed imagery. Int. J. Remote Sens. 2006, 27, 3025–3033.

14. Feyisa, G.L.; Meilby, H.; Fensholt, R.; Proud, S.R. Automated Water Extraction Index: A new technique for surface water mapping using Landsat imagery. Remote Sens. Environ. 2014, 140, 23–35.

15. Xie, H.; Luo, X.; Xu, X.; Tong, X.; Jin, Y.; Pan, H.; Tong, X. New hyperspectral difference water index for the extraction of urban water bodies by the use of airborne hyperspectral images. J. Appl. Remote Sen. 2014, 8, 5230–5237.

16. Yang, Y.; Liu, Y.; Zhou, M.; Zhang, S.; Zhan, W.; Sun, C.; Duan, Y. Landsat 8 OLI image based terrestrial water extraction from heterogeneous backgrounds using a reflectance homogenization approach. Remote Sens. Environ. 2015, 171, 14–32.

17. Ghrefat, H.A.; Goodell, P.C. Land cover mapping at Alkali Flat and Lake Lucero, White Sands, New Mexico, USA using multi-temporal and multi-spectral remote sensing data. Int. J. Appl. Earth Obs. 2011, 13, 616–625.

18. Guo, Y.L.; Li, Y.M.; Zhu, L.; Liu, G.; Wang, S.; Du, C.G. An Improved Unmixing-Based Fusion Method: Potential application to remote monitoring of inland waters. Remote Sens. 2015, 7, 1640–1666.

19. Hommersom, A.; Wernand, M.R.; Peters, S.; Eleveld, M.A.; van der Woerd, H.J.; de Boer, J. Spectra of a shallow sea-unmixing for class identification and monitoring of coastal waters. Ocean. Dyn. 2011, 61, 463–480.

20. Wang, Q.; Atkinson, P.M.; Shi, W. Fast subpixel mapping algorithms for subpixel resolution change detection. IEEE Trans. Geosci. Remote Sens. 2015, 53, 1692–1706.

21. Wang, Q.; Shi, W.; Atkinson, P.M.; Li, Z. Land cover change detection at subpixel resolution with a Hopfield neural network. IEEE J. Sel. Top. Appl. Earth Obs. Remote Sens. 2015, 8, 1339–1352.

22. Shi, C.; Wang, L. Incorporating spatial information in spectral unmixing: A review. Remote Sens. Environ. 2014, 149, 70–87.

23. Li, H.; Zhang, L. A hybrid automatic endmember extraction algorithm based on a local window. IEEE Trans. Geosci. Remote Sens. 2011, 49, 4223–4238.

24. Somers, B.; Asner, G.P.; Tits, L.; Coppin, P. Endmember variability in Spectral Mixture Analysis: A review. Remote Sens. Environ. 2011, 115, 1603–1616.

25. Zare, A.; Gader, P.; Bchir, O.; Frigui, H. Piecewise convex multiple-model endmember detection and spectral unmixing. IEEE Trans. Geosci. Remote Sens. 2013, 51, 2853–2862.

26. Sun, D.; Yu, Y.; Goldberg, M.D. Deriving water fraction and flood maps from MODIS images using a decision tree approach. IEEE J. Sel. Top. Appl. Earth Obs. Remote Sens. 2011, 4, 814–825.

27. Ma, B.; Wu, L.; Zhang, X.; Li, X.; Liu, Y.; Wang, S. Locally adaptive unmixing method for lake-water area extraction based on MODIS 250 m bands. IEEE J. Sel. Top. Appl. Earth Obs. Remote Sens. 2014, 33, 109–118.

28. Sethre, P.R.; Rundquist, B.C.; Todhunter, P.E. Remote detection of Prairie Pothole Ponds in the Devils Lake Basin, North Dakota. GISci. Remote Sens. 2005, 42, 277–296.

29. Applied Analysis Inc. Subpixel Classifier for IMAGINE: User's Guide; Applied Analysis Inc.: Billerica, MA, USA, 1997.

30. Ji, L.; Zhang, L.; Wylie, B. Analysis of dynamic thresholds for the normalized difference water index. Photogramm. Eng. Remote Sens. 2009, 75, 1307–1317.

31. Xie, H.; Luo, X.; Xu, X.; Pan, H.; Tong, X. Evaluation of Landsat 8 OLI imagery for unsupervised inland water extraction. Int. J. Remote Sens. 2016, 37, 1826–1844.

32. Otsu, N. A threshold selection method from gray-level histograms. IEEE Trans. Syst. Man Cybern. 1979, 9, 62–66.

33. Cleveland, W.S. Robust locally weighted regression and smoothing scatterplots. J. Am. Stat. Assoc. 1979, 74, 829–836.

34. Deng, C.; Wu, C. A spatially adaptive spectral mixture analysis for mapping subpixel urban impervious surface distribution. Remote Sens. Environ. 2013, 133, 62–70.

35. Powell, R.L.; Roberts, D.A.; Dennison, P.E.; Hess, L.L. Sub-pixel mapping of urban land cover using multiple endmember spectral mixture analysis: Manaus, Brazil. Remote Sens. Environ. 2007, 106, 253–267.

36. Roberts, D.A.; Quattrochi, D.A.; Hulley, G.C.; Hook, S.J.; Green, R.O. Synergies between VSWIR and TIR data for the urban environment: An evaluation of the potential for the Hyperspectral Infrared Imager (HyspIRI) Decadal Survey mission. Remote Sens. Environ. 2012, 117, 83–101.

37. Wu, C. Normalized spectral mixture analysis for monitoring urban composition using ETM+ imagery. Remote Sens. Environ. 2004, 93, 480–492.

38. Hu, X.; Weng, Q. Estimating impervious surfaces from medium spatial resolution imagery using the self-organizing map and multi-layer perceptron neural networks. Remote Sens. Environ. 2009, 113, 2089–2102.

39. Wu, C.; Murray, A.T. Estimating impervious surface distribution by spectral mixture analysis. Remote Sens. Environ. 2003, 84, 493–505.

40. Heinz, D.C.; Chang, C.I. Fully constrained least squares linear spectral mixture analysis method for material quantification in hyperspectral imagery. IEEE Trans. Geosci. Remote Sens. 2001, 39, 529–545.

41. Roberts, D.A.; Gardner, M.; Church, R.; Ustin, S.; Scheer, G.; Green, R.O. Mapping Chaparral in the Santa Monica Mountains using multiple endmember spectral mixture models. Remote Sens. Environ. 1998, 65, 267–279.

42. Franke, J.; Roberts, D.A.; Halligan, K.; Menz, G. Hierarchical Multiple Endmember Spectral Mixture Analysis (MESMA) of hyperspectral imagery for urban environments. Remote Sens. Environ. 2009, 113, 1712–1723.

43. Dennison, P.E.; Roberts, D.A. Endmember selection for multiple endmember spectral mixture analysis using endmember average RMSE. Remote Sens. Environ. 2003, 87, 123–135.

44. Gitelson, A.A.; Gao, B.C.; Li, R.R.; Berdnikov, S.; Saprygin, V. Estimation of chlorophyll-a concentration in productive turbid waters using a Hyperspectral Imager for the Coastal Ocean—The Azov Sea case study. Environ. Res. Lett. 2011, 6, 024023.

45. Song, K.; Li, L.; Tedesco, L.P.; Li, S.; Duan, H.; Liu, D.; Zhao, Y. Remote estimation of chlorophyll-a in turbid inland waters: Three-band model versus GA-PLS model. Remote Sens. Environ. 2013, 136, 342–357.

46. Mueller, N.; Lewis, A.; Roberts, D.; Ring, S.; Melrose, R.; Sixsmith, J.; Ip, A. Water observations from space: Mapping surface water from 25 years of Landsat imagery across Australia. Remote Sens. Environ. 2016, 174, 341–352.

47. Elmore, A.J.; Mustard, J.F.; Manning, S.J.; Lobell, D.B. Quantifying vegetation change in semiarid environments: Precision and accuracy of spectral mixture analysis and the normalized difference vegetation index. Remote Sens. Environ. 2000, 73, 87–102.

48. Weng, Q. Remote sensing of impervious surfaces in the urban areas: Requirements, methods, and trends. Remote Sens. Environ. 2012, 117, 34–49.

49. Liu, T.; Yang, X. Mapping vegetation in an urban area with stratified classification and multiple endmember spectral mixture analysis. Remote Sens. Environ. 2013, 133, 251–264.

50. Lu, D.; Weng, Q. Spectral mixture analysis of the urban landscape in indianapolis with Landsat ETM+ imagery. Photogramm. Eng. Remote Sens. 2004, 70, 1053–1062.

51. Song, C. Spectral mixture analysis for subpixel vegetation fractions in the urban environment: How to incorporate endmember variability? Remote Sens. Environ. 2005, 95, 248–263.

52. Keshava, N.; Mustard, J.F. Spectral unmixing. Proc. IEEE 2002, 19, 44–57.

CHAPTER 9

OPERATIONAL SURFACE WATER DETECTION AND MONITORING USING RADARSAT 2

Sandra Bolanos, Doug Stiff, Brian Brisco and Alain Pietroniro

Keywords: Hydrological modeling, water mapping, wetlands, flooding, SAR, Radarsat-2, RCM, potholes

INTRODUCTION

The National Hydrological Service (NHS) of Environment and Climate Change Canada is the national agency responsible for the collection, interpretation and dissemination of standardized data on water resources and information in Canada. Currently NHS's primary mechanism to collect and process data is by means of approximately 2100 active gauging stations that are employed to measure lake level or river stage used to calculate water discharge. This system has been in place for over 100 years, and although valuable, provides only a limited picture of water resources within Canada. As a part of the Government of Canada's strategy to leverage earth observation technology to further broaden the amount and value of environmental data and products [1], the NHS is working with the Canadian Space Agency on the development of technology and methods based on satellite imagery for water monitoring [2].

Water monitoring using Synthetic Aperture Radar (SAR) has been the object of study for many researchers [3,4,5,6,7,8,9,10,11,12,13,14,15,16,17,18,19]. Due to its all-weather capabilities, and its image acquisition capacity during day or night or in cloudy conditions, SAR imagery offers better alternatives for water mapping than optical imagery [5]. As a result of the radar's unique response to water, water mapping using intensity thresholding methods on SAR image had been extensively used [4,5,6,7,20,21]. In this method, water is separated from land in intensity images, but the accuracy of the results relies on the ability to differentiate land *vs.* water pixels in the intensity domain, which becomes especially difficult when the water backscatter is affected by wind-induced roughness. Since the values of backscatter vary depending on the incidence angle, image quality and wind-induced roughness, the threshold needs to be modified on a scene per scene basis [4,5,6]. In some other instances, image thresholding is combined with texture information [7] or region-growing segmentation algorithms used to determine the water pixels [14,16,21,22]. Segmentation algorithms are quite successful but computationally expensive. In many cases, statistical information is combined with digital elevation models [8,14,16], where pixels likely to contain water are first determined based on topographic data and the probability of water is based on histograms of water *vs.* land pixels in the image, but this requires a coarse water mask to determine the statistics of water pixels. This dependence on a pre-determined water mask and high-resolution digital elevation models makes these methods unusable for mapping small ephemeral water bodies. More recently, active contour algorithms, also called snakes [23], have been applied with some success for water mapping from radar images [17,22]. These algorithms, however, rely on ancillary data to determine candidate pixels for water as well as on morphological operators, which results in longer processing times.

This paper presents a novel threshold-based algorithm for automated water mapping based on Radarsat-2, which does not require ancillary data or *a priori* knowledge of the response of water bodies in each scene. It has been developed in such way that it will be transferable to the Radarsat Constellation

Citation: Sandra Bolanos, Doug Stiff, Brian Brisco and Alain Pietroniro, Operational Surface Water Detection and Monitoring Using Radarsat 2, doi:10.3390/rs8040285.

Mission (RCM) upon launch. The RCM will consist of three satellites that will provide daily worldwide coverage in lower resolution modes. The constellation will have a revisit period of 4 days, nominal resolution going from 3 m in very high resolution mode to 100 m in ScanSAR mode, and swaths ranging from 20 to 500 km. The system will provide imagery in single, dual and compact polarimetry modes and is expected to deliver 300,000 scenes per year [24]. RCM is owned and will be operated by the Government of Canada and will provide Earth Observation radar data continuity for Canada to fulfill the government priorities in maritime surveillance, disaster and natural resources management [25]. A combination of a thresholding technique with a texture indicator has been applied in a fully automated way on 82 Radarsat-2 images over the Canadian Prairies, to produce polygons delineating ephemeral water bodies (potholes) as well as permanent open water, which will be integrated into the Canadian Land Data Assimilation System (CaLDAS) [26], to improve numerical weather prediction [27]. The results were evaluated using nearly synchronous high resolution optical imagery for three areas of interest in the study site.

MATERIALS

Study Area and Data

Two study areas in the Canadian Prairies were established based on the temporal and spatial variability of open water in this region. The region has been affected by severe flooding after snowmelt, which will benefit from better land surface information for numerical weather prediction. The first study area is 494 km north to south stretching from Edmonton, Alberta, to the US/Canada border and is approximately 265 km wide. The second study area, over Saskatoon, Saskatchewan is approximately 150 km wide by 200 km north to south (Figure 1). Both areas are characterized by low relief, with elevations varying from 502 to 940 meters above sea level. Normal summer temperatures range from 12 to 27 degrees Celsius with winter temperatures remaining well below freezing. This region has the highest number of sunny days in Canada (2300 h of sunshine per year), an average annual precipitation of 450 mm and is primarily covered by grasslands and cropland. This area consists of numerous small potholes (also called sloughs) and is considered one of the largest wetland complexes in the world [28], supporting around 60% of the North American breeding bird populations of several species [29].

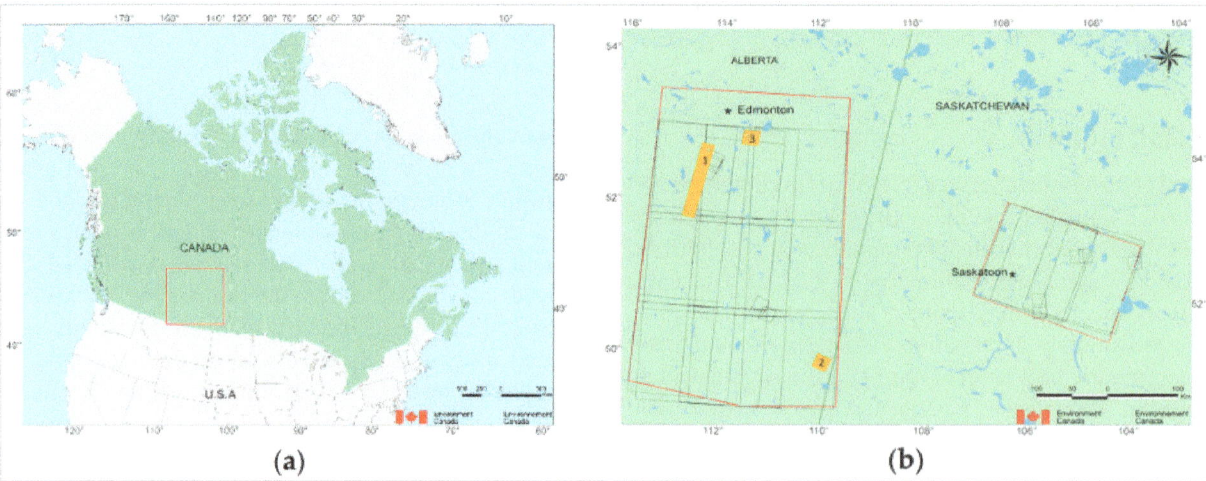

Figure 1. Study area: (**a**) Location of the study area in Canada (red square); (**b**) Areas of interest over Alberta and Saskatchewan (red polygons), radar image footprints (gray lines) and optical data (orange polygons) used for accuracy assessment. Numbers on the orange rectangles indicate the location of each of the three optical-SAR image pairs used for accuracy assessment, described in Section 4.

The Prairie potholes are glaciated depressions that store fresh water for which direct precipitation and spring runoff from snowmelt are the major sources of water supply [30]. Their water balance is influenced by redistribution of snow by wind from adjacent upland areas, precipitation, evapotranspiration, snowmelt runoff, groundwater exchange and soil type. Prairie potholes range in size from 1 m^2 to 100 km^2 and are also a significant storage of water that otherwise can oversaturate the soil during storm events [30]. Given the complexity of their vegetation structure, the minimum mapping unit identified for this project is defined as all open water features that are non-vegetated (or contain less than 15% vegetation areal cover) and are 1 ha or larger. This includes non-vegetated potholes, rivers and lakes.

Water Extents Dynamics of Prairie Potholes

Due to their dynamics and location, potholes are ephemeral and difficult to survey. Currently, there has not been a characterization of the potholes' size in the Canadian prairie region, and their physiological characteristics can vary significantly [31]. To get an idea of the spatial and temporal frequency required for water mapping, a time series analysis was conducted in our study area in Alberta. An area of 50 km by 100 km over the town of Olds, north of Calgary was selected and visually inspected using cloud-free orthorectified (and pan-sharpened) Landsat imagery acquired biweekly between May 2005 and October 2014 and available from the U.S. Geological Survey. This exploratory analysis on the pothole dynamics showed that:

- Potholes change very rapidly, especially in early spring (from May to beginning of June), then they become more stable—*i.e.*, the aerial extent of open water does not change significantly afterwards.
- The size of a pothole is not linked to its stability: inter-annual analysis showed that although the smaller potholes tend to disappear more rapidly (or become fully vegetated), large ones also could disappear (or be completely covered by vegetation) in less than 2 weeks.
- They have predictable locations. Over a 14 year period, potholes change the amount of water stored (they could shrink or expand) but for the most part, they're always in the same location.

This inspection helped us define the temporal scale of our image acquisition. Based on this analysis it was determined that imagery needs to be acquired in a period shorter than two weeks at the beginning of the season but the acquisitions can be switched to monthly from July through October without significant risk of changes.

Image Acquisition and Parameters

Surface water has been mapped from SAR in an operational way by the Emergency Geomatics Services of the Canada Centre for Mapping and Earth Observation (CCMEO) since 2006 during flooding events [5]. Based on their experience, the use of a beam mode with fine resolution and wide extent could offer the best trade-off between coverage and detail for water mapping. Eighty-two (82) Radasat-2 Wide Fine mode images with a nominal resolution cell size of 80 m and extents ranging from 100 km to 170 km were acquired bi-weekly over our study areas from May through October, 2014. Subsets of cloud-free pan-sharpened Landsat images were acquired on the same date (±1 day) and were used to help on the visual interpretation of water features in the SAR image and to define the temporal and spatial dynamics of open water features in the study area, as described in Section 2.2. For accuracy assessment, concurrent, orthorectified, pan-sharpened and cloud-free WorldView-2 imagery was used. To evaluate accuracy of the method at different scales, Fine Quad imagery acquired in 2012 was obtained and compared to a 5 m RapidEye Level 3A of the same day.

Radarsat-2 imagery can capture data in four polarizations: horizontal transmit, horizontal receive (HH); horizontal transmit, vertical receive (HV); vertical transmit, horizontal receive (VH); and vertical transmit, vertical receive (VV). Modes that have large extents (beyond 70 km) are available in one or two polarizations (HH and HV or VV and VH), due to the storage and downlink capacity of the satellite, while smaller swaths can provide all four polarizations. Previous studies [5,20] had shown that HH is the preferred polarization for mapping flooded vegetation, because it maximizes canopy penetration and enhances the contrast between forest and flooded vegetation. Also, HH has lower backscatter over rough open water (roughness induced by wind) than VV, and therefore is better to map open water under windy conditions. Noise limits the use of HV or VH, because water has lower radar-cross section in cross polarization than in co-polarized channels (HH or VV). Given this background, we chose a single-look complex dual-polarized Radarsat-2 image in Wide Fine mode with HH and HV polarizations to initiate our experiments.

METHODS

The main processing steps of the method are illustrated in Figure 2 and are fully described in the following sub-sections.

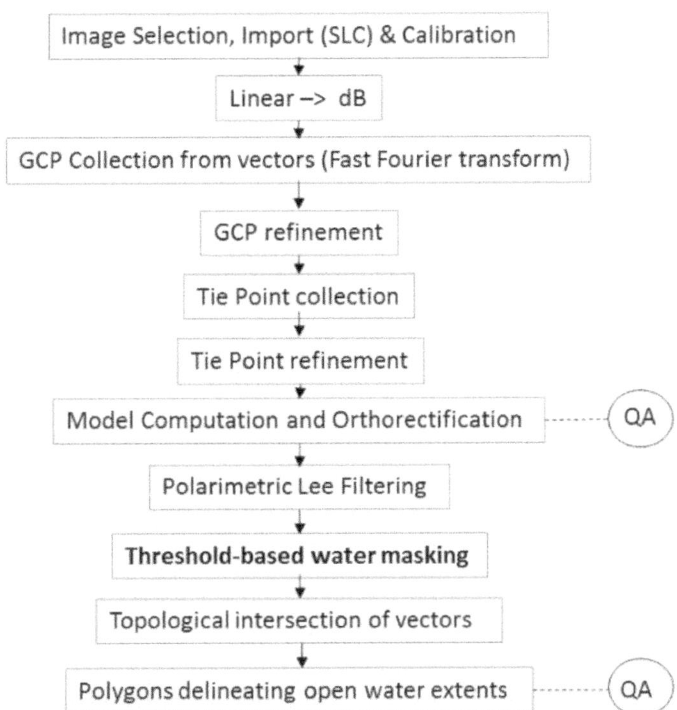

Figure 2. High-level workflow of the proposed water-mapping procedure consisting of geometric correction followed by novel threshold-based filtering and topological intersections. A quality assurance (QA) step should take place on the orthorectified imagery and on the final water polygons.

Image Preparation

Calibration of backscatter values is necessary to enable comparison of radar images acquired in different conditions [32]. Calibration of Radarsat-2 is applied using the information from the scene. The images were ordered from the Canadian Space Agency (CSA) with constant beta look-up table applied, and were converted to sigma-nought values based on the image's ancillary information. An intensity-based Enhanced Lee filtering [33] and a polarimetric Lee filtering of 5 × 5 pixels were applied to the images and compared. The Enhanced Lee filter was considered more appropriate to do this comparison than any other intensity-based filter because it divides the image into areas of three classes: homogeneous areas, such as the central area of open water bodies, where the speckle is eliminated with an average (low pass) filter; heterogeneous areas (e.g., land-water edges) on which an adaptive kernel is used; and areas containing isolated point targets (e.g., very small potholes), where the pixel value is preserved.

Unlike intensity-based filters, polarimetric filters use the intensity and phase information, hence capturing more dark water pixels and reducing the amount of false positives [34]. Then, the calibrated intensity values in the image are converted to physical units, decibels. The conversion to decibels increases the image contrast and minimizes the noise as well as sets all the images in the same physical units—so that a unique value or unique range of values can be used on multiple scenes. Once the image has been filtered and converted to physical units, a geometric correction process is applied.

All SAR images were orthorectified using a 3D rigorous geometric model [35] computed using the rational polynomial coefficients available with the image [36] and a 1:50,000 Digital Elevation model from the Canadian Digital Elevation Data (CDED) dataset [37]. The geometric correction algorithm uses the hybrid model for geometric processing of Radarsat-2 developed at the Canadian Center for Remote Sensing [38] and implemented in PCI Geomatics®. For the prairie region, with low topographic relief, the model provides sub-pixel positioning (mean Root Mean Square Error, RMSE = 0.35), when compared to roads at 1:10,000 scale from the National Road Network [39]—see Figure 3.

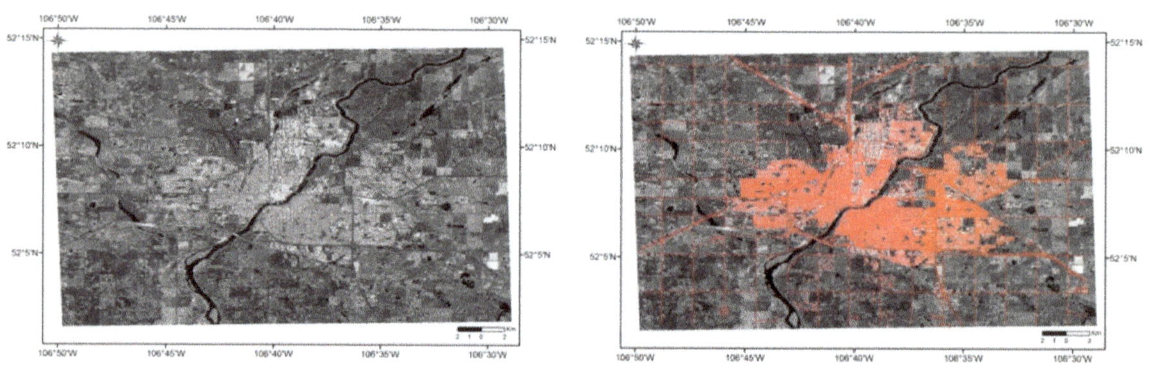

(a) (b)

Figure 3. Accuracy of automatic geometric correction using only ancillary data from Radarsat-2 in low terrain: (**a**) Wide Fine mode image over Saskatoon; (**b**) Wide Fine mode image over Saskatoon with overlaid roads at 1:10,000.

For areas of moderate to high relief, the collection of Ground Control Points (GCP) will be necessary to correct for relief displacement and a quality assurance (QA) step should take place to inspect the point accuracy and distribution as well as the overall RMSE prior to proceeding with the automatic tie point collection, orthorectification and mosaicking of scenes of the same swath.

Thresholding

Thresholding techniques on SAR image for water mapping had been covered at length in the literature [4,5,6,7,20,21]. In its most simplistic form, thresholding consists of establishing a value below or equal to which pixel values are selected. Unlike thresholding methods that depend on manual user intervention, the thresholding method presented here is less computationally intense and requires no user interaction. To avoid changing the threshold value for each scene depending on the wind speed or image viewing geometry, thresholding here is done in two steps. First, a low threshold has been pre-determined based on visual inspection of the lowest return of open calm water for each beam mode. This threshold is fixed for all the scenes in each run, but may differ for each beam mode because water has the lowest return of all the pixels in the image and this value changes depending on the noise level of each specific beam mode [40]. The pixels selected in this step constitute the "seed mask". Since the backscatter and noise change depending on the incidence angle [6], especially for images with wide extents, this seeding threshold is also modified based on the incidence angle value at every pixel, using an incidence angle raster created for each scene. This concurs with the work of other researchers on water mapping from SAR imagery [6,11,20]. For instance, Sokol *et al.* [11] found that a large decrease in C-HH backscatter occurs as the incidence angles increases and backscatter values from beam modes S1 to S7 (20°–49°) correspond to an average difference of 7 decibels.

In a second step, a more permissive threshold is used to create an "extended" mask based on a texture indicator (namely energy [41]), which is computed for each image channel (co and cross-polarized channels). In order to cope with wind-induced water roughness in either HH or HV, the highest value of this indicator is selected for each pixel. This texture indicator provides a measure of homogeneity of the signal by comparing the mean to the standard deviation of all the pixels in a 13 × 13 window [41]: the more homogeneous the section of the image is, the higher its value. Edges are regarded as boundaries between image objects and they are located where major changes in Digital Number (DN) values occur. This division of the image into a set of homogenous regions is the baseline concept of image segmentation, which has been widely used for object recognition from imagery [42,43]. Image segmentation algorithms can be region growing or edge detectors (among others). Edge detectors are the basis for active contour models (also called snakes), an object recognition technique that has been successfully used for SAR imagery to detect land-water boundaries in a semi-automated way [17,22]. In this technique, the energy in a moving window of group of pixels is minimized until the edge is found. Our extended mask uses the same concept: it compares the pixel values using local statistics in a window, to find the areas with the highest homogeneity.

The threshold defined to select the most homogeneous pixels in the image is based on the statistics of the highest value of this texture indicator for each pixel in the co-polarized and cross-polarized channels. The texture parameter is computed in both channels (HH and HV) in order to avoid a negative contribution of the lower signal-to-noise ratio in cross-polarized channels (HV). The statistics of the channel holding the highest (*i.e.*, most homogeneous) value of this texture indicator are computed and if a pixel is higher than the mean plus two standard deviations (chosen through experimental error) the pixel is considered very homogenous. This standard deviation multiplier could be adjusted to any number between 1 and 2, depending on the Noise Equivalent Sigma Zero (NESZ) of the employed beam mode and of the homogeneity of the surface water bodies (severely affected by roughness under windy conditions). Pixels meeting this criterion constitute the "extended mask". Figure 4 shows the logic used for thresholding.

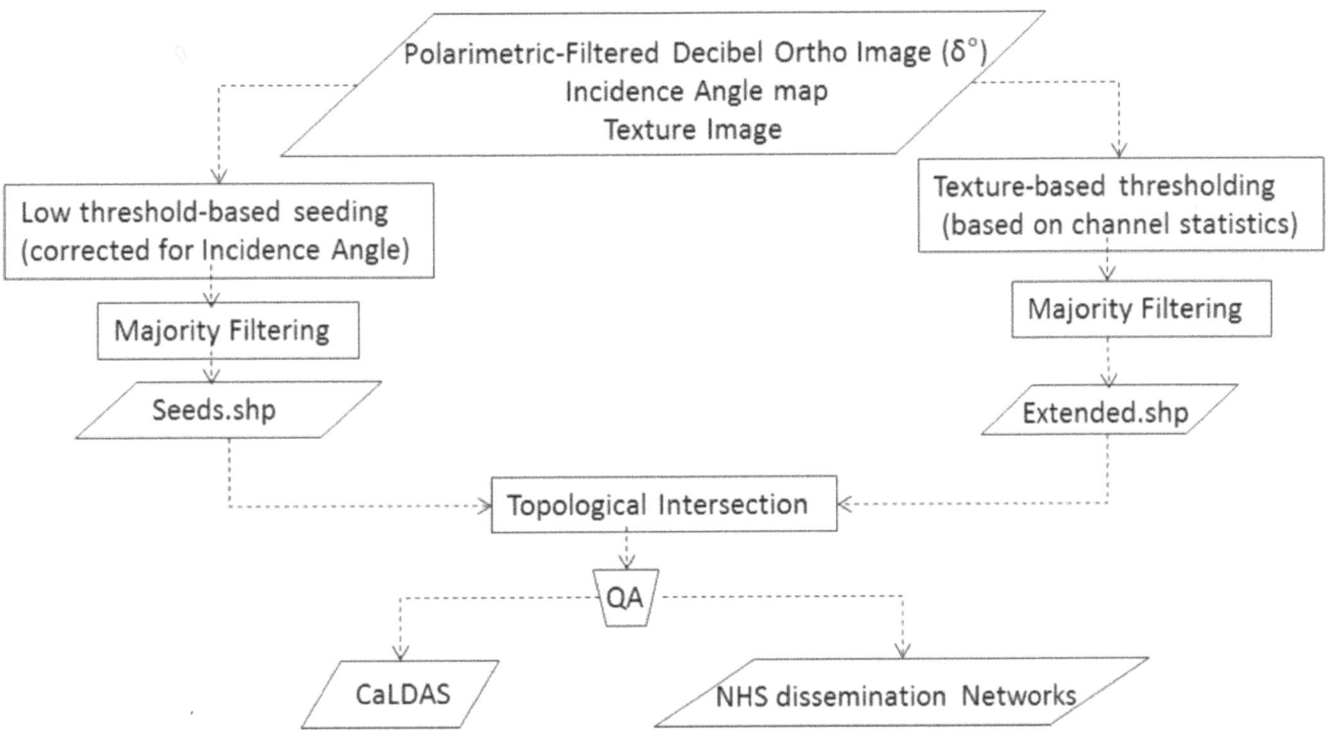

Figure 4. Threshold-based water masking consists of establishing a low threshold mask per beam mode and a texture-based mask per scene. After quality analysis of the polygons produced by the topological intersection, they are loaded into the dissemination networks of the National Hydrological Services of Environment and Climate Change Canada and will be used as input into the Canadian Land Data Assimilation System (CaLDAS).

Topological Intersection

A mode filter is applied to eliminate granularity of the extended mask (in a 7 by 7 kernel) and reduce the seed mask's false positives (in a 3 by 3 kernel). This is a particularly useful filter in classification procedures to eliminate noisy pixels, because each pixel value is replaced by its most common neighbor. The filter size on the seed mask is smaller than the one applied on the extended mask, because the seed pixels are always fewer and more clustered than the pixels in the extended mask. Both masks are then converted to polygons and exported to a GIS package, where the polygons are intersected using spatial criteria: if a polygon from the extended mask intersects a polygon from the seed mask, the extended polygon is considered water—see Figure 5. Polygons resulting from this intersection are aggregated based on a minimum area criteria and are visually inspected for dissemination to the weather networks.

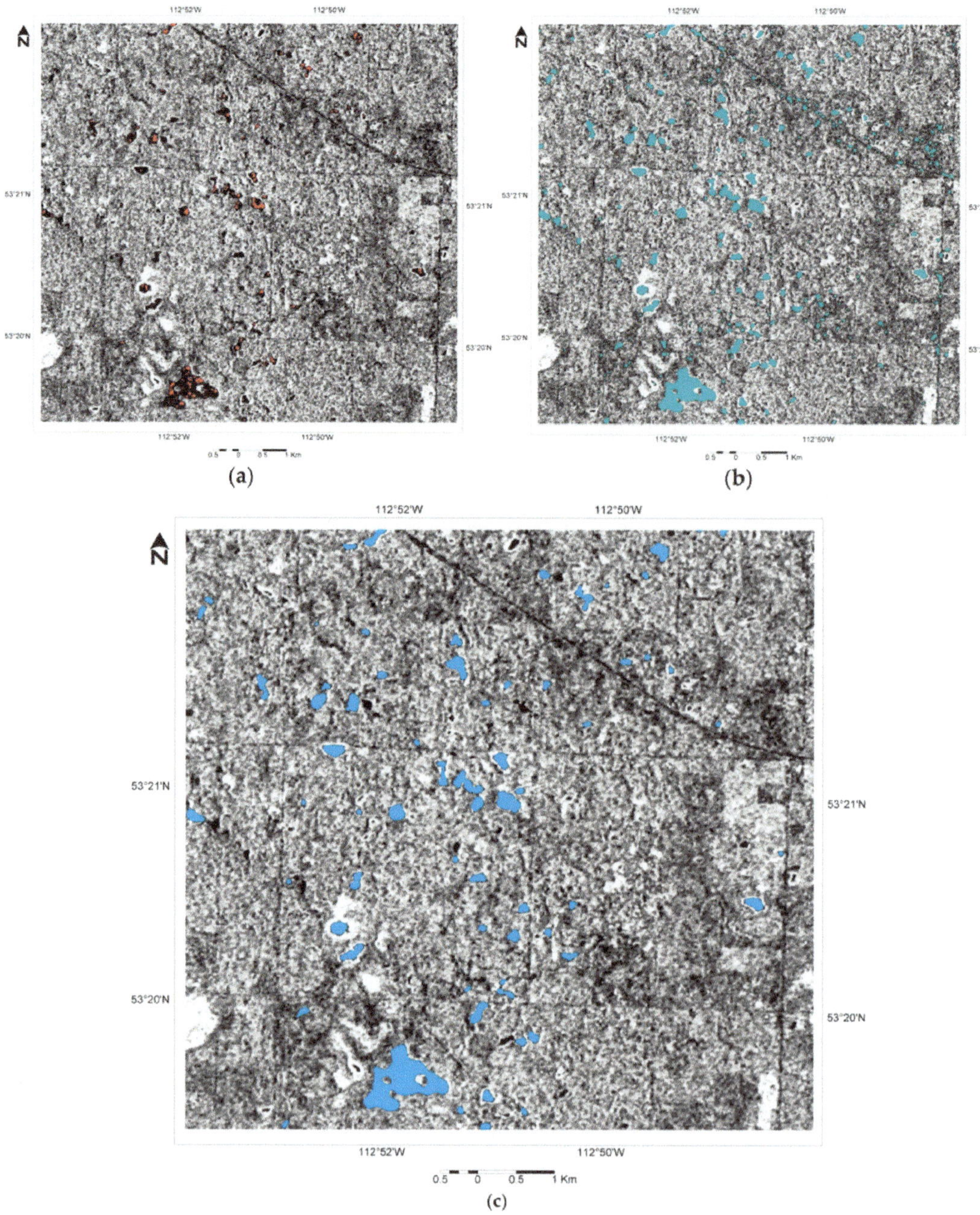

Figure 5. Topological Intersection: (**a**) Red: seed mask generated by using a unique threshold per beam mode; (**b**) Cyan: extended mask produced by the texture parameter –note the inclusion of false positives pixels on the road; (**c**) Blue: resulting water polygons produced by the intersection of the seed and extended masks.

RESULTS

Radarsat-2 derived polygons were evaluated against water polygons derived from cloud-free, temporally coincident high-resolution optical imagery over three areas of interest (AOIs) in Alberta, selected based on the availability of the imagery and landscape characteristics.

Due to the difficulty of accessing potholes' areas and their ephemeral nature, concurrent optical imagery with sub-meter spatial resolution was used whenever available. If this type of imagery was not available, then optical imagery with a pixel size comparable to the SAR image was employed. Although the evaluation against imagery instead of ground surveys is recognized as being limited to some degree, this was considered an appropriate proxy for an expensive field campaign that did not align well with the priorities of the Radarsat-2 acquisition plan.

In order to assess the veracity of the method to map surface water with different conditions, the three pairs of concurrent optical-SAR image were chosen because they represent a variety of conditions in which non-forested surface water areas can be found: open water, water with some vegetation and areas of very small (<1 ha) closed drainage flows, commonly found in the agricultural landscape of the Canadian Prairie region. The radar/optical image pairs were:

- Pair 1: WV2 and F0W3 (0.5 m and 8 m resolution respectively). Surface water polygons were derived from a cloud-free orthorectified and pan-sharpened WorldView2 (WV2) image taken on 12 May 2014 over the east side of the city of Red Deer and with a coverage 116 km long by 20 km wide. The polygons delineating open water were produced by thresholding of the near and short-wave infrared bands and manually edited. The resulting polygons were compared against the ones derived from a Wide Fine (F0W3) scene taken 1 day before (*i.e.*, 11 May 2014) and fully containing the WorldView-2 scene. The area covered by the optical image is cropland, which hydrological features are rivers, potholes and many shallow drainage flows that can be perceived mainly in sub-meter optical imagery, but are still considered surface water.

- Pair 2: WV2 and U76: (0.5 m and 2 m resolution respectively). Another WorldView-2 image taken on 10 May 2015 over an area 10 km west of the city of Schuler, Alberta was employed to derive water polygons using the same thresholding and manual editing procedure described above. The resulting polygons were compared against water polygons derived from an Ultra-Fine image (U76) taken 6 days after. The landscape in this area is mainly characterized by flooded vegetation.
- Pair3: RapidEye and FQ19 (5 m and 7 m resolution respectively). A RapidEye (RE) Image from 8 September 2012 over Elk Island National Pak in Alberta was employed to derive water polygons using thresholding and manual editing. The resulting polygons were compared against the water polygons obtained from a fully overlapping Fine Quad image (FQ19) taken the same day. Surface water in this area is mainly composed of open water bodies larger than 1 ha.

The minimum area, maximum area, number of shapes and cumulative area of the resulting water polygons from the optical and SAR images were derived. To facilitate the comparison, polygons bigger than 25 m^2 were selected, their statistics were computed using 5 area intervals (represented by the rows in the tables below) and the number of polygons as well as the contribution (expressed as percentage) of their cumulative area to the total area was calculated for each area range. Polygons smaller than 25 m^2 were excluded as they are not perceived by Radarsat-2 operational beam modes with repeated pass and continuous coverage. The comparison of surface water delineation based on area intervals was motivated because missing a high number of small polygons would not be detrimental for water quantification, as long as, together, they do not hold the biggest percentage of surface water in a particular area. Having an area for which missed small water polygons hold the majority of surface water will render the method or the data inappropriate for mapping ephemeral water bodies.Table 1, Table 2 and Table 3 show the accuracy in area quantification obtained from polygons, when surface water is mapped by manual vector editing on the optical image *vs.* applying our operational procedure on the SAR images:

Table 1. Water area quantification for Pair 1: WorldView-2 *vs.* Wide Fine (F0W3). The numbers on the second and third columns represent the total number of polygons in each area interval, while the numbers in brackets represent the contribution of the cumulative area in each range to total surface water area.

Area Range	WorldView2 (12 May 2014)	Radarsat-2 Wide Fine (F0W3) (11 May 2014)
[25 m^2–1000 m^2)	86576 (15.44%)	44 [1] (0.15%)
[1000 m^2–1 ha)	8513 (27.26%)	266 (6.19%)
[1–2 ha)	533 (8.76%)	75 (5.80%)
[2 ha–0.5 km^2)	488 (34.98%)	119 (56.45%)
[0.5–5 km^2]	6 (13.55%)	3 (31.41%)
Total number of polygons	96116	507
Total area	**84.825 km^2** (100%)	**17.68 km^2** (100%)

[1] The smallest water polygon detected in this Wide Fine mode scene is 147 m^2.

Table 2. Water area quantification for Pair 2: WorldView-2 *vs.* Ultra-Fine (U76). The numbers on the second and third columns represent the total number of polygons in each area range, while the numbers in brackets represent the contribution of the cumulative area in each range to total surface water area.

Area Range	WorldView2 (10 May 2014)	Radarsat-2 Ultra-Fine (U76) (16 May 2014)
[25 m^2–1000 m^2)	7634 (15.8%)	636 (23.27%)
[1000 m^2–1 ha)	401 (20.28%)	65 (25.69%)
[1–2 ha)	38 (10.16%)	2 (3.68%)
[2 ha–0.5 km^2)	43 (53.74%)	3 (47%)
[0.5 km^2–5 km^2]	0	0
Total number of polygons	8116	771
Total area	**5.52 km^2** (100%)	**0.64 km^2** (100%)

Table 3. Water area quantification for Pair 3: RapidEye-2 *vs.* Fine Quad (FQ19). The numbers on the second and third columns represent the total number of polygons in each area range, while the numbers in brackets represent the contribution of the cumulative area in each range to total surface water area.

Area Range	RapidEye L3A (8 September 2012)	Radarsat-2 Fine Quad (FQ19) (8 September 2012)
[25 m^2–1000 m^2)	5391 (3.29%)	72 [1] (0.20%)
[1000 m^2–1 ha)	674 (9.23%)	698 (7.85%)
[1–2 ha)	110 (5.87%)	99 (6.00%)
[2 ha–0.5 km^2)	155 (35.89%)	158 (48.90%)
[0.5 km^2–5 km^2]	8 (45.6%)	8 (36.93%)
Total number of polygons	6336	779
Total area	**26 km^2** (100%)	**23 km^2** (100%)

[1] The smallest water polygon derived from FQ19 was 62.5 m^2.

The accuracy achieved in quantifying the area of water bodies, when compared to high-resolution optical imagery for different beam modes, showed that:

- Our procedure fails to map open water bodies smaller than 1 ha when applied to Wide Fine mode. For the first paired optical-SAR dataset evaluated (Table 1), small water bodies were mostly missed by the algorithm. Also, the cumulative area contained in polygons smaller than 2 ha contributed to 52% of total surface water area in this particular AOI, which explains why the accuracy of polygons extracted from the SAR image drops significantly for this beam mode—see Figure 6.
- The quantification of the area on large water features were missed from Wide Fine mode due to fragmentation. This occurs due to discontinuity of polygons delineating water edges, when vegetation patches occur at the edges (e.g., riparian forest)—see Figure 7. On the other hand, two big water polygons that are seen as separate entities in the RapidEye image can be joined together and form one in the SAR image, due to the missing separation by small low vegetation patches (which are visible on the RapidEye image)—this changed the distribution of the area contribution for the water polygons derived from Fine Quad between 2 ha and 0.5 km² (Table 3).
- As expected, the algorithm also fails to detect flooded vegetation: a closer look at the polygons missed from the Ultra-Fine image *vs.* the polygons obtained from WolrdView 2 of the same time period showed that many polygons were larger than 1 ha and could have been seen in the Ultra-Fine image due its fine resolution, but were missed because of their high backscatter, which is characteristic of vegetation—see Figure 8 and Table 2.
- Fine mode imagery seems to provide the best results, as it quantified 88% of the total surface water area and picked up 97% of the total number of polygons larger than 1 ha when compared to polygons obtained from RapidEye (Table 3).
- Some water polygons that are selected by the algorithm from the SAR image are not shown as discernible open water bodies in the optical image (especially when the optical image pixels size is 5 m or more). They could be seen as false positives, but considering that SAR is more sensitive than optical imagery to water content, these areas could also be areas of low vegetation (where the vegetation cover is not high enough to influence backscatter but its chlorophyll content does influence reflectance) with particularly high water content—higher than its surroundings.

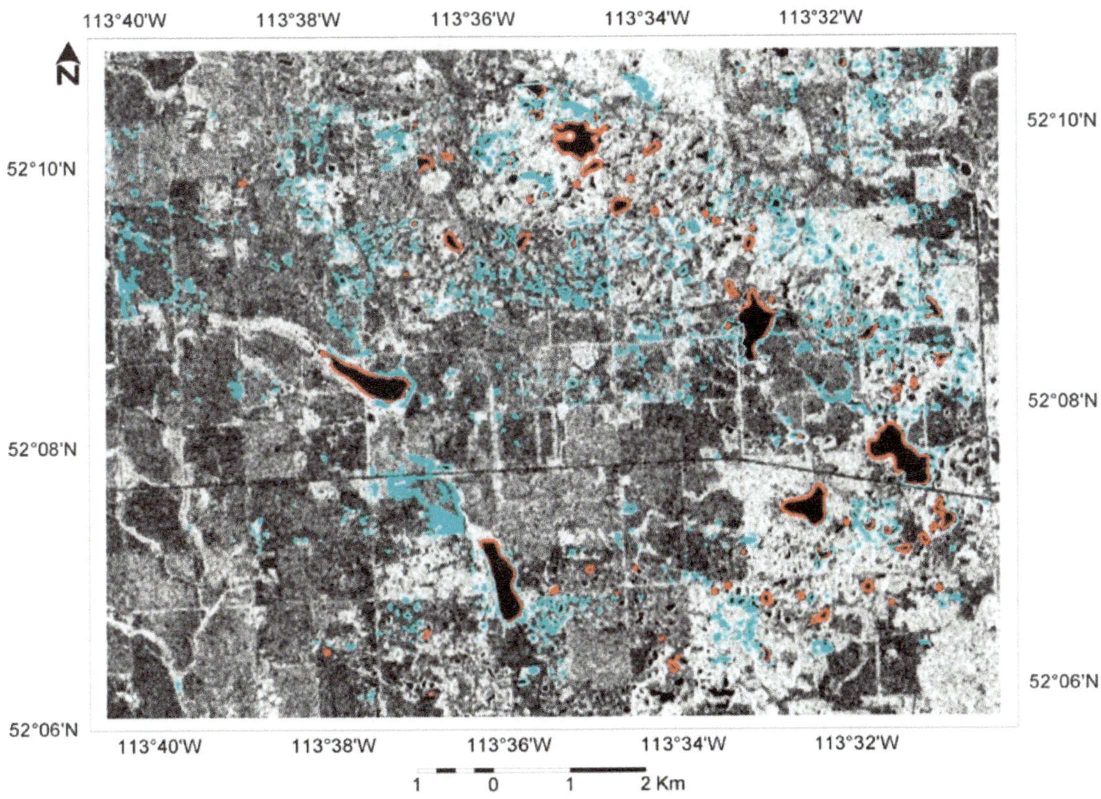

Figure 6. Results from the method applied on Wide Fine mode compared to 50 cm optical imagery taken 1 day apart. Blue: water polygons derived from pan-sharpened WorldView-2 (pixel size = 50 cm); red: water polygons derived from Wide Fine mode (F0W2) over our study area in Alberta.

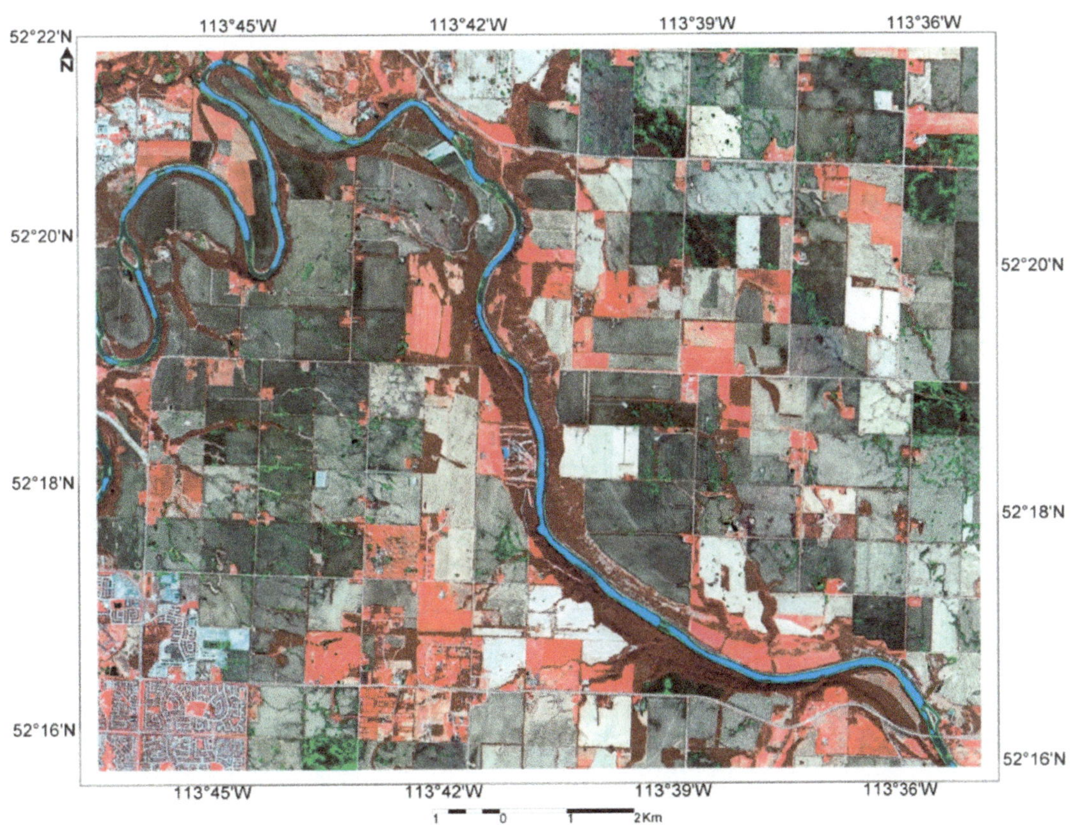

Figure 7. Effects of riparian forest on layer fragmentation. Subset over the Red Deer river in Alberta. The vectors shown are resulting polygons when the method is applied on Wide Fine mode. Background: Pan-sharpened Worldview2 imagery taken 1 day apart (Red = NIR band, Green = red band and Blue = green band). The fragmentation of SAR-derived water polygons (in blue) make them smaller than 0.5 km^2, excluding them from the last area interval in Table 1.

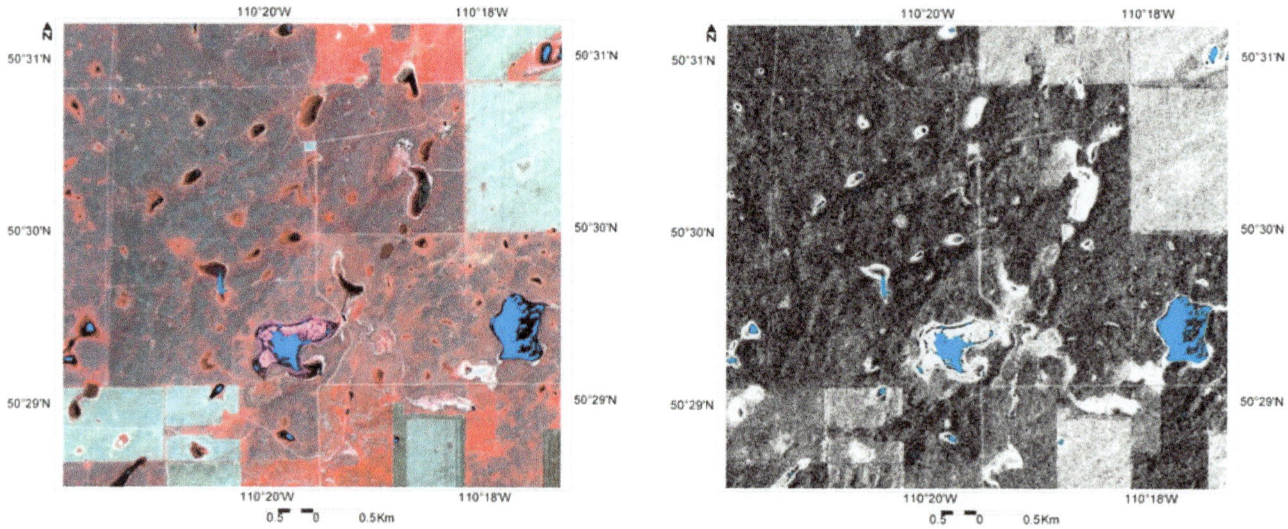

(a) (b)

Figure 8. Results on flooded vegetation: (**a**) Reference WorldView2 image displayed as NIR, red, green in RGB with overlaid water polygons from the Ultra-Fine image (in blue); (**b**) source Ultra-Fine image with overlaid derived polygons (in blue). Note that the images were taken 6 days apart, with changes in the aerial coverage of flooded vegetation between these two scenes.

In addition to this comparison of area estimates, randomly stratified points (50% of them on water pixels and 50% of them on non-water pixels) were generated and error matrices as well as kappa statistics [44] were computed using the same SAR-optical image pairs described above (see Table 4 and Table 5, respectively). The accuracy of the reference optical images to map water was evaluated through visual inspection, and their kappa values varied between 91.6% and 98.8%.

Table 4. Error matrices for all three pairs of images used for accuracy assessment.

Image	Water	Non-Water	Totals (Classifier)	Totals (Reference Data)
Pair 1 (F0W3 *vs.* WV2)				
Water	13	0	13	100
Non-water	87	99	187	99
Pair 2 (UF76 *vs.* WV2)				
Water	3	0	3	48
Non-water	45	52	97	52
Pair 3 (FQ19 *vs.* RE)				
Water	41	0	41	49
Non-water	8	51	59	51

Table 5. Kappa statistics for water classification from the three SAR images described above, using their paired optical image as reference.

Image	Number of Points	Overall Accuracy (%)	95% Confidence Interval (%)	Overall Kappa
Pair 1 (F0W3 *vs.* WV2)	199 [1]	56.5	49.38–63.621	0.125 ± 0.05
Pair 2 (UF76 *vs.* WV2)	100	55.0	44.75–65.25	0.065 ± 0.19
Pair 3 (FQ19 *vs.* RE)	100	92.0	86.18–97.82	0.839 ± 0.03

[1] This image contained the largest coverage of the optical images set (as can be seen in Figure 1); therefore a higher number of points was used to maintain a similar same sampling density in all three areas.

Considering the uniform spatial distribution, number and representation of each class in the sample points, we can state that the confidence interval provides a precise approximation of the estimated overall accuracy. The kappa statistic, as a measure of how closely the instances classified by the algorithm matched the reference data (compared to a random classifier) confirms the results presented in Table 1 through Table 3—despite the likelihood of large water bodies attracting more random points. Our method has low accuracy to map water from Wide Fine beam mode over areas with a high number of small potholes (less than 1 ha) or over flooded vegetation (*i.e.*, the dominant landscape in UF76). However, the method succeeds in automatically mapping non-vegetated water bodies larger than 1 ha from Fine Quad beam modes when compared to optical imagery of equivalent spatial resolution.

DISCUSSION

The areal extent of water bodies is important for a variety of reasons including a better understanding of the land–atmosphere boundary for meteorological and climate modeling [27]. The size and extent of water bodies across Canada vary both geographically and temporally throughout the year. This variation depends significantly on local climate and topography: the prairie region has many shallow water depressions (prairie potholes) that are filled through snow redistribution, snow melt, infiltration, and precipitation, and can evaporate quickly and exhibit "spill and fill" movement [45]. Due to local topography, small amounts of contributed water influence the areal extent of water bodies significantly. On the other hand, in the Canadian Shield, and mountainous regions, a large influx of water may significantly raise the stage, but areal extent does not significantly change.

The method presented in this paper includes procedures for geometric correction, calibration, filtering and thresholding of image values and derived texture. For surface water detection, a threshold-based algorithm that requires dual polarization is less restrictive than polarimetry-based algorithms that rely on fully polarimetric data, and therefore is easier to implement operationally, as more data can be obtained from scenes with larger coverage than fully polarimetric scenes.

Having accurate orbit and satellite ephemeris information as well as ancillary information (e.g., RPCs and GCPs) improves the efficiency of any operational system designed to retrieve data from satellite imagery: processing times were significantly lower when the system could skip the point collection routines, and eliminate manual quality control process on point collection. The nominal geometric model provided with Radarsat-2 was sufficient to perform orthorectification for the Prairie region. For the Radarsat Constellation Mission, we expect that the orbit accuracy as well as the rational polynomial coefficients are going to have accuracy that is at the same level or better than Radarsat-2, so that processing times remain as they are. Even for mountainous areas, where ground control point collection is required, the accuracy of the orbit and RPCs can significantly affect the processing time and quality of the results, as the image-matching algorithm uses this information to define the initial position of a searching window and to define a search radius [35].

A threshold-based technique offers an efficient and non-expensive way to do automatic mapping, but it has limitations. Work done by Manjusree *et al.* [20] proved that a water mapping threshold needs to be decreased from –8 to –12 decibels in the near range and from –15 to –24 in the far range. In our experiments, using a single threshold resulted in missing water bodies in the far range, because there were no seeds captured: the threshold value excluded more pixels on open water in the far range, which usually has lower signal to noise ratios and higher calibration uncertainty. Work by O'Grady *et al.*[21,46] also concluded that there are variations in the backscatter based on the incidence angle, given the varying noise levels of the scene. Another drawback of threshold-based methods for water mapping from SAR is that the threshold used for water mapping is severely affected by the noise (and noise floor) of the scene. Hardcoded threshold values are always "subject to change"—based on image quality and beam mode specifications. Wide beam modes with fine resolution (such as Wide Fine and Wide Ultra-Fine) are achieved by increasing the Noise Equivalent Sigma Zero (NESZ) which regulates the minimum signal that a SAR can measure, in order to maintain a constant resolution [36,40]. Also, to reduce data volumes when wide swaths modes with high resolution are acquired, a 2-bit block adaptive quantization (BAQ) compression technique is used [40]. Predominantly, the 2-bit BAQ does not provide enough detail to capture the difference in signal strength (and therefore compensate for it) when there is a large change in the radar return as a function of time. This results in artificial backscatter variability from the near to the far edge, and increased noise levels, especially in darker areas (calm water) that are close to bright features (vegetation), hence the necessity to vary the threshold used for water based on the incidence angle.

The traditional method of accuracy assessment based on a confusion matrix and random points (or stratified random points) did not adequately evaluate the success (or failure) of surface water mapping for different beam modes since an indication of the accuracy on obtaining the complete area of surface water is required for numerical weather prediction models [47]. Also, large water bodies that are easily identified in any SAR beam mode, have more pixels and therefore are more likely to get more random points, hence introducing some bias on the results of traditional accuracy assessment procedures. Furthermore, the elaboration of the confusion matrix requires a direct comparison of all points, without considering their contribution to the total surface water area. Quantifying water area based on polygons of a defined range (*i.e.*, 1 to 2 ha) better described the capacity of different SAR imaging modes for water mapping, as well the effectiveness of the method given the area distribution of ephemeral and permanent water bodies for the AOIs in our study region.

While omission errors in the SAR-derived polygons are relatively easy to evaluate, false positives (some of which could be wet surface or just homogeneous areas of low return, unrelated to water content) are much harder to assess. If we define the mapping unit as non-vegetated potholes that have a characteristic shape (normally high compactness), the commission errors can be assessed based on high-resolution optical image. However, if we were to define surface water as any standing water that might or might not contain any type of vegetation, the false positives are very difficult (if not impossible) to discriminate from high-resolution optical imagery, specially without a short-wave infrared band. Water areas with some vegetation are as important as open water for numerical weather prediction [26]. Nonetheless, doing fieldwork in these areas might neither be possible (due to accessibility) nor relevant if the field work campaign is not executed at the exact same time of the image acquisition, as water can evaporate very quickly, especially during the summer season. Radar is more sensitive to water content than optical imagery, and therefore it can detect water that cannot be easily discriminated in the visible or NIR part of the spectrum [48]. Furthermore, the chlorophyll activity and leaf structure can also affect the reflectance in optical imagery whereas for SAR these factors play a secondary role on the backscatter registered by the sensor. More work is underway to gather concurrent UAV and LiDAR data along with more SAR scenes, to obtain relative accurate ground truth that will serve to evaluate the commission errors. However, the validation of dynamic targets for multi-temporal, multi-source data remains a big challenge under a limited budget.

Further work is required to improve accuracy for small potholes and flooded vegetation. The upcoming Radarsat Constellation mission will offer compact polarimetry [24], which means that polarimetric data can be easily obtained in wider areas with more resolution than its predecessor, Radarsat-2. The acquisition of polarimetric data in the standard coverage that will be available with RCM, means that imagery will be acquired on a periodical basis, without the overhead of image acquisition planning and consolidation of conflicting acquisition requests from different users. If this type of imagery can be provided in fine resolution modes for inland applications, such as forestry, agriculture, moisture analysis, biomass estimation, flooding and landslides, it will open many new opportunities not only for more accurate and frequent water monitoring systems but also for better natural resource mapping, in general.

CONCLUSIONS

A threshold-based procedure to automatically extract surface water polygons from SAR imagery has been presented here. For the AOI covered by the Wide Fine scene, our analysis based on WorldView-2 imagery at 50 cm pixel size showed that 52% of the total water area was contributed by polygons smaller than 2 ha, whereas in the AOI covered by the Fine Quad scene, most of the potholes and permanent water bodies (87%) were larger than 1 ha. For areas where small potholes are dominant, mapping with Wide Fine can only reach 21% accuracy when estimating total surface water area, but for areas where potholes larger than 1 ha are dominant, Radarsat-2 Fine mode captured 88% of the total water area extracted from RapidEye. This provides much better information than currently available for regions where seasonal and ephemeral changes are expected on surface water. More efforts should be made to improve the performance and meet the requirements for operational applications on smaller water bodies and low-resolution modes.

Despite its limitations, our procedure is a fully automated method to derive polygons of open water from Radarsat-2 Fine beam modes, based on pixel values, which is, in terms of processing time, much faster than running more complex segmentation procedures. Processing time becomes a key factor when the information on the land surface needs to be derived to feed the models used in weather forecasting, which run as frequently as every 6 h [47]. The procedure has been implemented in Python as a processing chain, and it uses image processing software with multi-threat algorithms from PCI Geomatics, where only two steps for quality control and analysis are required: first, when a collection of points is required for geometric correction, and at the end, when water polygons are generated. The topological intersection of polygons to generate the final water mask is also being integrated into the same script using Python-based algorithms and open-source GIS software, which will make it easier to integrate into operational and/or web-based systems.

ACKNOWLEDGMENTS

This research was possible thanks to the Government Research Incentive Program of the Canadian Space Agency. We also would like to thank Stephan Belair and his team for their advice and support in the development of this project and revision of this manuscript. Thanks to Koreen Millard and Lori White for the feedback provided on this paper, and to Kevin Murnaghan for preparing and submitting the acquisition coverage plan for the 2014 Radarsat-2 imagery. The authors also would like to thank the AFOLU team in the Science and Technology branch of Environment and Climate Change Canada for providing the time to finalize the editing.

Conflicts of Interest

The authors declare no conflict of interest. The founding sponsors had no role in the design of the study; in the collection, analyses, or interpretation of data; in the writing of the manuscript, and in the decision to publish the results.

Abbreviations

The following abbreviations are used in this manuscript:

RCM	Radarsat Constellation Mission
DN	Digital Number
DB	decibels
RMSE	Root Mean Square Error
GCPs	Ground Control Points
RPC	Rational Polynomial Coefficients
GIS	Geographic Information System
AOI	Area Of Interest
SAR	Synthetic Aperture Radar
QA	Quality Analysis
NESZ	Noise Equivalent Sigma Zero
BAQ	Block Adaptive Quantization
NIR	Near-Infrared
Km	kilometers
RE	RapidEye imagery
WV2	WorldView-2 imagery
RGB	Red, Green, Blue guns used for visual display
UAV	Unmanned Aerial Vehicle
LiDAR	Light Detection and Ranging

REFERENCES

1. Canadian Space Agency. 2014 Report on Plans and Priorities. Available online: http://www.asc-csa.gc.ca/eng/publications/pr-2014-highlights.asp (accessed on 24 November 2015).
2. Pietroniro, A. Terrestrial Hydrology: Storage and Discharge of Water on Land: SWOT Hydrology for Canada. In Proceedings of the 4th SWOT Science Definition Team Meeting, Toulouse, France, 17–20 June 2014.
3. Brisco, B.; Touzi, R.; Van der Sanden, J.; Charbonneau, F.; Pultz, T.J.; D'Iorio, M. Water resource applications with RADARSAT-2 a preview. *Int. J. Digit. Earth* 2008, *1*, 130–147.
4. Wessels, J. Water extents mapping. In Proceedings of the LiDAR/SAR Wetlands and water monitoring workshop, Lethbridge, AB, Canada, 26–27 June 2014.
5. Brisco, B.; Short, N.; van der Sanden, J.; Landry, R.; Raymond, D. A semi-automated tool for surface water mapping with RADARSAT-1. *Can. J. Remote Sens.* 2009, *35*, 336–344.
6. White, L.; Brisco, B.; Pregitzer, M.; Tedford, B.; Boychuk, L. RADARSAT-2 Beam Mode Selection for Surface Water and Flooded Vegetation Mapping. *Can. J. Remote Sens.* 2014, *40*, 135–151.
7. Li, J.; Wang, S. An automated method for mapping inland surface waterbodies with Radarsat-2 imagery. *Int. J. Remote Sens.* 2015, *36*, 1367–1384.
8. Westerhoff, R.S.; Kleuskens, M.P.H.; Winsemius, H.C.; Huizinga, H.J.; Brakenridge, G.R.; Bishop, C. Automated global water mapping based on wide-swath orbital synthetic-aperture radar. *Hydrol. Earth Syst. Sci.* 2013, *17*, 651–663.
9. Töyrä, J.; Pietroniro, A. Towards operational monitoring of a northern wetland using geomatics-based techniques. *Remote Sens. Environ.* 2005, *97*, 174–191.
10. Touzi, R. Wetland characterization using polarimetric RADARSAT-2 capability. In Proceedings of the IEEE International Conference on Geoscience and Remote Sensing Symposium (IGARSS 2006), Denver, CO, USA, 31 July–4 August 2006; pp. 1639–1642.
11. Sokol, J.; NcNairn, H.; Pultz, T.J. Case studies demonstrating the hydrological applications of C-band multipolarized and polarimetric SAR. *Can. J. Remote Sens.* 2004, *30*, 470–483.
12. Song, Y.-S.; Sohn, H.-G.; Park, C.-H. Efficient water area classification using Radarsat-1 SAR imagery in a high relief mountainous environment. *Photogr. Eng. Remote Sens.* 2007, *73*, 285–296.
13. Pope, K.O.; Rejmankova, E.; Paris, J.F.; Woodruff, R. Detecting seasonal flooding cycles in marshes of the Yucatan Peninsula with SIR-C polarimetric radar imagery. *Remote Sens. Environ.* 1997, *59*, 157–166.
14. Martinis, S.; Twele, A.; Voigt, S. Towards operational near real-time flood detection using a split-based automatic thresholding procedure on high resolution TerraSAR-X data. *Nat. Hazards Earth Syst. Sci.* 2009, *9*, 303–314.
15. Chapman, B.; McDonald, K.; Shimada, M.; Rosenqvist, A.; Schroeder, R.; Hess, L. Mapping regional inundation with spaceborne L-Band SAR. *Remote Sens.* 2015, *7*.
16. Hahmann, T.; Twele, A.; Martinis, S.; Buchroithner, M. Strategies for the Automatic Extraction of Water Bodies from TerraSAR-X / TanDEM-X data. In *Geographic Information and Cartography for Risk and Crisis Management*; Konecny, M., Zlatanova, S., Bandrova, T.L., Eds.; Springer Berlin Heidelberg: Berlin, Germany, 2010; pp. 129–141.
17. Hahmann, T.; Wessel, B. Surface water body detection in high-resolution TerraSAR-X data using active contour models. In Proceedings of the 2010 8th European Conference on Synthetic Aperture Radar (EUSAR), Aachen, Germany, 7–10 June 2010; pp. 1–4.
18. Solbo, S.; Malnes, E.; Guneriussen, T.; Solheim, I.; Eltoft, T. Mapping surface-water with Radarsat at arbitrary incidence angles. In Proceedings of the Geoscience and Remote Sensing Symposium, 2003, IGARSS '03, Tolouse, France, 21–25 July 2003; Volume 2514, pp. 2517–2519.
19. Al-Bayati, M.; El-Zaart, A. Automatic thresholding techniques for SAR images. In Proceedings of the International Conference of Soft Computing, Dubai, United Arab Emirates, 18–19 May 2013.
20. Manjusree, P.; Prasanna Kumar, L.; Bhatt, C.; Rao, G.; Bhanumurthy, V. Optimization of threshold ranges for rapid flood inundation mapping by evaluating backscatter profiles of high incidence angle SAR images. *Int. J. Disaster Risk Sci.* 2012, *3*, 113–122.
21. O'Grady, D.; Leblanc, M.; Gillieson, D. Use of ENVISAT ASAR Global Monitoring Mode to complement optical data in the mapping of rapid broad-scale flooding in Pakistan. *Hydrol. Earth Syst. Sci.* 2011, *15*, 3475–3494.
22. Heremans, R.; Willekens, A.; Borghys, D.; Verbeeck, B.; Valckenborgh, J.; Acheroy, M.; Perneel, C. Automatic detection of flooded areas on ENVISAT/ASAR images using an object-oriented classification technique and an active contour algorithm. In Proceedings of the International Conference on Recent Advances in Space Technologies, 2003, RAST '03, Istanbul, Turkey, 20–22 November 2003; pp. 311–316.
23. Kass, M.; Witkin, A.; Terzopoulos, D. Snakes: Active contour models. *Int. J. Comput. Vis.* 1988, *1*, 321–331.
24. Vachon, P.; DeLisle, D. From Radarsat-2 to Radarsat constellation mission data continuity. In Proceedings of the Advanced SAR oceanography (SeaSAR) Symposium, Tromsø, Norway, 18–22 June 2012.
25. Canadian Space Agency. Radarsat Constellation. Available online: http://www.asc-csa.gc.ca/eng/satellites/radarsat/(accessed on 10 November 2015).
26. Carrera, M.L.; Bélair, S.; Bilodeau, B. The Canadian Land Data Assimilation System (CaLDAS): Description and synthetic evaluation study. *J. Hydrometeorol.* 2015, *16*, 1293–1314.
27. Belair, S. Better Weather Forecasts Resulting from Improved Land Surface Processes in Environment Canada's Numerical Prediction Systems. Available

online: http://gmao.gsfc.nasa.gov/seminars/archive/ (accessed on 29 November 2015).

28. Van der Valk, A.G. The prairie poholes of North America. In *The Wold's Largest Wetlands*; Fraser, L., Keddy, P., Eds.; Cambridge University Press: New York, NY, USA, 2005; pp. 393–423.
29. Nature Conservancy of Canada. Land Lines. Available online: http://www.natureconservancy.ca/en/blog/tweetable-facts-on-WWD.html#.VmSXKr-OshQ (accessed on 24 November 2015).
30. Pomeroy, J.; Fang, X.; Webrook, C.; Minke, A.; Guo, X.; Brown, T. *Prairie Hydrological Model Study Final Report*; University of Saskatchewan: Saskatoon, SK, Canada, 2010.
31. Wray, H.E.; Bayley, S.E. *A Review of Indicators of Wetland Health and Function in Alberta's Prairie, Aspen Parkland and Boreal Dry Mixedwood Regions*; Alberta Environment: Edmonton, AB, Canada, 2006.
32. Oliver, C.; Quegan, S. *Understanding Synthetic Aperture Radar Images*; Artech House: Norwood, MA, USA, 1998.
33. Lopes, A. Adaptive speckle filters and scene heterogeneity. *IEEE Trans. Geosci. Remote Sens.* 1990, *28*, 992–1000.
34. Lee, J.S.; Pottier, E. *Polarimetric Radar Imaging: From Basics to Applications*; CRC Press: Boca Raton, FL, USA, 2009.
35. Cheng, P.; Toutin, T. Radarsat-2 data: Automated high accuracy geometric correction and mosaicking without ground control points. *Geoinformatics* 2010, *13*, 22–27.
36. MacDonald Dettwiler and Associates (MDA). *Radarsat-2 Product Description*; RN-RP-51–2713; MacDonald Dettwiler and Associates: Richmond, BC, Canada, 2014.
37. Natural Resources Canada. *Canadian Digital Elevation Data: Level 1 Product Specifications*; Natural Resources Canada: Sherbrooke, QC, Canada, 2007.
38. Toutin, T.; Omari, K. A "new hybrid" modeling for geometric processing of Radarsat-2 data without user's GCP.*Photogr. Eng. Remote Sens.* 2011, *77*, 601–608.
39. Natural Resources Canada. *National Road Network, Canada: Level 1 Product Specifications*; Natural Resources Canada: Sherbrooke, QC, Canada, 2003.
40. Staples, G.; Cartwright, W. Radarsat-2 Imaging Modes and Aplications. In *Informative Webinar Series*; MDA Geospatial Services; Available online: http://gs.mdacorporation.com/includes/documents/webinar/RADARSAT-2 Imaging Modes and Applications Webinar.pdf (accessed on 23 April 2015).
41. Dekker, R.J. Texture analysis and classification of ERS SAR images for map updating of urban areas in The Netherlands. *IEEE Trans. Geosci. Remote Sens.* 2003, *41*, 1950–1958.
42. Blaschke, T.; Burnett, C.; Pekkarinen, A. Image segmentation methods for object-based analysis and classification. In*Remote Sensing Image Analysis: Including The Spatial Domain*; Jong, S.D., Meer, F.V.D., Eds.; Springer Netherlands: Houten, The Netherlands, 2004; Volume 5, pp. 211–236.
43. Blaschke, T. Object based image analysis for remote sensing. *ISPRS J. Photogr. Remote Sens.* 2010, *65*, 2–16.
44. Congalton, R.G. A review of assessing the accuracy of classifications of remotely sensed data. *Remote Sens. Environ.* 1991, *37*, 35–46.
45. Shook, K.; Pomeroy, J.W.; Spence, C.; Boychuk, L. Storage dynamics simulations in prairie wetland hydrology models: Evaluation and parameterization. *Hydrol. Process.* 2013, *27*, 1875–1889.
46. O'Grady, D.; Leblanc, M.; Gillieson, D. Relationship of local incidence angle with satellite radar backscatter for different surface conditions. *Int. J. Appl. Earth Observ. Geoinf.* 2013, *24*, 42–53.
47. Belair, S.; Canadian Meteorological Centre, Dorval, QC, Canada. Personal communication, 2015.
48. Klemas, V.; Pieterse, A. Using remote sensing to map and monitor water resources in arid and semiarid regions. In*Advances in Watershed Science and Assessment*; Younos, T., Parece, T.E., Eds.; Springer International Publishing: New York, NY, USA, 2015; Volume 33, pp. 33–60.

CHAPTER 10

FLOOD HAZARD MAPPING BY USING GEOGRAPHIC INFORMATION SYSTEM AND HYDRAULIC MODEL

Vahdettin Demir and Ozgur Kisi

INTRODUCTION

Flooding, as a major natural disaster, affects many parts of the world including developed countries. Due to this natural disaster, billions of dollars in infrastructure and property damages and hundreds of human lives are lost each year. These hazards and losses can be prevented and reduced by providing reliable information to the public about the flood risk through flood inundation maps [1]. Flood inundation maps are very essential for municipal planning, emergency action plans, flood insurance rates, and ecological studies [2]. Samsun is the largest and densely populated in the north of Turkey. This area is almost under threat of flooding in each year. In this region, the main reason of devastating flood is the influence of the Mert River especially during March, April, and July and due to seasonal rainfall which eventually makes the district vulnerable to flooding. In addition, the human based constructions and the collapse of water retaining structures are among the main causes of flooding.

Geographic Information Systems (GIS) are successfully used to visualize the extent of flooding and also to analyze the flood maps to produce flood damage estimation maps and flood risk map [3–5]. The GIS must be used together with a hydraulic method to estimate flood profile with a given return period. After 1970, Hydrologic Engineering Centers River Analysis System (HEC-RAS) software developed by United States Army Corps of Engineers (USACE) is widely used in Europe and America. In our country, it was first applied on Bartin River in 1998 by Yazıcılar and Önder [6]. GIS and HEC-RAS models were successfully used for obtaining flood maps of Waller River in Texas [7], Ohio Swan River Basin [3], Atrato River in Colombia [8], Vistula River in Warsaw, Poland [4], Gordon River in France [9], northwest of Colombia [8], mid-eastern Dhaka in Bangladesh [10], and Onaville in Haiti [11]. Çelik et al. analysed the 2004 flood of Kozdere Stream in Istanbul using HEC-RAS and GIS [12]. Sole et al. produced risk maps of Basilicata region (Italy) by acquiring water surface profiles according to different repetition flow in the main distributary (30, 200, and 500 years) [5]. Masood and Takeuchi used HEC-RAS and GIS for assessing flood hazard, vulnerability, and risk of mid-eastern Dhaka [10]. They obtained inundation map for flood of 100-year return period. Sarhadi et al. obtained flood inundation maps of ungauged rivers in southeastern Iran by using HEC-RAS and GIS [13]. Heimhuber et al. used HEC-RAS and GIS to perform one-dimensional, unsteady-flow simulations of design floods in the Ravine Lan Couline, which is the major drainage channel of the area [11]. To the knowledge of the authors, the HEC-RAS and GIS methods were not previously applied to Mert River Basin. Due to its proximity to numerous homes, businesses, and industrial area, the location of Mert River's flood plain is of great interest to city planners, developers, and property owners. To the knowledge of the authors, the GIS and HEC-RAS were not previously applied to this area where devastating floods happened.

The aim of this study is to obtain flood hazard maps of the Mert River Basin using GIS and HEC-RAS for floods

Citation: Vahdettin Demir and Ozgur Kisi, Flood Hazard Mapping by Using Geographic Information System and Hydraulic Model, doi.org/10.1155/2016/4891015.

of different return periods (10, 25, 50, 100, and 1000). First, topographical data were digitized and digital elevation model was prepared using ArcGIS. Then, flood flows of different return periods were simulated using a hydraulic model (HEC-RAS). Finally, flood risk maps were obtained by integrating the results of ArcGIS and HEC-RAS. The obtained flood map for 10-year return period was also tested by 2012 flood in which 12 people lost their lives.

STUDY AREA

Mert River is located in the center of Samsun. Geographic location of the study area is between Latitude 41.279 and Longitude 36.352. Samsun is the largest city in the Central Black Sea Region of Turkey. This district faces devastating floods which have a destructive effect on humans, buildings, and substructure systems. The Mert River which is about 8 kilometers long flows into the Black Sea. Mert River was selected for this study because it had a great loss of life and property in the recent floods (e.g., July 3, 2012). This river has five highway bridges and one pedestrian bridge. First, second, and third bridges of this river are located in the Black Sea coastline and provide ease of transport between cities. The study area is shown in Figure 1.

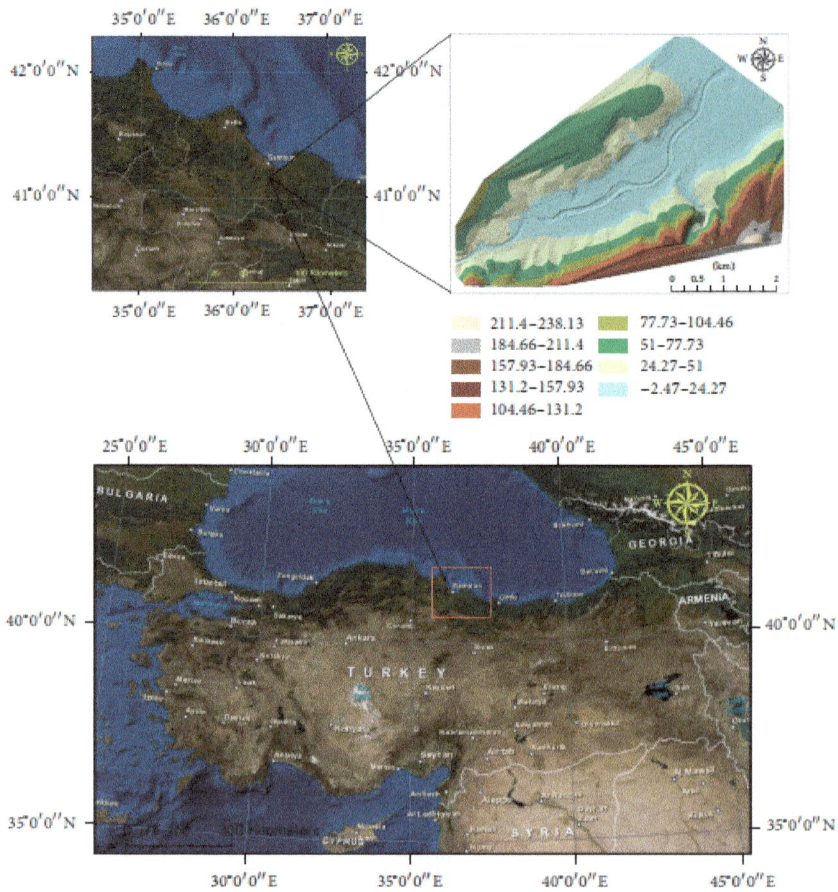

Figure 1: The location of the study area in Turkey.

Methodology

In the present study, flood hazard maps were obtained by using HEC-RAS, HEC-GeoRAS, and ArcGIS. The methodology for developing a flood hazard map can be explained by the following three phases: (i) preparing digital elevation model using ArcGIS, (ii) simulation of flood flows of different return periods using HEC-RAS hydraulic model, and (iii) preparing flood risk maps by integrating phases (i) and (ii). The flow chart of the methodology is illustrated in Figure 2. Next, brief information is provided for the HEC-RAS and HEC-GeoRAS. Detailed information for these methods can be obtained from related literature [14, 15].

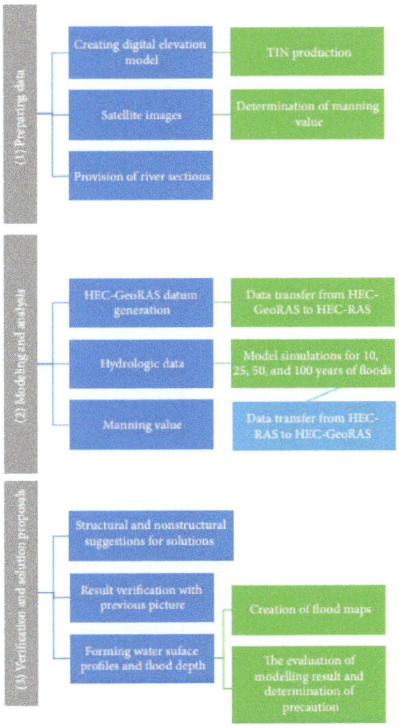

Figure 2: Flow chart of methodology.

HEC-RAS Model

HEC-RAS, a hydraulic model developed by the USACE, is extensively applied in calculating the hydraulic characteristics of rivers [16, 17]. It is an integrated program and uses the following energy equation for calculating water surface profiles [14, 18]:

$$Y_2 + Z_2 + \frac{\alpha_2 V_2^2}{2g} = Y_1 + Z_1 + \frac{\alpha_1 V_1^2}{2g} + h_e, \tag{1}$$

Where Y, Z, V, α, h_e, and g represent water depth, channel elevation, average velocity, velocity weighting coefficient, energy head loss, and gravitational acceleration; and subscripts 1 and 2, respectively, show cross sections 1 and 2.

This program provides user to input data, data correction, to receive output display and analysis. HEC-RAS model needs details of river cross sections and upstream flow rate. The water depth and mean velocity are calculated for a given cross section using the energy conservation equation [14].

HEC-RAS calculates the water levels' variation along the channel and the water level values are overlaid on a digital elevation model (DEM) of the area to get the extent and flood depth using GIS [19]. Spatial data like cross section, river reach, stream network, flow paths, and others have been obtained using HEC-GeoRAS (Arc-GIS extension) and these data then transferred to HEC-RAS [15].

HEC-GeoRAS Model (GIS)

HEC-GeoRAS is developed for the treatment of geographic data with the HEC-RAS and is working on an extension to ArcGIS (module). Other supplemental information with geometric data files is obtained from the Digital Terrain Models. This module can convert the format of HEC-RAS software and can read the obtained format. After analyzing the data with HEC-RAS, water surface profiles, water level, and water velocity can be obtained. The results obtained from hydraulic model can be converted to GIS format by using HEC-GeoRAS and thus flood mapping and flood depth map can be obtained [20].

The mixture of processing topographical information and other GIS data in ArcMap utilizing GeoRAS provides us with the capacity to create and export a geometry file to be investigated by RAS. The created geometry document holds information on river, catchment, and station cross section cut lines, bank stations, flow path. It achieves lengths for left and right overbanks and channel and roughness coefficients and furthermore can contain blocked obstructions. The results of RAS reproduction, for example, river profiles, can be sent specifically to a GIS environment, where they can be analyzed further by the assistance of the GeoRAS toolbar. A particularly arranged GIS information exchange document (. sdf) is utilized to perform the GIS data import and export between RAS and ArcMap [21].

APPLICATION AND RESULTS

In this study, HEC-RAS 4.10 was utilized for hydraulic analysis and ArcGIS 10.2 was used for mapping. First, 3D model of study area was prepared utilizing ArcGIS. Digital Elevation Model (DEM) was produced by 1/1000 scale topographical contour lines. Then, topographic data obtained from ArcGIS were transferred to HEC-RAS via Hec-GeoRAS module. Flood values of different return periods (10, 25, 50, and 100 years) and Manning roughness coefficient values were also entered into the HEC-RAS program for calculating water level for each cross section. Finally, the hydraulic analysis results were entered into the ArcGIS via Hec-GeoRAS module and flood hazard maps were obtained for each return period.

Manning roughness coefficients of 0.022, 0.026, and 0.045 were used for concrete, bush-wooded, and woodland river banks and 0.03 was utilized for the river base. Flood values of diverse return periods and annual instant maximum flows were obtained from the Turkish General Directorate of State Hydraulic Works. All these values are reported in Table 1. Table 2 gives the annual instant maximum flows of Mert River. As can be clearly seen from Table 2 a flood (near Q10, flood of ten-year return period) was seen in the studied area in 2012 and loss of life and property occurred.

Table 1: Flood values of different return periods of Mert River.

Return period	5	10	25	50	100	500	1000	10000
Flood (m^3/s)	508	641.8	839.7	1011.6	1207.6	1709.5	2028.5	3139.5

Table 2: Annual instant maximum flows of Mert River.

Year	2007	2008	2009	2010	2011	2012	2013
Flow (m^3/s)	158	102	66.3	87.1	73	570	66.1

Flood simulations were conducted using hydrodynamic program for the floods of 10, 15, 50, and 100 return periods. As an example, water surface profiles for the Q100 flood and the location of the bridges on Mert River are shown in Figure 3. Bridges were numbered according to their proximity to the Black Sea. Mert River flows into the sea after Mert River Bridge 1. It is clear from the figure that the last three bridges stay under water in the case of Q100 flood. 3D hazard maps of the Mert River acquired for the Q10, Q25, Q50, and Q100 floods are illustrated in Figure 4. As obviously seen from the figure, there are residential and industrial areas in the studied region which are significantly affected by flood disaster.

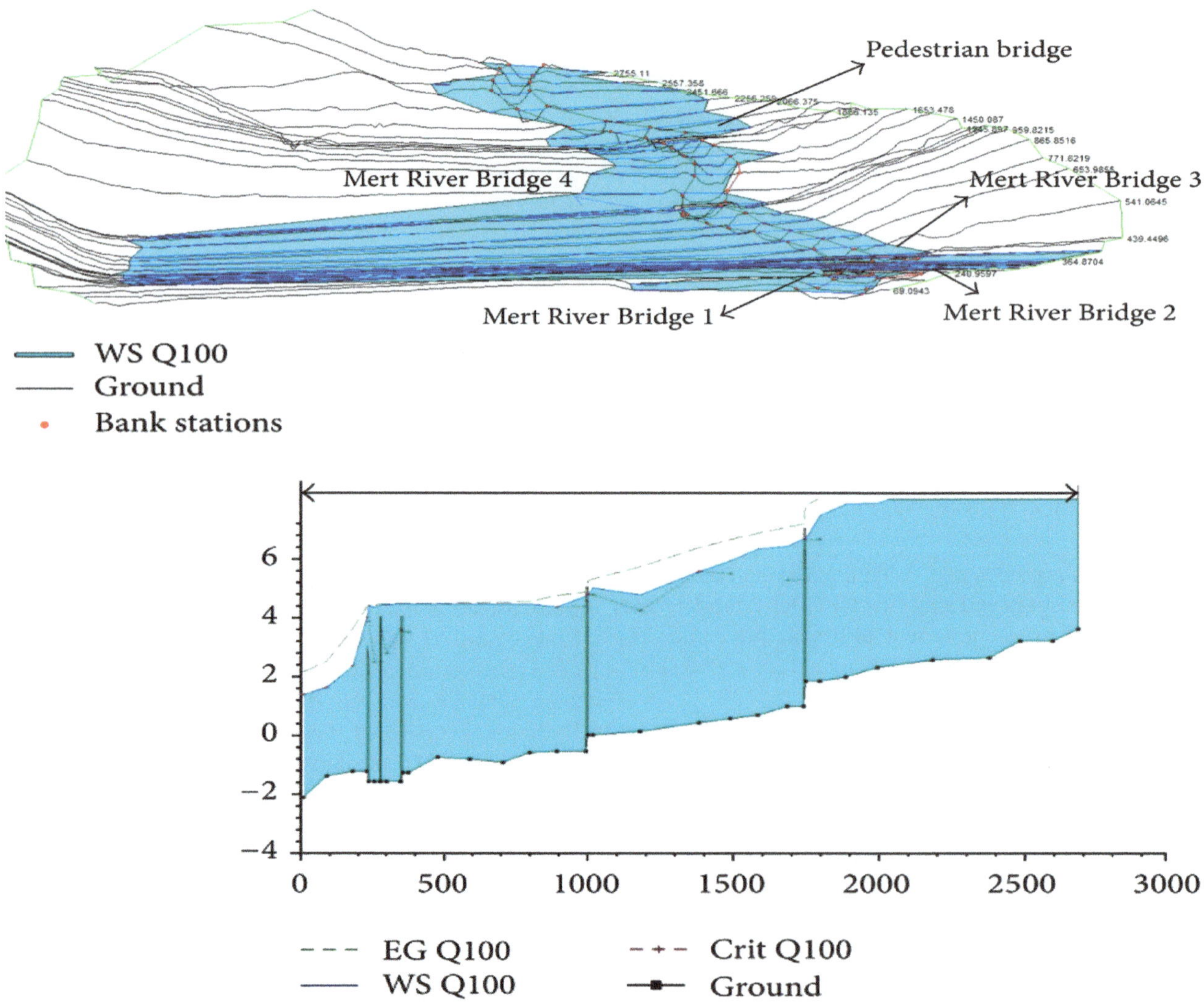

Figure 3: Water surface profile for the Q100 flood.

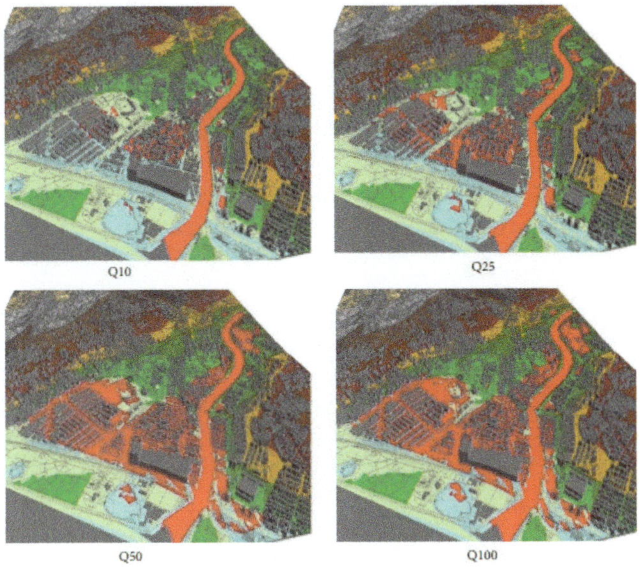

Figure 4: 3D hazard maps of the Mert River obtained for the Q10, Q25, Q50, and Q100 floods.

Flood depths for each return period were illustrated in Figure 5. The maps clearly demonstrates that when Q10 flood happens, the maximum depth is 6.2 m and affected area is approximately 30% (according to the urban area) in the downstream of the Mert River and the maximum depth and flooded area, respectively, increase to 7.6 m and 60% in the case of Q100 flood. This indicates the flatness of the study area. Concerning the quantity of affected residential area, 650 housings were affected by the 10-year event. This increases to 780, 840, and 960 housings in the case of Q25, Q50, and Q100 floods, respectively.

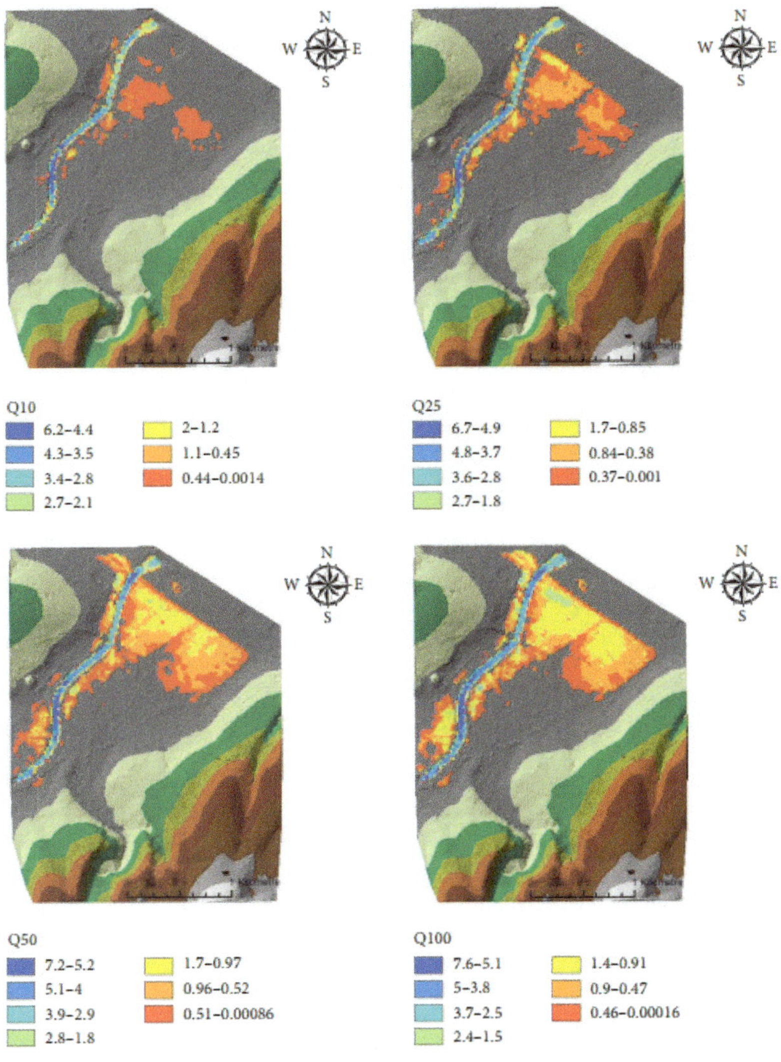

Figure 5: Water elevation maps of the studied area for the Q10, Q25, Q50, and Q100 floods.

2012 flood where loss of life and property occurred was also simulated in the present study. Flood hazard map and a photograph indicating a flood instant are outlined in Figure 6. It is clear from the figure that the influenced area is approximately 30% like the Q10 flood. The greatest hazards occur on the right side of the river which is mostly covered by industrial area. The flood magnitude alters a little on the left side of the river and the water reaches just a small number of houses near the river bank. It is clear from the hazard map prepared according to the 2012 flood which appeared in Figure 6 that the maximum depth is around 1 and 1.9 m in the residential area. A flooded building demonstrates that the water level in this area increased to 1–1.5 m when 2012 flood occurred. 619 housings were affected by the 2012 flood. This indicates that the simulation results obtained in this study correspond to the real flood hazard.

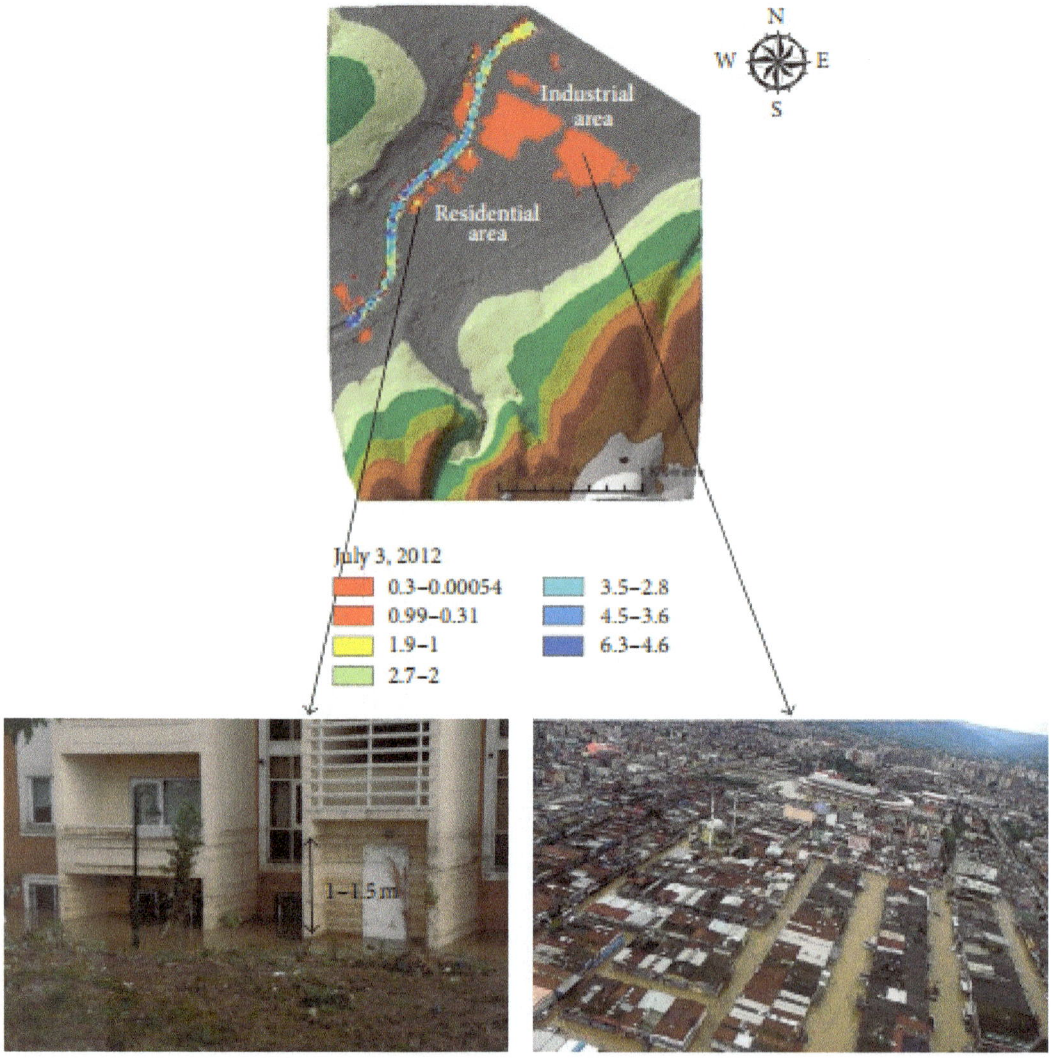

Figure 6: Flood hazard map and a photograph indicating the flood instants in industrial and residential area for the flood of July 3, 2012 [1, 22–24].

Flood of July 3, 2012, demonstrated that some areas (traffic roads and buildings surrounding the Mert River) are highly affected even though they have a low recurrence period (close to Q10). The flooded area is located in downstream of Mert River and includes industrial region and residential buildings. It ought to be noticed that the buildings are placed near watercourses. All these indicate a deficient urban planning which results in occupation of river and/or natural flooding areas [25].

The analyzed cross sections of Mert River and flooded area in the case of 2012 flood are represented in Figure 7. The flood impact additionally appeared for the selected section (red line) in this figure (see Figure 7(a)). Figure7(b) shows the prevention of flood by adding levee and regulation of river bottom. Dotted line in cross section indicates the swell height of the flood.

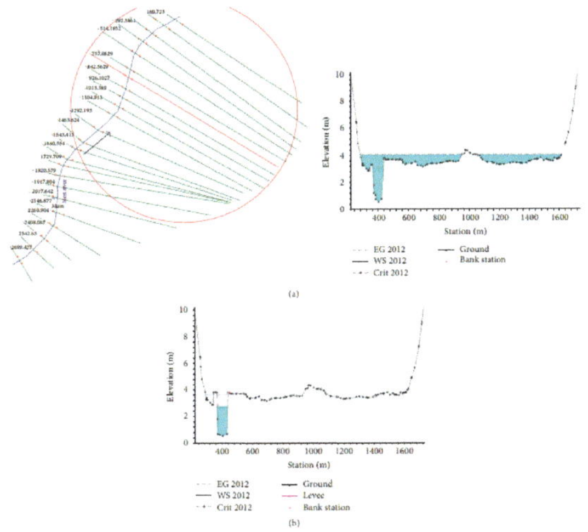

Figure 7: The analyzed cross sections of Mert River and flooded area in the case of 2012 flood: (a) flood effect for the selected section; (b) prevention of flood by adding levee and regulation of river bottom.

For the duration of an intense storm, real-time analysis includes using observed rainfall or gauged stage upstream as input for hydrologic modeling, utilizing output flow rates to hydraulic modeling, and finally mapping the output (flood hazard mapping) by a GIS program. Then, this information is utilized to manage flood warning activities such as voidances and road closures. However, the stream velocities are usually too great during a flood to make the flood hazard mapping practical. For solving this problem, the flood hazard mapping procedures employed in this study may be utilized to prepare a series of flood hazard maps taking into account diverse return periods. In the duration of an intense storm, the flood warning controller can choose the most appropriate digital flood hazard map that corresponds most closely to the real-time measured stream flow [26].

Numerous existing flood hazard maps require revision since they are outdated. The flood hazard mapping outlined in this study saves time and money versus traditional flood hazard delineation on paper maps. By this way, flood hazard maps can be regularly updated as variations in hydrologic and hydraulic conditions warrant [26].

CONCLUSIONS

Flood hazard mapping of Mert River Basin, Samsun, Turkey, was investigated using GIS and HEC-RAS in this study. 3D hazard maps were obtained for the Q10, Q25, Q50, and Q100 floods. The flood maps demonstrated that some areas are highly affected from flood for low return period (Q10) event. Through Q10 flood, the maximum depth reached 6.2 m and affected area was approximately 30% in the downstream of the Mert River. In addition, 650 housing were affected by this flood. All these indicated an insufficient urban planning in this area. Significant floods occurred for the 100-year return period on the downstream of the Mert River and three bridges out of five remained under flood. Flood hazard map of the 2012 flood where human life losses and a significant amount of property damages were experienced was additionally prepared utilizing GIS and HEC-RAS programs. The simulation results of the Q10 and 2012 floods were compared with each other and similarity was found between them. The studied area generally covers industrial and residential areas. It was seen that floods can be prevented in this region by adding levee and regulation of river bottom. Otherwise, the majority of this flooded area ought to be forested and/or kept as park area.

REFERENCES

1. V. Demir, Mert River (SAMSUN/TURKEY) the Flood Maps Determining the Help of Geographic Information Systems, Canik Basari University Institute of Natural Sciences, Samsun, Turkey, 2015.
2. C. Goodell and C. Warren, "Flood inundation mapping using HEC-RAS," Obras y Proyectos, pp. 18–23, 2006.
3. J. J. Wiles and N. S. Levine, "A combined GIS and HEC model for the analysis of the effect of urbanization on flooding; the Swan Creek watershed, Ohio," Environmental & Engineering Geoscience, vol. 8, no. 1, pp. 47–61, 2002.
4. M. Gutry-Korycka, A. Magnuszewski, J. Suchozebrski, W. Jaworski, M. Marcinkowski, and M. Szydlowski, "Numerical estimation of flood zones in the Vistula River valley, Warsaw, Poland," in Climate Variability and Change—Hydrological Impacts, pp. 191–195, IAHS-AISH Publication, 2006.
5. A. Sole, L. Giosa, and V. Copertino, "Risk flood areas, a study case: Basilicata region," in River Basin Management IV, vol. 1, pp. 213–228, WIT Press, Southampton, UK, 2007.
6. F. Yazıcılar and H. Önder, "Taşkın Yatakları Planlamasında HEC-RAS Bilgisayar Programı ile Su Yüzü Profili Hesaplanması-Bartın Nehrinde Bir Uygulama," in Su Mühendisliği Problemleri Semineri, Muğla, Turkey, September 1998.
7. E. C. Tate, D. R. Maidment, F. Olivera, and D. J. Anderson, "Creating a terrain model for floodplain mapping," Journal of Hydrologic Engineering, vol. 7, no. 2, pp. 100–108, 2002.
8. S. Mosquera-Machado and S. Ahmad, "Flood hazard assessment of Atrato River in Colombia," Water Resources Management, vol. 21, no. 3, pp. 591–609, 2007.
9. N. A. Sheffer, M. Rico, Y. Enzel, G. Benito, and T. Grodek, "The Palaeoflood record of the Gardon River,

France: a comparison with the extreme 2002 flood event," Geomorphology, vol. 98, no. 1-2, pp. 71–83, 2008.

10. M. Masood and K. Takeuchi, "Assessment of flood hazard, vulnerability and risk of mid-eastern Dhaka using DEM and 1D hydrodynamic model," Natural Hazards, vol. 61, no. 2, pp. 757–770, 2012.
11. V. Heimhuber, J.-C. Hannemann, and W. Rieger, "Flood risk management in remote and impoverished areas—a case study of Onaville, Haiti," Water, vol. 7, no. 7, pp. 3832–3860, 2015.
12. H. E. Çelik, G. Coskun, H. K. Cigizoglu, N. Ağıralioğlu, A. Aydin, and A. I. Esin, "The analysis of 2004 flood on Kozdere Stream in Istanbul," Natural Hazards, vol. 63, no. 2, pp. 461–477, 2012.
13. A. Sarhadi, S. Soltani, and R. Modarres, "Probabilistic flood inundation mapping of ungauged rivers: linking GIS techniques and frequency analysis," Journal of Hydrology, vol. 458-459, pp. 68–86, 2012.
14. C. Fan, C.-H. Ko, and W.-S. Wang, "An innovative modeling approach using Qual2K and HEC-RAS integration to assess the impact of tidal effect on River Water quality simulation," Journal of Environmental Management, vol. 90, no. 5, pp. 1824–1832, 2009.
15. A. A. Baky, A. M. Zaman, and A. U. Khan, "Managing flood flows for crop production risk management with hydraulic and GIS modeling: case study of agricultural areas in Shariatpur," APCBEE Procedia, vol. 1, pp. 318–324, 2012.
16. E. C. Carson, "Hydrologic modeling of flood conveyance and impacts of historic overbank sedimentation on West Fork Black's Fork, Uinta Mountains, northeastern Utah, USA," Geomorphology, vol. 75, no. 3-4, pp. 368–383, 2006.
17. F. Pappenberger, K. Beven, M. Horritt, and S. Blazkova, "Uncertainty in the calibration of effective roughness parameters in HEC-RAS using inundation and downstream level observations," Journal of Hydrology, vol. 302, no. 1–4, pp. 46–69, 2005.
18. G. Brunner and V. Bonner, HEC River Analysis System (HEC-RAS), US Army Corps of Engineers Institute for Water Resources Hydrologic Engineering Center, Davis, Calif, USA, 2010.
19. D. K. Gautam and R. G. Kharbuja, "Flood hazard mapping of bagmati river in kathmandu valley using geo-informatics tools," Journal of Hydrology and Meteorology, vol. 3, no. 2677, pp. 1–9, 2006.
20. HEC-GeoRAS, HEC-GeoRAS GIS Tools for Support of HEC-RAS using ARCGIS, HEC-GeoRAS, 2009.
21. A. Haghizadeh, L. Teang Shui, M. Mirzaei, and H. Memarian, "Incorporation of GIS based program intohydraulic model for water level modeling on river basin," Journal of Water Resource and Protection, vol. 4, pp. 25–31, 2012.
22. Samsunda-sel-felaketi-on-raporu-aciklandi-2012-07-08, 2012, http://www.ensonhaber.com/samsunda-sel-felaketi-on-raporu-aciklandi-2012-07-08.html.
23. NTV, "Samsunda-sel-9-can aldi, 2012," 2012, http://www.ntv.com.tr/galeri/turkiye/samsunda-sel-9-can-aldi,Qz9WsxxiRUuFyoEQ6BtJHg/.
24. 2012, http://www.samsunmimar.org/dosyalar/134218056132631.doc.
25. F. V. Silva, N. B. Bonumá, and P. K. Uda, "Flood mapping in urban area using HEC-RAS model supported by GIS," in Proceedings of the 6th International Conference on Flood Management (ICFM6 '14), São Paulo, Brazil, September 2014.
26. E. Tate and D. Maidment, "Floodplain mapping using HEC-RAS and ArcView GIS," CRWR Online Report 99-1, Center for Research in Water Resources, 1999.

CHAPTER 11

URBAN FLOOD MAPPING BASED ON UNMANNED AERIAL VEHICLE REMOTE SENSING AND RANDOM FOREST CLASSIFIER

Quanlong Feng, Jiantao Liu and Jianhua Gong

Keywords: UAV, flood mapping, urban landscape, random forest, texture analysis

INTRODUCTION

Flooding is among one of the most widespread and destructive natural disasters, which exerts a heavy toll on human life and property [1,2,3,4,5]. According to the United Nations, more people lose their lives due to floods than other natural hazards [1]. In addition, the damage caused by floods is much greater in highly-developed and densely-populated urban areas than in rural countrysides. Therefore, it is of vital significance to map flood extent areas rapidly and accurately to provide an overall view of the flood event in order to plan relief work efficiently, especially in urban districts.

Remote sensing has been recognized as a powerful tool to provide inundation maps in near real time according to a great deal of researches [1,3,4,5,6,7,8,9,10,11,12,13]. Compared with *in situ* measurement, remote sensing can provide synoptic and continuous coverage of flood events, which aids flood monitoring and damage assessment [5,6,7,8,9]. In general, remote sensing data used for flood mapping consists of space-borne and airborne imageries. As for the former, optical and radar data have been widely adopted to extract inundation areas with high accuracy [5,6,7,8,9,10,11,12,13]. The advantage of optical satellite remote sensing [5,6] is that it provides true color images through which visual interpretation and automatic classification can be done to provide first-hand information of flooded areas. Wang [8] used moderate resolution (30 m) Landsat 7 Thematic Mapper (TM) imagery to delineate the maximum flood event caused by Hurricane Floyd in North Carolina. Animi [5] proposed a model to generate a floodplain map based on high-resolution (1 m) Ikonos imagery and digital elevation model (DEM). Artificial neural network (ANN) was utilized in Animi's research [5] to classify the Ikonos imagery and the classification accuracy was improved by 15% compared to maximum likelihood (ML). However, the presence of extensive clouds during flood events often limits the feasibility of optical satellite imagery.

Space-borne radar enables data acquisition regardless of weather conditions because microwaves have the capability to penetrate clouds, which makes it an effective alternative for optical sensors in the differentiation of flooded from non-flooded areas [7,8,11]. Kuenzer *et al.* [1] employed the longest time series thus far of Environmental Satellite-Advanced Synthetic Aperture Radar (ENVISAT-ASAR) data to derive flood and inundation information for the Mekong Delta region, which contributes to a comprehensive understanding of Mekong Delta flood dynamics. Gstaiger *et al.* [7] used multi-sensoral data from Terra Synthetic Aperture Radar-X (TerraSAR-X) and ENVISAT-ASAR to automatically derive inundated areas also in the Mekong Delta. Although both optical and radar satellite remote sensing have been proven to extract flooded areas effectively, their disadvantages in urban flood mapping cannot be covered up due to the following reasons. First, satellite remote sensing imageries of the flooded areas are not always available because of revisit limitations. Second, the common used data (e.g., TM and ENVISAT-ASAR) are unable to capture the details of complex urban landscapes due to a relatively lower spatial resolution. High-

Citation: Quanlong Feng, Jiantao Liu, and Jianhua Gong, Urban Flood Mapping Based on Unmanned Aerial Vehicle Remote Sensing and Random Forest Classifier, doi:10.3390/w7041437.

resolution images, such as Ikonos and QuickBird, can provide richer information than TM and ENVISAT, yet the long revisit circle and exorbitant prices still limit their application in urban flood monitoring.

Compared with space-borne observations, aerial remote sensing is immune to extensive clouds and revisit limitations, which manifests itself as an ideal tool for flood monitoring [14,15,16,17]. In addition to this, aerial images are acquired at decimeter and sub-decimeter levels, representing much more abundant details than satellite data. In particular, piloted aircraft and unmanned aerial vehicle (UAV) are two main platforms for aerial remote sensing. The onboard sensors vary from off-the-shelf digital cameras and multispectral cameras to hyperspectral imager to Light Detection and Ranging equipment (LiDAR) [14,15]. The main drawback of piloted aircrafts in urban flood mapping is that it is always difficult to find ideal taking-off and landing places for the aircrafts, especially for fixed-wing airplanes. However, mini-UAVs provide a much safer and more convenient way to acquire data in disaster areas because they can be launched by hand and landed on a small piece of flat ground [15,16]. In addition, UAV can be deployed easily and frequently to satisfy the requirements of rapid monitoring and mapping in flooded areas at a user-defined spatio-temporal scale. However, research regarding the use of UAVs for urban flood mapping is still rare. Under this context, we are motivated to justify the performance of UAV remote sensing in extracting inundated areas under complex urban landscapes.

Commonly used methods [1,3,4,5,6,7,8,9,10,11,12,13] to extract flooded areas include thresholding, maximum likelihood, decision trees, *etc*. In this paper, we adopted a relatively novel classifier, Random Forest (RF) [18], for urban flood mapping. According to current research [18,19,20,21,22,23], the performance of Random Forest in flood mapping has not been well documented, thus, we are highly motivated to justify its performance. Meanwhile, texture analysis [24,25,26,27,28,29] was also used to provide enough shape and contextual information for the RF classifier to improve its accuracy.

The overall objective of this study is to develop a rapid and accurate method for urban flood mapping utilizing Random Forest and texture analysis of high-resolution UAV imagery (0.2 m). More specifically, this study's hypotheses are (i) to justify the feasibility of UAV remote sensing in urban flood mapping; (ii) to determine if Random Forest classifier shows good performance; and (iii) to justify whether the inclusion of texture features significantly improves classification accuracy in highly heterogeneous urban landscapes.

STUDY AREA

The study area is Yuyao City of Zhejiang Province in Eastern China (Figure 1). Yuyao City is located on the south shore of Hangzhou Bay. The downtown area is located in a relatively open and flat floodplain. The Yuyao River and its tributary, the Zuiliang River, flow through the whole city from west to east. Influenced by Typhoon Fitow, Yuyao experienced extreme precipitation on 7 October 2013, which led to the most serious floods in the last 60 years. Most downtown areas were inundated for more than five days and the direct economic losses were more than 69.61 billion RMB (about 11.33 billion USD). Figure 1b also depicts the overall Red-Green-Blue (RGB) image acquired by UAV, which covers the worst-hit areas of downtown Yuyao. It has a high spatial resolution of 0.2 m, which allows for an accurate extraction of the inundated areas. The size of the whole image is 19,965 × 12,555 covering an area of about 10 km^2.

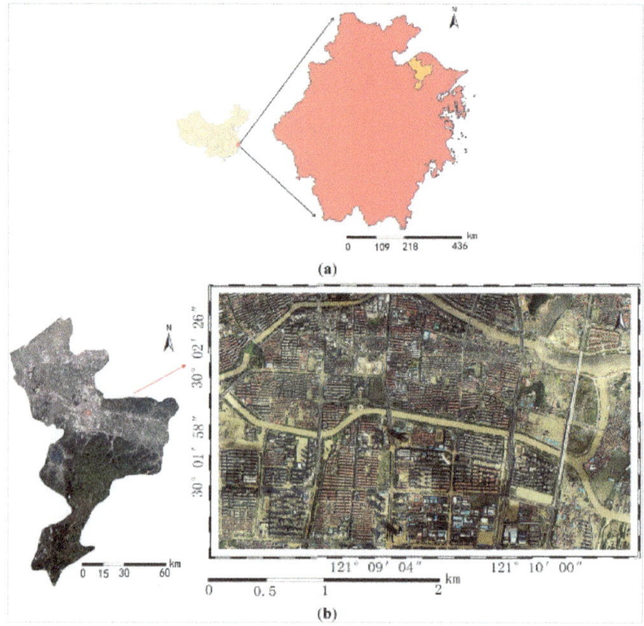

Figure 1. Study area. (**a**) Zhejiang Province; and (**b**) Yuyao City.

METHODS

Workflow

The overall workflow of the proposed method for urban flood mapping is depicted in Figure 2. There are four major steps in the workflow: (i) UAV data acquisition and preprocessing (Section 3.2); (ii) feature selection and texture analysis (Section 3.3); (iii) image classification (Section 3.4); and (iv) accuracy assessment (Section 3.5). Data preprocessing was a prerequisite for urban flood mapping, which involved downloading data from UAV digital camera, image registration, orthorectification, and automatic mosaicking. Feature selection was based on gray-level co-occurrence matrix (GLCM). The extracted texture features were combined with raw RGB images to construct a multi-dimensional feature space. Training samples were randomly picked to train Random Forest, and the trained classifier was then utilized to extract the inundated areas. Accuracy assessment was done based on a confusion matrix derived from validation samples to test the performance of the proposed method.

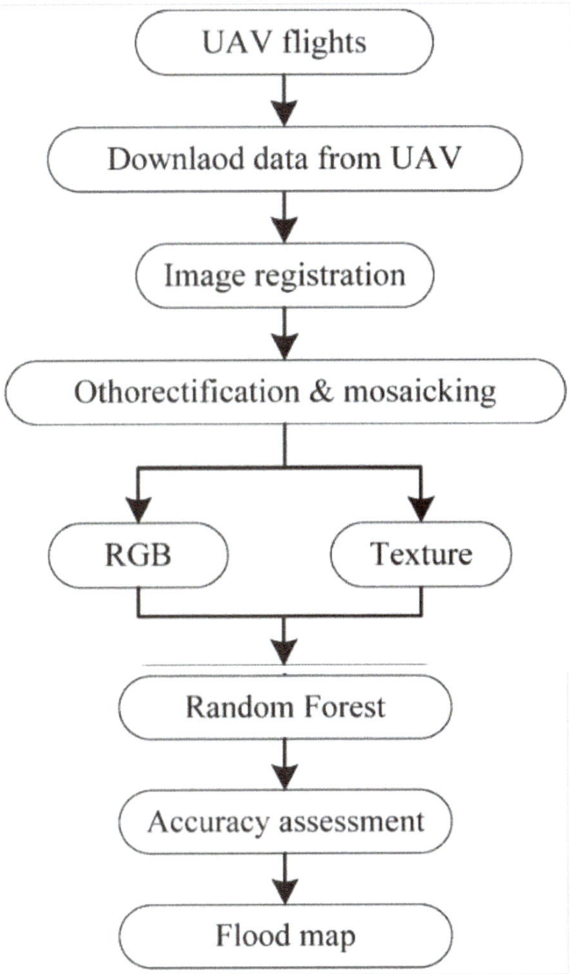

Figure 2. Workflow of urban flood mapping.

Data Acquisition and Preprocessing

In order to monitor the floods and provide the inundation map for emergency relief workers, a mini-UAV called River-Map was deployed twice on 10 October 2013. A small, wide-field camera, GoPro, was onboard during the first flight to provide real time surveillance and an RGB digital camera was used in the second flight to acquire images for further mapping of the inundated areas. The video images are shown in Figure 3, from which we can see clearly that Yuyao City had experienced serious and extensive flooding. Most roads and bottomlands were submerged, which caused great difficulty for relief work.

The River-Map UAV (Figure 4) belongs to small electric UAVs. It has good flexibility and mobility, which facilitates its application in urban cartography, precision agriculture, and water pollution surveys [15]. River-Map UAV can be launched by hand and landed on a small piece of flat ground, which makes it particularly suitable in monitoring urban flooding. It has a wingspan of 2.5 m and a length of 1.58 m. The payload capacity of River-Map is 4.5 kg and its endurance is about 1.5 h. It has an auto pilot for the control of the entire craft, a GPS for navigation, and an IMU (Inertial Measurement Unit) for inertia measurements. Additionally, mission planning and information interaction are done through the ground control station. The flight altitude was set to be 350 m, resulting in a high resolution of 0.2 m per pixel. To satisfy the requirements of aerial photography, the forward lap was set to be 60% and the side lap 30% during mission planning. During the fight, it took about one hour for the UAV to fly over an area of 10 km^2 in this study. Because the River-Map UAV is battery-powered and has an endurance of 1.5 h, it has a maximum mapping area of 15 km^2 during one flight.

Figure 3. Typical flood depictions acquired by wide field camera on board the UAV: (**a**) flooded residential areas; (**b**) flooded commercial areas.

Figure 4. River-Map UAV.

Raw data acquired by River-Map UAV were a series of RGB images with a central projection, which needed orthorectification to generate orthophotos. We chose Pix4D software [15] as the preprocessing tool due to its high efficiency and good accuracy. Raw images were then orthorectified and mosaicked into a single image using Pix4D with a root mean square error of 0.2 pixels.

Texture Analysis

Feature selection is of great importance in image classification [25]. Good features can increase the between-class separability while they decrease the within-class variance. Due to the limitation of payload capacity, an off-the-shelf digital camera was on board the River-Map UAV. Raw images acquired had only three bands (RGB), which resulted in a lack of spectral features. It was difficult to yield a high classification accuracy using only the RGB features because different ground objects may have similar colors. The details of submerged roads, bare soil, inundated grasslands, and vegetation are shown in Figure 5. Because of the high levels of muds and sediments, the submerged roads have the similar color as the yellow bare soil. In addition, the inundated grasslands are rather green, which are difficult to separate from non-flooded vegetation. However, the flooded areas (both submerged roads and grasslands) are much more homogenous than the fragmented bare soil and vegetation, which demonstrates that the inclusion of texture features may increase the between-class separability.

Figure 5. (**a**) Submerged roads; (**b**) bare soil; (**c**) submerged grassland; (**d**) vegetation.

$$\mathrm{MEA}i = \sum_{i=0}^{N-1}\sum_{j=0}^{N-1} i \times P(i,j) \tag{1}$$

$$\mathrm{STD}\,i = \sqrt{\sum_{i=0}^{N-1}\sum_{j=0}^{N-1} P(i,j) \times (i - \mathrm{MEA}\,i)^2} \tag{2}$$

$$\mathrm{HOM} = \sum_{i=0}^{N-1}\sum_{j=0}^{N-1} \frac{P(i,j)}{1+(i-j)^2} \tag{3}$$

$$\mathrm{DIS} = \sum_{i=0}^{N-1}\sum_{j=0}^{N-1} P(i,j) \times |i-j| \tag{4}$$

$$\mathrm{ENT} = \sum_{i=0}^{N-1}\sum_{j=0}^{N-1} -P(i,j) \times \ln(P(i,j)) \tag{5}$$

$$\mathrm{ASM} = \sum_{i=0}^{N-1}\sum_{j=0}^{N-1} P(i,j)^2 \tag{6}$$

where N is the number of grey levels; P is the normalized symmetric GLCM of dimension $N \times N$; $P(i, j)$ is the normalized grey level value in the cell i, j of the co-occurrence matrix, such that the sum of $P(i, j)$ equals to 1 [15]. The calculations of texture statistics were based on a moving window around each pixel and the statistics were then attributed the pixel itself [24]. In general, the bigger the window size, the coarser the information that can be provided by texture features. The contribution of texture features to classification accuracy depends on both the texture scale and the ground objects scale. Experiments were done to find that textures calculated at a 5 × 5 moving window yielded the highest accuracy. Hence, the six texture features derived at a 5 × 5 window were added as additional ancillary bands to the RGB images, which were ready for Random Forest classification.

Random Forest Classifier

After feature selection, a classifier should be utilized to classify different ground objects in order to extract flooded areas. A robust and high-performance classifier has always been the key issue to yield a high precision in image classification. As a relatively novel machine learning based classifier, Random Forest has rarely been used in urban flood mapping [18,19,20,21,22,23] and its performance needs to be studied and justified.

Random Forest is an ensemble learning method proposed by Breiman in 2001 [18], which is a collection of independent individual classification and regression tree (CART) classifiers, and can be defined as Equation (7):

$$\{h(x, \theta_k), k = 1, 2...i...\} \tag{6}$$

where h represents Random Forest classifier, x stands for input variable, and $\{_k\}$ are independently identically distributed (i.i.d.) random predictor variables, which are used for generating each CART tree. The final response of Radom Forest is calculated based on the output of all the decision trees involved [18]. Compared with other machine learning methods, such as support vector machine and artificial neural network, Random Forest has a relatively lower computational burden, and it is insensitive to multivariate linear variables and outliers [19]. In addition, Random Forest shows outstanding performance in high dimensional feature space, which provides great potential for classifying complex and texture-abundant UAV images. Another important advantage of Random Forest is that it can measure the importance of input variables, which enables researchers to better understand the contribution of each variable to the overall classification accuracy. The schematic diagram of classification using Random Forest is illustrated in Figure 6.

As depicted in Figure 6, training samples were firstly selected using a stratified sampling method. Bootstrap strategy was then adopted to draw about 2/3 of the total training samples with replacement to reduce generalization error. The remaining 1/3 of the samples are called out-of-bag (OOB) data, which are used for cross-validation to evaluate the performance of Random Forest. Gini index [18] was used as a measure of heterogeneity to split each individual CART tree.

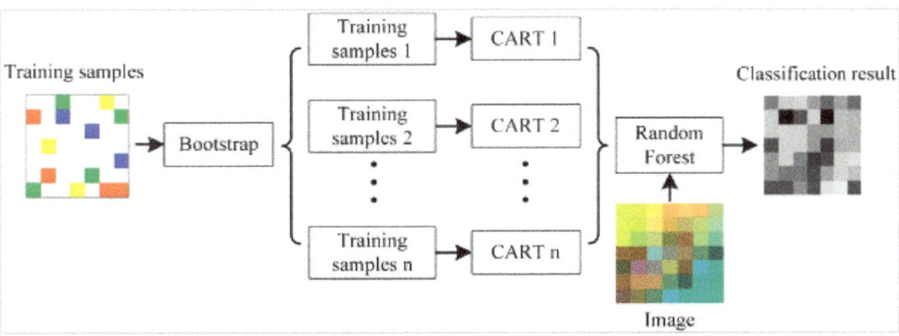

Figure 6. Schematic diagram of Random Forest for image classification.

Only two parameters are utilized to construct a Random Forest classifier: *ntree*, the number of trees to grow into a whole forest, and *mty*, the number of randomly selected predictor variables [20]. With the increment of *ntree*, OOB error decreases gradually. In addition, according to the Law of Large Numbers, OOB error is convergent when *ntree* is bigger than a certain threshold. A smaller *mty* indicates a weaker prediction ability of each individual tree, but also a smaller correlation between different trees, which contributes to reduce the generalization error. There are two methods to calculate *mty*, *i.e.*, 1/3 or square root of the number of input variables.

Because Random Forest belongs to supervised classification method, training samples for each ground object should be selected separately. According to field survey and visual interpretation, four classes were initially chosen as follows: water, impervious surface, vegetation, and bare soil, and the number of training sample for each class was set to be 500. Training samples were then randomly chosen in a small polygon block, and it was assumed that all pixels within each polygon belonged to the same category [15]. After the classification, impervious surface, vegetation, and bare soil were merged into the non-flooded class. Since the classified water consisted of both flooded areas and persistent water (rivers, lakes, *etc.*), the latter were then eliminated according to the city map of Yuyao given by the official authority.

Accuracy Assessment

In the accuracy assessment of the proposed urban flood mapping method, a confusion matrix was calculated based on the validation samples derived from visual interpretation of the high resolution UAV orthophotos. These additional samples were selected independently from the training samples, and the number of the validation samples was 5000. Overall accuracy and Kappa index can be derived from the confusion matrix to quantify the performance of Random Forest.

However, the field measurements of acquiring validation data were not performed in this study. Validation samples were selected from the UAV images by visual inspection. Two reasons accounting for this are as follows. Firstly, the image acquired by the UAV has a very high resolution of 0.2 m, which implies that we can see clearly whether certain areas had been inundated or not. Secondly, most roads were severely submerged after the urban water logging of Yuyao City, and the actual traffic conditions made it difficult to carry out the field measurements. Some errors in selecting validation samples due to the blocking effect of dense vegetation and tall buildings may exist. Additionally, the shadows caused by tall buildings may also degrade the accuracy. The lack of actual field measurements was a defect of this paper and may cause errors in accuracy assessment.

RESULTS

Parameterization of Random Forest

As mentioned above, two important parameters *ntree* (the number of individual trees) and *mty* (the number of randomly selected predictor variables) should be tuned to improve the performance of the Random Forest classifier. The input image consisted of three spectral bands (e.g., RGB) and six texture bands, *i.e.*, the number of input variables was 9, thus *mty* was set to be 3 (the square root of 9). As for *ntree*, an optimal value can be achieved when the OOB error began to converge. To obtain the best estimation of *ntree*, a relatively large value of 600 was chosen to construct the Random Forest model and the relationship between *ntree* and OOB error is shown in Figure 7.

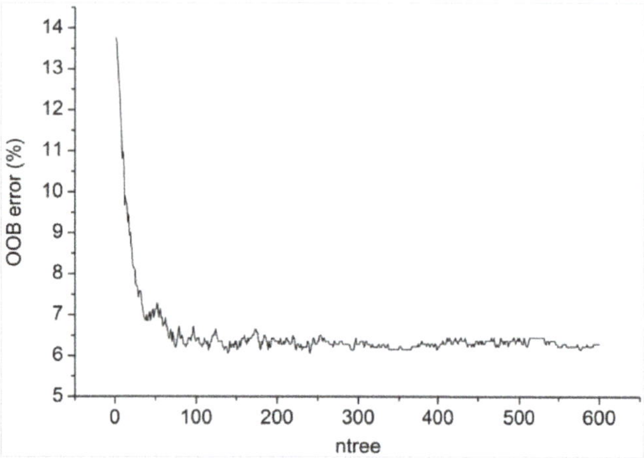

Figure 7. OOB error of Random Forest.

It can be seen from Figure 7 that OOB error decreases sharply from 13.7% to 7.2% as *ntree* increases from 1 to 35. Additionally, OOB error continues to decrease with minor fluctuations at a much slower speed until *ntree* reaches 200. OOB error stays rather stable at 6.4% when *ntree* is after 200. Therefore, *ntree* was set to be 200, which has a relatively higher accuracy and reduces calculation amounts at the same time.

Classification Results of UAV Imagery

A Random Forest with 200 decision trees was utilized to classify the original UAV RGB image (RGB-only) and image added with texture features (RGB + texture). The classification results for RGB-only and RGB + texture are depicted inFigure 8.

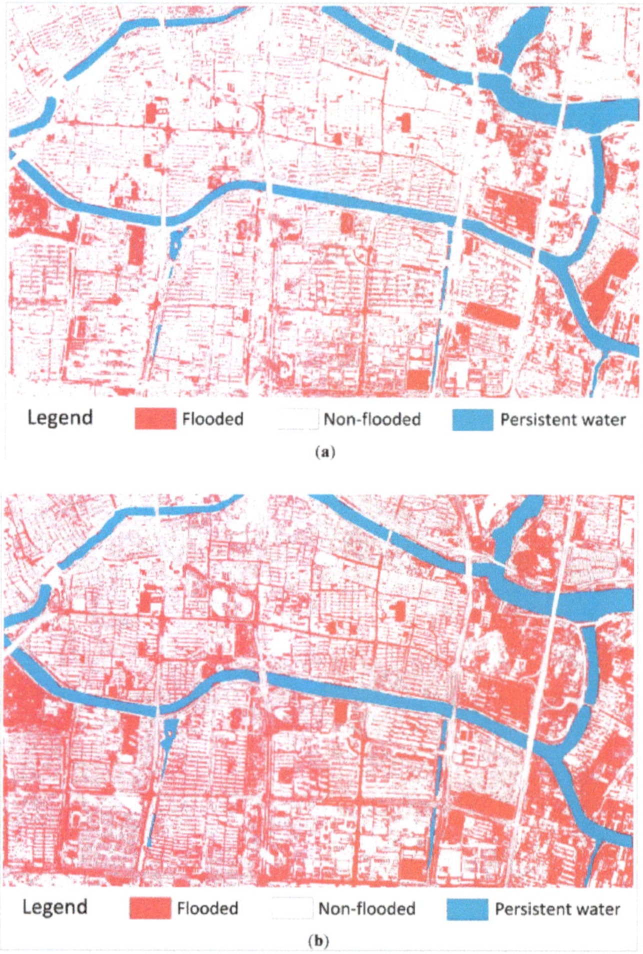

Figure 8 illustrates that significant differences of classification results, before and after the inclusion of texture features, exist. When classifying using the RGB-only image, large amounts of flooded areas were misclassified into non-flooded areas. The area statistics of RGB-only and RGB + texture are shown in Table 1.

Table 1. Class statistics for RGB-only and RGB + texture.

Type	RGB-Only		RGB + Texture	
	Area (km²)	Area (%)	Area (km²)	Area (%)
Flooded	1.845	18.40	3.367	33.58
Non-flooded	7.444	74.26	5.922	59.07
Persistent water	0.737	7.34	0.737	7.34
Total	10.026	100	10.026	100

Table 1 indicates that classification using the RGB-only image greatly underestimated the flooded areas. The extracted flooded areas only account for 18.4%, which is less than that of RGB + texture (33.58%). This is mainly due to the spectral similarity between flooded areas and other ground objects. The submerged roads show in yellow because the flood water consists of a large amount of mud and sediments. In addition, the flooded grasslands are in green due to the grass beneath the water. All these factors lead to the difficulty in separating flooded areas from bare soil and vegetation, which, in turn, results in the underestimation of inundated areas. However, the inclusion of texture features can improve this situation. This is mainly because flooded waters show rather homogeneous textures compared to those of bare soil and grassland. The inclusion of texture features can increase the between-class separability, hence to improve the urban flood mapping results.

In order to demonstrate the inundation map in detail, two typical regions, together with digitized results and the extracted flooded areas of RGB-only and RGB + texture, are shown in Figure 9.

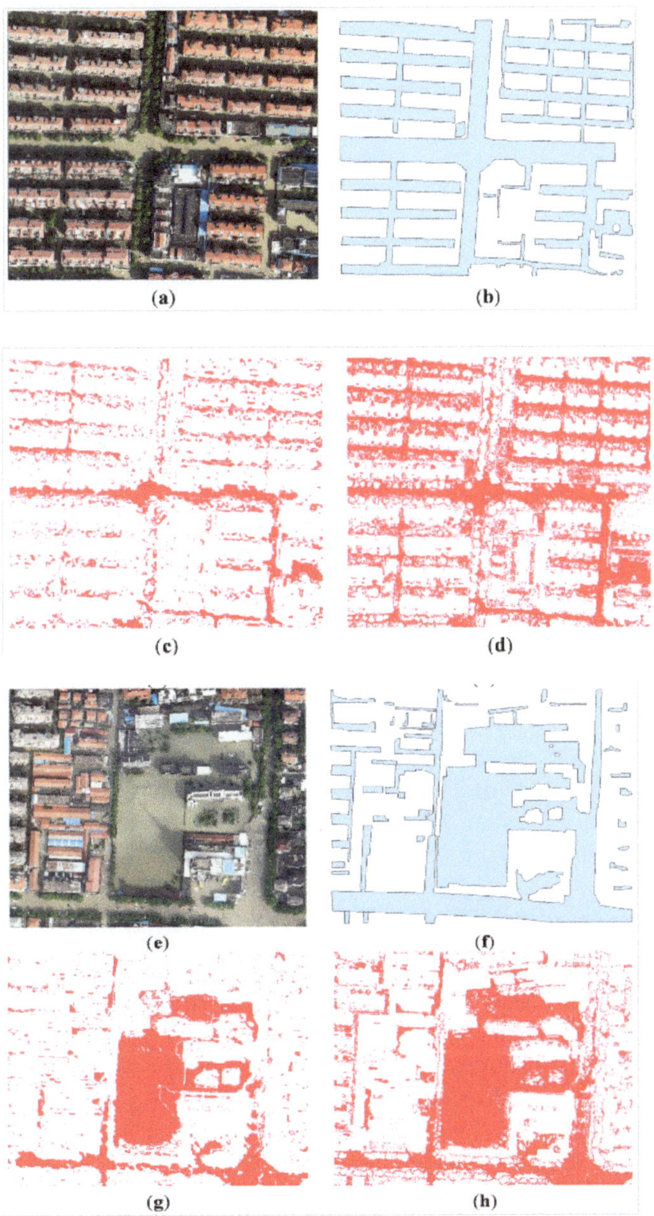

Figure 9. Image-A: (**a**) original RGB image; (**b**) digitized flooded areas; (**c**) extraction result of RGB-only; (**d**) extraction result of RGB + texture; Image-B: (**e**) original RGB image; (**f**) digitized flooded areas; (**g**) extraction result of RGB-only; (**h**) extraction result of RGB + texture.

Figure 9 clearly depicts that the proposed method (RGB + texture) outperforms "RGB-only" and can extract more inundated areas. However, the former still underestimates some inundated areas, including areas under the trees and areas in shadows, when compared to UAV orthophotos and the digitized flooded areas. This underestimation is caused by the following reasons: First, the fact that visible light can hardly penetrate the dense tree-crowns makes it difficult to detect the flooded areas under trees. Second, the shadows of tall buildings can change the spectral signature of the flooded areas, which in turn accounts for the underestimation.

Results of Accuracy Assessment

To quantitatively assess the classification accuracy before and after the inclusion of texture features, confusion matrix derived from validation samples was calculated for RGB-only and RGB + texture images, the results are shown in Table 2and Table 3.

Table 2. Confusion matrix for RGB-only.

Classification Results	Ground Truth		UA
	Flooded	Non-flooded	-
Flooded	3321	713	82.3%
Non-flooded	1679	4287	71.9%
PA	66.4%	85.7%	-
OA	76.1%	Kappa index	0.522

Table 3. Confusion matrix for RGB + texture.

Classification Results	Ground Truth		UA
	Flooded	Non-flooded	-
Flooded	3823	92	97.7%
Non-flooded	1177	4908	80.7%
PA	76.5%	98.2%	-
OA	87.3%	Kappa index	0.746

Notes: PA, Producer Accuracy; UA, User Accuracy; OA, Overall Accuracy.

Overall Accuracy and Kappa index for RGB-only and RGB + texture increased from 76.1% to 87.3% and 0.552 to 0.746, respectively. The significant increments of 11.2% and 0.194 were observed to verify that the inclusion of texture features can greatly improve classification accuracy.

Variable Importance

The importance of input variables given by Random Forest can be used to measure their contribution to classification accuracy [15]. The importance of the nine input variables is depicted in Figure 10. It can be seen that spectral features are more important than texture features from an overall perspective. Red band is the most important variable with a value of 5.56 followed by Blue band (5.15). The third most important variable is MEA, which exceeds Green band. The remaining variables all belong to texture features with relatively lower importance values. MEA indicates that, although the inclusion of texture features can improve the classification accuracy of Random Forest classifier, spectral features still remain the most important variables in urban flood mapping.

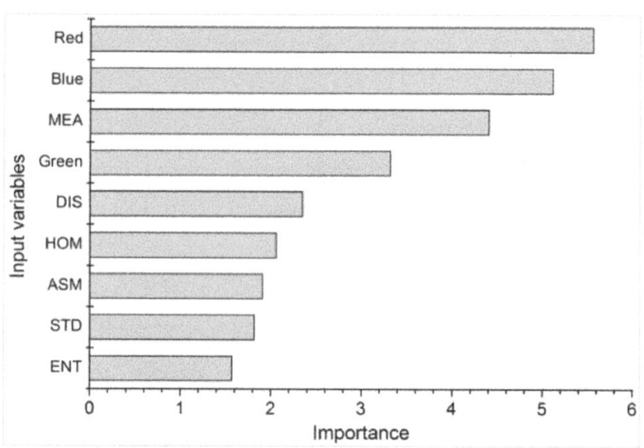

Figure 10. Importance of input variables.

Comparison with Other Classifiers

To further verify the performance of the proposed method, other classifiers including maximum likelihood (ML), artificial neural network (ANN), and support vector machine (SVM) were adopted to classify the RGB + texture image. The comparison results are described in Table 4.

Table 4. Accuracy comparison results of RF and other classifiers.

Method	Overall Accuracy (%
ML	81.2
ANN	83.6
SVM	87.8

Table 4 indicates that Random Forest shows an outstanding performance when compared to other classifiers. Specifically, the overall accuracy of RF (87.3%) is higher than ML (81.2%) and ANN (83.6%), but slightly lower than SVM (87.8%). However, SVM needs many experiments to search for the optimal combination of kernel function type, punishment coefficient, and kernel parameter Gamma. The complicated parameterization process of SVM cannot match that of RF, which needs only two parameters to be tuned. Based on this fact, RF shows great potential for urban flood mapping due to its high accuracy and simplicity.

DISCUSSION

The results of urban flood mapping indicate that UAVs can be an ideal platform for data acquisition during a flood event. UAVs have low dependence on launching and landing conditions, which makes them safer and more applicable than piloted aircrafts in urban flood monitoring. Additionally, a UAV is immune to extensive clouds due to its low-flight altitude, which makes it a feasible alternative of optical satellite remote sensing. The high-resolution images acquired by UAV can provide abundant details of ground objects, which show great potential for the precise mapping of inundated areas in heterogeneous urban landscapes. In addition, legal requirements when deploying UAV in China must be obeyed. According to Civil UAV Air Traffic Management Measures, enacted by the Civil Aviation Administration of China (CAAC), one should apply to the local air traffic control department for airspace and flight plans before the implementation of UAV flying activities.

Whether or not UAVs can provide real-time information mainly depends on the sensor used on board. When a digital video camera, such as GoPro, is on board the UAV, real-time images can be transferred back to the ground control station and be shown on the screen, which can provide first hand data for decision makers. When an off-the-shelf camera is on board, the RGB images acquired by a UAV can only be accessible to decision-makers after the landing of the UAV. In this case, UAV can still provide near real-time information, which is also very valuable for rescue operations after the urban flooding event.

Experimental results indicate that the inclusion of texture features can improve the classification accuracy, which verify the conclusion in many previous studies [24,25,26,27,28,29]. Since shape and contextual relationship are into consideration in texture analysis, between-class separability can be increased in the spectral-textural feature space, which contributes to the improvement of classification accuracy.

Random Forest shows high performance for urban flood mapping. The value of *ntree* should be optimized to decrease OOB error and the computational burden at the same time. Random Forest outperformed ML and ANN in the comparison experiments. Since ML assumes that all training samples are normally distributed [5], and it is difficult to meet this assumption in reality this results in a limitation of ML. Unlike ML, Random Forest does not need this assumption. The ensemble learning mechanism of Random Forest guarantees the good performance, regardless of the distribution pattern [18]. As for ANN, the main drawback is its low generalization capability due to over-fitting of the training data [29]. In contrast, Random Forest uses a bootstrap method to generate independent training samples to tackle the issue of over-fitting. In the end, Random Forest is slightly inferior to SVM in terms of classification accuracy. This is not surprising since SVM has been proven to be the best statistic learning theory, which is based on Vapnik-Chervonenkis (VC) dimension theory and structural risk minimization [30]. However, it is time consuming to test the optimal parameters of SVM, such as kernel function type, punishment coefficient, and so on [30]. Taking this into consideration, it can be inferred that a simpler classifier with a low computation burden and satisfactory accuracy will be the best choice for urban flood mapping. Random Forest is such a classifier for which performance has already been verified in this study.

However, the approach used in this paper can only extract the submerged areas based on two-dimensional UAV orthophotos and cannot provide depth information at the same time. This may be a demerit of the presented method. In order to get accurate vertical depth information, a Digital Elevation Model (DEM), before and after the flood event, is always necessary. Light Detection and Ranging equipment (LiDAR) could be used onboard the UAV to obtain those DEMs, and the submerged depth could be calculated by subtracting the DEM after the flood from the DEM before the flood invent.

In terms of the processing time of the presented approach, it took Pix4D about eight hours to process about 400 raw images to generate the final orthorectified image, covering an area of 10 km^2. Inundated area extraction consumed about one hour in this study, including texture calculation, training sample selection, and Random Forest classification. Therefore, the whole process of extracting the inundated areas from UAV images took about nine hours. From the experience, a period of nine hours to interpret the UAV remotely-sensed data can be acceptable, although not perfect, for decision makers.

CONCLUSIONS

This paper proposed a hybrid method for urban flood mapping by combining Random Forest and texture analysis based on high-resolution UAV imagery. Six least correlated GLCM texture features were calculated and combined with the original RGB image to construct a multi-dimensional spectral-textural feature space. A Random Forest consisting of 200 decision trees was utilized to extract the inundated areas. Experimental results indicated that Random Forest showed good performance in urban flood mapping with an overall accuracy of 87.3% and a Kappa index of 0.746. The inclusion of texture features can significantly improve classification accuracy with an increase of 11.2%. The comparison with other classifiers indicated that Random Forest outperformed maximum likelihood and artificial neural network, while showed similar performance to support vector machine.

Above all, the results demonstrate that a UAV is an outstanding platform for urban flood monitoring, and that the hybrid method proposed in this paper can provide accurate extraction results under complex urban landscapes. As Object Based Information Analysis (OBIA) is widely used in classifying high-resolution remote sensing images, future studies should be focused on incorporating OBIA into Random Forest to further increase the accuracy of urban flood mapping.

ACKNOWLEDGMENTS

This study is funded by the European Space Agency (ESA) and National Remote Sensing Center of China (NRSCC) Dragon 3 Program (DRAGON 3 Project ID. 10668) and the National Key Technology R&D Program of China (2012BAH33B04). Particular thanks to the anonymous referees, Academic Editor, and Assistant Editor for their very useful suggestions and comments of our paper.

AUTHOR CONTRIBUTIONS

Quanlong Feng proposed the main structure of this study and contributed to experiments, data analysis and manuscript writing. Jiantao Liu contributed to UAV data preprocessing, map making and manuscript revision. Jianhua Gong mainly contributed to manuscript revision.

Conflicts Of Interest

The authors declare no conflict of interest.

REFERENCES

1. Kuenzer, C.; Guo, H.; Huth, J.; Leinenkugel, P.; Li, X.; Dech, S. Flood mapping and flood dynamics of the Mekong delta: ENVISAT-ASAR-WSM based time series analyses. *Remote Sens.* **2013**, *5*, 687–715.
2. Smith, M.W.; Carrivick, J.L.; Hooke, J.; Kirkby, M.J. Reconstructing flash flood magnitudes using "Structure-from-Motion": A rapid assessment tool. *J. Hydrol.* **2014**, *519*, 1914–1927.
3. Gianinetto, M.; Villa, P. Mapping Hurricane Katrina's widespread destruction in New Orleans using multisensor data and the normalized difference change detection (NDCD) technique. *Int. J. Remote Sens.* **2011**, *32*, 1961–1982.
4. Wang, Y. Using Landsat 7 TM data acquired days after a flood event to delineate the maximum flood extent on a coastal floodplain. *Int. J. Remote Sens.* **2004**, *25*, 959–974.
5. Amini, J. A method for generating floodplain maps using IKONOS images and DEMs. *Int. J. Remote Sens.* **2010**, *31*, 2441–2456.
6. Mallinis, G.; Gitas, I.Z.; Giannakopoulos, V.; Maris, F.; Tsakiri-Strati, M. An object-based approach for flood area delineation in a transboundary area using ENVISAT ASAR and LANDSAT TM data. *Int. J. Digit Earth.* **2013**, *6*, 124–136.
7. Gstaiger, V.; Huth, J.; Gebhardt, S.; Wehrmann, T.; Kuenzer, C. Multi-sensoral and automated derivation of inundated areas using TerraSAR-X and ENVISAT ASAR data. *Int. J. Remote Sens.* **2012**, *33*, 7291–7304.
8. Henry, J.B.; Chastanet, P.; Fellah, K.; Desnos, Y.L. Envisat multi-polarized ASAR data for flood mapping. *Int. J. Remote Sens.* **2006**, *27*, 1921–1929.
9. Wang, Y.; Colby, J.D.; Mulcahy, K.A. An efficient method for mapping flood extent in a coastal floodplain using Landsat TM and DEM data. *Int. J. Remote Sens.* **2002**, *23*, 3681–3696.
10. Osorio, J.D.G.; Galiano, S.G.G. Development of a sub-pixel analysis method applied to dynamic monitoring of floods.*Int. J. Remote Sens.* **2012**, *33*, 2277–2295.
11. White, L.; Brisco, B.; Pregitzer, M.; Tedford, B.; Boychuk, L. RADARSAT-2 beam mode selection for surface water and flooded vegetation mapping. *Can. J. Remote Sens.* **2014**, *40*, 135–151.
12. Schnebele, E.; Cervone, G.; Kumar, S.; Waters, N. Real time estimation of the calgary floods using limited remote sensing data. *Water* **2014**, *6*, 381–398.
13. Gerl, T.; Bochow, M.; Kreibich, H. Flood damage modeling on the basis of urban structure mapping using high-resolution remote sensing data. *Water* **2014**, *6*, 2367–2393.
14. Colomina, I.; Molina, P. Unmanned aerial systems for photogrammetry and remote sensing: A review. *ISPRS J. Photogramm.* **2014**, *92*, 79–97.
15. Feng, Q.; Liu, J.; Gong, J. UAV Remote sensing for urban vegetation mapping using Random Forest and texture analysis. *Remote Sens. 201* **2015**, *7*, 1074–1094.
16. Gong, J.; Yue, Y.; Zhu, J.; Wen, Y.; Li, Y.; Zhou, J.; Wang, D.; Yu, C. Impacts of the Wenchuan Earthquake on the Chaping River upstream channel change. *Int. J. Remote Sens.* **2012**, *33*, 3907–3929.
17. Hunt, E.R.; Hively, W.D.; Fujikawa, S.J.; Linden, D.S.; Daughtry, C.S.T.; McCarty, G.W. Acquisition of NIR-Green-Blue digital photographs from unmanned aircraft for crop monitoring. *Remote Sens.* **2010**, *2*, 290–305.
18. Breiman, L. Random forests. *Mach. Learn.* **2001**, *45*, 5–32.
19. Rodriguez-Galiano, V.F.; Chica-Olmo, M.; Abarca-Hernandez, F.; Atkinson, P.M.; Jeganathan, C. Random Forest classification of Mediterranean land cover using multi-seasonal imagery and multi-seasonal texture. *Remote Sens. Environ.* **2012**, *121*, 93–107.
20. Rodriguez-Galiano, V.F.; Ghimire, B.; Chica-Olmo, M.; Rigol-Sanchez, J.P. An assessment of the effectiveness of a random forest classifier for land-cover classification. *ISPRS J. Photogramm.* **2012**, *67*, 93–104.
21. Immitzer, M.; Atzberger, C.; Koukal, T. Tree species

classification with Random Forest using very high spatial resolution 8-band WorldView-2 Satellite data. *Remote Sens.* **2012**, *4*, 2661–2693.

22. Hayes, M.H.; Miller, S.N.; Murphy, M.A. High-resolution landcover classification using Random Forest. *Remote Sens. Lett.* **2014**, *5*, 112–121.
23. Xu, L.; Li, J.; Brenning, A. A comparative study of different classification techniques for marine oil spill identification using RADARSAT-1 imagery. *Remote Sens. Environ.* **2014**, *141*, 14–23.
24. Anys, H.; He, D.C. Evaluation of textural and multipolarization radar features for crop classification. *IEEE Trans. Geosci. Remote Sens.* **1995**, *33*, 1170–1181.
25. Szantoi, Z.; Escobedo, F.; Abd-Elrahman, A.; Smith, S.; Pearlstine, L. Analyzing fine-scale wetland composition using high resolution imagery and texture features. *Int. J. Appl. Earth Obs.* **2013**, *23*, 204–212.
26. Haralick, R.M.; Dinstein, I.; Shanmugam, K. Textural features for image classification. *IEEE Trans. Syst. Man Cybern.* **1973**, *3*, 610–621.
27. Laliberte, A.S.; Rango, A. Texture and Scale in Object-Based Analysis of Subdecimeter Resolution Unmanned Aerial Vehicle (UAV) Imagery. *IEEE Trans. Geosci. Remote Sens.* **2009**, *47*, 761–770.
28. Aguera, F.; Aguilar, F.J.; Aguilar, M.A. Using texture analysis to improve per-pixel classification of very high resolution images for mapping plastic greenhouses. *ISPRS J. Photogramm.* **2008**, *63*, 635–646.
29. Zhai, Y.; Thomasson, J.A.; Boggess, J.E., III; Sui, R. Soil texture classification with artificial neural networks operating on remote sensing data. *Comput. Electron. Arg.* **2006**, *54*, 53–68.
30. Mountrakis, G.; Im, J.; Ogole, C. Support vector machines in remote sensing: A review. *ISPRS J. Photogramm.* **2011**, *66*, 247–259.

CHAPTER 12

FLOOD MAPPING AND FLOOD DYNAMICS OF THE MEKONG DELTA: ENVISAT-ASAR-WSM BASED TIME SERIES ANALYSES

Claudia Kuenzer, Huadong Guo, Juliane Huth, Patrick Leinenkugel, Xinwu Li and Stefan Dech

Keywords: Flood, flood dynamics, flood progression, water detection, inundation, radar, Envisat, ASAR, WSM, feature extraction, time series, Mekong Delta, Vietnam

INTRODUCTION: REMOTE SENSING BASED FLOOD MONITORING

Floods occur in many areas of the world and bring with them a range of negative but also many positive side effects. Every year floods cause numerous casualties [1]. In the last decade of the 20th century alone, about 1.4 billion people were directly or indirectly affected by floods and about 100,000 lost their lives [2]. According to the United Nations, more people lose their lives due to floods than due to any other natural hazard [3]. Unwanted inundation leads to the destruction of infrastructure and harvests and limits access to natural resources.

At the same time, floods and frequent inundation can be accompanied by positive side effects. Floodwaters contain nutrient-rich sediment fertilizing floodplain soils. Furthermore, inundated areas are habitat for fish and other aquatic animals. Floodwaters—when controlled, channelled, or stored—support long term water supply, irrigation activities, or even electricity generation. The monitoring of flood events, therefore, is a crucial task in earth observation. *In situ*assessments of flooded areas are usually difficult to undertake, and airborne or space-borne observations are the only realistic choice for continuous monitoring of large flooded areas [4].

Citation: Claudia Kuenzer, Huadong Guo, Juliane Huth, Patrick Leinenkugel, Xinwu Li and Stefan Dech, Flood Mapping and Flood Dynamics of the Mekong Delta: ENVISAT-ASAR-WSM Based Time Series Analyses, doi:10.3390/rs5020687.

According to numerous studies, e.g., [5,6], Synthetic Aperture Radar (SAR) provide the most suitable data for this endeavour, as cloud cover during the rainy season obscures optical observations. According to Leinenkugel *et al.*[7], cloud cover for the Mekong region reaches approximately 85%–95% in the rainy season, thereby making remote sensing analyses on the basis of optical image data a very challenging task. Radar systems, in contrast, enable data acquisition independent of weather conditions and time of day, as microwaves pass through most clouds, rain fields, water vapour, and aerosols. Flooded areas have very characteristic backscattering properties in radar images. As undisturbed water surfaces act as specular reflectors for incident microwaves, flooded areas yield no or a very low return signal, and can thus clearly be distinguished from open soil, vegetated agricultural land, forest land, or settled land [8].

However, influences of wind (leading to waves and hence increased surface roughness), extreme sediment load, or vegetative components in or covering the water can have impacts on the signal as well. A standard literature reference for the impact of forest canopies above flooded surfaces is Richards *et al.*[9]. They underline that C- and X-band data will usually not penetrate a tree canopy, and the return of a suitable L-band signal from a water surface below forest also depends on incidence angle, trunk geometry, and stand density. However, even with L-band data flood detection under canopy is usually not fully satisfactory. Hess *et al.*[10,11] employed SIR-C [10] and 100 m resolved JERS-1 L-band data and achieved flood detection user accuracies of 62% in the low water stage of the Amazon, and of 86% during the high water stage. The area covered in this latter study was extensive.

Furthermore, very small observation angles can hamper water detection [4], and polarization of the incident radiation also plays an important role, as elaborated by Henry *et al.*[12]. They compared flood mapping results of Envisat ASAR Image Mode and Alternating Polarization Mode with ERS-2 derived flood maps and Landsat ETM+ derived ground truth. They found that HH polarized data is best suited for flood area discrimination and superior to HV and VV data. However, if HV is available jointly with HH data, the result can be further refined. VV polarized data is least suitable, as it is strongly influenced by surface roughness. This was confirmed in numerous studies, amongst others by Gstaiger *et al.*[13].

The above effects are also reviewed by Pierdicca *et al.*[14], who employed Italian COSMO SkyMed data for semi-automated flood detection in numerous case study sites located in Italy, Pakistan, Albania, Thailand, and Australia. These authors also underline the impact of microwave frequency on water detectability. While X-band as well as C-band data cannot penetrate into a forest canopy, L-band data has an increased capability to penetrate down to water surfaces which might exist below forest. Here the chance for a water-related return signal (if the stand is not too dense) is definitely higher than for X- or C-band data. Pierdicca *et al.*[14] also state that detailed local knowledge of an area (land cover, terrain, typical flood patterns) definitely helps to improve and interpret the flood mapping result and addresses the trade-off between semi-automated methods involving more human interaction (maybe even manual threshold selection) yielding very high accuracy *versus* fully automated rapid mapping approaches.

Methods of flood area extraction—or so called "water mask" derivation—are numerous and have been presented by several authors. Matgen *et al.*[15] have been working in the field for many years, and conclude that hybrid methodologies combining radiometric thresholding with region-growing approaches are the future pathway towards SAR based automated flood monitoring. Pulvirenti *et al.*[16] presented a fuzzy-logic approach considering three scattering models, simplified hydrologic assumptions and context information. Based on available pre-flood data their algorithm has the capability to detect not only open water areas, but also flooded areas beneath vegetation senescence. Martinis *et al.*[17,18] developed an automatic near-real-time flood detection tool for very high resolution TerraSAR-X SAR data combining split-based histogram thresholding and segmentation based classification. Although results are convincing, a challenge to this approach is that costly licensed software (Definiens Developer) is needed for the segmentation of the radar images. Mason *et al.*[19,20] also focus on high resolution TerraSAR-X data, presenting a semi-automated algorithm for flood detection in urban areas, addressing the problem of floods in settled regions which might be obscured by layover and foreshortening or radar shadow. This specific issue—while in general of high interest—is not so relevant for the data and observation scale addressed in this paper.

With respect to applications, the observation of flooded areas is usually undertaken based on multi-temporal data. Kuehn*et al.*[21] employed three Radarsat-1 data sets as well as optical Aster data for flood mapping in the Ganges Delta in Bangladesh. Hoque *et al.*[6] utilized multi-temporal optical and SAR data for flood monitoring in Bangladesh, Henry *et al.*[12] analysed one multi-polarized ASAR data set, and Schumann *et al.*[22] focus on future flood modelling using SAR data for surface roughness derivation. Pulvirenti *et al.*[23] analysed the flood situation after the 2011 Tsunami hitting Japan using COSMO SkyMed data. Some of the numerous examples for published flood related studies based on SAR data appear in [24–31]. However, for method studies as well as for application oriented studies larger time series are hardly ever studied. For method studies focussing on novel or improved tools for flood detection, usually a pre-flood image, a further image acquired during the flood event, as well as (possibly) a post-flood image are sufficient. In many application studies only two data sets (pre-flood, flood) are compared. Publications on flood dynamics and floodwater progression over a longer time period are rare, and for the Mekong Delta in Vietnam little is available.

In the Mekong Delta, Sakamoto *et al.*[32] detected temporal changes in the extent of annual flooding based on mainly optical 500 m resolved MODIS data spanning the years 2000–2004. In the field of radar based analyses, Long *et al.*[3] used five ERS-2 SAR data sets to map flood occurrence. However, so far flood patterns in the Mekong Delta have never been analysed with a radar based time series at a resolution of 150 meters, as we present here. A challenge of the delta is its more or less completely flat topography. Ancillary data like a DEM cannot support the refinement of flood detection. The questions which we aim to answer are the following:

- Which spatial patterns exist with respect to flood occurrence?
- Which areas in the Mekong Delta are most frequently flooded and why?
- Which areas in the Mekong Delta are not affected by floodwaters and why?
- How do floodwaters in the Mekong Delta progress over time?
- Which similarities or differences in floodwater dynamics exist between the observed years?
- How do artificial dikes impact flood patterns in the delta?
- What are the advantages and disadvantages of Envisat ASAR WS based flood monitoring?

STUDY AREA: THE MEKONG DELTA

The Mekong Delta in southern Vietnam with 39,000 km^2 is one of the largest river deltas in the world and home to over 18 million inhabitants. It is located between 8°30′–11°30′N and 104°30′–106°50′E and is the vulnerable end of the Mekong river [33] (see Figure 1). The Mekong originates at the Tibetan plateau and crosses China, Myanmar, Laos, Thailand, and Cambodia before entering the territory of Vietnam. The Mekong's discharge of 475 km^3/year makes it to one of the world's largest single-peak pulsing rivers [34]. Severe flood-pulse changes are expected to occur in the future due to regulatory measures (e.g., hydropower dams) upstream [33].

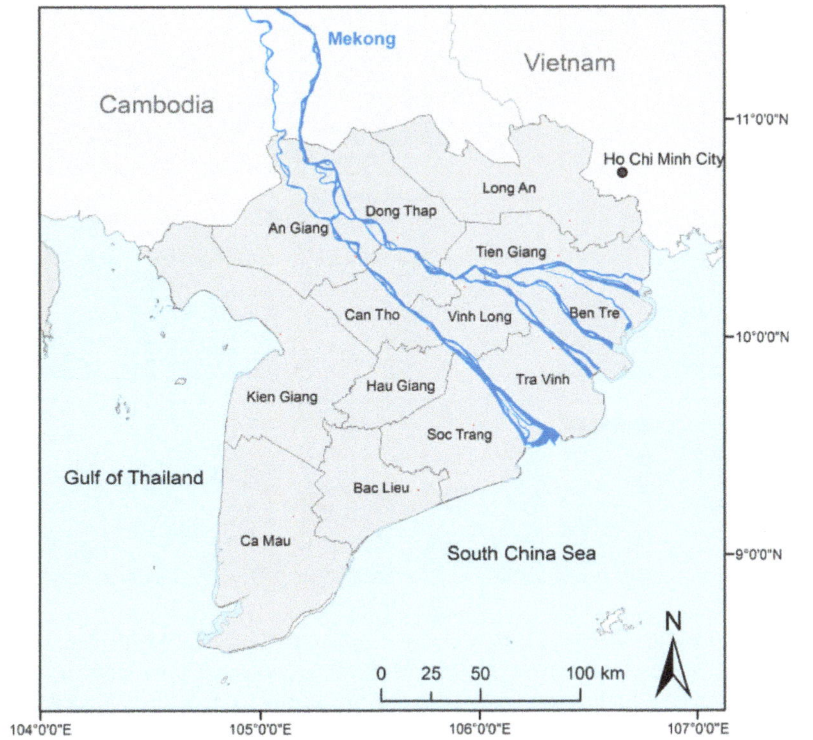

Figure 1. The study area: the Mekong Delta and its 13 Vietnamese provinces in southern Vietnam. Here the Mekong splits into nine branches before reaching the South China Sea.

The delta—mostly located at an elevation well below 10 m above sea level—developed via thousands of years through the accumulation of fertile sediments [33,35]. Over 90% of the 13 Vietnamese provinces comprising the delta are located well below 3 m above sea level. The flat and open delta is often termed the 'rice bowl of Vietnam'. Rice production yields two, and in some areas up to three harvests per year [32,36,37].

Just recently, a detailed work examining rice cropping patterns in the delta based on hyper-temporal SPOT NDVI data was presented by Thi Thu Ha Nguyen *et al.*[36]. However, especially the coastal areas of the Mekong Delta are increasingly being converted to aquaculture areas. Shrimp ponds extend all the way to the coast and lead to the depletion of the last remaining forest fragments of coastal mangroves or other natural vegetation protecting the coast [38,39]. Overall, the delta is mainly dominated by seasonal agricultural land use, and tree cover is very limited. Tree cover occurs as slim belts of mangroves along the south-western coastline, and as fruit tree orchards in the southern centre of the delta. The region experiences prominent dry and rainy seasons influenced by the south-west Indian and north-west Pacific monsoon. The dry season lasts from late December to May, while the rainy season (causing most of the 1,800 mm of precipitation in the area) usually lasts from early June to December. During the rainy season, large areas of the delta are flooded and fertilized by about one billion m^3 of sediment [36]. Thus, the people living in the delta call the annual flood 'the beautiful flood' [40]. They have learned to live with frequent flooding, as floodwaters are indispensable for the delta's agriculture [41]. Furthermore, inundation allows for seasonal fishing. Extreme events leading to fatalities are relatively rare, but occurred in 1961, 1978, 2000, 2001, and 2002, whereby each event was accompanied by the death of several hundred people, many of them unable to swim [40]. The hydrodynamic conditions in the delta are strongly altered by over 55,000 km of man-made canals [42]. Figure 1 depicts an overview of the area.

Major challenges in the delta in recent times can be attributed to socio-economic transformation and urbanisation processes [43] leading to the degradation of the last natural forest and wetland areas [38,39], accompanied by increasing water pollution. Furthermore, climate change is a threat. The delta is one of the most endangered places on Earth with respect to sea level rise. Rising sea levels are forecasted to lead to intensified saltwater intrusion into the Mekong main stems and canals, as well as into aquifers and soils, as tidal saltwater influences progress further inland [40,41,44,45]. According to an IPCC prognosis (2007) of sea level rise between 75 and 100 cm by the end of this century, about 20%–50% of the low lying delta will be affected if no countermeasures are taken. Currently, ocean-induced flooding mainly plays a role during tropical storms and typhoons, which are expected to increase in frequency [40]. In the Mekong Delta, the flood regime is defined by a complex mix of four influencing factors. Type 1 floods are mainly induced by the flood pulse of the Mekong as well as overland flow. Type 2 floods are artificial floodwaters distributed via a dense network of canals and controlled by dykes and sluice gates. Type 3 floods are short-term floods due extreme local precipitation events. Type 4 floods are mainly related to high tides during storms. Usually, no single flood occurrence in the Mekong can be solely assigned to one of the four factors. Instead, flood occurrences are always triggered by a combination of several of the four sources (see Figure 2). This complexity has already been introduced and discussed by Nguyen Viet Dung [35], Vo Khac Tri [46], Apel *et al.*[47], and Delgado *et al.*[48].

Figure 2. Typical flood regime depictions of the Mekong Delta, Vietnam. **Upper left**: Small canal with permanent water used for transport. **Upper centre**: Banks of canals are usually vegetated, and water may occur beneath vegetation. **Upper right**: Typical small dyke, which borders a large flooded area. **Middle left**: Overland flow. Large areas are flooded. **Middle centre**: At this stage flooded fields are used for fishing. **Middle right**: pumping excess water from field into the canal at a higher level. **Lower left**: Floodwaters have mostly retreated from this agricultural field. **Lower centre**: Large rice fields have been freshly planted. **Lower right**: The rice is heading, fields are slightly irrigated. (All photographs taken by C. Kuenzer in the Mekong Delta, November 2011).

AVAILABLE SATELLITE AND ANCILLARY DATA

On 1 March 2002 ESA's Envisat satellite was launched, and data from sensors like ASAR (Advanced Synthetic Aperture Radar)—operating at C-band—were available until Envisat operational services ceased in April 2012. Envisat ASAR data come in different modes with different spatial resolutions and coverage. Data is available in Image, Wave, and Alternating Polarisation modes at approximately 30 m × 30 m in the highest resolution, approximately 150 m × 150 m in Wide Swath Mode (WS), and at 1 km pixel resolution in Global Monitoring (GMM) mode. Data from Envisat ASAR can be obtained via ESA's science programme. Since ESA approved a new data policy in May 2010, ESA's Earth Observation Data could be accessed mostly free of charge via the submission of a project proposal for research applications. We obtained all data (archived as well as scheduled, whereas scheduling often collided with other requests) in ASAR WS mode (WSM), which has a swath width of 400 km, and 150 m pixel spacing. Not all of the data were exclusively available for the rainy season. In the context of the present study, the following Envisat ASAR WSM data sets were available for the consecutive years 2007, 2008, 2009, 2010, and 2011 (see Table 1):

Table 1. Envisat ASAR WSM data sets available for monitoring inundation in the Mekong Delta. Scenes acquired during the rainy season appear in bold letters.

2007	2008	2009	2010	2011
2007-06-14	2008-06-01	2009-06-02	2010-01-14	2011-01-03
2007-07-03	2008-06-17	2009-06-18	2010-01-17	2011-01-14
2007-07-10	2008-06-24	2009-07-04	2010-02-18	2011-02-02
2007-07-19	2008-07-03	2009-08-27	2010-03-25	2011-03-04
2007-08-07	2008-07-22	2009-10-01	2010-04-29	2011-03-15
2007-08-14	2008-08-14	2009-12-10	2010-05-02	2011-04-03
2007-08-23	2008-08-23	2009-12-13	2010-06-03	2011-06-21
2007-09-11	2008-08-26		2010-07-08	
2007-09-18	2008-09-11		2010-08-12	
2007-10-16	2008-09-30		2010-08-15	
2007-10-23	2008-10-07		2010-09-16	
2007-11-01	2008-10-16		2010-10-08	
2007-11-20	2008-11-04		2010-10-21	
2007-11-27	2008-11-11			
2007-12-06	2008-11-20			
	2008-11-23			
	2008-12-16			
	2008-12-25			

ASAR has a repetition rate of 35 days. However, more than one path can be used to cover the Mekong Delta. We used mainly archive data for our analyses and ordered all the data available in the archive, which led to approximately two to four coverages per month. As a window for our data search we defined June to December (rainy season in the Mekong Delta), however, in 2010 and 2011 other dates were also retrieved from the archive. Unfortunately, in 2009 and 2011 less data were available in the archive due to the acquisition of alternative ASAR modes (WSM cannot be tasked in parallel). However, despite these restrictions, we could build up a comprehensive time series of over 60 ASAR scenes located over the Mekong Delta. Additional data available for this study included the MEKONG LC2010 map, a MODIS time-series-derived land cover/land use classification generated by Leinenkugel *et al.*[49], as well as different vector data sets, such as province boundaries.

METHODS AND DATA ANALYSES

All radar amplitude data were firstly geo-corrected using the open source software package BEAM, provided by the European Space Agency ESA. The flooded/inundated areas were then derived from the radar amplitude data, as briefly explained below, and in depth in Huth *et al.*[50] and Gstaiger *et al.*[13].

One main goal of water mask processing is the rapid and repeated delivery of Mekong Delta flood information over the course of several years in the context of the large German-Vietnamese WISDOM project (Water related Information System for the Sustainable Development of the Mekong Delta, [51,52]). Therefore, a main requirement for the algorithm used for data processing is simplicity and speedy performance, applicability to different types of radar data (not relevant for this paper, but a determinant for algorithm choice), independence of expensive licensed software, and availability as an embedded web processing service (WPS).

A simple threshold method enables the separation of water pixels, represented by low backscatter values, from non-water pixels, which usually have higher values in the image histogram depicting backscatter distribution. As laid out in Gstaiger *et al.*[13], to firstly reduce the typical speckle typical for SAR imagery, the first step of the algorithm is to apply a standard convolution median filter with a kernel size of 5 × 5 pixels, resulting in a filtered and speckle reduced image P1. After this pre-processing two empirically chosen thresholds divide water from non-water pixels (processed image P5). Here, the first threshold, T1, which has a lower value than the final water threshold, defines confident water areas, leading to P2. The second threshold T2, which has a higher value than the land threshold, classifies confident land areas, leading to image product P3. Then buffer zones of two pixels, which are only generated via dilatation, are applied to P", which results in product P4. The buffers define the transition zone from water to land, also represented by mixed pixels. The second threshold now enables the inclusion of the water pixels within this zone in the initial binary water mask. The temporary results P3 and P4 are now compared, and if coincidence occurs the value (water or land, 0 or 1) is written to P5. Otherwise, the value from P2 is written to P5 (P4 & P3 || P2). In this way overestimated water pixels are excluded. Proceeding further, isolated pixels are removed via morphological image closing (P6) [13] (see Figure 3). The removal of so-called 'islands' and 'lakes' according to a defined maximum size (T3,T4) is mainly of relevance for higher resolution SAR data (e.g., TerraSAR-X data also processed in the course of the project), but does not affect Envisat ASAR derived results at 150 m resolution.

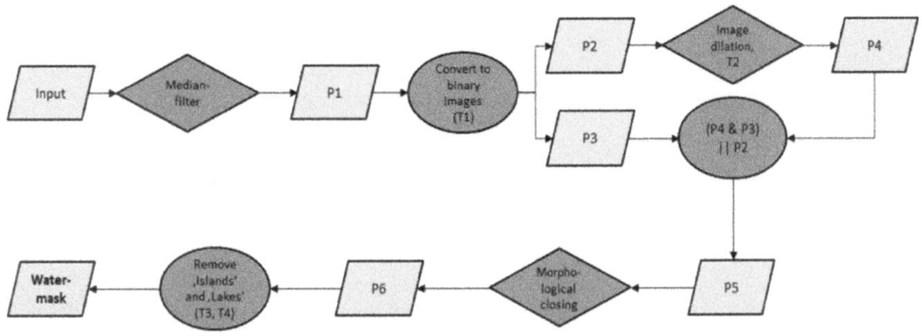

Figure 3. Processing chain of the histogram based approach for water mask derivation (modified from Gstaiger *et al.*[13]).

The results of this algorithm have been compared with other flood mask derivation approaches—e.g., a histogram texture based approach, and a very interactive object based approach—and it was found that the accuracies of the approach which we use here (97.6%) and the very interactive object based approach (97.8%) were more or less the same, and exceeded the accuracy of a further texture based approach tested (95.6%) [13].

Furthermore, our algorithm enables automatic processing of the water masks right at data delivery via email by the data provider and up until integration of the product in a web-based information system for the Mekong Delta [53]. The water mask processing chain combines all processing steps from data retrieval to product generation, and data ingestion and all steps have been encapsulated in a Web Processing Service, WPS, allowing for fully automatic processing [13,54].

The algorithm for water surface extraction yields products of high accuracy, and derived flooded areas have been validated indirectly via comparison with high resolution optical and TerraSAR-X data and *in situ* ground surveys of water-land boundaries [13]. Considering that the ASAR WS data have a rather coarse resolution of 150 metres, the land-water boundary cannot be determined anyway down to metre accuracy. We employed the algorithm to derive so-called 'water masks' for our study region. Processing yielded the water masks for the individual time steps available as binary 0/1 files (no water/water) for each individual scene (see Figure 4).

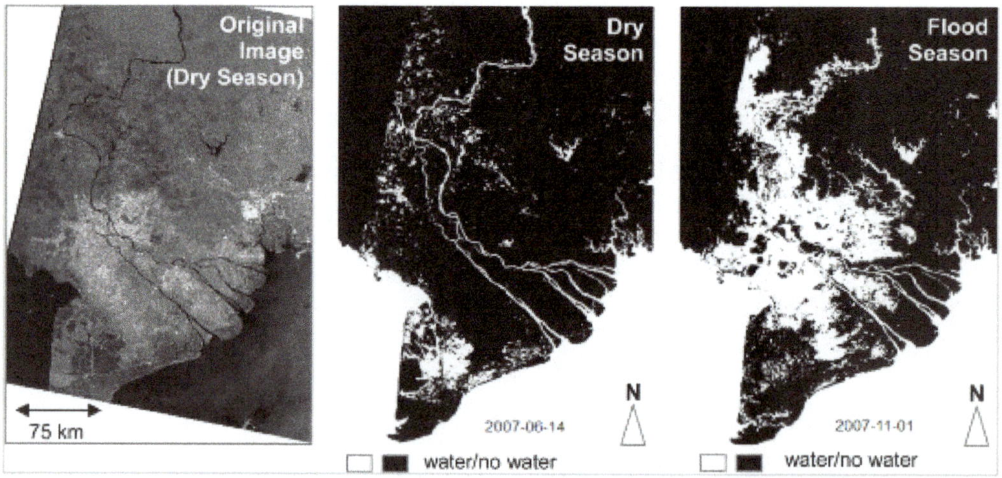

Figure 4. Original ENVISAT ASAR WSM radar image for 2007-06-14 (**left**), and water mask derived for 2007-06-14 (**centre**). The differing extent of flooding between the start of the rainy season (centre) and the flood peak around the end of the rainy season (**right**) is evident. Extent: UL: 12°N, 104°15′E, LR: 8°30′N, 106°50′E.

Water masks for each year (e.g., 2008) were summed up to obtain an image product depicting how many times an area was covered with water during the respective rainy season within that year. This enables an understanding of flood length for different areas. Furthermore, all sum products for the individual years (*i.e.*, ASAR 2007, 2008, 2009, 2010, 2011) were added to retrieve a long time series of water covered (flooded) *versus* non-flooded areas, which happens to be the longest SAR-based time series of flood occurrence for the delta.

Since the footprints of many scenes did not completely overlap, but only shared a certain region in common, a so-called 'largest common coverage', LCC, footprint was generated. This LCC footprint was derived for each individual year, as well as for all years, to enable later statistical analyses within the years, as well as for the five-year time-span in regions of comparable coverage only. This LCC approach ensures that we do not compare areas which were covered by a varying amount of data acquisitions over the year. This process flow is depicted in Figure 5.

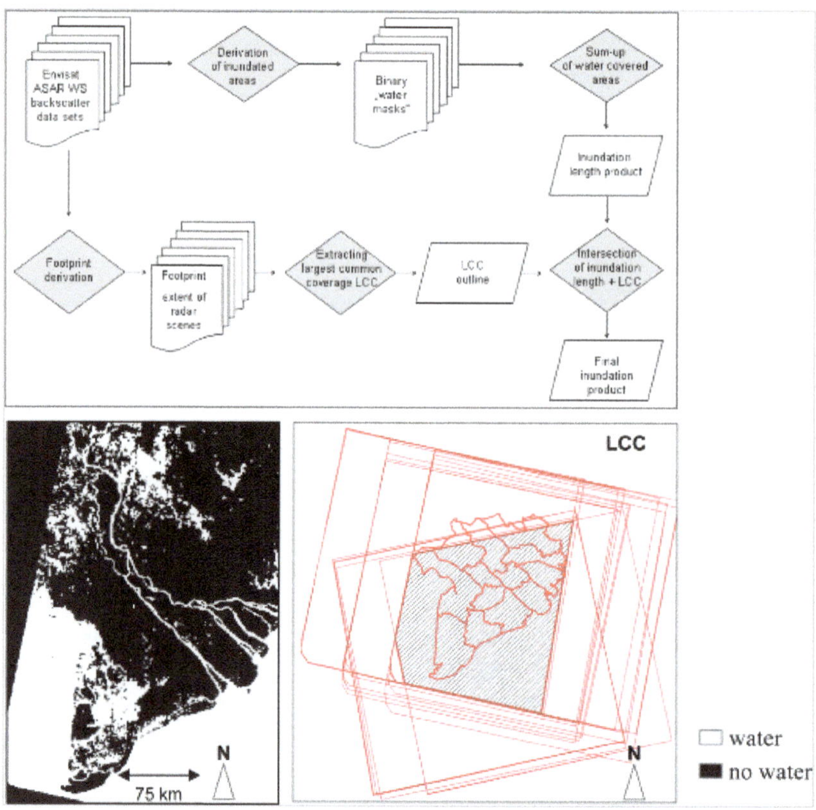

Figure 5. Derivation of flood occurrence (summed) for individual years or all years, considering the largest common coverage (LCC) of all scenes. Data processing procedure, exemplary binary water mask for one time step, creation of the largest common coverage based on radar scene footprints. Extent of lower left image: UL: 12°N, 104°15′E, LR: 8°30′N, 106°50′E.

RESULTS

Floodwater Occurrence and Progression for One Selected Year (2007)

Figure 6 presents the flooding regime over the rainy season for the year 2007. This year was chosen for display since the largest common coverage for this year (within red boundary) is larger than those for 2008, 2009, 2010 and 2011 and therefore depicts typical patterns best. The year 2007 was neither an extreme flood year nor a drought year. As listed in Table 1 and depicted in the figure, the time span between 14 June and 6 December is covered.

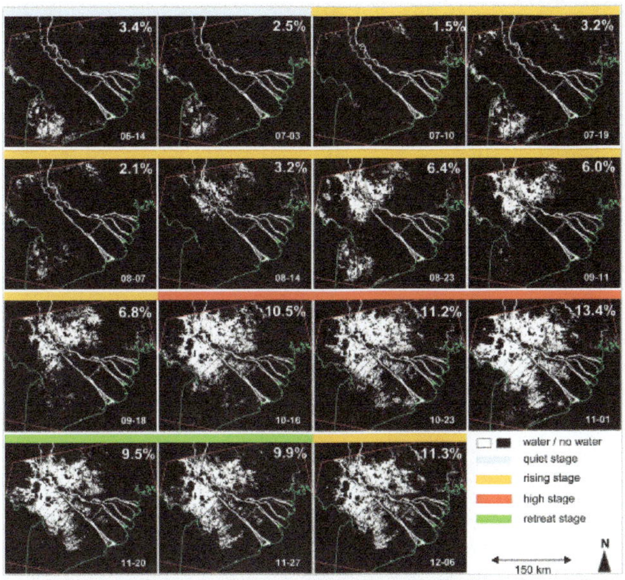

Figure 6. Floodwater distribution during the 2007 rainy season, depicted only for the largest common coverage. Each subset contains the exact date and flood percentage. Image extent: LL: 8°46′50″N, 104°35′05″E, UR: 10°55′50″N, 106°29′26″E.

We can observe a clear rising stage of floodwaters (orange), where water progresses from north to south. River-induced and overland flow (type 1 flood) as well as human control (type 2 flood) are the main causes for the spread of inundation (orange). During the high stage (red) the flood area covers more than 10% of the observed area and a slight progression is still recognizable. From November onwards floodwater retreated back, from south to north, and radially from areas further away from the main stem to areas closer to the river. However, in December floodwaters unexpectedly rose again. Flood variability in the south-western tip of the delta (Ca Mau province and neighbours) is not related to the type 1 flood dominating the northern part. Instead, in the very lowest lying part of the delta flooding is mainly induced by the tide (type 4 floods) and human control (type 2 floods) [35,42,55]. Extreme rain events in the delta (type 3 flood) can occur at any time in the rainy season, and lead to short-term patches of shallow water [47]. According to Nguyen Nghia Hung *et al.*[42] and Delgado *et al.*[48] the flood hydrograph (as measured at *In situ* gauge stations at the northern border) in the delta is characterized by two peaks: the first between mid-July to mid-August, and the second between September and October (note the sharp increase between 08–14 and 08–23, as well as between 09–18 and 10–16).

Flood Frequencies for 2007, 2008, 2009, and 2010

Figure 7 below presents the results of the added flood occurrence (sum) products for the years 2007, 2008, 2009, and 2010. As only one acquisition during the rainy season (see Table 1) was available for the year 2011, this year is not presented inFigure 7. However, the years 2007 (already presented as an individual observation in Figure 6) and 2008, as well as 2009 and 2010 are covered with an about equal amount of rainy season data and are thus comparable—at least for the spatial areas covered by both of the respective data sets. The flood occurrence sum images are generated according to the flow chart depicted in Figure 5. Very bright (white) areas were flooded numerous times, while lower grey values indicate less frequent flooding. Images would look identical if they were normalized to per cent—however, in the legend we prefer to keep the number of available observations for better transparency. The figures have slightly different extents due to the varying largest common coverage, LCC. The long time series for 2007 and 2008 show a similar flood distribution: the areas most often flooded are the northern areas of the Mekong Delta, whereas the coastal areas are less affected by flooding. This is confirmed by the spatial profile, which cuts the delta along the orange line (2007) from north-northwest to south-southeast. A clear trend towards shorter flood durations when approaching the coast can be noted. This can also be observed for the 2008 subset. Since 2010 was an extreme drought year for the delta [56] it stands out that the northern flood plains are not as strongly flooded as in the previous years.

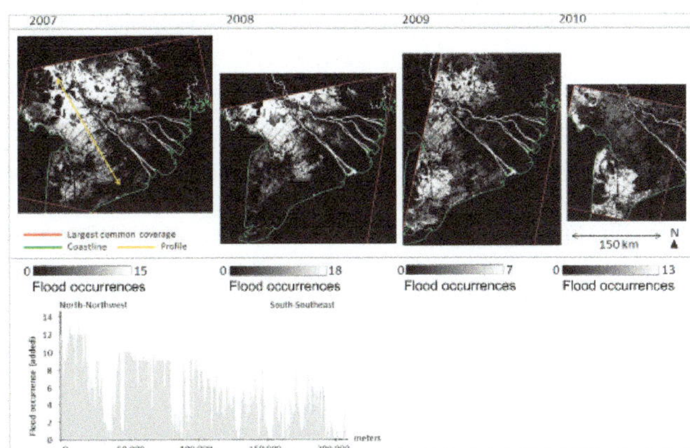

Figure 7. Added flood occurrence (sum) images for 2007, 2008, 2009, and 2010. According to the largest common coverage, LCC, the subsets differ in extent, but are displayed at the same scale. The profile presents inundation frequency along the profile line depicted in the upper left figure of 2007.

Figure 8 presents a more detailed illustration of the 2007 flood occurrence data set shown in Figure 6. Displayed at enlarged extent and in color, we can clearly depict flood occurrence patterns during the rainy season of 2007. Floodwaters remain for nearly all the rainy season in the northern delta, mainly in the provinces of An Giang, Dong Thap, northern Long An, and Can Tho. Within the northern region, areas which are not flooded are either elevated or strongly dyked, as will be presented later. Subset b) on the right side of Figure 8 depicts flooded fields in northern Can Tho province. The boundaries between the fields, usually paved road or dirt roads constructed on top of a dyke, are rarely flooded. Flooding here only occurs a very few times per year, when over-bank flow or very heavy rains lead to a sudden increase in the water level of canals and on fields. The coastal province of southern Long An, Tien Giang, Ben Tre, Tra Vinh, and Soc Trang (see Figure 1) are hardly ever affected by type 1 or type 2 flooding.

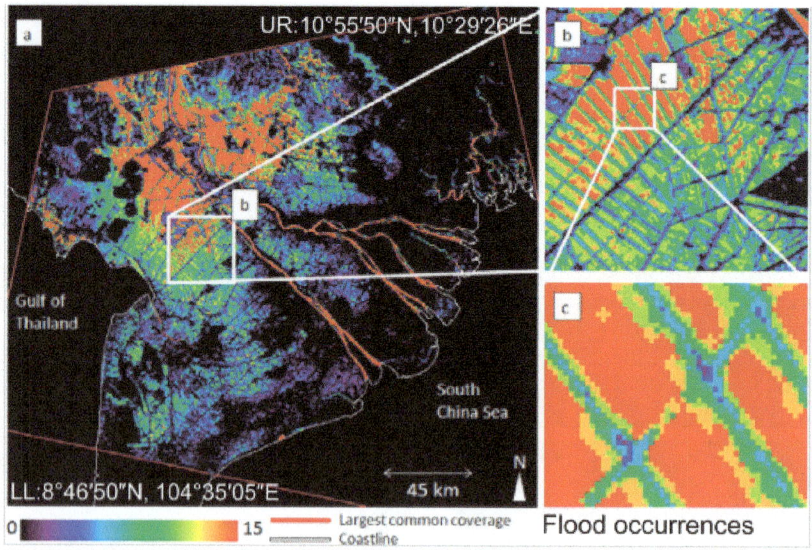

Figure 8. Flood dynamics in the Mekong Delta for the 2007 rainy season, based on 15 Envisat ASAR WSM derived water masks. Areas which are nearly always flooded are shown in reddish tones. The two zooms ((**a**) and (**b**)) depict flooded fields which are separated by dykes (elevated roads, pathways)

Figure 9 shows the inundation patterns in the Mekong Delta over the course of five years. Dark areas (except the ocean) were never, or only very rarely, detected as being flooded or inundated during the entire observation period, while coloured regions were inundated up to 58 times. Areas most frequently flooded during the rainy season are located in the northern and south-western Mekong Delta. The southern coastal provinces and the south-eastern areas, where the main stem splits up and reaches the south China Sea, are rarely flooded or inundated. It is furthermore obvious that the areas close to the main river arms are well protected by dykes, and extensive large-scale flooding and inundation rarely occur in the vicinity of the river banks. Another reason for limited flood detection is the fact that river banks as well as canals banks are usually vegetated with palm trees and other trees—with Envisat ASAR WSM data it is not possible to detect flooded areas beneath dense tree cover. In contrast, the main Mekong branches are depicted as water covered areas in nearly all observations. As the Mekong never runs dry, the river's main branches should actually be displayed as water-covered for the maximum time span (58 observations). However, a few times probably not all pixels within the river were declared as water—this can occur due to the extremely high sediment load, or strong wind (wavy water surface). Strictly speaking, only the area within the white boundary is covered by all 58 data sets. However, we present the whole delta, as general patterns are still valid. Several areas are rarely or even never detected as being flooded (see optical data snapshots in Figure 8). Except for the mangrove area, where floodwaters often exist below a dense canopy, this is reasonable.

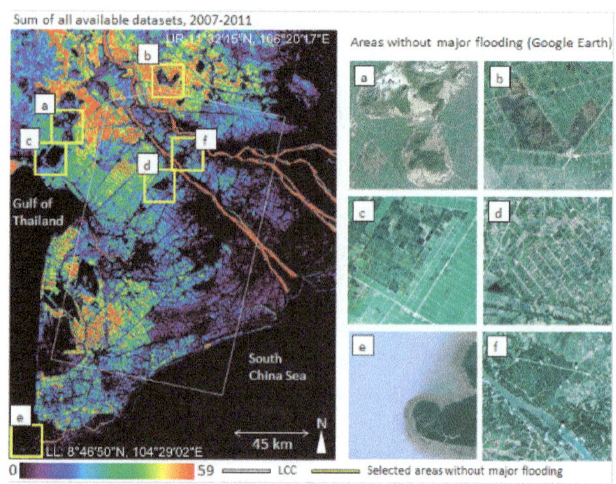

Figure 9. Inundation in the Mekong Delta from 2007 to 2011 derived from all available Envisat ASAR WSM data. (**a**) hills in the Tri Ton District north of Chau Lang, (**b**) Tram Chim National Park, (**c**) well-dyked agricultural area north of Hon Dat, (**d**) the Can Tho Cuu Long Rice Research Institute test field, (**e**) the tip of Ca Mau province with dense mangroves, and (**f**) fruit tree orchards.

All distinct areas within the Mekong Delta which are rarely or never detected as being covered with water can be explained with local land cover and land elevation. For example, regions (a), (b), (c), and (d) are in fact never flooded (because located at too high elevation, or too well protected). However, tree covered areas (e) and (f) need some additional explanation. Regions where mixed fruit tree plantations dominate land use are usually also slightly elevated, and flooding in this area is very rare [46]. However, where *melaleuca*, *nypa* palms, and mangroves are prevalent, radar beams cannot pass through the dense canopy, and also here inundation below the trees cannot be observed.

The three diagrams in Figure 10 depict how much of the fully covered (LLC) provinces for the respective years are flooded. The figure enables an understanding of flood progression.

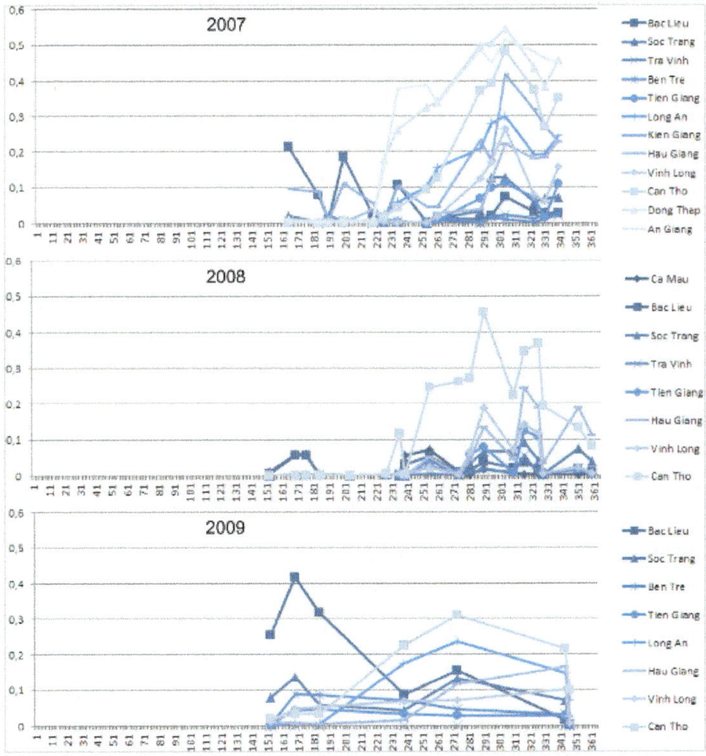

Figure 10. Flood patterns during observed days of the year (DOY) for different fully covered provinces. Provinces displayed in darker blues are located on the coast, provinces in lighter blue further inland (the lighter the blue tone, the further north the location).

Overall Flood Progression and Comparison of Flood Progression for Selected Years

The diagrams below depict how much of the provinces were flooded during the respective day of the year when we observed them. Compared to—for example—Figure 6, Figure 10 enables a more quantitative comparison of flood progression. The x-axis presents the day of the year (DOY), and the y-axis displays how much (what percentage) of the province is flooded (1 = 100%). In 2007 Ca Mau was not included in the LCC. In 2008 Ben Tre, Long An, Kien Giang, Dong Thap, and An Giang are not fully covered and thus missing. In 2009 Ca Mau, Tra Vinh, Kien Giang, Dong Thap, and An Giang are not included. However, a comparison of the province areas fully covered during all three years is possible, and we do not want to withhold the information for the other provinces, as clear patterns can be observed.

Provinces located near the coast are displayed in darker blue, while the light blue plots represent provinces further away from the coast. For all three years provinces further away from the coast are flooded to a larger extent than were provinces near the coast. The only exception is the large flooded area at the very beginning of the rainy season in 2009 in Bac Lieu province. This flood event cannot be attributed to river or overland flooding (type 1), as in this case the northern provinces should show larger flood extents as well. The flooded area here may stem from coastal inundation after a very high tide, but this is only speculation.

Except for this outlier event, coastal provinces are usually flooded over not more than 0%–15% of their territory. In flood prone provinces such as An Giang, Dong Thap, and Can Tho above 40% of the province area can be flooded during the high stage. During the observation for DOY 305 in 2007 the three provinces were covered with floodwater to over 50%. In 2008 An Giang and Dong Thap were unfortunately not included in the largest common coverage, but floods in Can Tho again covered over 40% of the province during DOY 290.

Generally, all three graphs show that the flood area steeply increases after DOY 215 in 2007 and 2008. In 2009 floodwater increase starts earlier and does not seem to be as extreme, but in that year we have far fewer observations available. Both in 2007 and 2008 the high flood stage occurs roughly between DOY 275 and DOY 320. In 2008 the flood furthermore seemed to arrive via three strong pulses starting around DOY 236, DOY 274 and DOY 308.

However, precise interpretation is difficult. The fact that four flood types (river/overland, human control, local rain events, as well as tidal influence) impact inundation in the delta makes it nearly impossible to separate the components—especially as we unfortunately had no detailed gauge station or hydrologic station data at hand. The time series could be interpreted at even greater detail if we had information available on typhoon landfalls (and the time lags for the accompanying rain events), extraordinary high tides and storms, as well as dyke control schemes. However, interpretation would then also exceed the scope of this remote sensing related paper.

Figure 11 presents patterns of flood progression for the two comparable time series of 2007 and 2008. In 2007 a total of 15 observations were available. In the color display of Figure 11 three composites were generated which show flood progression over the course of three consecutive observations. Please note that near exact dates can be compared. The three binary data sets are stacked, and displayed in RGB (red, green, blue) mode. In this way, areas which are only flooded during the first observation appear red, areas which are exclusively flooded during the second observation appear green, and areas flooded only during the last observation appear blue. According to additive color theory, areas which are flooded during the first two observations (red + green) appear yellow, areas flooded during the last two observations (green + blue) appear cyan, and areas flooded during the first and last (red + blue) observation appear magenta. If an area is flooded three times consecutively it appears white (red + green + blue). For 2007 three triplets are shown: covering the month of August, mid-September to mid-October, and the second half of October, respectively. We can depict how floodwaters progress from north to south and expand radially from the main river. The northern areas of the delta are always flooded. Similar patterns can be observed in three selected triplets of 2008. Comparing 2007 and 2008 it is obvious that in 2007 floodwater extent (especially at the onset in August) was larger, and flooding occurred slightly earlier.

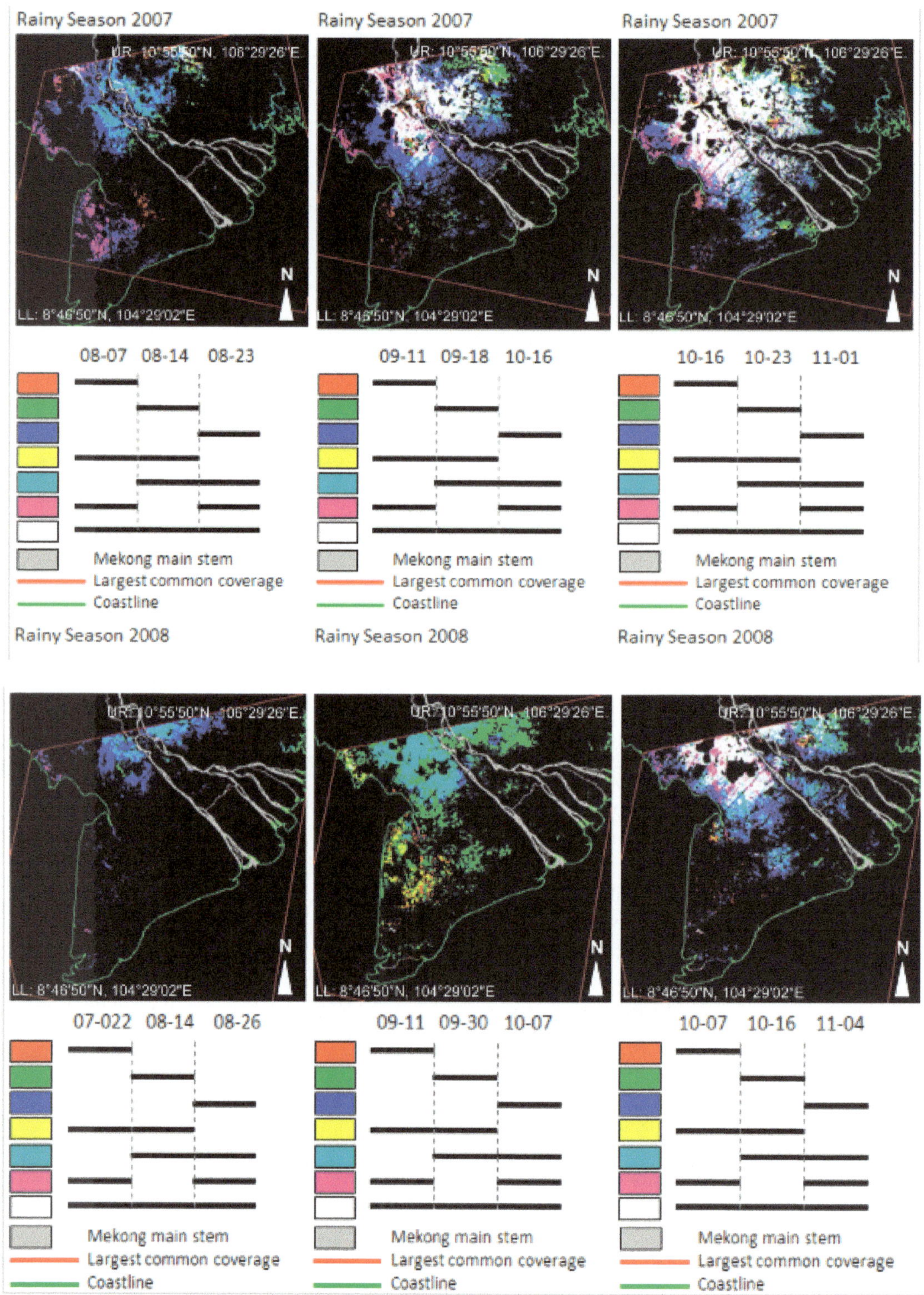

Figure 11. Flood progression in the delta in 2007 (**upper**) and 2008 (**lower**).

Please note that in each of the triplets dates are chosen that enable comparison between 2007 (upper row) and 2008 (lower row).

Impact of Dykes on Floodwater Distribution and Progression

As mentioned before, floodwater distribution in the Mekong Delta is influenced by anthropogenic forcing. Farmers control the floodwaters via dykes, and the operation of sluice gates and pumps.

Figure 12 shows the effectiveness of different types of dykes. In An Giang semi-protected and fully protected dykes exist. They vary in height and stability. Fully protected dykes are dykes which managed to fend off water during the century flood in the year 2000, while semi-protected dykes are lower and not as solid. The results show that areas located behind fully protected dykes (pink) are much less often flooded than areas behind semi-protected dykes (green). An Giang is famous for rice production and has converted many areas to three-crop farms. Strict inundation and floodwater control within the dyke systems led to a fourfold increase in rice yield within only the past 40 years [40]. Especially triple rice crop areas are located within high ring dykes. High ring dyke areas are usually provided with higher pumping capacities. According to their findings, areas extensively flooded for longer periods are customarily used for growing two rice crops per year, while strictly controlled fields with shorter flood durations are used for cultivating triple-rice crop systems. For the latter areas farmers prefer varieties with shorter growing periods (≈90 days) compared to varieties of longer growing period in two-season crop areas.

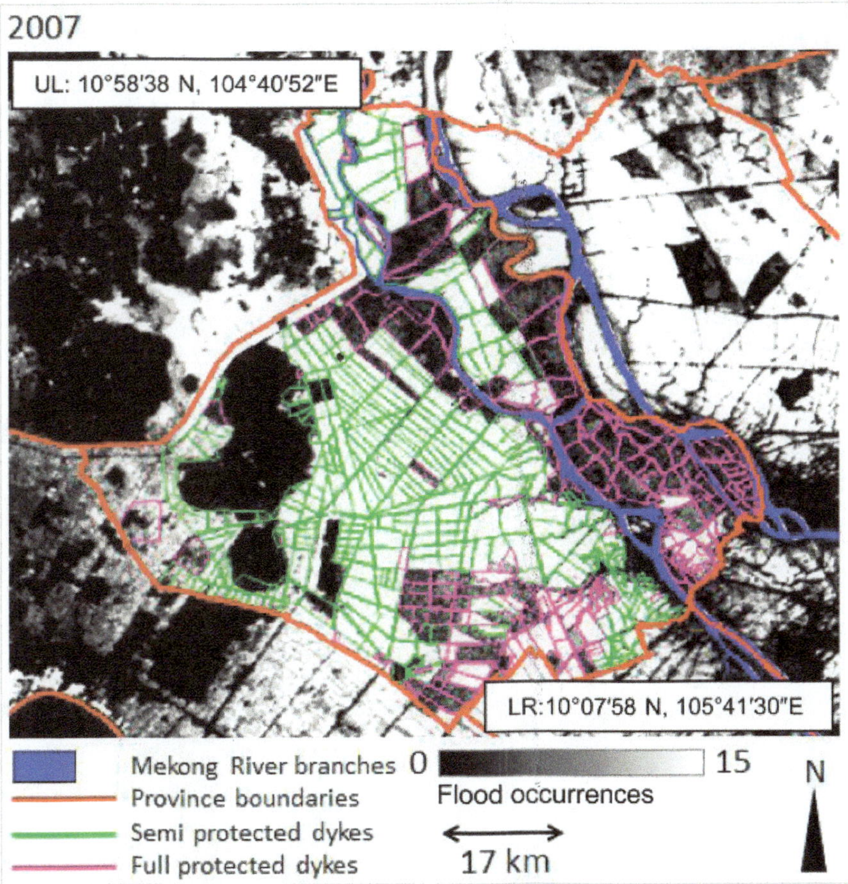

Figure 12. An Giang province in the very north of the Mekong Delta. Fully protected and semi protected dykes (kindly provided by Nguyen Viet Dung) are superimposed on the inundation frequency.

In Figure 13 we depict floodwater progression over the course of the 2007 rainy season within this fully dyked province (Figure 11). Thi Thu Ha *et al.*[40] already underlined that in flood controlled areas gates of dykes are opened about every 3–4 days during the rainy season to flood rice paddy fields—usually before planting the winter-spring rice crop. Based on the results displayed in Figure 13 we can observe that the inundation of compartments with semi-protected dykes or no dykes is controlled by type 1 river floods and overland flow and dyke overflow. In the subsets of time steps 6 and 7 (14 August, 23 August) the flooded area extends rapidly and uncontrollably outside of the protected dyke areas (pink) within un-dyked, and semi-protected (green) dyke areas. Taking a closer look at time steps 7 and 8 (23 August and 11 September), within the turquoise box we can witness how fully protected areas (pink boundary) were intentionally flooded sometime before 23 August, but are non-inundated only two weeks later. Within time steps 9 and 10 (18 September and 16 October) it is depicted how overland and over-dyke flows reach the area of An Giang province, which is furthest away from the Mekong main stem, in the very western corner, between the small hills (pink box). In the same sequence, floodwater control within fully protected dykes (green box) can be observed.

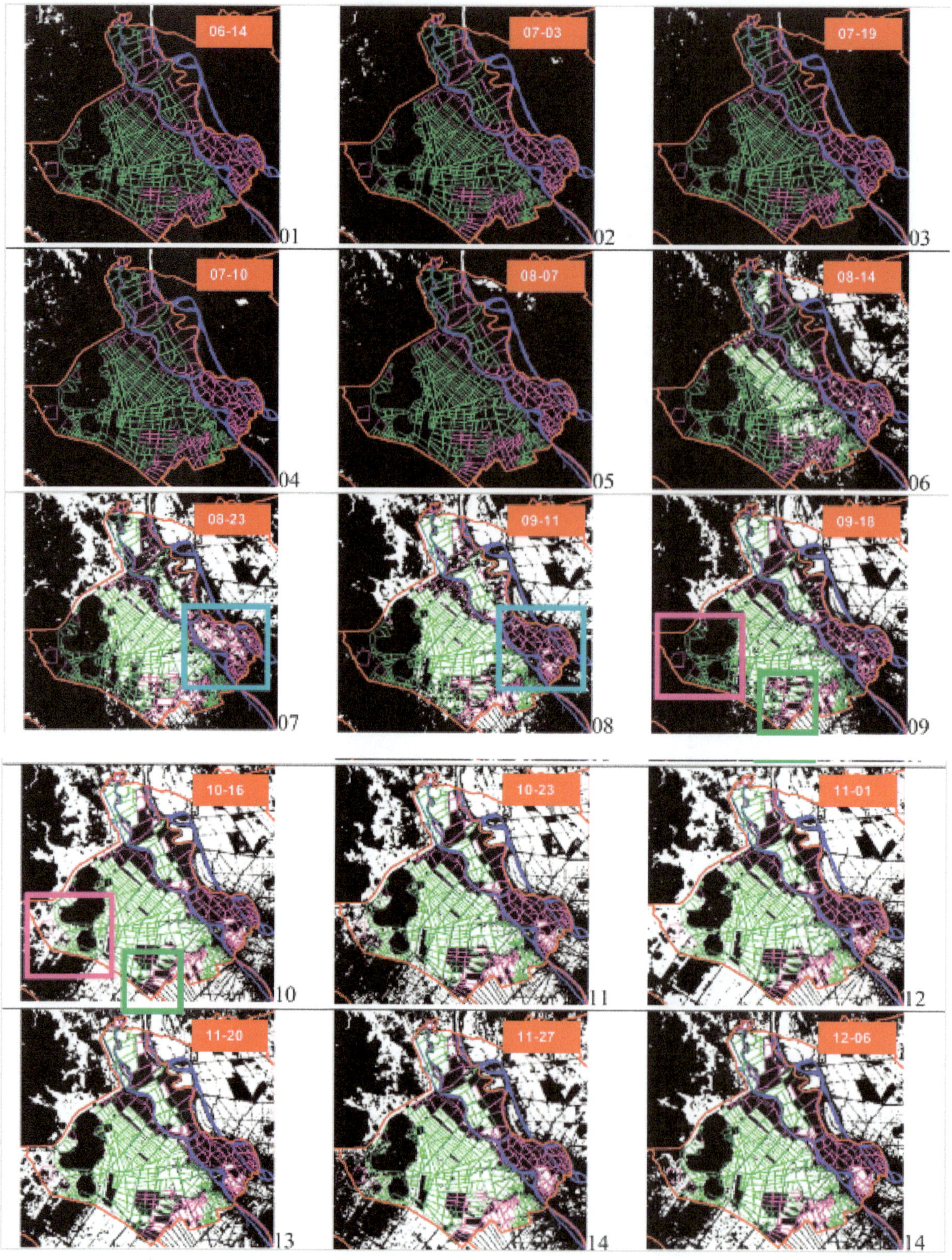

Figure 13. Floodwater progression influenced by the dyke system. Am Giang province, 2007.

A heated upstream/downstream discussion is not only being led by all Mekong riparians with respect to flood pulse changes induced by hydropower construction or upstream water diversion [37], this upstream-downstream discourse even takes place within the Vietnamese Mekong Delta. Currently, there is a trend towards areal expansion of triple rice crop systems. However, if more fully protected ring dykes are built this will lead to higher flood hazards downstream in the delta, as a larger amount of water will be pumped out at the falling flood stage. Furthermore, despite the flushing of fields to battle algae, diseases, or accumulated agro-chemicals, the normally stagnant water within dyked fields leads to a complete sedimentation of fertile nutrients. Less sediment reaches downstream localities.

The close relationship of flood patterns, dyke distribution and land use is demonstrated in the following figures. Figure 13presents a land use classification based on a comprehensive MODIS EVI and reflectance time series at 500 m resolution [56]. As the classification was derived based on the utilisation of over 45,000 MODIS scenes for the whole Mekong Basin, the accuracy and thematic differentiation for the delta might not be as precise as—for example—a manual classification based on higher resolution data that focuses solely on the delta. However, only phenologic time series analysis really allows the derivation of classes such as one-, two-, and three-season rice crops (yellow, orange, and red), fruit tree orchards (purple), and aquaculture mixed with vegetation (turquoise). There is a clear relationship between land use and flood patterns. The upper part of Figure 14—where the whole delta is presented—depicts that fruit tree orchards (purple) are clearly the regions that are never flooded in the frequency image of 2007 (largest extent of all years). Two-season rice crop regions have the highest flood frequencies and are more often inundated than three-season rice crop areas. The enlarged subset for An Giang province (Figure 14 lower part) underlines this finding. We can show that in this province all areas controlled by full protection dykes (pink lines) are classified as triple rice crop fields, whereas most of the semi-protected dykes encircle two-rice-crop paddies.

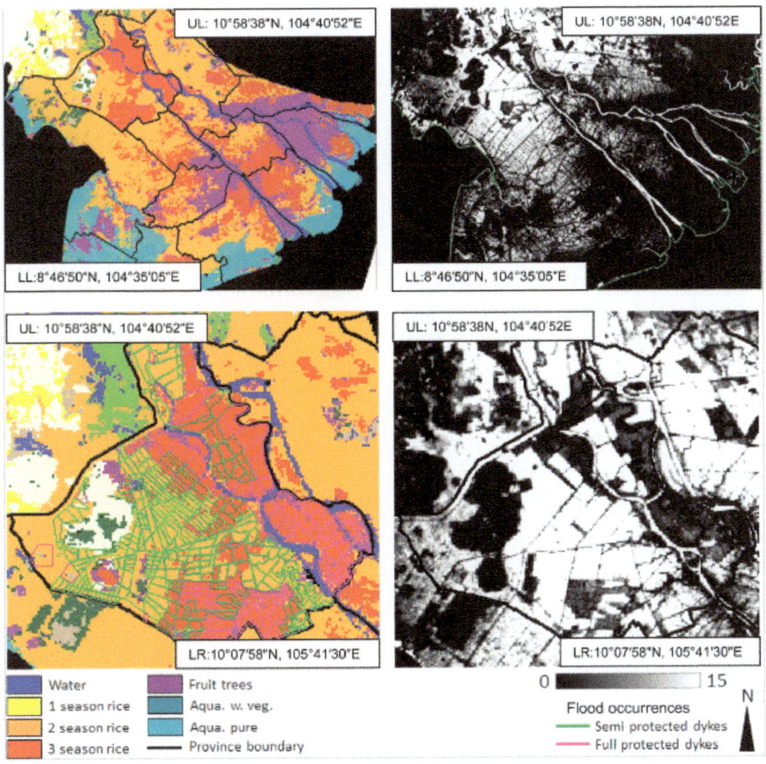

Figure 14. Relationship between land use and flood occurrence, and land use and dykes.

DISCUSSION

The Mekong Delta has an extremely complex flood regime, and a large amount of earth observation data and ancillary data is needed to visualize, interpret, and understand the driving forces of floodwater occurrence, progression, retreat, and control. So far, only few recent papers have addressed the interwoven topics of Mekong Delta hydrology, diking, and land use. We had the chance to analyse a SAR based time series spanning 60 observations over the course of five years and could shed some light on overall patterns and relations in the Mekong Delta. However, our findings lead to a large set of new questions which cannot all be addressed in the context of this paper. We discuss in the following some major critical points with respect to this study.

Choice of Algorithm

As presented in the introduction numerous publications on water detection algorithms are available. Several algorithms have been compared by different authors, and most water detection algorithms achieve high accuracies and thus—although differing in technique and ancillary data used—do not lead to strongly varying results. For our study it was important to use an algorithm which was available to us (so we used one developed in our group, sources are presented in the introduction). Furthermore, an important criterion for us was the use of an approach not depending on expensive licensed software, so that mapping can also be undertaken by project partners in Vietnam or other relevant countries. The flood maps derived based on our flood detection tool have been validated extensively via comparison with high resolution TerraSAR-X data, with Landsat and Quickbird data, as well as with water land boundary maps. As accuracy of water detection (water surface within the observation scale, and uncovered, compare also below) exceeds 95% [13], we refrained from presenting this data again.

Reliability of Presented Flood Maps

As already stated in the 'state of the art' section of the introduction, several external factors influence floodwater detection. Water detection in SAR data is hampered by unfavorable incidence angles (which we did not encounter in this study), by strong wind leading to large waves (also not a major problem for the flooded areas in the delta), and by vegetation cover above water covered areas. Whereas L-band might provide the chance to assess floodwater beneath tree cover, SAR data acquired at C-band does not allow for the detection of such water surfaces. Luckily, the Mekong Delta has relatively little tree cover. Trees occur as fruit tree orchard plantations especially in the southern part of Can Tho and provinces further to the south. However, these orchard regions are usually located at slightly higher elevation, and floodwaters in the plantation areas are rare. However, there are two other land cover classes where floodwater may occur below tree cover and thus remain un-detected. Firstly, dense mangrove belts exist along the southern and especially south-western and western coast of the delta. Mangroves need brackish water and can survive on flooded ground. However, as the mangrove canopy in these coastal regions is very dense, it is not possible to detect floodwater in these regions. However, as the mangrove areas in the delta are (unfortunately) very small and declining, this does not significantly impair the overall flood detection result. Secondly, floodwater can occur alongside canals whose banks might be overgrown with water hyacinth or *nypa* palms. Also below such vegetation water will not be detectable in SAR data. Due to such areas, where no flood detection algorithm using C-band data can usually recognize water, it is likely that the flood extents presented are rather under- than over-estimated. In many other regions of the world additional DEM data can help to improve flood mapping (simply assuming that a certain terrain height is evenly flooded if not intersected by impenetrable barriers). However, for delta regions of this world this is not feasible, as long as no highly accurate DEM derived from detailed topographic maps or LIDAR air campaigns is available. Global DEMs like SRTM are too coarse in vertical accuracy to be useful for the analysis of flat delta regions. Here, the TanDEM DEM derived from TerraSAR-X data, which is expected to be completed in 2013, might provide one additional data layer to support flood mapping in flat areas.

Observation Scale (Spatial Resolution) of the SAR Data

One topic which urgently needs to be addressed for the Mekong Delta concerns the observation scale for flood mapping. A disadvantage of the—still relatively low—resolution of 150 m is the fact that canals and rivers, which are usually equal or less than 150 m in width, cannot necessarily be extracted. The same applied to flooded areas and flooded fields which are distinctly smaller than 150 m × 150 m. In such areas the radar pixel yields a return signal which represents a mix of different surface properties. If water surfaces are mixed with too many other surfaces (bare soil, infrastructure, vegetation *etc.*) the return signal will be too 'bright' for the flood detection algorithm to accurately retrieve water surfaces. This is, for example, the case in the south-western and southern shoreline areas of the Mekong Delta, which are dominated by aquaculture, as observable in high resolution optical data and determined by *in situ* observations. However, in Figures 5, 6, and 7, the aquaculture ponds located very near the coast of most coastal provinces could not be extracted, as they are much smaller (on average 50 m × 50 m, with other uses, such as orchards, housing, and dry pools, side by side) than the minimum mapping unit of the Envisat ASAR WSM data. The same holds true for small flooded patches within urban areas. In the data presented here, the south-eastern tip of the province of Can Tho, where Can Tho city is located, appears black (never flooded). However, we do know from numerous visits that floodwaters can hamper daily life in Can Tho, and flooded areas of smaller extent occur within and also surrounding the city. These occurrences, though, are also too small to be mapped by the flood detection algorithm when using ASAR WSM data. As the scope of this study was to understand general flood patterns of the Mekong Delta, this is not of much concern here. It is clear that occurrences which are smaller than the minimum mapping unit can hardly be resolved.

This also affects the mapping of aquaculture. Aquaculture in the delta occurs in natural ponds or artificial pools which are usually far below 30 m × 30 m in size (this would already be a very big aquaculture pond). Aquaculture ponds and pools—even if dominating an area—are separated by smaller agricultural areas, settlements (bright in radar data), mangroves, or shrubland. A 150 m resolved ASAR pixel will thus contain a mix of backscatter from different surface types, and aquaculture is only one part within such a pixel. This is why (see Figure 14) not all aquaculture dominated areas appear as such in the water masks—this is simply a consequence of the size of aquaculture ponds, their density, the types of other surfaces in the area, and the SAR resolution.

As in any remote sensing application there is a limit to the observation scale, and objects which are much smaller than this scale cannot be detected—be they small flooded patches within urban areas, small flood patches within agricultural areas, or aquaculture ponds. Therefore, it is very likely that the flooded area for the Mekong Delta derived from ENVISAT ASAR WSM data is rather under- than over-estimated. To account for this and analyse the degree of this underestimation, in the future we plan to investigate selected areas of the delta based on TerraSAR-X derived inundation time series at 8.25 m and 2.5 m resolutions.

CONCLUSION

Our main findings based on four to five years of Envisat ASAR-WSM derived flood occurrence and flood progression analyses in combination with further ancillary data are:

- Flooding in the Mekong Delta progresses radially from the Mekong branch main stems outwards, as well as from north-west to south-east. In the rainy season floodwaters of type 1 (river flood and overland flow) arrive via Cambodia and spread into the delta. Within the delta, the northern floodplains are inundated the longest, while coastal areas are rarely affected by type 1 flooding. The floods related to the flood pulse of the Mekong, as well as to overland flow in the five years observed, arrive in the Vietnamese Mekong Delta roughly after DOY 215. The high flood stage occurs between DOY 275 and DOY 320. Flood prone provinces such as An Giang, Dong Thap, and Can Tho are flooded to about 40%–50% of the province area during the high flood stage.
- The many different flood types in the delta (river and overland flow, human control, larger rain events, and tidal influences) seriously complicate the interpretation of flood patterns derived from SAR data. Floodwater distribution and timing is not only a complex mix of those four components, but may also be influenced via upstream impacts on the flood pulse. Based on Envisat ASAR WSM time series alone, it is not fully possible to differentiate the individual components, and it is also

not possible to distinguish between 'good flooding' and 'bad flooding'. Natural *versus*human induced flooding cannot always be distinguished. However, if ancillary data such as information on land use and dykes are available, interpretation can be more reliable in this respect. A large human-induced control component via dykes and pumping activities can be observed. This fact has to be cautiously considered in all Mekong Delta related flood modeling. The fused data sets generated here are suitable and have been used for flood model validation and calibration, as it is quite rare to have model evaluation data available that is multi-temporal in nature [35,42].

- Areas which are most often flooded coincide with two season rice crop areas in the delta. They are often located behind semi-protected lower dykes, and also experience overland flooding, such as depicted for An Giang Province. Triple rice crop areas show lower inundation frequencies than double crop areas, and are usually located behind fully protected dyke systems. Concerning the overall delta (dyke data were not available everywhere, but land use data were available) we observed the same pattern: double rice crop areas show a higher inundation frequency than triple crop areas. Regions in the delta which are hardly ever detected as flooded coincide with extensive fruit tree orchards. It is uncommon for orchard regions to flood. However, the problem of water detection below forest canopy exists, and this definitely applies to some small coastal mangrove belt areas (e.g., south-western Ca Mau province), where crown closure of mangroves is too dense for the C-band spectrum of Envisat ASAR to penetrate. However, these really 'forested' areas in the delta are relatively rare—most of the delta is a flat, open agricultural plain. Nevertheless, the fact that some water surfaces lead us to the conclusion that the flood area we present here is rather under- than over-estimated.
- Envisat ASAR WSM analysis is limited by insufficient data availability for true time series analysis. Even though all available data for five consecutive years were derived from the archives of the European Space Agency, ESA, actually only a small fragment covering only two of 13 delta provinces remains for a real time-series analysis of five years. Nevertheless, a current large challenge for the earth observation community is SAR data availability. With the recent failure of the European Envisat in April 2012, the foregoing failure of the Japanese 'Advanced Land Observation satellite', ALOS, in April 2011, the fact that Canadian Radarsat data is quite costly, and that Italian COSMO-SkyMed and German TerraSAR-X data are usually not available at large areal coverage for scientific purposes (also have only been in orbit only since 2007), complicates the current observation of flooded areas. While daily optical data from NOAA-AVHRR and MODIS might yield a chance to circumvent the cloud cover problem in cloud-prone areas, the resolution of 1 km to a maximum of 250 m often has proven to be too low for sufficient flood mapping. The launch of ESA's Sentinel 1 is therefore urgently awaited.

The data and data products derived for this study are available in a web based information system [51,53] to which German and Vietnamese project partners of the WISDOM project [52] have full access, and to which access can be given to further interested scientists upon request and individual evaluation.

ACKNOWLEDGMENTS

The authors thank the German Ministry of Education and Research, BMBF, for funding the WISDOM project. Further thanks go to the European Space Agency, ESA, for the free provision of Envisat ASAR WS data. We acknowledge the provision of dyke data for the province of An Giang by Nguyen Viet Dung. Very special thanks go to Trin Thi Long and Vo Khac Tri of the Southern Institute of Water Resources Research, SIWRR, in Vietnam for their support with numerous field campaigns and for sharing their extensive knowledge on the Mekong Delta. We are also thankful for the constructive comments of four anonymous reviewers.

REFERENCES

1. Werle, D., Martin, T.C., Hasan, K. Flood and coastal zone monitoring in Bangladesh with Radarsat ScanSAR: Technical experience and institutional challenges. *John Hopkins APL Tech. Digest* **2001**, *21*, 148–154.
2. Jonkman, S.N. Global perspectives on loss of human life caused by floods. *Nat. Hazards* **2005**, *34*, 151–175.
3. Long, N.T., Trong, B.D. Flood Monitoring of Mekong River Delta, Vietnam using ERS SAR Data. Presented at the 22nd Asian Conference on Remote Sensing, Singapore, 5– 9 November 2001.
4. Kussul, N., Shelestov, A., Skakun, S. Flood Monitoring from SAR Data. In *Use of Satellite and In situ Data to Improve Sustainability*, Springer: Heidelberg, Germany, 2011, pp. 19–29.
5. Tholey, N., Clandillon, S., De Fraipont, P. The contribution of spaceborne SAR and optical data in monitoring flood events. Examples in northern and southern France. *Hydrolog. Process* **1997**, *11*, 1409–1413.
6. Hoque, R., Nakayama, D., Matsuyama, H., Matsumoto, J. Flood monitoring, mapping and assessing capabilities using RADARSAT remote sensing, GIS and ground data for Bangladesh. *Nat. Hazards* **2010**, *58*, 525–548.
7. Leinenkugel, P., Kuenzer, C., Dech, S. Comparison and optimisation of MODIS cloud mask products for South East Asia. *Int. J. Remote Sens.* **2013**. accepted.
8. Zhou, C., Luo, J., Yang, C., Li, B., Wang, S. Flood monitoring using multi-temporal AVHRR and RADARSAT imagery.*Photogramm. Eng. Rem. Sensing* **2000**, *66*, 633–638.
9. Richards, J.A., Woodgate, P.W., Skidmore, A.K. An explanation of enhanced radar backscattering from flooded forests. *Int. J. Remote Sens* **1987**, *8*, 1093–1100.
10. Hess, L.L., Melack, J.M., Filoso, S., Wang, Y. Delineationg of Inundated Area and Vegetation along the Amazon Floodplain with the SIR-C Synthetic Aperture Radar. *IEEE Trans. Geosci. Remote Sens.* **1995**, *33*.

11. Hess, L.L., Melack, J.M., Novo, E.M.L.M., Barbosa, C.C.F., Gastil, M. Dual-season mapping of wetland inundation and vegetation for the central Amazon basin. *Remote Sens. Environ* **2003**, *87*, 404–428.

12. Henry, J.-B., Chastanet, P., Fellah, K., Desnos, Y.-L. Envisat multi-polarized ASAR data for flood mapping. *Int. J. Remote Sens* **2006**, *27*, 1921–1929.

13. Gstaiger, V., Gebhardt, S., Huth, J., Wehrmann, T., Kuenzer, C. Multi-sensoral and automated derivation of inundated areas using TerraSAR-X and ENVISAT ASAR data. *Int. J. Remote Sens* **2012**, *33*, 7291–7304.

14. Pierdicca, N., Pulvirenty, L., Chini, M., Guerriero, L., Candela, L. Observing floods from space: Experience gained from COSMO-SkyMed observations. *Acta Astronautica* **2013**, *80*, 122–133.

15. Matgen, P., Hostache, R., Schumann, G., Pfister, L., Hoffman, L., Svanije, H.H.G. Towards and automated SAR based flood monitoring system: Lessons learned from two case studies. *Phys. Chem. Earth* **2011**, *36*, 241–252.

16. Pulvirenti, L., Pierdicca, N., Chini, M., Guerriero, L. An algorithm for operational flood mapping from Synthetic Aperture Radar (SAR) data using fuzzy logic. *Nat. Hazards Earth Sci* **2011**, *11*, 529–540.

17. Martinis, S., Twele, A., Voigt, S. Towards operational near-real time flood detection using a split-based automatic thresholding procedure on high resolution TerraSAR-X data. *Nat. Hazards Earth Syst. Sci* **2009**, *9*, 303–314.

18. Martinis, S., Twele, A., Voigt, S. Unsupervised extraction of flood-induced backscatter changes in SAR data using Markov image modeling on irregular graphs. *IEEE Trans. Geosci. Remote Sens* **2011**, *49*, 251–263.

19. Mason, D.C., Speck, R., Devereux, B. Flood detection in Urban Areas using TerraSAR-X. *IEEE Trans. Geosci. Remote Sens* **2010**, *48*, 882–893.

20. Mason, D.C., Davenport, I.J., Neal, J.C., Schumann, G.J-P., Bates, P.D. Near Real-Time Flood Detection in Urban and Rural Areas Using High-Resolution Synthetic Aperture Radar Images. *IEEE Trans. Geosci. Remote Sens* **2012**, *50*, 3041–3052.

21. Kuehn, S., Benz Hurley, J., Hurley, U. Efficient Flood Monitoring Based on RADARSAT-1 Images Data and Information Fusion with Object-Oriented Technology. Proceedings of IEEE International Conference on Geoscience and Remote Sensing Symposium (IGARSS '02), Toronto, ON, Canada, 22– 27 July 2002, 5, pp. 2862–2864.

22. Schumann, G., Matgen, P., Hoffman, L., Hostache, R., Pappenberger, F., Pfister, L. Deriving distributed roughness values from satellite radar data for flood inundation modelling. *J. Hydrol* **2007**, *344*, 96–111.

23. Pulvirenti, L., Chini, M., Pierdicca, N., Guerriero, L., Ferrazzoli, P. Flood monitoring using multi-temporal COSMO-SkyMed data: Image segmentation and signature interpretation. *Remote Sens. Environ* **2011**, *115*, 990–1002.

24. Chaouch, N., Temimi, M., Hagen, S., Weishampel, J., Medeiros, S., Khanbilvardi, R. A synergetic use of satellite imagery from SAR and optical sensors to improve coastal flood mapping in the Gulf of Mexico. *Hydrolog. Process* **2011**,*26*, 1617–1628.

25. Kasischke, E.S., Smith, K.B., Bourgeau-Chavez, L.L., Romanowicz, E.A., Brunzell, S., Richardson, C.J. Effects of seasonal hydrologic patterns in south Florida wetlands on radar backscatter measured from ERS-2 SAR imagery.*Remote Sens. Environ.* **2003**, *88*, 423–441.

26. Kiage, L.M., Walker, N.D., Balasubramanian, S., Babin, A., Barras, J. Applications of Radarsat-1 synthetic aperture radar imagery to assess hurricane-related flooding of coastal Louisiana. *Int. J. Remote Sens* **2005**, *26*, 5359–5380.

27. Lang, M.W., Kasischke, E.S., Prince, S.D., Pittman, K.W. Assessment of C-band synthetic aperture radar data for mapping and monitoring Coastal Plain forested wetlands in the Mid-Atlantic Region, USA. *Remote Sens. Environ* **2008**,*112*, 4120–4130.

28. Martinez, J.M., Le Toan, T. Mapping of flood dynamics and spatial distribution of vegetation in the Amazon floodplain using multitemporal SAR data. *Remote Sens. Environ* **2007**, *108*, 209–223.

29. Oberstadler, R., Hoensch, H., Huth, D. Assessment of the mapping capabilities of ERS-1 SAR data for flood mapping: a case study in Germany. *Hydrolog. Process* **1997**, *11*, 1415–1425.

30. Siqueira, P., Chapman, B., McGarragh, G. The coregistration, calibration, and interpretation of multiseason JERS-1 SAR data over South America. *Remote Sens. Environ* **2004**, *90*, 536–550.

31. Townsend, P.A. Mapping seasonal flooding in forested wetlands using multi-temporal radarsat SAR. *Photogramm. Eng. Rem. Sens* **2001**, *67*, 857–864.

32. Sakamoto, T., Van Nguyen, N., Kotera, A., Ohno, H., Ishitsuka, N., Yokozawa, M. Detecting temporal changes in the extent of annual flooding within the Cambodia and the Vietnamese Mekong Delta from MODIS time-series imagery.*Remote Sens. Environ* **2007**, *109*, 295–313.

33. Kuenzer, C., Campbell, I., Leinenkugel, L., Roch, M., Dech, S. Understanding upstream-downstream relations in the Mekong Basin in the context of hydropower developments. *Sustainability Science* **2012**, *11*.

34. Västilä, K., Kummu, M., Sangmanee, C., Chinvanno, S. Modeling climate change impacts on the flood pulse in the Lower Mekong floodplains. *J. Water Clim. Chang* **2010**, *1*, 67–86.

35. Nguyen, V.D. Multi-Objective Automatic Calibration of Hydrodynamic Models—Development of the Concept and an Application in the Mekong Delta. Ph.D. Thesis, University of Stuttgart, Stuttgart, Germany. 2011, 177.

36. Ha Nguyen, T.T., De Bie, C.A.J.M., Ali, A., Smaling, E.M.A., Chu, T.H. Mapping the irrigated rice cropping patterns of the Mekong delta, Vietnam, through hyper-temporal SPOT-NDVI image analysis. *Int. J. Remote Sens* **2012**, *33*, 415–434.

37. Kuenzer, C., Knauer, K. Remote sensing of rice crop areas—A review. *Int. J. Remote Sens.* **2012**. in print.

38. Kuenzer, C., Bluemel, A., Gebhardt, S., Vo Quoc, T., Dech, S. Remote sensing of mangrove ecosystems: A review.*Remote Sens* **2011**, *3*, 878–928.

39. Vo, Q.T., Kuenzer, C., Vo Quang, M., Moder, F., Oppelt, N. Review of valuation methods for mangrove ecosystem services. *Ecol. Indic* **2012**, *23*, 431–446.

40. Huu, N.N. Flooding in the Mekong River Delta, Viet Nam. Human Development Report 2007/2008, UNDP Occasional Paper No. 53 within the Program 'Fighting Climate Change'. 2007, 23.

41. Renaud, F., Kuenzer, C. The Water-Development Nexus: Importance of Knowledge, Information and Cooperation in the Mekong Delta. In *The Mekong Delta System—Interdisciplinary Analyses of a River Delta*, Renaud, F., Kuenzer, C., Eds., Springer: Heidelberg, Germany, 2012, pp. 445–458.

42. Nguyen, N.H., Delgado, J.M., Vo, K.T., Le, M.H., Merz, B., Bardossy, A., Apel, H. Floodplain hydrology of the Mekong Delta, Vietnam. *Hydrolog. Process* **2011**.

43. Leinenkugel, P., Esch, T., Kuenzer, C. Settlement detection and impervious surface estimation in the Mekong delta using optical and SAR data. *Remote Sens. Environ* **2011**, *115*, 3007–3019.

44. Kuenzer, C., Liu, G., Renaud, F., Ottinger, M., Dech, S. Asian River Deltas Experiencing Slow-Onset Hazards: Vulnerability, Resilience and Adaptation to Environmental Degradation and Climate Change. Proceedings of the International Risk and Disaster Reduction Conference, Beijing, China, 31 October–2 September 2011.

45. Kuenzer, C., Renaud, F. Climate Change and Environmental Change in River Deltas Globally. In *The Mekong Delta System—Interdisciplinary Analyses of a River Delta*, Renaud, F., Kuenzer, C., Eds., Springer: Heidelberg, Germany, 2012, pp. 7–48.

46. Vo, K.T. Hydrology and Hydraulic Infrastructure Systems in the Mekong Delta, Vietnam. In *The Mekong Delta System—Interdisciplinary Analyses of a River Delta*, Renaud, F., Kuenzer, C., Eds., Springer: Heidelberg, Germany, 2012, pp. 49–38.

47. Apel, H., Nguyen, N.H., Trinh, T.L., Vo, K.T. Flood Hydraulics and Suspended Sediment Transport in the Plain of Reeds, Mekong Delta. In *The Mekong Delta System—Interdisciplinary Analyses of a River Delta*, Renaud, F., Kuenzer, C., Eds., Springer: Heidelberg, Germany, 2012, pp. 221–232.

48. Delgado, J.M., Merz, B., Apel, H. Monsoon Variability and the Mekong Flood Regime. In *The Mekong Delta System—Interdisciplinary Analyses of a River Delta*, Renaud, F., Kuenzer, C., Eds., Springer: Heidelberg, Germany, 2012, pp. 233–245.

49. Leinenkugel, P., Kuenzer, C., Oppelt, N., Dech, S. Characterization of land surface phenology and land cover based on moderate resolution satellite data in cloud prone areas—A novel product for the Mekong Basin. *Remote Sens. Environ.***2013**. accepted.

50. Huth, J., Gebhardt, S., Wehrmann, T., Schettler, I., Kuenzer, C., Schmidt, M., Dech, S. Automated Inundation Monitoring Using TerraSAR-X Multi-Temporal Imagery. Proceedings of the European Geosciences Union General Assembly, Vienna, Austria, 19–24 April 2009.

51. Kuenzer, C. WISDOM: Water related Information System for the Sustainable Development of the Mekong Delta—Thoughts on Sustainability and Long Term Implementation. Proceedings of the 3rd Indo-German Conference on Research for Sustainability: Water and Waste Management, New Delhi, India, 3–4 February 2010.

52. WISDOM. *A German-Vietnamese Initiative*, Available online: www.wisdom.eoc.dlr.de (accessed on 1 February 2012).

53. Klinger, V., Wehrmann, G., Gebhardt, G., Kuenzer, C. A Water Related Web-Based Information System for the Sustainable Development of the Mekong Delta. In *The Mekong Delta System—Interdisciplinary Analyses of a River Delta*, Renaud, F., Kuenzer, C., Eds., Springer: Heidelberg, Germany, 2012, pp. 423–444.

54. Gebhardt, S., Wehrmann, T., Klinger, V., Schettler, I., Huth, J., Kuenzer, C., Dech, S. Improving data management and dissemination in web based information systems by semantic enrichment of descriptive data aspects. *Comput. Geosci.***2010**, *36*.

55. Vo, Q.T., Oppelt, N., Leinenkugel, P., Kuenzer, C. Remote sensing in mapping mangrove ecosystems—An object-based approach. *Remote Sens* **2013**, *5*, 183–201.

56. Stone, R. Severe drought puts spotlight on Chinese dams. *Science* **2010**, *327*, 1311.

CHAPTER 13

FLOOD MAPPING BASED ON MULTIPLE ENDMEMBER SPECTRAL MIXTURE ANALYSIS AND RANDOM FOREST CLASSIFIER

Quanlong Feng, Jianhua Gong, Jiantao Liu and Yi Li

Keywords: Flood mapping, spectral mixture analysis, random forest, medium resolution imagery

INTRODUCTION

Flooding is among one of the most destructive disasters which results in tremendous economic and human losses worldwide [1,2,3,4]. It is of great significance to map accurately the extent of inundated areas and the land cover types under water [4,5], which can assist in flood monitoring, relief works planning and damage assessment.

Due to its synoptic view and continuous coverage of flooding events, remote sensing has been recognized as a powerful and effective tool to provide inundation maps in near real time according to many researches [1,2,3,4,5,6,7,8,9,10,11]. Generally, remotely sensed data used for flood monitoring are mainly collected from radar and optical satellites. The advantage of radar remote sensing is that it enables data acquisition regardless of weather conditions and time of day [7]. Space-borne sensors such as Synthetic Aperture Radar (SAR) [8] are capable of penetrating the cloud, and can provide views of the extent of inundation even when thick clouds exist above the disaster-stricken areas [1,2,6,7,8]. Brivio *et al.* [9] utilized visual interpretation and thresholding algorithms for multi-temporal ERS-1 (European Remote Sensing satellite) SAR data in the determination of inundated areas at the peak of the flood. Results showed that only 20% of the flooded areas were determined due to the time delay between the flood peak and the satellite overpass [9]. To tackle this limitation, Brivio *et al.*[9] proposed a new procedure with the synthetic use of SAR data and digital topographic data from a Geographical Information System (GIS) technique and a high proportion, 96.7%, of the flooded areas was detected [9].

Although optical sensors are unable to penetrate thick clouds which is a major drawback in flood monitoring, the images acquired under cloud free conditions can still be utilized to extract flooded areas with high accuracy [10,11,12]. Besides, optical remote sensing can provide true color images which are much easier for visual interpretation than radar data. Widely used optical remote sensing data are mainly from Landsat series, *i.e.*, Thematic Mapper (TM), Enhanced Thematic Mapper Plus (ETM+), SPOT (Systeme Probatoire d'Observation de la Terre) series (*i.e.*, SPOT-4, SPOT-5), and Ikonos due to their ideal combination of spatial and temporal resolutions, easy access, and ease in data processing and analysis [4]. Wang*et al.* [10] used medium resolution (30 m) TM images to delineate the maximum flood event caused by Hurricane Floyd in North Carolina. Animi [11] proposed a model based on Artificial Neural Network to generate a floodplain map using high-resolution (1 m) Ikonos imagery and digital elevation model (DEM). Gianinetto *et al.* [12] utilized multisensor data (TM and SPOT-4) to map Hurricane Katrina's widespread destruction in New Orleans and adopted a change detection method to extract the land cover types under water. Above all, optical remote sensing can also play an important role in flood mapping given cloud free conditions.

Nevertheless, mixed pixels are common in medium spatial resolution data such as TM and ETM+, and such pixels have been recognized as a problem for remote sensing applications

Citation: Quanlong Feng, Jianhua Gong, Jiantao Liu and Yi Li, Flood Mapping Based on Multiple Endmember Spectral Mixture Analysis and Random Forest Classifier, doi:10.3390/rs70912539.

[13,14,15,16,17,18,19,20]. In terms of flood monitoring using medium resolution data, flooded and non-flooded features may co-exist within one pixel, resulting in the challenge to extract these small and fragmented flooded patches especially under complex urban landscapes. To deal with the mixed pixel problem, several approaches such as linear spectral mixture analysis (LSMA) [13], fuzzy-set possibilities [14], and Bayesian possibilities [15] have been developed to partition the proportions of each pixel between classes. Among these methods, LSMA appears to be the most promising and has been widely used to extract sub-pixel information with physical meaning [13,16,17,18,19]. LSMA assumes that the reflectance of each pixel can be modeled as a linear combination of a few spectrally pure land cover components, known as endmembers [13]. The aim of LSMA is to decompose a mixed pixel into the spectra of endmembers and estimate the fractional abundance of each endmember within a pixel. However, LSMA uses an invariant set of endmembers to model the entire landscape while the spectrum of each endmember is assumed to be constant across the image scene [20,21,22,23]. It neglects the fact that the same material may have different spectral curves, which cannot account for within-class spectral variability. As an extension of LSMA, multiple endmember spectral mixture analysis (MESMA) allows the type and number of endmembers to vary for each pixel, which takes into account the spectral and spatial variability of the real complex landscapes [20,21,22,23,24,25,26,27]. MESMA has been used in remote sensing fields especially in urban vegetation mapping [20,21,22,23,24,25,26,27], but to our knowledge, it has been rarely used in flood monitoring. Thus, this study aims to justify the performance of MESMA in extracting flooded areas from medium resolution remote sensing data.

Given the fact that most LSMA and MESMA applications have focused on the extraction of fractional information rather than thematic land cover types [21,22,23,24,25,26,27], this study attempted to introduce an image classifier, Random Forest (RF) [28], to classify the fraction maps generated from MESMA into different land cover categories. In general, Random Forest classifier has several advantages over other classification methods [29,30,31,32,33,34,35,36,37,38]. It is easy to parameterize, it is non-sensitive to over-fitting and is good at dealing with outliers in the training data [29]. When compared to statistical methods such as the maximum likelihood classifier (MLC), Random Forest requires no assumptions of data distribution [29], which can improve the classification performance. When compared to other machine learning methods such as support vector machine (SVM), RF has the advantage of easier parameterization and better generalization capability [30]. The successful use of SVM depends on several experiments to search for the optimal combination of kernel function type, punishment coefficient, and the kernel parameter Gamma [30]. However, the parameterization of SVM can be time consuming and cannot match that of RF. Meanwhile, RF has shown similar or even higher classification accuracy than SVM according to several studies [29,30]. RF has been widely used in remote sensing fields due to the above advantages. Rodriguez-Galiano *et al.* [32] utilized Random Forest for the Mediterranean land cover classification using multi-seasonal imagery and multi-seasonal texture, and the results indicated a high Kappa index of 0.92. Feng *et al.* [29] adopted Random Forest and texture analysis for urban vegetation mapping using high resolution UAV images and the results also showed a high classification accuracy of 90.6%. However, the usage of Random Forest in flood mapping has not been well documented and we are motivated to justify its performance in this study.

Overall, the main objective of this study is to propose a hybrid method based on MESMA and RF classifier for flood mapping using medium resolution optical satellite data. Specifically, this paper aims to (i) extend the application of MESMA to the field of flood mapping and verify its performance; (ii) justify whether Random Forest classifier can show good performance in flood mapping; (iii) discuss the merits and demerits of the proposed method based on the comparison with other state of art studies.

STUDY AREA AND DATASETS

The study area is Yuyao City in Zhejiang Province in eastern China (Figure 1), which is located at the south shore of Hangzhou Bay whose coordinates are 29°–30°N, 120°–121°E. Yuyao City is located in a relatively open and flat plain while the Yuyao River flows through the middle of the whole city from west to east. The study region has a total area of 521.2 $km^2$with an elevation range of 1 to 331 m and a slope range of 0° to 39.4°. The main land cover types include woodland (e.g., broad-leaved forest and coniferous forest), cropland (e.g., paddy rice, oilseed rape, leaf mustard), built up area (e.g., urban and suburban regions), bare soil (e.g., non-vegetated bare ground) and water (e.g., rivers, lakes, and reservoirs). Besides, the study area has an annual temperature of 16.2 °C and an average annual precipitation of 1547 mm. There is a typhoon season between May and October with an average precipitation of 5.5 mm/day in the last 60 years. Influenced by Typhoon Fitow, Yuyao experienced extreme precipitation on 7 October 2013, which led to the most serious floods in the last 60 years [30]. Typhoon Fitow brought about an accumulated rainfall of 496.4 mm in three days and the mean precipitation is about 165.5 mm/day. Most downtown areas were inundated for more than 5 days while about 833 thousand people were impacted by the disaster and the direct economic losses were more than 69.61 billion RMB (about 11.33 billion USD).

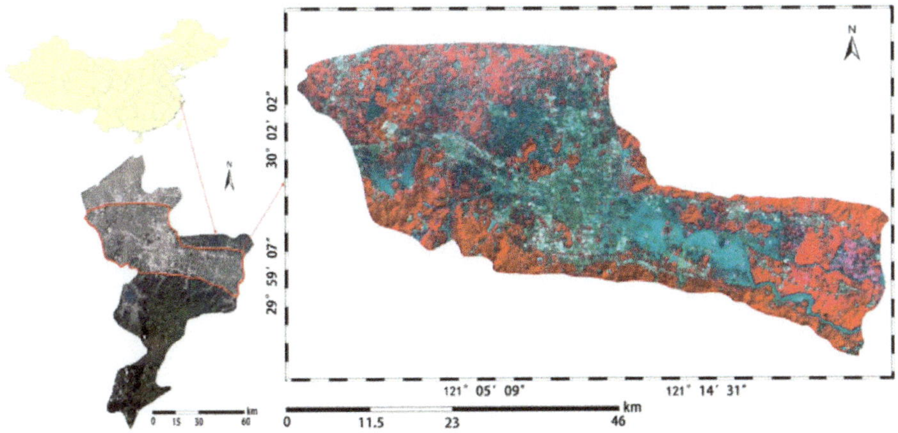

Figure 1. Study area. Red-Green-Blue composition: near-infrared, red and green bands of the multispectral charge coupled device camera (HJ-CCD) after the flood.

Remote sensing data used in this study were acquired by HJ-1B satellite of China on 11 October 2013. HJ-1B belongs to a small satellite constellation (HJ-1A/1B), was launched in September 2008 and is targeted for rapid mapping and monitoring of natural hazards and disasters [39]. Sensors on board HJ-1B consist of a multispectral charge coupled device camera (HJ-CCD) and an infrared camera (HJ-IRS) which have similar wavelengths to that of Landsat TM. The revisit time is four days which can meet the requirements of dynamically monitoring the flood events. The basic parameters of HJ-1B data can be viewed in Table 1.

Table 1. Parameters of HJ-1B data.

Sensor	Band	Wavelength/μm	Spatial Resolution/m	Image Breadth/km
CCD	1	0.43–0.52	30	
	2	0.52–0.60	30	360
	3	0.63–0.69	30	
	4	0.76–0.90	30	
IRS	5	0.75–1.10	150	
	6	1.55–1.75	150	720
	7	3.50–3.90	150	
	8	10.5–12.5	300	

Since Band-4 and Band-5 have similar wavelengths, we only utilized Band-4 in the analysis due to its higher spatial resolution. Besides, the data quality of Band-7 is quite low due to severe stripe noises and Band-8 belongs to a thermal infrared band; they were discarded in the following research. Thus, we only focused on the visible, near-infrared and short wave infrared wavelengths and Band-1–4 and Band-6 were selected for further analysis.

METHOD

Workflow

The workflow of this study is depicted in Figure 2 which comprises the following steps: image preprocessing, spectral unmixing using MESMA, image classification via Random Forest, and accuracy assessment. First, images acquired by both HJ-CCD and HJ-IRS after the flood event were preprocessed including radiometric calibration, geometric rectification and atmospheric correction (Section 3.2). Second, MESMA was employed to decompose the multispectral images into four fraction maps consisting of water, vegetation, impervious surface, and soil (Section 3.3). Third, both the fraction maps and the five reflectance bands together with the Normalized Difference Water Index (NDWI) and topographical parameters (DEM, slope and aspect) were included for the Random Forest classifier to determine thematic land cover types after the flood (Section 3.4). Finally, accuracy assessment was done based on the confusion matrix derived from testing samples to verify the performance of the proposed approach. Meanwhile, the importance of input variables was analyzed in Section 3.5. Variable importance is similar to variable ranking, which is provided by the Random Forest classifier to identify the contribution of each variable to the classification accuracy.

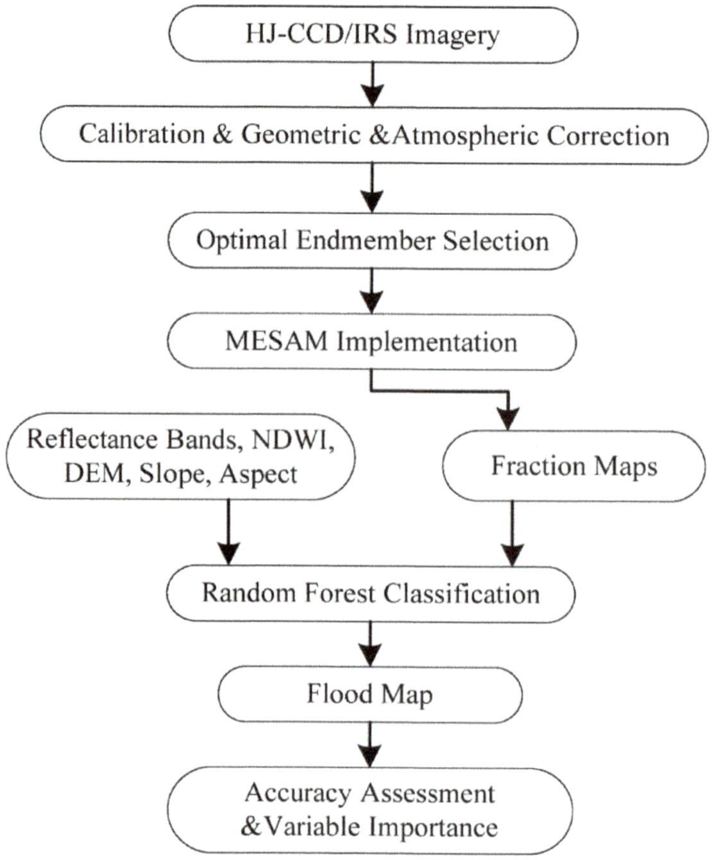

Figure 2. Workflow of this study.

Image Preprocessing

Image preprocessing was a prerequisite which involved radiometric calibration, geometric rectification, and atmospheric correction. Radiometric calibration was to convert the digital numbers recorded by the sensor to radiance. Because HJ-IRS data has a spatial resolution of 150 m, it needs to be resampled at 30 m in order to match HJ-CCD data before geometric correction. Nearest Neighbor was selected as the resampling method due to its capacity in preserving the original spectral characteristics. The resampled HJ-IRS data were layer-stacked with HJ-CCD data for further processing.

The calibrated and resampled multispectral images were then geometrically rectified to a standard Landsat-8 Operational Land Imager (OLI) image, which had already been geo-corrected with a topographic map with a scale of 1:50000. The resulting root mean square error (RMSE) of the rectified image was less than 0.2 pixels which showed good accuracy.

As for atmospheric correction, by-band 6S model [40] was adopted to retrieve remote sensing reflectance from HJ-1B multispectral data. A Mid-Latitude Summer atmospheric model was used while the aerosol type was set to be continental. Other parameters such as the altitude and wavelength of the image sensor and the imaging time can be acquired through the header files of HJ-1B data. The visibility for the 6S model was set to be 36.8 km which was obtained from the local meteorological bureau. The retrieved remote sensing reflectance was then used for MESMA to derive sub-pixel fraction maps.

Multiple Endmember Spectral Mixture Analysis

As stated above, LSMA assumes that the spectrum measured by a sensor is a linear combination of the spectra of all endmembers within the pixel [13]. The mathematical model of LSMA can be expressed as Equation (1),

$$R_i = \sum_{k=1}^{N} f_k R_{ik} + \varepsilon_i \tag{1}$$

where R_i is the reflectance of band i of a pixel, k is the number of endmembers, f_k is the proportion of the kth endmember within the pixel, R_{ik} is the reflectance of the kth endmember within the pixel in band i, ε_i is the residual of band i which indicates the unmodeled portions of the spectrum.

However, two major drawbacks exist in the LSMA method. Firstly, the number of endmembers is invariant when modeling the entire landscape using LSMA, regardless of whether the ground components represented by the endmembers are present within the pixel [20,21,22,23]. This may result in a decreased accuracy of the fractional abundance. Secondly, each endmember has only one fixed spectrum, neglecting that the same ground component may have different spectral characteristics [21,22,23]. Hence, LSMA is unable to account for with-in class spectral separability.

To address the defects of LSMA, MESMA allows the number and type of endmembers to vary on a per pixel basis to better represent the spectral variability of the complex landscapes [20], which can increase the accuracy of the estimated fractions. To be specific, MESMA consists of endmember selection and spectral unmixing modeling which is documented in the following sections.

Endmember Selection

Careful selection of a high quality set of endmembers is an important step in MESMA implementation [20,21,22,23,24,25,26,27]. Spectrally "pure" endmembers can be obtained from either a spectral library (reference endmember) or the remote sensing image itself (image endmember) [16,20]. In this study, image endmembers were utilized due to the following advantages: (i) they can be easily obtained and (ii) they can represent spectra measured at the same scale as the image data [16,21,22,23]. All the image endmembers were organized into four groups in this study: water, vegetation, impervious surface, and soil. Each group consisted of several subsets to represent the spectral variations for each material, e.g., vegetation (woodland, cropland), impervious surface (white material, dark material, blue roof, red roof), water (clear water, turbid water, submerged urban and suburban area), and soil. The shade endmember which has a zero value in all bands was also included here to account for the illumination variation [21,22,23].

In order to select spectrally "pure" endmembers, the pixel purity index (PPI) image was firstly calculated based on several thousand iterations of the PPI algorithm [21,22]. A higher value in the PPI image indicated a relatively purer pixel or endmember. The PPI results were also dynamically linked in an N-dimensional visualizer together with the original multispectral images to select the candidate endmembers by searching for a set of vertices on a convex hull [21].

The selected endmembers were re-selected using the following three techniques to obtain the most appropriate ones: (i) count-based endmember selection (CoB); (ii) endmember average root mean squared error (EAR) and (iii) minimum average spectral angle (MASA) [21,22,23,41]. The optimal set of endmembers was selected iteratively by adding high CoB, low EAR or low MASA endmembers to the library and assessing the model performance using root mean square error images and a visual comparison with high resolution orthoimagery [21,22,23].

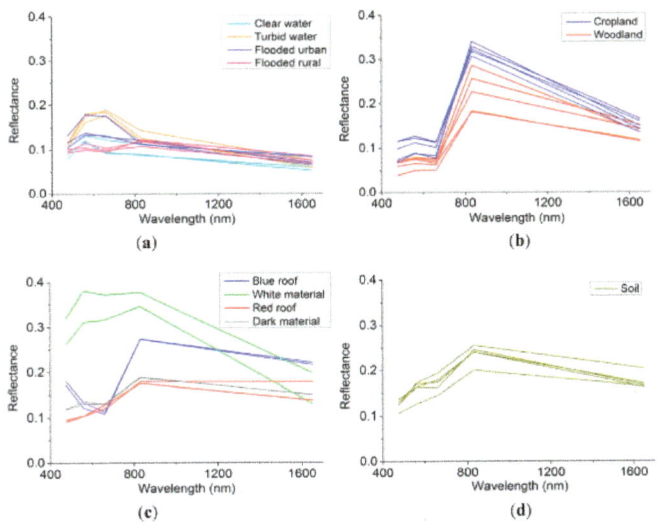

Figure 3. Spectral reflectance of endmembers including (**a**) water; (**b**) vegetation; (**c**) impervious surface; (**d**) soil.

A total of 35 optimal endmember spectra were selected using the above method from 2500 candidates. Specifically, it included twelve spectral curves for water, ten for vegetation, eight for impervious surface and five for bare soil, which are illustrated in Figure 3. Water and impervious surface spectra showed greater variation than that of vegetation and bare soil. The variation of water spectra mainly lies in that the persistent water and the submerged regions have different spectral signatures.

The whole procedure of MESMA is available in Visualization and Image Processing for Environmental Research (VIPER) tools software [41] which was adopted in this study.

Spectral Unmixing Modeling

After the selection of the optimal spectra for each endmember, the combinations of all the possible endmember spectra were considered to run the MESMA procedure (Table 2). Specifically, we tested 35 two-endmember models, 446 three-endmember models and 2640 four-endmember models for each pixel. The process of running the MESMA models based on the established spectral library is as follows. Take the three-endmember models as an example. Table 2 shows that the three-endmember models consisted of six combinations, *i.e.*, water + veg + shade, water + imp + shade, water + soil + shade, veg + imp + shade, veg + soil + shade, imp + soil + shade. Because the water category had 12 spectral curves while the vegetation category had 10 spectral curves, the combination of water + veg + shade would have a total of 120 (12 × 10 × 1) SMA models. Similarly, we could calculate model numbers of the other five combinations. Hence, the three-endmember models would consist of 446 (120 + 96 + 60 + 80 + 50 + 40) SMA models. During the modeling process, several constrains must be satisfied according to previous studies [21,22,23]: (a) the non-shade fractions should be constrained between −0.05 and 1.05; (b) the maximum allowable shade fraction should be 0.8; (c) the maximum RMSE was set to be 0.025.

Table 2. Combinations of endmember models by land cover classes.

Two-Endmember (35)	Three-Endmember (446)	Four-Endmember (2640)
water + shade	water + veg + shade	water + veg + imp + shade
veg + shade	water + imp + shade	water + veg + soil + shade
imp + shade	water + soil + shade	water + imp + soil + shade
soil + shade	veg + imp + shade	veg + imp + soil + shade
	veg + soil + shade	
	imp + soil + shade	

Note: veg, vegetation; imp, impervious surface.

How the variation of the main endmembers was handled in MESMA is as follows. As for LSMA, only a few SMA models can be established due to the limited combinations of different endmembers. However, MESMA can account for the variation by considering all the combinations of the endmember spectra in order to build the candidate SMA models [21]. Meanwhile, the SMA model with the lowest RMSE was chosen as the final output for each pixel in the MESMA procedure [21]. Pixels that cannot be modeled were left unmodeled in the output. From another perspective, the variation within each endmember was an objective existence; the MESMA method allows for the establishment of all the possible combinations of SMA models to account for the variation. The mechanism in that the SMA model with the lowest RMSE was chosen as the final output makes MESMA outperform LSMA. Besides, as shade fractions were added initially to account for illumination variation and they did not belong to the interested land cover types, we generated the shade normalized maps by dividing each land cover fraction by the sum of total non-shade fractions [22].

After the implementation of MESMA, thematic land cover types should be extracted from the fraction images. However, considering that with the simple threshold method it is difficult to discriminate different land cover types from fraction maps [21,22], a machine learning based classifier, Random Forest [28] was employed to identify the land cover types using the sub-pixel fractional information. This part of the work is documented in detail in the next section.

Random Forest Classifier

Random Forest classifier was used to classify the fraction maps derived through MESMA into thematic land cover types. This supervised classifier was chosen due to its robustness and efficiency in remote sensing applications [29,30,31,32,33,34,35,36,37,38]. Besides, Random Forest has been rarely used in flood mapping and its performance needs to be studied and verified.

Random Forest is a machine learning based method proposed by Breiman [28] in 2001. It can be viewed as an ensemble of many decision trees and can be defined as Equation (2) [28],

$$\{h(x, \theta_k), k = 1, 2 \ldots i \ldots\} \tag{2}$$

where h represents Random Forest, x represents input variables, and $\{\theta_k\}$ stands for independently identically distributed random predictor variables which are used to split each decision tree. The final response of Random Forest is determined on the output of all the decision trees involved [29].

Two steps involving random selection are used to create each decision tree. First, a bootstrap strategy [28] is utilized to randomly select only 2/3 of the training samples with replacement to build each decision tree. The remaining 1/3 of the training samples are defined as out-of-bag (OOB) data [28], which are used for inner cross-validation to evaluate the performance of Random Forest. Second, only a subset of the predictor variables is randomly selected to split each tree using the Gini index [28]. Actually, the Gini index is the most frequently used attribute selection measure for splitting each decision tree, which measures the impurity of a given attribute with respect to the rest of the classes [28]. Both of the two random selection steps result in less correlation between the decision trees and thus a higher generalization capability of the Random Forest classifier. Besides, the importance of input variables can be measured, which indicates their contributions to the classification accuracy [28,29].

The parameterization of Random Forest classifier is simple and can be carried out using the package *random Forest* in *R*language [42,43]. Only two parameters need to be specified including *ntree* and *mty*. *ntree* is the number of decision trees making up the whole forest. In general, the OOB error decreases with the growth of *ntree* and the plot of OOB error *vs. ntree*is always necessary to see whether the number of trees is sufficient in the grown forest [34]. *mty* is the number of randomly selected predictor variables. Two methods [28] are commonly used to calculate *mty*, *i.e.*, one-third or square root of the number of input variables.

In this research, the four shade-normalized fraction images generated from MESMA, together with the five spectral bands were included as input for the Random Forest classifier. Besides, since the main interest of this study was with the extraction of inundated extent, an additional image of the normalized difference water index [44] was also included for image classification. In general, the Normalized Difference Spectral Indices (NDSIs) is suitable for detecting the open water surface and the inundated areas. Boschetti *et al.* [45] provided a useful review which summarized and compared dozens of NDSIs for detecting surface water in flooded rice fields, which also shed light on the inundated area detection.

Moreover, the water related spectral indices are proposed as the combination of shortwave infrared and near infrared or visible spectral regions [45]. Due to the band positions of the HJ satellite data, we incorporated a normalized difference water index as shown in formula (3):

$$\mathrm{NDWI} = \frac{\rho(\mathrm{Green}) - \rho(\mathrm{NIR})}{\rho(\mathrm{Green}) + \rho(\mathrm{NIR})} \tag{3}$$

where ρ(Green) and ρ(NIR) refer to the reflectance of the green band (Band-2) and near infrared band (Band-4), respectively.

Meanwhile, to improve the separability of different land cover types, the topographical parameters were also added here including DEM, slope, and aspect. The DEM, with a spatial resolution of 30 m, was derived from the Advanced Spaceborne Thermal Emission and Reflection Radiometer Global Digital Elevation Model (ASTER GDEM) products [11].

According to visual interpretation through high resolution unmanned aerial vehicle photos [30], five classes were chosen for the entire image as follows, water, built-up, woodland, cropland, and bare soil. The selection procedure of training and testing samples is of great significance in supervised image classification. According to previous studies [29,46], there are several selection strategies including single pixel selection, similar contiguous pixels through seeding, and polygon-block selection. In this study, a polygon-block sampling procedure was utilized according to Chen *et al.* [46]. All the training and sampling were selected by visualizing and digitizing small polygons or blocks of pixels, and it was assumed all the pixels within each polygon belong to the same land cover category [46]. Usually there were two to six pixels in a polygon utilized in this research. Besides, all the training and testing samples were derived from the remote sensing image based on high resolution areal image, Google Earth history image and *a priori* knowledge. Each land cover type had 200 training points to train the Random Forest classifier in order to avoid any under- or over-estimation of the spectral patterns. The trained Random Forest was then utilized to classify the input images. Next, the non-water classes (woodland, cropland, built up area, and bare soil) were merged into the non-flooded class to simplify the problem while the flooded class was extracted by subtracting the persistent water. The "persistent water" was derived from the "water" class of the classification result of the pre-flood image (2013.09.10). Because the "real" flooded area and the persistent water were both included in the "water" class of the classification result of the flooding image (2013.10.11), the persistent water was then utilized to separate the "real" flooded areas.

Accuracy Assessment

In order to quantitatively assess the accuracy of the proposed method, the confusion matrix was calculated from the validation samples. Overall accuracy, producer accuracy, user accuracy, and the Kappa index can be derived from the confusion matrix to justify the performance of the hybrid approach.

There were only two land cover types, *i.e.*, non-flooded class and flooded class in the accuracy assessment. Each of the non-flooded and flooded classes had 200 testing points to derive the confusion matrix. Besides, the testing points were selected independently from the training points, using the polygon-block selection procedure [46]. The spatial distribution of training and testing samples of the "water" class is depicted in Figure 4. Actually, the "point" in Figure 4 is a small polygon containing several pixels which is assumed to belong to the same land cover category. Besides, all the testing points were derived manually based on *a priori* knowledge and reference images, therefore, under this context, testing points can be distributed almost evenly among the whole study area through manual selection. Although it can generate some bias when compared to the statistical sampling method, the above sampling method can be feasible and acceptable in the remote sensing field [29,46].

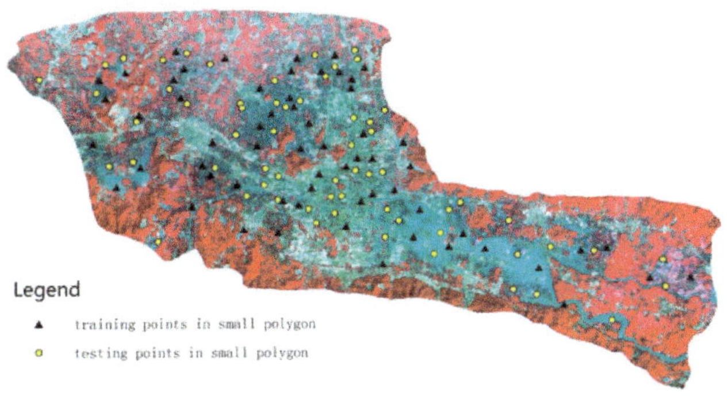

Figure 4. Spatial distribution of training and testing points for "water" class.

RESULTS

Fraction Maps Derived from MESMA

After the procedure of MESMA, four fraction maps including water, vegetation, impervious surface, and soil can be derived and are shown in Figure 5. Red areas represent high fractions while blue areas mean low fractions.

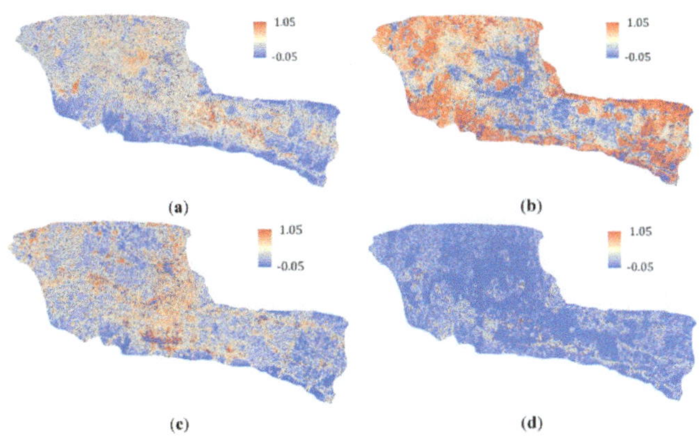

Figure 5. Fraction maps derived from multiple endmember spectral analysis (MESMA) including (**a**) water; (**b**) vegetation; (**c**) impervious surface; (**d**) soil.

Figure 5 indicates that MESMA can generate precise fraction maps for the entire post-flood regions. The peripheral regions of the reservoir (on the left side of the image) and the Yuyao River, which also belong to the flooded areas, have a high value of water fractions (Figure 5a). The extent of Yuyao City, which manifests itself with high values in the impervious surface fraction map, is clearly depicted in Figure 5c. The high vegetation fractions are mainly located outside the city core (Figure 5b) while the soil fractions are relatively lower and scattered among the vegetated areas (Figure 5d). Above all, the fraction maps derived from MESMA shows distinct differences between each other, which can improve the between-class separability of different land cover types.

Parameterization of Random Forest

As stated in Section 3.4, two important parameters including *mty* (the number of randomly selected predicable variables) and *ntree* (the number of individual trees) should be determined. Since the input image consisted of 13 bands, *i.e.*, five spectral bands, four fraction bands, one NDWI band and three topographical bands, thus *mty* was set to be 3.61, which was the square root of the number of input variables. As for *ntree*, the optimal value should make the OOB error of RF converge. We set it to be 200 according to our previous studies [29,30], and the plot of *ntree vs.* OOB error is depicted in Figure 6 to justify whether 200 trees is enough for the classifier.

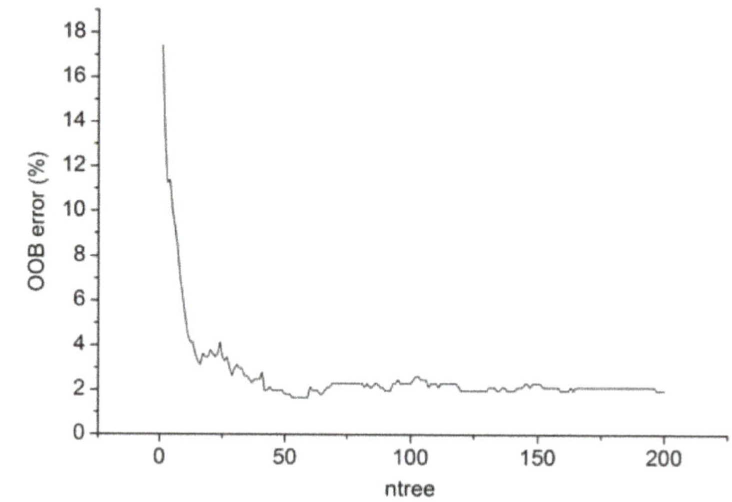

Figure 6. Out-of-bag (OOB) error *vs. ntree*.

Figure 6 depicts clearly that the OOB error has already been convergent when *ntree* reaches 200. Specifically, the OOB error decreases sharply from 17.4% to 3.1% while *ntree* increases from 1 to 15. After that, the OOB error drops slowly and begins to show convergence. Besides, Figure 6 also indicates that *ntree* = 50 or 100 witnessed a similar OOB error to that of*ntree* = 200, however, the OOB error curve showed greater fluctuations at *ntree* = 50 or 100 than *ntree* = 200. Therefore, we rendered that a number of 200 decision trees shows more robustness and *ntree* was then set to be 200 in the Random Forest classifier.

Flood Mapping Results

The Random Forest classifier with 200 decision trees was incorporated to classify the post-flood image containing 13 input bands into five land cover types: water, woodland, cropland, impervious surface, and bare soil. The final flood map was generated through the merging of non-water land cover types into the non-flooded class and the subtracting of persistent water from the water class (Figure 7).

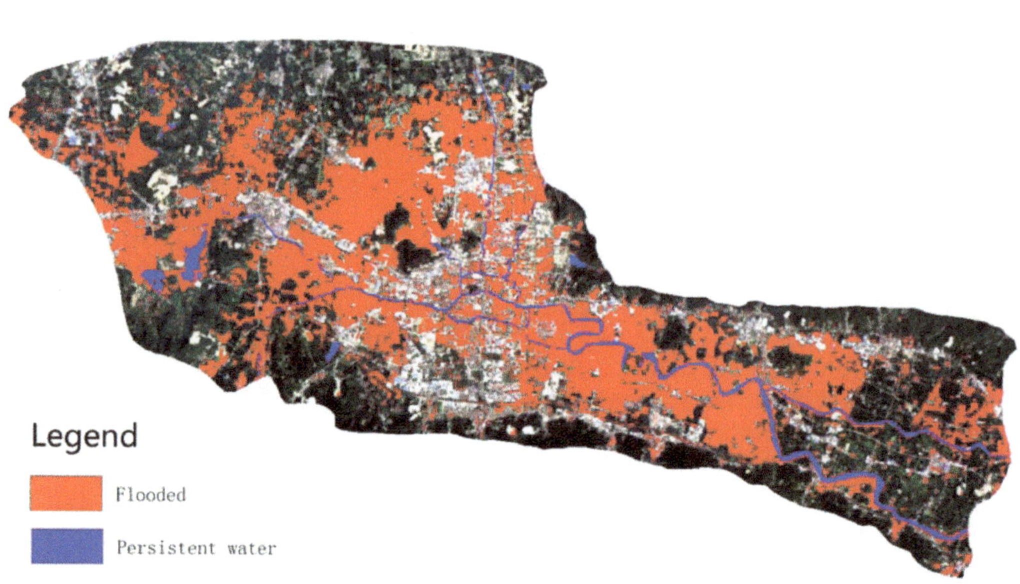

Figure 7. Flooded areas generated through RF classifier.

As depicted in Figure 7, the red areas represent the flooded extent while the blue areas stand for the persistent water. In order to manifest the inundated areas more obviously, the original true color remote sensing image is set to be the background on which the flood mapping result is overlaid. Figure 7 also illustrates that the inundated areas show distinct spatial patterns. Large flat areas along the branches of the Yuyao River were submerged due to the overflow of the rivers. The downstream area of the reservoir which lies on the left side of the image was also inundated due to the flood discharge through the dam. Besides, vast areas of low-lying cropland, adjacent to the northwestern part of the city core, were submerged.

Moreover, land cover maps just before a flood event should be useful to help extract submerged land cover types and areas to assist disaster loss evaluation and post-disaster reconstruction. HJ-CCD and HJ-IRS images acquired on 10 September 2013 were selected to generate the land cover map before the flood, also using the proposed method in this study. Specifically, MESMA together with Random Forest classifier were utilized to derive the land cover types of the pre-flood image. First, 32 optimal spectral curves (nine for water, ten for vegetation, eight for impervious surface and five for bare soil) were selected for the MESMA model to derive the fractional information. Second, NDWI of the pre-flood image, together with DEM, slope, aspect, original spectral bands, and fractional bands were layer stacked for the Random Forest classifier with 200 decision trees. A number of 200 testing points for each land cover type were selected to calculate the confusion matrix and the result showed high performance with an overall accuracy of 91.20% and a Kappa index of 0.89, which can provide accurate and reliable data source to derive the inundated land cover types and areas. The pre-disaster land cover map is shown in Figure 8.

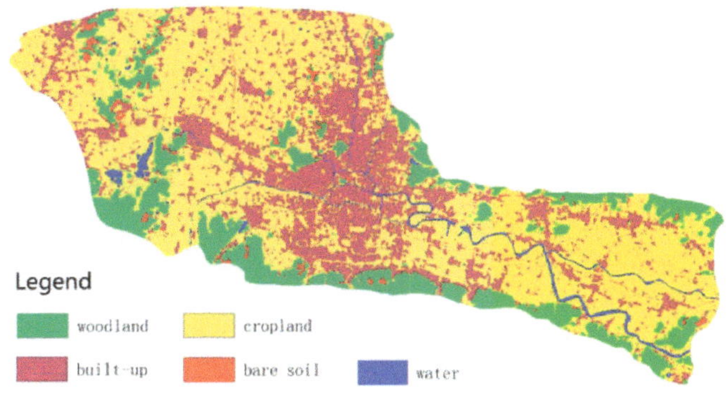

Figure 8. Land cover types before the flood event.

Figure 8 illustrates that the city core is centered in the study area while the suburban areas possess large areas of cropland and scattered villages. The woodland is mainly located on the mountains adjacent to the floodplain while the bare soil is mainly located where the forest had been cut down. The area (km^2) and proportion (%) of pre-flood land cover types are listed in Table 3.

Table 3. Area and proportion of the land cover types before the flood.

Pre-Flood Land Cover Type	Area (km^2)	Proportion
Woodland	76.023	14.59%
Cropland	294.823	56.56%
Built up area	136.145	26.12%
Bare soil	4.100	0.79%
Water	10.124	1.94%
Total	521.215	100.00%

It can be observed that cropland was the dominant land cover type with a percentage of 56.56%, followed by built up area (26.12%) and woodland (14.59%). Water and bare soil accounted for a small proportion of 1.94% and 0.79%, respectively. The pre-disaster land cover map and the flood map were sent for a GIS analysis to compute the land cover types and areas intersected by flooding (Table 4).

Table 4. Inundated land cover types and area statistics.

Inundated Land Cover Type	Area (km^2)	Proportion
Woodland	0.075	0.07%
Cropland	76.133	75.66%
Built up area	24.373	24.23%
Bare soil	0.039	0.04%
Total	100.620	100.00%

Table 4 indicates that a total area of about 100 km² was inundated after the flood event. Among this, about 76 km² was cropland and 24 km² was built-up areas which altogether account for about 99% of all the inundated areas. There existed very little woodland and bare soil that had been submerged. Furthermore, the fact that large cropland and built-up areas were flooded indicates that great economic losses had occurred, including the reduction of agriculture yield, the stagnation of industrial production, *etc.*

Results of Accuracy Assessment and Variable Importance

In order to quantitatively justify the performance of the proposed method, accuracy assessment was carried out based on the confusion matrix derived from validation data (Table 5). It indicates that the proposed method can extract the inundated areas accurately with an overall accuracy of 94% and a Kappa index of 0.88. Besides, more flooded testing points were misclassified as non-flooded (17) than the situation in which non-flooded were misclassified as flooded (7), meaning that the proposed method has the tendency to underestimate the flooded areas.

Table 5. Confusion matrix for two land cover types.

Classification Results	Validation Data		
	Flooded	Non-flooded	UA (%)
Flooded	183	7	96.3
Non-flooded	17	193	91.9
PA (%)	91.5	96.5%	
OA (%)	94	Kappa	0.88

Notes: PA, Producer Accuracy; UA, User Accuracy; OA, Overall Accuracy.

It is necessary to show the accuracy assessment of the overall five land cover classes. To calculate the confusion matrix, 200 testing points were selected for each class and the following Table 6 shows the accuracy assessment results for the entire five land cover types.

Table 6. Confusion matrix for five land cover types.

Class	Validation Data					
	Woodland	Cropland	Built up	Bare soil	Water	UA (%)
Woodland	158	0	0	0	0	100.00
Cropland	31	173	0	0	0	84.80
Built up	1	8	200	9	16	85.47
Bare soil	1	0	0	191	0	99.48
Water	9	19	0	0	184	86.79
PA (%)	79.00	86.50	100.00	95.50	92.00	
OA (%)	90.60		Kappa	0.8825		

Table 6 indicates that the classification errors mainly occurred between woodland and cropland, water and cropland, water and built up area. The spectral similarity between forests and crops accounts for the classification errors between woodland and cropland. Meanwhile, the errors between water (mainly flooded areas), cropland, and built up area lie in the fact that a large amount of croplands and built up areas were submerged by the flood water, which resulted in the spectral confusion that accounts for the classification errors.

Another important characteristic of Random Forest classifier is that it can provide the importance of input variables, which can be used to measure their contribution to the classification accuracy. Actually, the variable importance can play a role in feature selection in choosing the more important variables for further classification [32]. The feature selection process is significant especially when several dozens of variables are used and the amount of calculations are correspondingly heavy [32]. Therefore, it is possible to re-run the RF classifier without the less important variables to reduce the amount of calculations. However, we did not discard any input variables because the number of variables was small in this study and the amount of calculation was acceptable. The importance of all the 13 input variables is depicted in Figure 9. In general, the original reflectance bands are more important than fraction bands and topographical bands. The near-infrared band (Band-4) is the most important followed by the short wave infrared band (Band-6). This is partly because the reflectance of water is very low in those two bands, which helps to increase the difference between flooded and non-flooded regions. As for the fraction bands, the water fraction is more important and this is predictable since water fraction can help provide sub-pixel information of the flooded areas.

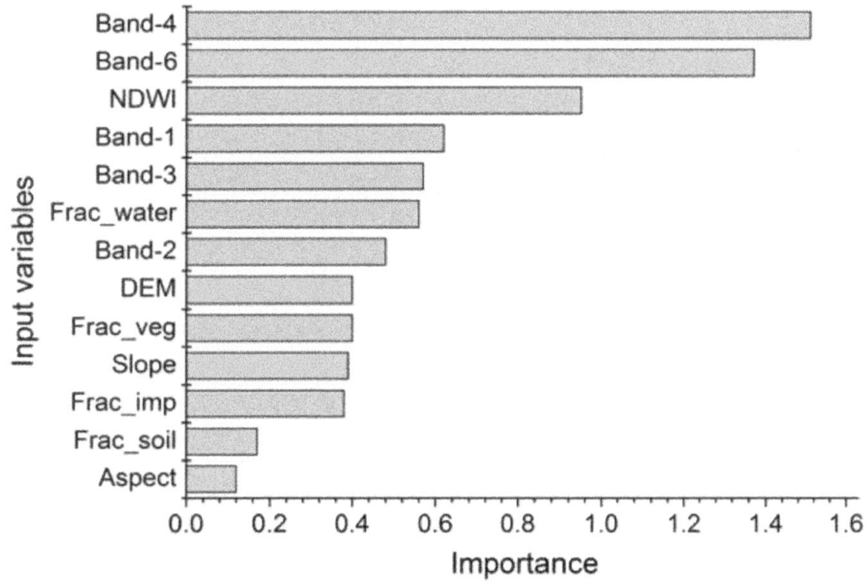

Figure 9. Importance of input variables.

Comparison with Other Methods

To further verify the performance of the proposed method, a series of comparison experiments was carried out. First, Random Forest was utilized to classify the input image without fractions to verify whether the inclusion of fraction information can improve the classification accuracy. Next, maximum likelihood classifier (MLC) was incorporated as a benchmark to classify both the input image, with and without fraction information. Also, the same training and validating data were used during all the classification to ensure comparability. Finally, the traditional NDWI thresholding method was utilized to extract the flooded areas. A trial and error approach was employed to determine the optimal threshold of NDWI and the final threshold was set to be –0.18. The results of the above comparison experiments are listed in Table 7.

Table 7. Comparison of the classification results by different approaches.

	MESMA + RF	RF-Only	MESMA+MLC	MLC-Only	NDWI-Thresholding
OA (%)	94	91.5	92	89	82.3
Kappa index	0.88	0.83	0.84	0.78	0.65

Notes: MESMA+RF, the proposed method; RF-only, using RF to classify the input image without fraction bands; MESMA+MLC, using MLC to classify the input image with fraction bands; MLC-only, using MLC to classify the input image without fraction bands. Maximum likelihood classifier (MLC). Multiple endmember spectral analysis (MESMA). Random Forest (RF). Normalized Difference Water Index (NDWI).

Table 7 indicates that the proposed method in this study yields the highest accuracy. The inclusion of fraction information increases the overall accuracy by 2.5% when comparing MESMA+RF to RF-only. Similarly, when using MLC, the overall accuracy rises by 3% after the addition of fraction bands, indicating that the inclusion of sub-pixel fractions can improve the classification accuracy regardless of the type of classifiers. Besides, RF outperformed MLC with an increase of 2% and 2.5% in classifying the images with and without fractions bands, respectively. The traditional NDWI thresholding method shows the lowest accuracy of 82.3%, indicating that it is inferior to these classification methods. Although the inclusion of MESMA based fraction images only increased the classification accuracy by 2.5% for RF classifier, this was derived from the classification results of only one study area during the flooding period. However, when considering the classification results of the pre-flood image, the inclusion of MESMA improved the accuracy by 4.7% of RF, which was higher than that of the flood image. Therefore, future studies should include more study cases to further verify the role of MESMA in the improvement of classification accuracy.

DISCUSSION

Experimental results demonstrated that both the flood map and pre-flood land cover map can be derived accurately from the proposed method, which provides reliable and valuable information for flood management and hazard assessment. The advantage of the proposed method lies in the integration of MESMA and RF classifier. Specifically, due to its capability to account for the spectral and spatial variability of complex landscapes [21], the adoption of MESMA can tackle the mixed pixel problem of flood mapping when using medium resolution remote sensing data. In previous studies [20,21,22,23,24,25,26,27], MESMA was successfully utilized in several application fields, such as mapping vegetation types under complex urban environment [21], mapping burn severity in Mediterranean countries from moderate resolution satellite data [22], and mapping urban land cover types using HyMap hyperspectral data [26], *etc.* This study introduced MESMA into the field of flood mapping, which broadens its range of application and provides a new clue on deriving inundated areas accurately from medium resolution data. Besides, the adoption of the robust and efficient RF classifier also contributes to the high performance of this research. In previous studies of flood mapping [8], statistical classifier such as MLC was widely used. However, MLC is inferior to RF classifier according to several studies [29,30,31], therefore, we introduced RF classifier to improve the classification accuracy in the field of flood mapping.

Meanwhile, it is necessary to compare with other literature on flood mapping to further verify the performance of the proposed approach. These state of art studies include water index (WI) [44,45], linear spectral mixture analysis [47], artificial neural network (ANN) [11] and object based image analysis (OBIA) [8]. To begin with, water index (e.g., NDWI) has been widely used for the detection of open surface water and inundated areas. The advantage of the water index lies in its simplicity and practicality, which can enhance open water features and suppress built up, vegetation and soil noise at the same time [44]. Memon *et al.* (2012) [48] used three water indexes for delineating and mapping of surface water using MODIS (Terra) near real time images during the 2012 floods in Pakistan. The three water indexes included NDWI, Red and Short Wave Infra-Red (RSWIR) water index [49] and Green and Short Wave Infra-Red (GSWIR) water index [50]. Experimental results indicated the accuracy of NDWI, RSWIR, and GSWIR was 73.12%, 85.80%, and 81.54%, respectively, which was lower than the proposed MESMA+RF approach (94%). This was mainly due to the fact that the vegetation has relatively high reflectance in the NIR region, so the water indexes could not take water under vegetation into account [48]. Therefore, the co-existence of flooded water and vegetation within one pixel would lead to the underestimation of flooded areas when using water indexes. However, MESMA can derive the sub-pixel water information from the mixed pixels, which accounts for a higher accuracy than with the water index thresholding methods.

Juan *et al.* (2012) [47] utilized linear spectral mixture analysis as part of the proposed sub-pixel analysis methodology to identify flooded areas from MODIS remote sensing data. The proposed methodology was demonstrated to be effective for mapping the flood extent with an accuracy of 80% [47]. However, the proposed approach of our paper outperformed that of Juan *et al.* One possible reason lies in that Juan *et al.* only adopted the LSMA model, which used an invariant set of endmembers (water, vegetation, *etc.*) to model the complex flooded landscape and could not account for the within-class variance [21]. On the other hand, MESMA allowed the type and number of endmembers to vary on each pixel [22], which could yield more accurate fractional information than LSMA.

Artificial neural network was employed in Amini's study (2010) [11] to classify the high-resolution Ikonos image to determine the inundated classes after flooding. A multi-layer perceptron neural network was chosen due to its ability to implement nonlinear decision functions and the fact that no prior conjectures needed to be made for the input data [11]. Results showed that the utilization of ANN achieved an accuracy of 70% and outperformed MLC with an accuracy increase of 15% [11]. Meanwhile, the proposed MESMA+RF method outperformed that of Amini. One possible interpretation is that ANN has the drawback of a low generalization capability due to over-fitting of the training data, which leads to the decline of performance in predicting new datasets. However, RF uses a bootstrap strategy to generate independent training samples to tackle the problem of over-fitting, which accounts for the higher accuracy when compared with ANN.

In addition, as one important method in remote sensing image classification, object based image analysis has also been applied in detecting inundated areas. Mallinis *et al.* (2013) [8] utilized Geographic Object-Based Image Analysis (GEOBIA) and Landsat TM data for flood area delineation. Results indicated that the proposed GEOBIA based method showed high performance and attained an overall accuracy of 92.67% in inundated-areas detection [8]. The advantage of OBIA in flood mapping is the ability to incorporate semantic knowledge in the classification process, thus restricting limitations resulting from imagery characteristics and temporal availability [8]. The RF classifier used in this paper is a pixel-based method, although it has a similar accuracy to that of OBIA, it may still cause a "salt and pepper" effect in the classification results. Therefore, future study should be focused on incorporating OBIA into the RF classifier to further increase the accuracy of flood mapping.

Although the above discussion verifies the high performance of the proposed approach of the study, there are still some limitations which can be stated as follows. First, the optical sensor's inability to penetrate the canopies of forests can cause the underestimation of flooded areas in the woodland region. The fact that only 0.075 km^2 of woodland was detected as flooded in this study supported this point of view. This is in accordance with Wang's study [10], which found that scattered "holes" or "islands" exited in the submerged forest areas. To tackle this issue, radar data should be incorporated and future study should focus on the fusion of multi-sensor (optical and radar) data to increase flood mapping accuracy. Second, the underestimation of flooded areas in the built up regions was also a problem. This is due to the fact that some submerged roads are too narrow to be detected in the medium resolution HJ data (30 m). The buildings adjacent to the submerged roads usually have a very high reflectance and will cover the signal of the flooded roads within one pixel. Although MESMA can tackle the mixed pixel problem to some degree, however, high resolution remote sensing data should be used in order to generate more accurate flood maps in built up areas such as city blocks and factories.

Meanwhile, the proposed MESMA+RF method can be extended to other remote sensing applications such as the extraction of imperious surface from medium resolution satellite data and land cover mapping using airborne hyperspectral remote sensing data.

CONCLUSIONS

This paper proposed a hybrid method for flood mapping using medium resolution optical remote sensing data based on multiple endmember spectral mixture analysis and Random Forest classifier. A spectral library was first established using four image endmembers, including water, vegetation, impervious surface, and soil. A total of 35 optimal endmembers were selected and 3111 SMA models were constructed for each pixel to derive the fractional information. In order to increase the separability between different land cover types, the original reflectance bands, NDWI, DEM, slope, and aspect together with the fraction maps derived through MESMA were combined to construct the multi-dimensional feature space. A Random Forest classifier consisting of 200 decision trees was utilized to extract the inundated areas of Yuyao City, China. Experimental results indicated that the proposed hybrid method showed good performance with an overall accuracy of 94% and a Kappa index of 0.88. The inclusion of fractions from MESMA can improve the classification accuracy with an increase of 2.5%. Comparison experiments with other methods including maximum likelihood classifier and NDWI thresholding verified the effectiveness of the proposed method.

Above all, the hybrid method of this paper can extract inundated areas accurately using medium resolution multispectral optical data. Meanwhile, the proposed method can be expanded to the field of hyperspectral image analysis. Future studies should include more study cases to further verify the role of MESMA in the improvement of classification accuracy. A statistical sampling method should also be considered to further increase the reliability of the results of the accuracy assessment. Additionally, multi-sensor and high resolution remote sensing data are required to increase flood mapping accuracy.

ACKNOWLEDGMENTS

Special thanks to the anonymous referees, Academic Editor and Assistant Editor for very useful comments and suggestions to help improve the quality of this paper. Besides, this study is funded and supported by the European Space Agency (ESA) and National Remote Sensing Center of China (NRSCC) Dragon 3 Program (10668), National Key Technology R&D Program of China (2012BAH33B04), the National Natural Science Foundation of China (41471341); "135" Strategy Planning (Y3SG1500CX) of the Institute of Remote Sensing and Digital Earth, CAS; the Young Scientists Foundation of RADI (Y5SJ1000CX).

AUTHOR CONTRIBUTIONS

Quanlong Feng proposed the main structure of this study, and contributed to data analysis and manuscript writing. Jiantao Liu contributed to MESMA data processing and manuscript revision. Jianhua Gong and Yi Li mainly contributed to manuscript revision.

CONFLICTS OF INTEREST

The authors declare no conflict of interest.

REFERENCES

1. Kuenzer, C.; Guo, H.; Huth, J.; Leinenkugel, P.; Li, X.; Dech, S. Flood mapping and flood dynamics of the Mekong Delta: ENVISAT-ASAR-WSM based time series analyses. *Remote Sens.* **2013**, *5*, 687–715.
2. Kuenzer, C.; Guo, H.; Schlegel, I.; Tuan, V.Q.; Li, X.; Dech, S. Varying scale and capability of Envisat ASAR-WSM, TerraSAR-X Scansar and TerraSAR-X Stripmap data to assess urban flood situations: A case study of the Mekong Delta in Can Tho Province. *Remote Sens.* **2013**, *5*, 5122–5142.
3. Schnebele, E.; Cervone, G.; Kumar, S.; Waters, N. Real time estimation of the Calgary floods using limited remote sensing data. *Water* **2014**, *6*, 381–398.
4. Wang, Y.; Colby, J.D.; Mulcahy, K.A. An efficient method for mapping flood extent in a coastal floodplain using Landsat TM and DEM data. *Int. J. Remote Sens.* **2002**, *23*, 3681–3696.
5. Ticehurst, C.; Guerschman, J.P.; Chen, Y. The strengths and limitations in using the daily MODIS open water likelihood algorithm for identifying flood events. *Remote Sens.* **2014**, *6*, 11791–11809.
6. Martinis, S.; Twele, A. A hierarchical spatio-temporal Markov model for improved flood mapping using multi-temporal X-band SAR data. *Remote Sens.* **2010**, *2*, 2240–2258.
7. Gstaiger, V.; Huth, J.; Gebhardt, S.; Wehrmann, T.; Kuenzer, C. Multi-sensoral and automated derivation of inundated areas using TerraSAR-X and ENVISAT ASAR data. *Int. J. Remote Sens.* **2012**, *33*, 7291–7304.
8. Mallinis, G.; Gitas, I.Z.; Giannakopoulos, V.; Maris, F.; Tsakiri-Strati, M. An object-based approach for flood area delineation in a transboundary area using ENVISAT ASAR and LANDSAT TM data. *Int. J. Digit. Earth* **2013**, *6*, 124–136.
9. Brivio, P.; Colombo, R.; Maggi, M.; Tomasoni, R. Integration of remote sensing data and GIS for accurate mapping of flooded areas. *Int. J. Remote Sens.* **2002**, *23*, 429–441.
10. Wang, Y. Using Landsat 7 TM data acquired days after a flood event to delineate the maximum flood extent on a coastal floodplain. *Int. J. Remote Sens.* **2004**, *25*, 959–974.
11. Amini, J. A method for generating floodplain maps using IKONOS images and DEMs. *Int. J. Remote Sens.* **2010**, *31*, 2441–2456.
12. Gianinetto, M.; Villa, P. Mapping Hurricane Katrina's widespread destruction in New Orleans using multisensor data and the normalized difference change detection (NDCD) technique. *Int. J. Remote Sens.* **2011**, *32*, 1961–1982.
13. Adams, J.B.; Sabol, D.E.; Kapos, V.; Filho, R.A.; Roberts, D.A.; Smith, M.O.; Gillespie, A.R. Classification of multispectral images based on fractions of endmembers:

Application to land cover change in the Brazilian Amazon.*Remote Sens. Environ.* **1995**, *52*, 137–154.

14. Foody, G.M. Estimation of sub-pixel land cover composition in the presence of untrained classes. *Comput. Geosci.* **2000**,*26*, 469–478.
15. Foody, G.M.; Campbell, N.A.; Trodd, N.M.; Wood, T.F. Derivation and applications of probabilistic measures of class membership from the maximum-likelihood classification. *Photogramm. Eng. Remote Sens.* **1992**, *58*, 1335–1341.
16. Lu, D.; Weng, Q. Spectral mixture analysis of the urban landscape in Indianapolis with Landsat ETM+ imagery. *Photogramm. Eng. Remote Sens.* **2004**, *70*, 1053–1062.
17. Lu, D.; Batistella, M.; Moran, E.; Mausel, P. Application of spectral mixture analysis to Amazonian land-use and land-cover classification. *Int. J. Remote Sens.* **2004**, *25*, 5345–5358.
18. Tooke, T.R.; Coops, N.C.; Goodwin, N.R.; Voogt, J.A. Extracting urban vegetation characteristics using spectral mixture analysis and decision tree classifications. *Remote Sens. Environ.* **2009**, *113*, 398–407.
19. Zhang, Y.; Lu, D.; Yang, B.; Sun, C.; Sun, M. Coastal wetland vegetation classification with a Landsat Thematic Mapper image. *Int. J. Remote Sens.* **2011**, *32*, 545–561.
20. Roberts, D.A.; Gardner, M.; Church, R.; Ustin, S.; Scheer, G.; Green, R.O. Mapping chaparral in the Santa Monica Mountains using multiple endmember spectral mixture models. *Remote Sens. Environ.* **1998**, *65*, 267–279.
21. Liu, T.; Yang, X. Mapping vegetation in an urban area with stratified classification and multiple endmember spectral mixture analysis. *Remote Sens. Environ.* **2013**, *133*, 251–264.
22. Quintano, C.; Fernández-Manso, A.; Roberts, D.A. Multiple Endmember Spectral Mixture Analysis (MESMA) to map burn severity levels from Landsat images in Mediterranean countries. *Remote Sens. Environ.* **2013**, *136*, 76–88.
23. Gan, M.; Deng, J.; Zheng, X.; Hong, Y.; Wang, K. Monitoring urban greenness dynamics using multiple endmember spectral mixture analysis. *PLoS ONE* **2014**, *9*, e112202. [PubMed]
24. Michishita, R.; Jiang, Z.; Xu, B. Monitoring two decades of urbanization in the Poyang Lake area, China through spectral unmixing. *Remote Sens. Environ.* **2012**, *117*, 3–18.
25. Powell, R.L.; Roberts, D.A.; Dennison, P.E.; Hess, L.L. Sub-pixel mapping of urban land cover using multiple endmember spectral mixture analysis: Manaus, Brazil. *Remote Sens. Environ.* **2007**, *106*, 253–267.
26. Franke, J.; Roberts, D.A.; Halligan, K.; Menz, G. Hierarchical multiple endmember spectral mixture analysis (MESMA) of hyperspectral imagery for urban environments. *Remote Sens. Environ.* **2009**, *113*, 1712–1723.
27. Myint, S.W.; Okin, G.S. Modelling land-cover types using multiple endmember spectral mixture analysis in a desert city. *Int. J. Remote Sens.* **2009**, *30*, 2237–2257.
28. Breiman, L. Random forests. *Mach. Learn.* **2001**, *45*, 5–32.
29. Feng, Q.; Liu, J.; Gong, J. UAV remote sensing for urban vegetation mapping using random forest and texture analysis. *Remote Sens.* **2015**, *7*, 1074–1094.
30. Feng, Q.; Liu, J.; Gong, J. Urban flood mapping based on unmanned aerial vehicle remote sensing and random forest classifier—A case of Yuyao, China. *Water* **2015**, *7*, 1437–1455.
31. Immitzer, M.; Atzberger, C.; Koukal, T. Tree species classification with random forest using very high spatial resolution 8-band WorldView-2 satellite data. *Remote Sens.* **2012**, *4*, 2661–2693.
32. Rodriguez-Galiano, V.F.; Chica-Olmo, M.; Abarca-Hernandez, F.; Atkinson, P.M.; Jeganathan, C. Random forest classification of mediterranean land cover using multi-seasonal imagery and multi-seasonal texture. *Remote Sens. Environ.* **2012**, *121*, 93–107.
33. Ghosh, A.; Sharma, R.; Joshi, P.K. Random forest classification of urban landscape using Landsat archive and ancillary data: Combining seasonal maps with decision level fusion. *Appl. Geol.* **2014**, *48*, 31–41.
34. Rosa, D.L.; Wiesmann, D. Land cover and impervious surface extraction using parametric and non-parametric algorithms from the open-source software R: An application to sustainable urban planning in Sicily. *GISci. Remote Sens.* **2013**, *50*, 231–250.
35. Hayes, M.H.; Miller, S.N.; Murphy, M.A. High-resolution landcover classification using Random Forest. *Remote Sens. Lett.* **2014**, *5*, 112–121.
36. Mishra, N.B.; Crews, K.A. Mapping vegetation morphology types in a dry savanna ecosystem: Integrating hierarchical object-based image analysis with random forest. *Int. J. Remote Sens.* **2014**, *35*, 1175–1198.
37. Puissant, A.; Rougier, S.; Stumpf, A. Object-oriented mapping of urban trees using random forest classifiers. *Int. J. Appl. Earth Obs.* **2014**, *26*, 235–245.
38. Xu, L.; Li, J.; Brenning, A. A comparative study of different classification techniques for marine oil spill identification using RADARSAT-1 imagery. *Remote Sens. Environ.* **2014**, *141*, 14–23.
39. Feng, Q.; Gong, J. Destriping remotely sensed data using anisotropic diffusion in wavelet domain. In Proceedings of 2014 4th IEEE International Conference on Information Science and Technology (ICIST), Shenzhen, China, 26–28 April 2014; pp. 316–319.
40. Vermote, E.F.; Tanre, D.; Deuze, J.L.; Herman, M.; Morcrette, J. Second simulation of the satellite signal in the solar spectrum, 6S: An overview. *IEEE Trans. Geosci. Remote Sens.* **1997**, *35*, 675–686.
41. Roberts, D.A.; Halligan, K.; Dennison, P.E. VIPER Tools User Manual (Version1.5). Available online: http://www.vipertools.org/ (assessed on 12 February 2015).
42. American Museum of Natural History, Center for Biodiversity and Conservation. Available online: http://

biodiversityinformatics.amnh.org/ (assessed on 12 February 2015).

43. Random Forest Classification. Available online: http://bitbucket.org/rsbiodiv/randomforest classification/ (assessed on 12 February 2015).

44. McFeeters, S. The use of the normalized difference water index (NDWI) in the delineation of open water features. *Int. J. Remote Sens.* **1996**, *17*, 1425–1432.

45. Boschetti, M.; Nutini, F.; Manfron, G.; Brivio, P.; Nelson, A. Comparative analysis of normalised difference spectral indices derived from MODIS for detecting surface water in flooded rice cropping systems. *PLoS ONE* **2014**, *9*, e88741. [PubMed]

46. Chen, D.; Stow, D. The effect of training strategies on supervised classification at different spatial resolutions. *Photogramm. Eng. Remote Sens.* **2002**, *68*, 1155–1162.

47. Juan, D.; Sandra, G. Development of a sub-pixel analysis method applied to dynamic monitoring of floods. *Int. J. Remote Sens.* **2012**, *33*, 2277–2295.

48. Memon, A.; Muhammad, S.; Rahman, S.; Haq, M. Flood monitoring and damage assessment using water indices: A case study of Pakis tan flood-2012. *Egypt. J. Remote Sens. Space Sci.* **2015**, *18*, 99–106.

49. Rogers, A.; Kearney, M. Reducing signature variability in unmixing coastal marsh thematic mapper scenes using spectral indices. *Int. J. Remote Sens.* **2004**, *25*, 2317–2335.

50. Ji, L.; Zhang, L.; Wylie, B. Analysis of dynamic thresholds for the normalized difference water index. *Photogramm. Eng. Remote Sens.* **2009**, *75*, 1307–1317.

CHAPTER 14 FLOOD RISK MAPPING USING LIDAR

Tim L. Webster

Keywords: LiDAR, flood inundation, flood risk, water levels, DEM

INTRODUCTION

Generally, Digital Elevation Models (DEMs) derived from aerial photography or other remote sensing systems such as the Shuttle Radar Topography Mission (SRTM) have degraded accuracies in forested areas. In Maritime Canada, previous topographic maps derived from aerial photo surveys were hampered in forested areas where the ground was obscured by trees and accurate measurements of the terrain were difficult (http://www.gov.ns.ca/snsmr/land/programs/post/manual/default.asp). Light Detection and Ranging (LiDAR) is a remote sensing technology to derive very accurate elevation measurements of the earth's surface. Flood *et al.* and Wehr *et al.* [1,2] provide a general overview of airborne laser scanning (LiDAR) technology and principles. The benefit of LiDAR is that a narrow laser beam is directed from the aircraft towards the earth's surface and reflected back in order to measure the range or distance from the aircraft to the ground. The beam divergence is typically very small (0.3 rmad), resulting in a laser footprint diameter of 20–30 cm on the ground, depending on flying height. Only a portion of that beam has to make it through the gaps in the forest canopy and hit the ground in order to be reflected back to the aircraft. Thus in the forest, if one can see patches of sky above them, there is a good chance the laser beam will partially make it through a gap and make it to the ground or near ground features such as dense shrubs. The LiDAR sensor records a series of points that represent what the laser pulse was reflected off and contains "hits" from the vegetation, buildings, and bare ground targets, for example. Sensors in the 1990s could record the first or last returning laser pulse and eventually evolved to record first and last pulse and the intensity of the backscattered energy. Sensors today are capable of recording intermediate pulse returns in addition to the first and last return.

In order to derive an accurate DEM, the LiDAR points are classified or filtered into 'ground' and 'non-ground' target classes. A critical review of ground classification is presented by [3] where they discuss the criteria for selecting the most appropriate filtering algorithm. They also describe where all of the algorithms are suboptimal: rough terrain or discontinuous slope, dense forest areas where the beam cannot penetrate to the ground, and low vegetation being confused with the ground. The classification methods are categorized into the following groups: segmentation or clustering, morphological, directional scanning, contour, Triangular Irregular Network (TIN), and Interpolation [3]. Kraus *et al.* [4,5] utilized a least squares interpolation method to iteratively remove trees and generate a ground surface. Vosselman [6] applied an approach that compares the slope between a LiDAR point and its neighbor to determine if it is ground or not where a threshold is used to evaluate the slope angle. Axelsson [7] described a method where seed low points are selected within a grid of user-defined size greater than the largest non-ground features. These points are used to construct an initial TIN. Then each point is added to the TIN if the parameters are below the threshold settings. The problem is that different thresholds are required for different terrain conditions. Zhang *et al.* [8] converted the LiDAR points to a raster and then use mathematical morphology filter on the grayscale image to extract ground and non-ground points. They progressively increase the size of the filter window where

Citation: Tim L. Webster, Flood Risk Mapping Using LiDAR, doi:10.3390/rs2092060.

different threshold parameters are applied to detect ground and non-ground features. They first generate a minimum elevation surface from the LiDAR points. In the next iteration, they increase the window and apply threshold parameters to form a smoother surface which is subtracted from the previous to identify non-ground points. The current industry standard for processing LiDAR is TerraScan from Terrasolid to process and classify the points. This method is based on the TIN construction from a set of low seed points initially and then iteratively evaluates additional points based on a set of parameters including slope and triangle edge length. Raber *et al.* [9] optimized the parameter settings of this software for different land covers and also used traditional parameter settings and then compared the accuracy of the derived Digital Terrain Models (DTMs). They found the differences were small although statistically significant. Hodgson *et al.* [10] compared LiDAR points and interpolated surfaces from TINs with validation points for different land cover types. Their results indicated that errors varied by land cover type and found that the TIN process did not introduce additional error; in fact it reduced the error in some cases.

LiDAR has been used in a variety of flood inundation applications involving rivers [11,12,13,14,15,16]. The research has mainly focused on assessing the impact of LiDAR derived surface model resolution on model results [12,14,16] or on the impact of vegetation and building extraction for urban areas [13,15]. The application of LiDAR to coastal areas has focused on modeling inundation from storms and sea-level rise [17,18,19,20,21,22,23,24,25]. A good review of LiDAR calibration, validation and processing and a summary of how LiDAR can be used in coastal research to assess geohazards were provided by [17]. They found that the rate and extent of coastal inundation were dependent on LiDAR horizontal grid resolution which also affected assumptions about connectivity of the ocean to low lying inland areas. Many countries that have a significant coastline are using LiDAR to monitor and model surface processes including shoreline mapping and change detection [26,27,28,29,30,31,32]. The Atlantic Coast of Canada is considered vulnerable to sea-level rise [33]. The communities of Charlottetown and the north shore of Prince Edward Island (PEI) were mapped for vulnerability to storm surge flooding and long-term sea-level rise [34]. A large storm surge impacted the area in January of 2000 and was used as a benchmark to test the inundation model developed within the ArcGIS system [18,22]. LiDAR was instrumental in facilitating accurate flood risk maps in the PEI study [18,21,34]. The PEI study was followed by a project in southeast New Brunswick that involved the acquisition of LiDAR along 175 km of shoreline for flood risk mapping [20]. The LiDAR surveys were conducted in 2003 and 2004 ensuring that the areas were captured during leaf-off conditions and prior to significant snow cover. A large coastal dune system, La dune de Buctouche, was surveyed in both years and was impacted by a storm in February of 2004. Following similar methods described in [26,27,28,29,30,31,32] this multi- temporal LiDAR dataset was used to measure coastal erosion [20].

The third assessment report from the Intergovernmental Panel on Climate Change (IPCC) indicates that there will be an increase in mean global sea-level from 1990 to 2100 between 0.09 m and 0.88 m [35]. Previous flood risk mapping projects from climate change sea-level rise in the region have used the central value of 0.5 m/century to project global sea-level rise into the future [18,20,21,22,23]. The latest IPCC Assessment Report 4 (AR4) has projected global mean sea-level to rise between 0.18 and 0.59 m from 1990 to 2095 [36,37]. However, as Forbes *et al.* [38] pointed out, these projections do not account for the large ice sheets melting as not enough literature on the subject was available at the time of the AR4 report. Recent sea-level observations were compared with the IPCC projections from 2001 [39]. The results indicated that sea-level may be responding quicker than the climate models predicted. Observations were based on tide gauge measurements and satellite altimetry. The satellite data show a linear trend (1993–2006) of 33 cm/century. The effects of fresh melt water from ice sheets on the thermohaline circulation (THC) pattern, changes sea-levels regionally and modeling results indicate the North American Atlantic coast could see an increase in sea-level between 0.5 and 1 m as the THC shuts off [40]. Comparison of altimetry data from the Topex-Poseidon satellite between 1993 and 2003 with tide gauge measurements is consistent with a weakening of the THC [40]. Sea-level changes cannot yet be predicted with confidence using models based on physical processes, because the dynamics of ice sheets and glaciers and, to a lesser extent, that of oceanic heat uptake is not sufficiently understood. The correlation between sea-level and temperature is >0.99 for observations from 1880 to 2000 [41]. Climate models predict global mean temperature with confidence and these results were used to estimate sea-levels [41,42]. The future global temperature scenarios based on the PICC AR4 report were used to estimate global sea-level rise between 75 and 190 cm for the period of 1990–2100 [41]. Due to the uncertainty in the global sea-level predictions, [38] used the upper limit of 1.3 m as a precautionary approach to sea-level rise projections in the Halifax region.

Communities around the Bay of Fundy, including Annapolis Royal, are vulnerable to coastal flooding from storm surge events such as the Saxby Gale of 1869 [44,46,47,48,49]. The Saxby Gale storm surge could have reached land elevations of between 6.2 and 6.5 m if it were to occur today on a perigean tide, without even considering the relative sea-level rise that has occurred since 1869 [44]. The more recent Groundhog Day storm on February 2, 1976 has been used as the benchmark storm to assess the model results and estimate wave run-up [45,48,50]. The tide gauge at Digby, 50 km southwest of Annapolis Royal, was not operating during the Groundhog Day storm and could not be used to assess the height of the storm surge of this event. However, water levels from the tide gauge at Saint John, New Brunswick, directly across the Bay of Fundy, were used to determine the height of the storm surge during this event and are discussed later in the methods section (Figure 1). The effects of the storm could have been much worse if it had occurred at the highest predicted astronomical tide, which was one meter higher than the tide predicted during the Goundhog Day storm. The tide gauge data from Saint John, NB, were analyzed for this event by [47], who reported a 1.46 m storm surge on top of a mean higher water level of 7.71 m (above chart datum), and stated that the damage would have been worse if the storm had occurred on a perigean spring tide, which recorded water levels of 8.84 m on April 16th, 1976.

Figure 1. Location of the study area, Annapolis Royal, and the nearest long-term tide gauge; Saint John. Digby is directly west of the Annapolis Royal label on the figure.

A previous study [44] indicated that there are several key locations around the town of Annapolis Royal, Nova Scotia, that are susceptible to flooding as a result of storm surges. As a result of that study, flood risk mapping was integrated in a recent Climate Change and Adaptation project, funded by Natural Resources Canada (http://adaptation.nrcan.gc.ca/projdb/178_e.php). In this study, a LiDAR derived DEM of Annapolis Royal, Nova Scotia, was used to construct flood inundation maps based on sealevel rise and climate change (Figure 1). Similar to other flood risk mapping projects in the region [18,19,20,21,22,23], still water levels were used within the GIS to inundate the land. In these past studies, the return period of the inundation water levels was calculated using 40 years worth of tide gauge records for the region [43]. Bernier *et al.* [43] recently compared the return period of storm surges and total water level from tide gauge records for the region, to the Dalhousie storm surge model driven by 40 years of wind observations and inferred atmospheric pressure fields. In this study, the flood inundation mapping has been coupled with a risk assessment tool, the Water Modeler, to determine return periods of water levels under current conditions and with sea-level rise rates from climate change. The return periods generated by the Water Modeler are compared to those produced by [43]. As a result of uncertainties in future sea-level rise rates, flood inundation maps have been constructed at 10 cm increments so that as sea-level rise estimates become refined in the future, the appropriate water level inundation map is available. As with previous regional studies [18,19,20,21,22,23], a benchmark storm from the past has been used to assess the model. In addition to the increased water levels associated with a storm surge (regional scale effect), strong winds associated with a storm blowing on shore can increase wave heights and cause water levels to increase. As waves push toward the shore, the effect of breaking waves produces wave setup, and individual breaking waves produce wave run-up. Wave run-up is more of a local phenomenon that is controlled by the wind speed, wind direction, and the local offshore bathymetry depth and shoreline geometry. Thus, in addition to modeling still water levels, the wave run-up has been estimated for the benchmark storm by utilizing the SWAN model [45] with a seamless DEM which integrated the LiDAR and bathymetric chart data of the basin. The objectives of this study were: (1) utilize a LiDAR derived DEM for flood inundation mapping and assess its accuracy; (2) construct flood risk maps that the community can use to adapt to rising sea-levels; (3) merge the LiDAR DEM with bathymetric data to generate wave run-up predictions based on a historic benchmark storm; and (4) construct return periods of water levels for current and projected sea-level rise conditions and map the water inundation using the LiDAR DEM.

METHODS

Airborne LiDAR

LiDAR data were acquired for the study area on April 20–25, 2004 and covered an area of 595 km^2 as part of the AGRG study of coastal processes along the Nova Scotia shoreline of the Bay of Fundy [19]. The survey was conducted under contract to Terra Remote Sensing Inc. utilizing their Mark II LiDAR system. The area was surveyed at low tide in order to capture the elevation of the exposed inter-tidal zone. The LiDAR system recorded the first and last laser returns with a posting every 1 m on the ground with vertical precision of 15 cm (Figure 2). The Mark II system was mounted on a pod that was fixed to the underside of a Bell 206L helicopter (Figure 2). Since the Mark II could only record intensity of one of the returns (first or last), it was decided to acquire the intensity on alternating returns; thus, every other first and last return would have the intensity recorded. The system utilizes a laser with a wavelength of 1,064 nm and beam divergence of 0.45 mrad, and operated at a 30,000 Hz laser repetition rate, along with the scanning mirror operating at 17 Hz to direct the laser pulses across the swath. The survey was flown at an altitude of 800 m which resulted in a laser footprint diameter of 36 cm (Figure 2). The point spacing was approximately 1 m in the along track and across track direction (Figure 2). The point density increased to near 0.5 m where the flight lines overlapped.

The point cloud was classified utilizing the TerraScan™ software into 'ground' and 'non-ground' points by the service provider who did not provide the parameters used in the ground classification process. The LiDAR points were processed to have heights relative to the Canadian Geodetic Vertical Datum of 1928 (CGVD28) which is a close approximation of mean sea-level (MSL). The data were delivered in 2 km × 2 km tiles as ASCII files where each point contained the following attributes: UTM easting, UTM northing, ellipsoidal height (relative to GRS80), orthometric height (referenced to CGVD28), GPS time, echo code, flight line number, and intensity. The data were processed into ArcInfo™ GIS point coverages, and different surfaces were constructed. In this study, four surfaces were constructed using a triangular irregular network algorithm with linear interpolation to a 1 m grid: (1) a "bare-earth" DEM; (2) a digital surface model (DSM) incorporating 'ground' and 'non-ground' points *i.e.* trees and buildings; (3) a normalized height model which is derived from the subtraction of the DEM from the DSM (height of trees and buildings); (4) an intensity image derived from the 'ground' and 'non-ground' hits. These surfaces have been used to classify the land cover of the area using an object-orientated classification method [51].

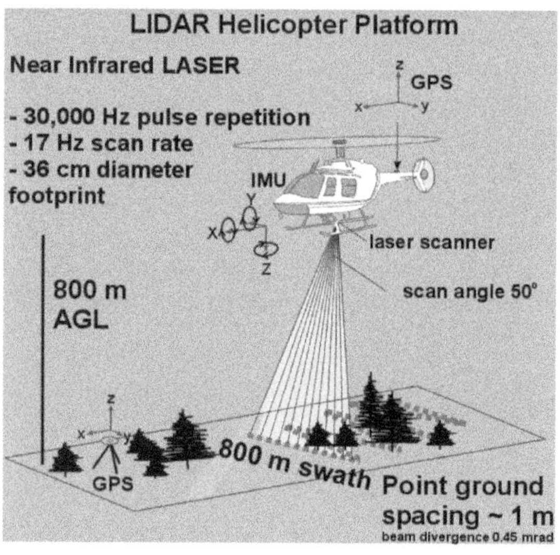

Figure 2. LiDAR configuration for the April 2004 survey.

LiDAR Validation

The LiDAR points and surfaces were expected to have an absolute vertical error of less than 30 cm, and typically less than 15 cm. Independent survey-grade GPS checkpoints were collected over paved surfaces to validate the LiDAR points and derived surface models. The real-time kinematic (RTK) GPS surveys were conducted using a Leica GPS System 500, where baselines were kept below 15 km in length, allowing the RTK GPS rover to achieve a vertical precision of 3 cm or better. The GPS data were post-processed and orthometric heights were computed and used to compare with the LiDAR points and surfaces following methods described in [52,53]. The LiDAR ground classification was assessed visually and by comparing the DSM and DEM surface models and found to be satisfactory. Extra attention was paid along the coastal zone as these areas are prone to classification errors as a result of the steep slopes and some anthropogenic features (e.g., wharfs).

Seamless DEM Construction & Wave Modeling

Bathymetric soundings for the Annapolis Basin and river south of Annapolis Royal from the Canadian Hydrographic Service (CHS) Chart 4396 were used to construct a grid that was transformed from chart datum (typically a value lower than the average lowest tide over a 19 year tidal astronomical cycle) to CGVD28 and merged with the LiDAR DEM and DSM to form "seamless" elevation models. Chart datum is 4.39 m below CGVD28 for Digby. In order to model wave heights, the wind direction and speed must be known as well as local bathymetry. Wave heights were predicted with the SWAN (Simulating WAves Nearshore) model [45]. The seamless DEM (bathymetry + land) and the predicted tide height (orthometric height of 4.1 m) during the Groundhog Day storm was used as the initial water level. Hourly wind information recorded by Environment Canada at Greenwood, NS (approximately 100 km northeast along the Annapolis Valley) was used to drive the SWAN model [45]. The model was run for 15-minute increments from February 1–3, 1976, and hourly wave height predictions were output at a horizontal grid resolution of 40 m that have been resampled to 10 m and imported into the GIS. The significant wave height that occurs along the shore was added to the storm surge water level in order to determine the area inland affected by wave run-up.

Observed and Predicted Sea-Levels

Observed hourly water levels were acquired from the Marine Environmental Service (MEDS) website (http://www.meds-sdmm.dfo-mpo.gc.ca) for Saint John, NB and Digby, NS (located within Annapolis Basin south of Annapolis Royal) and predicted tide water levels for these sites were acquired (http://tbone.biol.sc.edu/tide/). The tide gauge records at Digby only cover 15 months from 1969 to 1970, while the Saint John records span over 30 years from 1966 to 2004. In order to generate return periods for high water levels, 30 years of observations are recommended [23]. The water level record at Digby was not of sufficient length to calculate return periods with any confidence, so the record at Saint John was used in Water Modeler to generate return periods. In order to evaluate the relationship of water levels observed in Saint John with those in the Annapolis Basin (Digby), the tide gauge water level records were examined to find times when both sites were operating simultaneously and the residuals (storm surge events) were calculated for each site by subtracting the predicted water level from the observed water level. The difference in storm surges between the two sites was calculated and statistics generated in order to quantify how similar the sites were. This was used to estimate the uncertainty of the return period of high water events in the Annapolis Basin.

Sea-Level Rise from Climate Change Estimates

The Saint John tide gauge information indicates that the sea-level has been rising at a rate of 36 cm per century [47]. The International Panel on Climate Change (IPCC) predicts global sea-level to continue to rise [35,36,37]. In this study, flood inundation maps were calculated up to the 9 m contour level on land as a precautionary approach. In order to calculate return periods on different water levels in the future, a modest rate of 0.5 m per century was used to estimate long-term sea-level rise. As the estimates of sea-level rise from climate change are revised and refined in the future, the return period of any given water level can easily be calculated using the new estimate.

In addition to global sea-level rise, local crustal dynamics also affect relative sea-level. The major influence on crustal motion for this region relates to the last glaciation that ended *ca.* 10,000 years ago [33,54]. The areas where the ice was thickest were depressed the most, and peripheral regions were uplifted, termed the "peripheral bulge". The ice was thickest over Hudson Bay in central Canada, where the crust was most depressed; even today this area is still rebounding from the load of the ice and continues to uplift. The Maritimes represent part of the peripheral bulge and southern New Brunswick and Nova Scotia are subsiding [54]. Subsidence rates vary across the region with this area of Nova Scotia sinking at a rate of ~20 cm per century [54]. The subsidence of the crust is important for coastal communities in that it compounds the problem of local sea-level rise and must be considered when projecting future flood risk.

The Bay of Fundy tidal range is expected to increase by 10 cm in the future with an increase in sea-level [55]. When all of these factors are combined, global sea-level rise, crustal subsidence, and tidal amplitude, a potential increase in relative sea-level of 80 cm in the next century is probable. This number will be used to determine the water level of storms, such as the Groundhog Day storm of 1976, should they occur in the future. If one considers the estimates by [40,41,42], global sea-level could rise by as much as 190 cm over the next century, thus decreasing the return periods of any given water level. With the high level of uncertainty involved in the estimates of global sea-level rise over the next century, flood inundation levels were mapped at every 10 cm up to 9 m. However, instead of determining the return period for several different future sea-level rise estimates which are highly uncertain, I have calculated it for the present relative sea-level rise rate (36 cm/century), and for the estimated rate of 80 cm/century based on the conservative global estimate of 50 cm/century, and for the extreme estimated rate of global sea-level rise of 190 cm/century [41], thereby producing a relative sea-level rise rate of 220 cm/century.

Flood Inundation Mapping

Water levels were converted from chart datum to CGVD28 (DEM datum), and used to raise the water level of the Annapolis Basin to inundate low-lying areas connected to the basin. In order to ensure that only low-lying areas are flooded that are connected to the basin, care must be taken to determine the extent of culverts and bridges that allow sea water to flow landward. A manmade barrier such as a raised roadbed or railway will be represented on the LiDAR DEM in a similar way as a dyke or levee. The main differences between these features relates to the control and direction of the drainage. Streams and ditches landward of a dyke are often drained by a one-way culvert, or aboiteau, thus not allowing the ocean in at high water levels. However, roads and railways are typically equipped with culverts or bridges to drain the land and allow the ocean to flow landward at high water levels. In order to effectively automate mapping the flood extent in a GIS, the DEM was modified to allow hydraulic pathways for flow through the road and rail beds, in order to connect upstream areas with the ocean following the method of [20]. Once the DEM had the appropriate pathways to allow the flow of water, a water level associated with a tide and storm surge was mapped, based on land elevations and the flood water extent determined. This assumed a flat planar water surface (still water) when projecting a given water level landward. An Arc script has been developed to raise the water level in 10 cm increments, ensuring only areas are flooded that are connected to the basin [22]. The range of water levels that have been used in this study are between −4 m and 9 m above CGVD28. The upper limit of 9 m was selected to accommodate the highest astronomical tide of 4.8 m, plus relative sea-level rise of 2.2 m in the next century (0.2 m crustal subsidence + 0.1 m tidal amplitude + 1.9 m extreme estimate of global sea-level rise [41]), plus a 2 m storm surge and wave-setup combination. The flood inundation maps were then assigned a probability of occurrence or a return period in years for each water level to assess the risk of flooding using the method outlined in [23].

RESULTS

LiDAR Validation and Surface Construction

The LiDAR 'ground' and 'non-ground' points were used to construct triangulated irregular networks (TINs) based on the orthometric height that was linearly interpolated to a 1-m resolution digital surface model (DSM) (Figure 3). The ground points were used to construct a second grid using the same method that represented a "bare-earth" digital elevation model (DEM) (Figure 3).

The RTK GPS points were compared to the LiDAR points within a 2 m horizontal distance resulting in a mean difference of 10 cm and a standard deviation of 12 cm, indicating the LiDAR data was within the vertical accuracy expected. The LiDAR DEM was also compared to the GPS points resulting in a mean difference of 4 cm and a standard deviation of 32 cm. The chart bathymetric information was converted to CGVD28 and merged with the LiDAR to construct a seamless DEM for wave modeling with SWAN [54] (Figure 3).

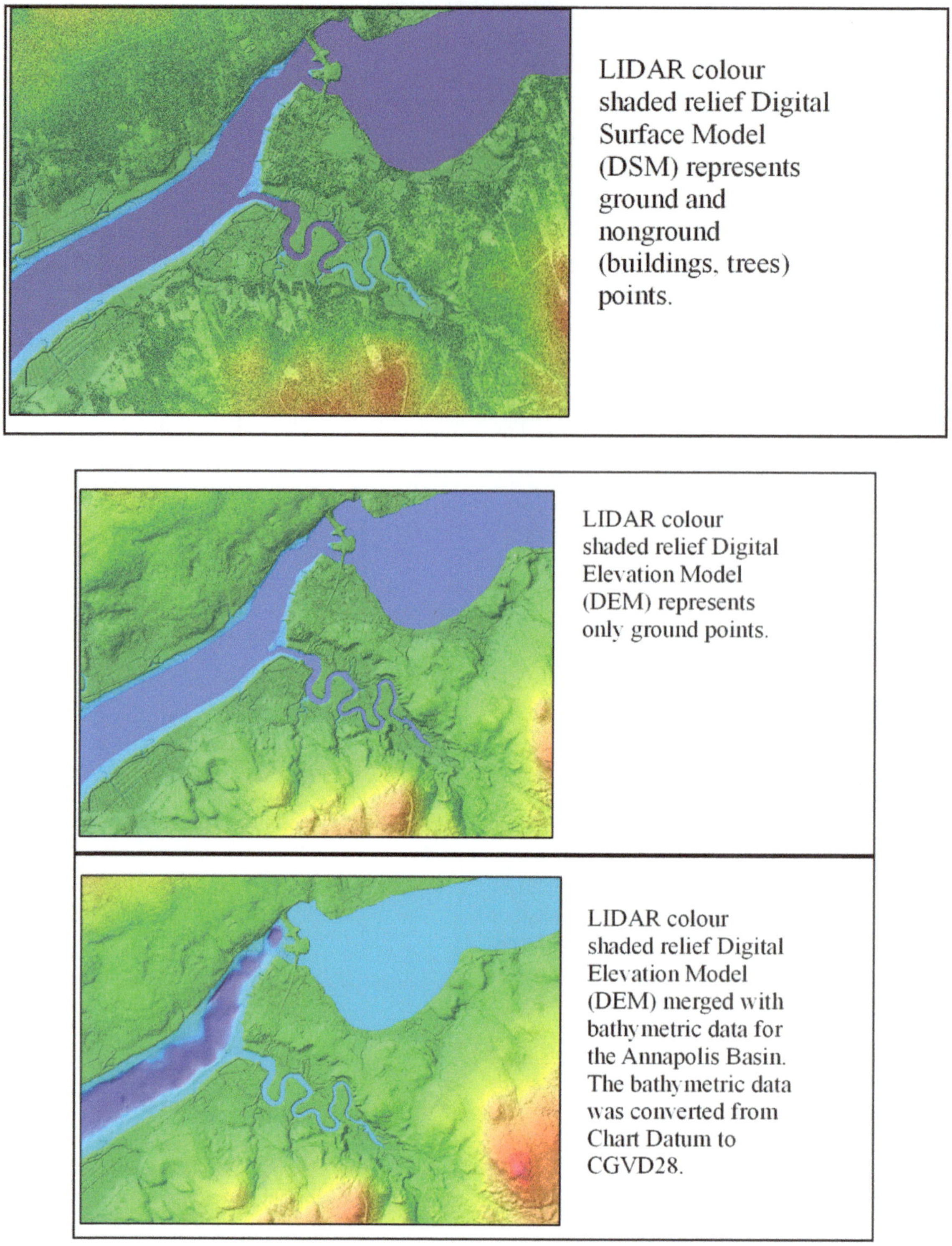

Figure 3. LiDAR surface models. Top: DSM; Middle: DEM; Bottom: seamless DEM comprised of LiDAR for the land and intertidal zone and bathymetric soundings from a chart. Blue shades are below mean sea level and elevation increases from green to red in color.

Wave Modeling of the Benchmark Storm

The results of the wave modeling for the Groundhog Day storm for Annapolis Royal show that significant waves were generated in the Annapolis Basin that impacted the shoreline. The water level used for this simulation corresponded to an orthometric height of 4.1 m which represents the predicted tide. Modeled significant wave height along the shore of Annapolis Royal resulted in approximate heights of 2 m, with the largest waves occurring in the channel, calculated to have peak heights greater than 3 m (Figure 4). With the water level set to 4.1 m, much of the low lying area landward of the dykes was flooded and waves generated. This model does not consider connectivity of the ocean to low lying areas separated by dyke structures. Thus the wave heights landward of the dykes were not considered to be accurate because according to the LiDAR-based flood inundation mapping, the dykes on the east and west side of the basin, southwest of Annapolis Royal, were not overtopped by a 4.1 m water level. However, the significant wave heights near Annapolis Royal appear to agree with newspaper reports of waves overtopping the causeway.

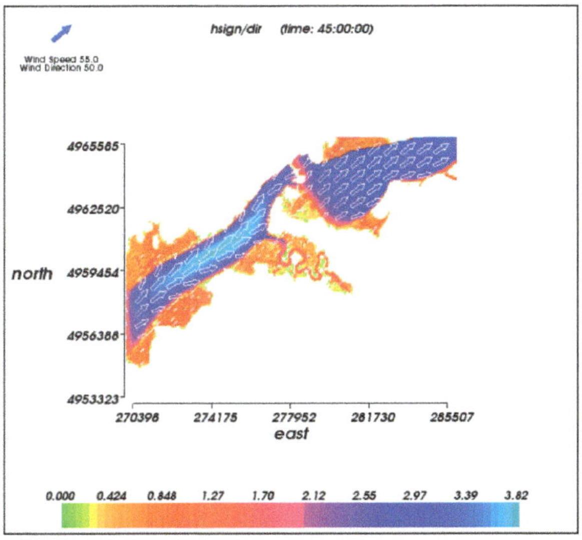

Figure 4. Results from the SWAN wave model. The model represents the maximum significant wave heights generated for winds from Greenwood, NS, February 1–3, 1976 and a water level set to 4.1 m, which represents the predicted tide. The colors represent the significant wave height (m) and white arrows represent the direction and size of the waves. The blue arrow in the upper left represents the speed (km/hr) and direction of the wind. Coordinates along the side are UTM Zone 20 NAD83 easting and northing values.

Observed Water Level Residuals (Storm Surges) between Sites

In order to examine the relationship between water levels in Saint John and those in the Annapolis Basin, the observed water level records at Saint John and Digby for the period of January to March, 1970, have been analyzed. The predicted water level for each site has been determined and combined with the observed water levels and residuals (storm surge) calculated during this period (Figure 5). Generally, the storm surges observed in Saint John are observed in Digby nearly concurrently. This is not surprising, considering the scale of the weather systems (low pressure and associated wind) that drive storm surges and the proximity of the two sites. The storm surge levels at each site were then compared which resulted in a mean difference of 1 cm and a standard deviation of 39 cm (Figure 5). A 48 hour moving average of storm surge levels was calculated, which clearly shows that Saint John experiences slightly higher storm surge events (approximately 1 cm on average) than Digby (Figure 5). This is probably due to the fact that Digby is situated in the Annapolis Basin and is somewhat sheltered. Based on these results, it was deemed reasonable to use the Saint John water level record to estimate water levels and return periods for storm surge events in the Annapolis Basin. The significant wave height is not the same as the wave run-up however.

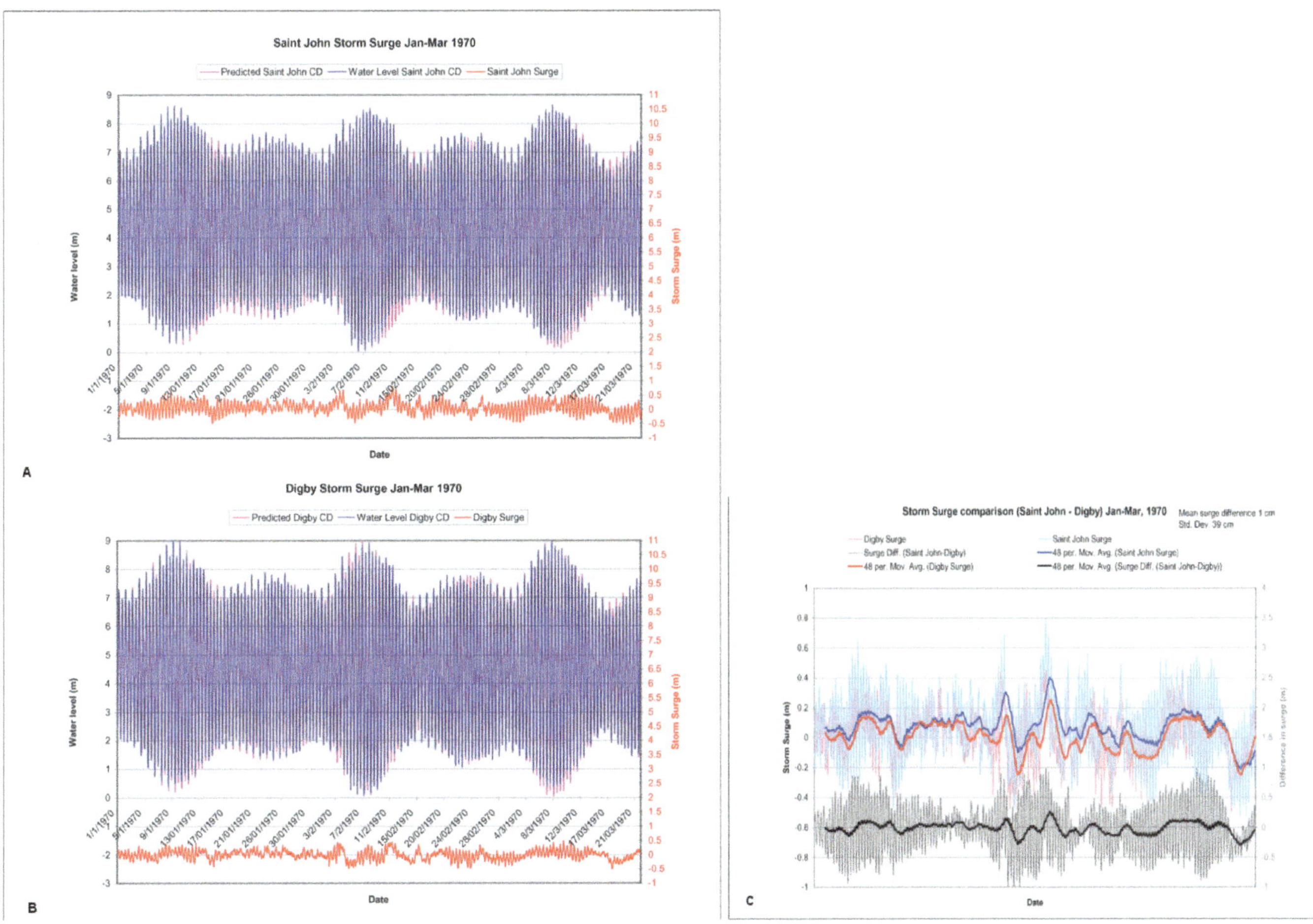

Figure 5. Comparison of water levels at Saint John and Digby for the period of January to March, 1970. Top panel: the observed and predicted water level and calculated storm surge (m) for Saint John, NB. Middle panel: the observed and predicted water level and calculated storm surge (m) for Digby, NS. Bottom panel: comparison of the storm surges for Saint John (shades of blue) and Digby (shades of red) with a 48 hour mean (heavy lines) to reduce high frequency noise and the difference between them (heavy black line).

Flood Inundation Mapping, Sea-Level Rise and Return Periods

It is difficult to validate any flood inundation model, especially when using future predictions. In order to test if the water levels translated from Saint John to the Annapolis Basin were accurate and the DEM was sufficient, the benchmark Groundhog Day storm was used to test the model. The observed water level was subtracted from the predicted tide water level for Saint John, resulting in storm surge of approximately 1.3 m. This value is close to the reported value of 1.5 m by [50] for Saint John. To translate this information from Saint John across the Bay of Fundy to the Annapolis Basin, the value 1.5 m was added to the predicted high tide level for Digby during the Groundhog Day storm in order to determine the total water level. This value was used to estimate storm surge and wave run-up within the Annapolis Basin, considering the Annapolis Basin storm surges are usually less than those experienced in Saint John (Figure 6). The predicted high tide water level of 8.5 m above Chart Datum (CD) or 4.1 m orthometric (using Digby CGVD28-CD offset of 4.39 m) was used for Annapolis Royal with an increased water level of 1.5 m, producing a total water level of 5.6 m. The difference between storm surges in Saint John and the Annapolis Basin indicated a standard deviation of 40 cm, thus a water level of 5.2 m was also generated to give a minimum estimate of the flood extent. The resultant water levels inundated some key areas within the town including lower George Street (inset map Figure 6), and overtopped the government wharf (Figure 6). The upper level of 5.6 m predicts that Highway 1 between the causeway and the town would have been inundated; however the newspaper reports on the extent of the flood did not mention this. The lower water level of 5.2 m partially covers the wharf and approaches the landmarks described in the newspaper reports. These results are consistent with the variance (standard deviation of 0.4 m) in storm surge levels observed between Digby and Saint John. Thus, a water level of between 5.2 and 5.6 m orthometric height appears to accurately represent the range of flood conditions observed in Annapolis Royal during the Groundhog Day storm. A median value of 5.4 m was chosen to predict flooding from a similar storm in 2076 under different sea-level rise conditions.

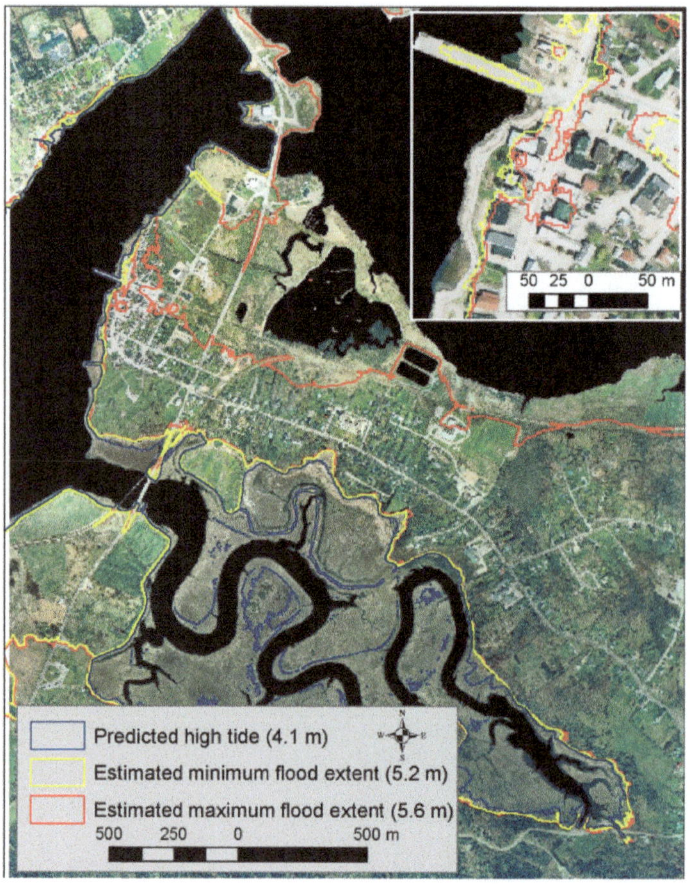

Figure 6. Modeled flood inundation area for a predicted tide (4.1 m) and storm surge; possible water level ranges (5.2–5.6 m) of the Groundhog Day storm, 1976. The background map is a color orthophoto acquired in May 2003.

The results from the Groundhog Day storm model are in sufficient agreement with reported flood extents to confidently predict future flood extents. Water levels were generated in 10 cm increments from −4 m to 9 m orthometric height. A select set of water levels have been used to generate perspective views of the town to demonstrate potential future flooding conditions. These include the current low tide limit during the LiDAR flight (−4 m), the Groundhog Day storm in 100 years at a RSL rate of 80 cm/century (6.2 m), in 100 years at a upper estimate RSL rate of 220 cm/century (7.6 m), and finally the highest astronomical tide plus a 2 m storm surge and the RSL rate of 220 cm/century in 100 years (9.0 m) (Figure 7 andFigure 8). The perspective views were generated by overlaying the May 2003 orthophoto on the LiDAR DSM and superimposing the flood inundation level in blue. The viewing angle for the perspectives is toward the southeast (Figure 7).

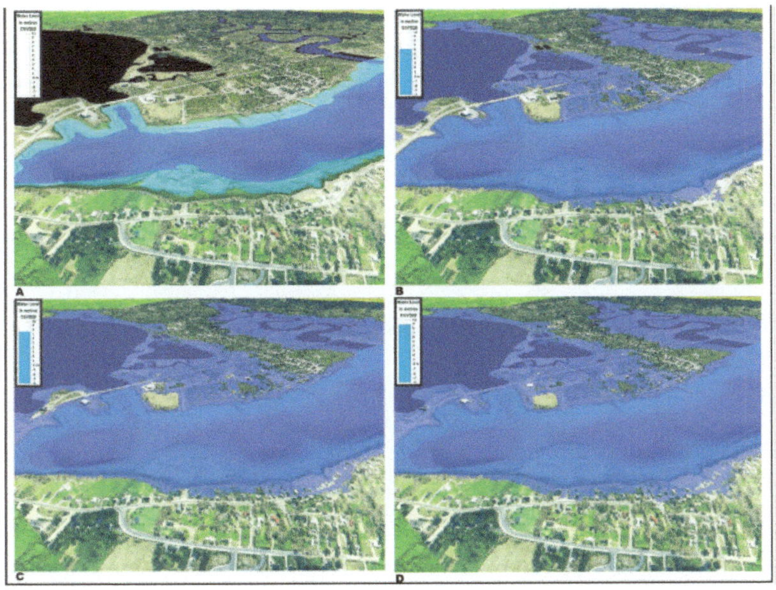

Figure 7. Perspective views of Annapolis Royal looking southeast. May 2003 color orthophoto and bathymetric data overlayed on the LiDAR DSM. (**A**) low tide (−4 m); (**B**) Groundhog Day storm in 100 years with a conservative RSL of 80 cm, (6.2 m); (**C**) Groundhog Day storm in 100 years with a potential RSL of 220 cm, (7.6 m); (**D**) highest predicted tide with a 2 m storm surge in 100 years with a potential RSL of 220 cm, (9.0 m).

The perspective views in Figure 7 are very useful to demonstrate to the public and municipal officials where the water may inundate with the rising sea-level rise in the future. These views have been extracted from a series of frames in which the sea-level was increased in 10 cm increments. The entire set of frames were used to construct animations that are useful for demonstrating the path the water will take as the sea-level rises. In the case of Annapolis Royal, the water moves up a ditch system and around an abandoned protective dyke along the water front. The ditch is located to the right of the isolated land area in the middle of the scene in Figure 7(D). As the sea-level increases, eventually the water inundates the land behind the dyke and crosses the road, separating some key emergency services such as the fire department from the hospital. However, these views are not suitable for mapping or planning purposes and a series of maps were constructed defining the extent of the flood inundation (Figure 8).

Figure 8. Different possible sea-level rise projections of the Groundhog Day storm in 100 years. Digital orthophoto (May 2003) of Annapolis Royal with a vector outline of flood inundation layers based on the LiDAR DEM.

The risk of these water levels was calculated using the Water Modeler software and the time series of water levels from the Saint John tide gauge record. As a result of the analysis of coincident water level records from Saint John and Digby (Annapolis Basin) (Figure 5), the long time series at Saint John can be used to estimate the water level and return periods in Annapolis Royal. The benchmark storm was used to illustrate the possible effects of climate change and sea-level rise on storm return periods (Table 1). Municipal planners are often interested in the 50 and 100 year flood level for restrictions on development. Water Modeler was used to assess the water level for the 10 and 100 year return period events based on the Saint John records. However, the water levels calculated for return periods of 50 and 100 years based on the Saint John records should be considered as estimates with a possible uncertainty of ±39 cm at Annapolis Royal (Table 2).

Table 1. Return periods (years) and probabilities of occurrence of the Groundhog Day storm water level (4.95 m) under current and projected sea-level rise conditions (from [35,41,50]). The return period has been rounded to the nearest year.

Probability	Return period (years) RSL = 0.36 cm/year (current)	Return period (years) RSL = 0.80 cm/year (climate change)	Return period (years) RSL = 2.20 cm/year (climate change)
Average (~65–75 %)	30	23	14
100 %	66	44	22

Table 2. Water levels (m) for return periods of the 50 and 100 year events with different sea-level rise projections (from [35,41,50]). The water level has been rounded to the nearest decimeter.

Return Period (Years)	Water level (m) with RSL = 0.36 cm/year (current)	Water level (m) with RSL = 0.80 cm/year (climate change)	Water level (m) with RSL = 2.20 cm/year (climate change)
50	4.8	4.9	5.6
100	5.1	5.5	6.8

DISCUSSION AND CONCLUSIONS

Annapolis Royal is vulnerable to storm surges and a rising sea-level. Although the Bay of Fundy has some of the largest tides in the world and the probabilities of a significant storm surge occurring on a high tide may be lower than some other regions of the Maritimes, past events such as the Groundhog Day storm of 1976 prove that this region is not immune to such events. This region of Canada experiences storm surges of 1–2 m periodically. Prior to the collection of LiDAR and the construction of high-resolution DEMs with accuracies better than 30 cm, existing digital mapping was not sufficient to accurately map flood risk for most communities, including Annapolis Royal. The LiDAR DEM provides sufficient detail to construct accurate flood inundation maps and when combined with probability statistics from long–term tide-gauge records, flood risk maps were produced. Climate change will impact sea-levels on a global scale and the local crustal conditions in this region will add to the problem as a result of subsidence, thus increasing relative sea-level rise. Currently the lower sections of the town of Annapolis Royal, particularly lower George St. and the government wharf, experience flooding during extreme high tide events in the absence of storms. When storm surges and high winds and waves occur with higher than usual tides, this area of the town is most vulnerable to flooding. The occurrence of these extreme high tides can be predicted based on astronomical conditions. The most significant high tides occur on Soras cycles, of which the most famous storm to cause coastal flooding in the Bay of Fundy region was the "Saxby Gale" of 1869. The predicted water level at Annapolis Royal during the next Soras cycle will peak at 9.14 m CD (4.75 m orthometric) on May 7, 2012. Other time periods where the predicted water level will be above 9 m CD will occur on April 8–9, May 6–7, June 5, October 16, November 14–16, and December 14–15 of that year. If coastal development is planned to take place in any of the risk areas mapped when simulating the Groundhog Day storm, then significant protective structures such as dykes must be constructed to prevent water from the basin or Annapolis River from flooding these areas.

ACKNOWLEDGEMENTS

I would like to thank several individuals and groups for their assistance during this project. Angela Templin, research intern who initially worked on the Annapolis Royal flood risk project. Roger Mosher from the Applied Geomatics Research Group, for providing assistance and guidance with the Water Modeler software to compute return period probabilities and the SWAN wave modeling. Natacha Bernier and Keith Thompson, Dalhousie University Oceanography Dept. for providing a copy of Bernier's PhD thesis. The Odell House for providing historical photos of the town and providing access and copies of the Annapolis County Spectator for the Groundhog Day storm coverage. Charles Hanna for providing information and advice on sea-level rise predictions for the Bay of Fundy. George Parkes for providing information and insights into the Groundhog Day storm and storm surges in the Bay of Fundy. David Mossman for providing reprints on past storm events in the Bay of Fundy. Charles O'Reilly for providing the chart datum—CGVD28 relationships for the different regions in the study. Denise Sullivan, Andy Sharp and Steve Hawbolt from CARP and other residents of Annapolis Royal for providing insights into the flood extent of the Groundhog Day storm. Tam Hill & Jeff Parks for their project management and involving us in the project.

REFERENCES

1. Flood, M.; Gutelius, B. Commercial implications of topographic terrain mapping using scanning airborne laser radar.*Photogramm. Eng. Remote Sensing* 1997, *4*, 327–366.

2. Wehr, A.; Lohr, U. Airborne laser scanning—An introduction and overview. *ISPRS J. Photogramm. Remote Sens.* 1999,*54*, 68–82.
3. Meng, X.; Currit, N.; Zhao, K. Ground filtering algorithms for airborne LiDAR data: A review of critical issues. *Remote Sensing* 2010, *2*, 833–860.
4. Kraus, K.; Pfeifer, N. A new method for surface reconstruction from laser scanning data. In *International Archives of Photogrammetry, Remote Sensing and Spatial Information Sciences*; ISPRS: Vienna, Austria, 1997; Volume 32, Part 3-2W3; pp. 80–86.
5. Kraus, K.; Pfeifer, N. Determination of terrain models in wooded areas with airborne laser scanner data. *ISPRS J. Photogramm. Remote Sens.* 1998, *53*, 193–203.
6. Vosselman, G. Slope based filtering of laser altimetry data. In Proceedings of Technical Commission III, Amsterdam, The Netherlands, July 2000.*International Archives of Photogrammetry, Remote Sensing and Spatial Information Sciences*; ISPRS: Vienna, Austria, 2000; Volume 33, Part B4; pp. 958–064.
7. Axelsson, P. DEM generation from laser scanner data using adaptive tin methods. In Proceedings of Technical Commission III, Amsterdam, The Netherlands, July 2000.*International Archives of Photogrammetry, Remote Sensing and Spatial Information Sciences*; ISPRS: Vienna, Austria, 2000; Volume 33, Part B3; pp. 85–92.
8. Zhang, K.; Chen, S.; Whitman, D.; Shyu, M.; Yan, J.; Zhang, C. A progressive morphological filter for removing nonground measurments from airborne LIDAR data. *IEEE Trans. GeoSci. Remote Sens.* 2003, *4*, 872–882.
9. Raber, G.; Jensen, J.R.; Schill, S.R.; Schuckman, K. Creation of digital terrain models using an adaptive LiDAR vegetation point removal process. *Photogramm. Eng. Remote Sensing* 2002, *68*, 1307–1315.
10. Hodgson, M.E.; Bresnahan, P. Accuracy of airborne LiDAR-derived elevation: Empirical assessment and error budget.*Photogramm. Eng. Remote Sensing* 2004, *3*, 331–339.
11. Lane, S.; James, T.D.; Pritchard, H. Photogrammetric and laser altimetric reconstruction of water levels for extreme flood event analysis. *Photgramm. Record* 2003, *18*, 293–307.
12. Bates, P.D.; Marks, K.J.; Horritt, M.S. Optimal use of high-resolution topographic data in flood inundation models.*Hydrol. Process.* 2003, *17*, 537–557.
13. Cobby, D.M.; Mason, D.C.; Horritt, M.S.; Bates, P.D. Two-dimensional hydraulic flood modeling using a finite-element mesh decomposed according to vegetation and topographic features derived from airborne scanning laser altimetry. *Hydrol. Process.* 2003, *17*, 1979–2000.
14. Haile, A.T.; Rientjes, T.H.M. Effect of LiDAR DEM resolution in flood modeling: A model sensitivity study for the city of Tegucigalpa, Honduras. In Proceedings of ISPRS WG III/3, III/4, V/3 Workshop "Laser scanning 2005", Enschede, The Netherlands, September 12–14, 2005; pp. 168–173.
15. Mason, D.C.; Horritt, M.S.; Hunter, N.M.; Bates, P.D. Use of fused airborne scanning laser altimetry and digital map data for urban flood modeling. *Hydrol. Process.* 2007, *21*, 1436–1447.
16. Raber, G.; Jensen, J.R.; Hodgson, M.E.; Tullis, J.A.; Davis, B.A.; Berglund, J. Impact of LiDAR nominal post-spacing on DEM accuracy and flood zone delineation. *Photogramm. Eng. Remote Sensing* 2007, *73*, 793–804.
17. Gutierrez, R.; Gibeaut, J.C.; Smyth, R.C.; Hepner, T.L.; Andrews, J.R. Precise airborne LiDAR surveying for coastal research and geohazards applications. In Proceedings of the ISPRS Workshop "Land Surface Mapping and Characterization Using Laser Altimetry", Annapolis, MD, USA, October 22–24, 2001.International Archives of Photogrammetry, Remote Sensing and Spatial Information Sciences; ISPRS: Vienna, Austria, 2001; Volume 34-3/W4, pp. 22–24.
18. Webster, T.L.; Forbes, D.L.; Dickie, S.; Shreenan, R. Using topographic LiDAR to map flood risk from storm surge events for Charlottetown, Prince Edward Island, Canada. *Canad. J. Remote Sens.* 2004, *30*, 54–76.
19. Webster, T.L.; Christian, M.; Sangster, C.; Kingston, D. High-resolution elevation and image data within the Bay of Fundy Coastal Zone, Nova Scotia, Canada. In *GIS for Coastal Zone Management*; Bartllet, D., Smith, J., Eds.; CRC Press: Boca Raton, FL, USA, 2005; pp. 195–218.
20. Webster, T.L.; Forbes, D.L.; MacKinnon, E.; Roberts, D. Floodrisk mapping for storm surge events and sea-level rise in Southeast New Brunswick. *Canad. J. Remote Sens.* 2006, *32*, 194–211.
21. Webster, T.L.; Forbes, D.L. Airborne laser altimetry for predictive modelling of coastal storm surge flooding. In *Remote Sensing of Aquatic Ecosystem Processes, Science and Management Applications*; LeDrew, E., Richardson, L., Eds.; Springer: Dordrecht, The Netherlands, 2006; pp. 157–180.
22. Webster, T.; Stiff, D. The prediction and mapping of coastal flood risk associated with storm surge events and long-term sea level changes. In *Risk Analysis VI Simulations and Hazard Mitigation*; Brebbia, C.A., Beriatos, E., Eds.; WIT Press: Wessex, UK, 2008; pp. 129–139.
23. Webster, T.L.; Mosher, R.; Pearson, M. A coastal zone decision support system to generate flood risk maps from storm surge events and sea level rise. *Geomatica* 2008, *62*, 393–406.
24. Poulter, B.; Halpin, P.N. Raster modeling of coastal flooding from sea-level rise. *Int. J. Geogr. Inf. Sci.* 2008, *22*, 167–182.
25. Klemas, V. The role of remote sensing in predicting and determining coastal storm impacts. *J. Coast. Res.* 2009, *6*, 1264–1275.
26. Krabill, W.B.; Wright, C.; Swift, R.; Frederick, E.; Manizade, S.; Yungel, J.; Martin, C.; Sonntag, J.; Duffy, M.; Brock, J. Airborne laser mapping of assateague national seashore beach. *Photogramm. Eng. Remote Sensing* 1999, *66*, 65–71.
27. Brock, J.C.; Wright, C.W.; Sallenger, A.H.; Krabill, W.B.; Swift, R.N. Basis and methods of NASA airborne

topographic mapper LIDAR surveys for coastal studies. *J. Coast. Res.* 2002, *18*, 1–13.

28. Stockdon, H.F.; Sallenger, A.H.; List, J.H.; Holman, R.A. Estimation of shoreline position and change using airborne topographic LIDAR data. *J. Coast. Res.* 2002, *18*, 502–513.
29. Sallenger, A.H., Jr.; Krabill, W.B.; Swift, R.N.; Brock, J. Evaluation of airborne topographic LiDAR for quantifying beach changes. *J. Coast. Res.* 2003, *19*, 125–133.
30. Robertson, W.V.; Whitman, D.; Zhang, K.; Leatherman, S.P. Mapping shoreline position using airborne laser altimetry. *J. Coast. Res.* 2004, *20*, 884–892.
31. Morton, R.A.; Miller, T.L.; Moore, L.J. Historical shoreline changes along the US Gulf of Mexico: A summary of recent shoreline comparisons and analysis. *J. Coast. Res.* 2004, *4*, 704–709.
32. Zhou, G.; Xie, M. Coastal 3-D morphological change analysis using LiDAR series data: A case study of Assateague Island National Seashore. *J. Coast. Res.* 2009, *2*, 435–447.
33. Shaw, J.; Taylor, R.B.; Forbes, D.L.; Ruz, M.-H.; Solomon, S. *Sensitivity of the Canadian Coast to Sea-Level Rise*; Geological Survey of Canada, Bulletin 505; Natural Resources Canada: Ottawa, ON, Canada, 1994; p. 114.
34. McCulloch, M.M.; Forbes, D.L.; Shaw, R.W.; the CCAF A041 Scientific Team. *Coastal Impacts of Climate Change and Sea-Level Rise on Prince Edward Island*; Open File 4261; Geological Survey of Canada, Natural Resources Canada: Ottawa, ON, Canada, 2002.
35. Church, J.A.; Gregory, J.M.; Huybrechts, P.; Kuhn, M.; Lambeck, K.; Nhuan, M.T.; Qin, D.; Woodworth, P.L. Changes in sea level. In *Climate Change 2001: The Scientific Basis. Contribution of Working Group I to the Third Assessment Report of the Intergovernmental Panel on Climate Change*; Cambridge University Press: Cambridge, UK, 2001; pp. 639–693.
36. Meehl, G.A.; Stocker, T.F.; Collins, W.D.; Friedlingstein, P.; Gaye, A.T.; Gregory, J.M.; Kitoh, A.; Knutti, R.; Murphy, J.M.; Noda, A.; Raper, S.C.B.; Watterson, I.G.; Weaver, A.J.; Zhao, Z.-C. Global climate projections. In *Climate Change 2007: The Physical Science Basis. Contribution of Working Group I to the Fourth Assessment Report of the Intergovernmental Panel on Climate Change*; Solomon, S., Qin, D., Manning, M., Chen, Z., Marquis, M., Averyt, K.B., Tignor, M., Miller, H.L., Eds.; Cambridge University Press: Cambridge, UK and New York, NY, USA, 2007; pp. 1–844.
37. Bindoff, N.L.; Willebrand, J.; Artale, V.; Cazenave, A.; Gregory, J.; Gulev, S.; Hanawa, K.; Le Quéré, C.; Levitus, S.; Nojiri, Y.; Shum, C.K.; Talley, L.D.; Unnikrishnan, A. Observations: Oceanic climate change and sea level. In *Climate Change 2007: The Physical ScienceBasis. Contribution of Working Group I to the Fourth Assessment Report of the Intergovernmental Panel on Climate Change*; Solomon, S., Qin, D., Manning, M., Chen, Z., Marquis, M., Averyt, K.B., Tignor, M., Miller, H.L., Eds.; Cambridge University Press: Cambridge, UK and New York, NY, USA, 2007; pp. 1–429.
38. Forbes, D.L.; Manson, G.K.; Charles, J.; Thompson, K.R.; Taylor, R.B. *Halifax Harbour Extreme Water levels in the Context of Climate Change: Scenarios for a 100-Year Planning Haorizon*; Open File 6346; Geological Survey of Canada, Natural Resources Canada: Ottawa, ON, Canada, 2009; pp. 1–21.
39. Rhamstorf, S.; Cazenave, A.; Chruch, J.A.; Hansen, J.E.; Keeling, R.F.; Parker, D.E.; Somerville, R.C.J. Recent climate observations compared to projections. *Science* 2007, *316*, 709. [PubMed]
40. Levermann, A.; Griesel, A.; Hofmann, M.; Montoya, M.; Rahmstorf, S. Dynamic sea level changes following changes in the thermohaline circulation. *Clim. Dynam.* 2005, *24*, 347–354.
41. Vermeer, M.; Rahmstorf, S. Global sea level linked to global temperature. *Proc. Nat. Acad. Sci. USA* 2009. [PubMed]
42. Rahmstorf, S. A semiempirical approach to projecting future sea-level rise. *Science* 2007, *315*, 368–370. [PubMed]
43. Bernier, N.B.; Thompson, K.R. Predicting the frequency of storm surges and extreme sea levels in the Northwest Atlantic. *J. Geophys. Res.* 2006.
44. Belbin, J.; Clyburn, D. Tidal Surge Project: The Coastal Flooding Component of the Annapolis Climate Change Outreach Project; Unpublished Report for the Clean Annapolis River Project. 1998. Available online: http://c-ciarn.dal.ca/workshops/3/presentations/Belbin.pdf (accessed on 20 August 2010).
45. Booij, N.; Ris, R.C.; Holthuijsen, L.H. A third-generation wave model for coastal regions 1. Model description and validation. *J. Geophys. Res.* 1999, *104*, 7649–7666.
46. Abraham, J.; Parkes, G.; Bowyer, P. The transition of the "Saxby Gale" into an extratropical storm. In Proceedings of the 23rd Conference on Hurricanes and Tropical Meteorology, American Meteorological Society, Dallas, TX, USA; 1999; pp. 795–798.
47. Desplanque, C.; Mossman, D. Tides and their seminal impact on the geology, geography, history, and socio-economics of the Bay of Fundy, eastern Canada. *Atlantic Geol.* 2004, *40*, 1–130.
48. MacDonald, A. Winds of Change Severe Weather in The Annapolis Valley. Unpublished Thesis, Bachelor of Arts with Honours in History, Acadia University, Wolfville, NS, Canada, 2006.
49. Parkes, G.; Ketch, L.; O'Reilly, C. Storm surge events in the Maritimes. In Procedures of the Canadian Coastal Conference, Guelph, ON, Canada; 1997; pp. 115–129.
50. Desplanque, C.; Mossman, D. Storm tides of the fundy. *The Geogr. Rev.* 1999, *89*, 23–33.
51. Brennen, R.; Webster, T.L. Object oriented land cover classification of LiDAR derived surfaces. *Canad. J. Remote Sens.*2006, *32*, 167–172.
52. Webster, T.L.; Dias, G. An automated GIS produce for comparing GPS and proximal LIDAR elevations. *Comput.Geosci.*2006, *32*, 713–726.

53. Webster, T.L. LIDAR validation using GIS: A case study comparison between two LIDAR collection methods. *GeoCarto Int.* 2005, *20*, 11–19.

54. Peltier, W.R. Global glacial isostasy and the surface of the ice-age earth: The ice-5G (VM2) model and Grace. *Ann. Rev. Earth Planetary Sci.* 2004, *32*, 111–149.

55. Godin, G. Possibility of rapid changes in the tide of the Bay of Fundy, based on a scrutiny of the records from Saint John. *Continental Shelf Res.* 1992, *12*, 327–338.

CHAPTER 15

LAND COVER MAPPING USING THE HC-MMK APPROACH

Guangbin Lei, Ainong Li, Jinhu Bian, Zhengjian Zhang, Huaan Jin, Xi Nan, Wei Zhao, Jiyan Wang, Xiaomin Cao, Jianbo Tan, Qiannan Liu, Huan Yu, Guangbin Yang and Wenlan Feng

Keywords: Land cover mapping, hierarchical classification, knowledge, quality control, multi-source data, HC-MMK approach, CLC-SW2010 product, Southwestern China

INTRODUCTION

Land cover data is essential for a variety of studies such as global change [1,2], ecological environments [3,4] and resource management. It plays a critical role in improving the performance of hydrological, ecological, biogeochemical and atmospheric models [5]. Satellite remote sensing has been widely applied and recognized as a powerful and effective tool for producing land cover products [6]. A series of land cover products at global and regional scales, such as the IGBP-DIS global 1-km land cover product [7], MODIS global 500-m land cover product (MCD12Q1) [8], GlobCover 2009 at 300 m-resolution [9], global 30-m land cover products (FROM-GLC [5,10] and GlobeLand30 [11]) and Chinese land use datasets at 1:100,000 scale (CLUDs) [12,13], have been derived from remotely sensed data in response to the policy- and science-driven need for land cover data. Currently, land cover products at 30 m-resolution over large regions or even the globe are highlighted, since most significant human activities on the land surface can be captured at this scale [11,14]. However, the spatial agreement of multiple existing land cover products tends to be low in regions with complex, heterogeneous land cover [15]. Especially in mountainous areas, severe challenges arise for land cover mapping, such as the rugged terrain, the high spatial heterogeneity of land surfaces and the frequent cloud contamination of satellite imagery [16], having an enormous impact on scientific research and applications related to land cover over mountainous areas.

Multi-temporal satellite imagery can improve our ability to distinguish between vegetation types by leveraging variability in phenological patterns across the landscape [17,18], such as distinguishing between evergreen forests and deciduous forests [19] and identifying rubber plantations from natural forests [20]. It is natural to consider the use of multi-temporal satellite imagery in enhancing the quality of land cover products over mountainous areas with a wide variety of vegetation heterogeneities. The Landsat TM image is ideal input data for 30 m-resolution land cover mapping due to its long-term archive (over four decades) [21] and free availability. The native HJ-CCD sensors have a distinct advantage to acquire more high-quality satellite images due to the wide range (700 km × 700 km) and short repeat cycle (2–4 days) [22,23]. Meanwhile, they have the same spatial resolution (30 m) and similar spectrum range in the visible and near-infrared band [24]. Therefore, the combination of Landsat TM images and native HJ-CCD images offers a feasible approach for the collection of multi-temporal satellite images needed for 30 m-resolution land cover mapping over largely mountainous areas.

The spatial distribution of some land cover classes in mountainous areas often follow certain geographical rules and is constrained by various topographical factors (such as elevation, aspect and slope). For example, the distribution of natural vegetation displays distinct vertical stratification in mountainous areas, while the distribution of cropland is constrained by terrain slope, and the tree line represents the highest elevation that forests can grow. These rules and topographical factors play an important role in improving

Citation: Guangbin Lei, Ainong Li, Jinhu Bian, Zhengjian Zhang, Huaan Jin, Xi Nan, Wei Zhao, Jiyan Wang, Xiaomin Cao, Jianbo Tan, Qiannan Liu, Huan Yu, Guangbin Yang and Wenlan Feng, Land Cover Mapping Using the HC-MMK Approach, doi:10.3390/rs8040305.

the accuracy of land cover mapping in mountainous areas [16]. Meanwhile, some existing thematic maps (such as land use maps, vegetation maps and soil maps), field photos and records, as well as relevant research results and statistical data could also provide some useful information for enhancing the quality of land cover products from different perspectives. We refer to these rules and information as geo-knowledge in this paper. Although geo-knowledge is central to decision-making and discovering classification errors for land cover mapping in mountainous areas, how to translate this geo-knowledge into computer language and maximize its potential is a practical problem. Additionally, due to the high spatial heterogeneity and rugged terrain, the geographical rules vary from region to region, and it is advisable to try to construct an optimal rule for each region. Moreover, uncertainties and errors are inevitable for various auxiliary data, and how to make best use of them and avoid any negative impacts caused by such uncertainties needs to be considered.

Pixel-based and object-oriented classification are two different approaches for land cover mapping. The object-oriented classification method was found to be able to avoid the "Salt and Pepper Noise" often existing in pixel-based classification results [25], taking fully into consideration spectral, shape and texture features [26] as well as topographical features, and produced more accurate land cover products in highly heterogeneous areas. Most importantly, it will be the optimal classification method for the next generation of global land cover products [5,14]. This study should be treated as a reference for other object-oriented land cover mapping projects over large areas or even the globe in the future. Currently, many artificial intelligence algorithms have been applied in the field of land cover mapping, such as Support Vector Machine [27,28], decision tree [29], neural network [30] and genetic algorithms [31]. In these algorithms, decision tree is a "white box" model, and this helps to understand the relationship between the inputs and outputs. Meanwhile, it provides a convenient way to represent the geographical rules used in classification. Therefore, the object-oriented classification approach and the decision tree algorithm were chosen for land cover mapping in mountainous areas.

Since 2010, the Chinese government has launched the Strategic Priority Research Program "Climate Change: Carbon Budget and Relevant Issues" (hereinafter referred to as "Carbon Special Program") [32] and "National Ecosystem Survey and Assessment of China (2000–2010)" Project (hereinafter referred to as "Ecological Decade Project") [33], aiming to answer relevant scientific questions about the terrestrial carbon budget and to assess the quality of eco-environments in China. Southwestern China, where more than 90% of land is mountains and hills, is one of the core regions in the Carbon Special Program and Ecological Decade Project due to its enormous carbon storage capacities and a fragile ecological environment. The specific objectives of this research were to (1) develop an operable object-oriented land cover hierarchical classification method based on the multi-source and multi-temporal satellite images for the largely mountainous area; (2) provide an interactive quality control approach based on a variety of geo-knowledge to further improve the quality of land cover products over mountainous areas; and (3) generate a new 30 m-resolution land cover product in Southwestern China (CLC-SW2010) based on the proposed approach according to a pre-defined time schedule.

STUDY AREA

Southwestern China (78°25′–110°21′E, 21°09′–36°32′N) in this study includes the Chongqing municipality, the Sichuan, Yunnan and Guizhou provinces and the Tibet autonomous region (Figure 1), with an area of 2.33 million km^2, accounting for about a quarter of the total land area of China. It is bordered by Vietnam, Laos, Myanmar, India and Bhutan to the south, India and Nepal to the west, Xinjiang, Qinghai, Gansu and Shanxi provinces to the north, and the Guangxi, Hunan and Hubei provinces of China to the east. Southwestern China is one of most complicated terrain sections, including plateaus, mountains, hills, basins and plains, and can be divided into four geomorphic units: Qing-Tibet plateau, Hengduan mountain, Yunnan-Guizhou plateau and Sichuan basin [34]. The elevation ranges from less than 500 m (Yangtze River) to 8844 m (Mt. Everest in Qing-Tibet plateau) in this region. There are various types of climates in this area, mainly subtropical, temperate and alpine climates. A variety of land cover types and distinct vertical stratification of vegetation classes can be found in this place.

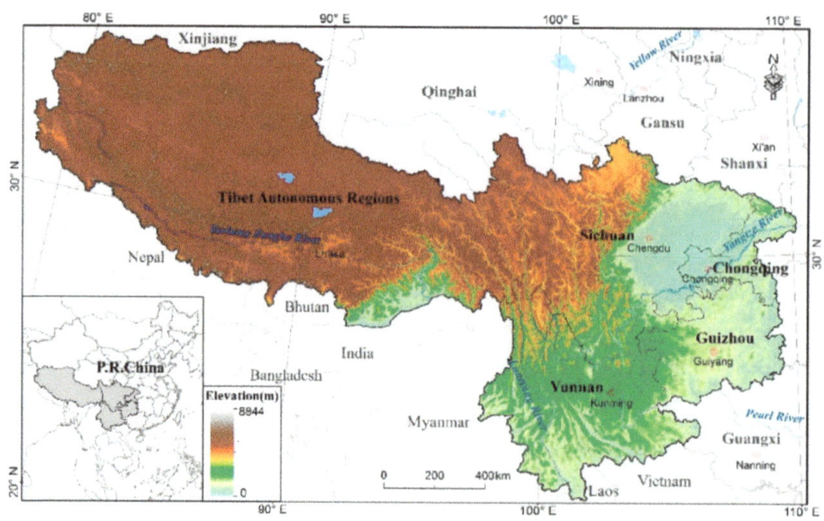

Figure 1. The location of Southwestern China.

METHODOLOGY

Overall Description

A new HC-MMK approach (Hierarchical Classification based on Multi-source and Multi-temporal data and geo-Knowledge) was proposed in this study to produce the 30 m-resolution CLC-SW2010 product, which is essentially an object-oriented classification method. The multi-source and multi-temporal satellite images and field data were firstly collected and preprocessed. After object-oriented multiresolution segmentation, a series of hierarchical classification trees for each mapping unit were built by decision tree algorithm and were then conducted to produce the preliminary classification results. Then, five-step knowledge-based interactive quality control was used to further improve the quality of classification results. The CLC-SW2010 product was independently validated by a third-party group. The construction of hierarchical classification trees and the five-step knowledge-based interactive quality control are the major components of this proposed approach. The flowchart of this proposed approach can be seen in Figure 2.

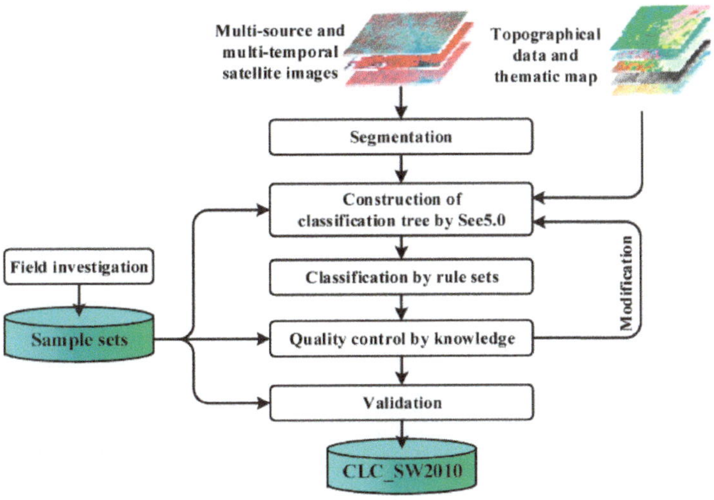

Figure 2. The flowchart of the HC-MMK (Hierarchical Classification based on Multi-source and Multi-temporal data and geo-Knowledge) approach.

Land Cover Classification System

The classification system of China land cover for carbon budget [35] was adopted in this study, which includes two levels (Table 1). The first level has six land cover classes, including woodlands, grasslands, wetlands, croplands, artificial surface lands and bare lands. The basic 38 classes at the second level are formed by taking into full consideration the interpretation abilities of satellite imagery and the application demands of the Carbon Special Program and Ecological Decade Project. The land cover classes in this system are able to easily reorganize according to application requirements, and also are compatible with international land cover classification systems, such as FAO LCCS [35]. For more details on each land cover class, refer to the article by Zhang *et al.* [35].

Table 1. Land cover classification system of the CLC-SW2010 product.

Code I	Category I	Code II	Category II
1	Woodlands	101	Evergreen Broadleaf Forests
		102	Deciduous Broadleaf Forests
		103	Evergreen Needleleaf Forests
		104	Deciduous Needleleaf Forests
		105	Mixed Forests
		106	Evergreen Broadleaf Shrubs
		107	Deciduous Broadleaf Shrubs
		108	Evergreen Needleleaf Shrubs
		109	Woody Plantations
		110	Shrub Plantations
		111	Woody Greenland
		112	Shrub Greenland
2	Grasslands	21	Meadows
		22	Steppes
		23	Herbosa
		24	Herbaceous Greenland
3	Wetlands	31	Woody Wetlands
		32	Shrub Wetlands
		33	Herbaceous Wetlands
3	Wetlands	34	Lakes
		35	Reservoirs
		36	Rivers
		37	Canals
4	Croplands	41	Paddy Fields
		42	Dry Lands
5	Artificial Surface Lands	51	Residential Lands
		52	Industrial Lands
		53	Transportation Lands
		54	Mineral Land
6	Bare Lands	61	Sparse Forests
		62	Sparse Shrubs
		63	Sparse Grasslands
		64	Lichens/Mosses
		65	Bare Rocks
		66	Bare Soils
		67	Deserts
		68	Saline Lands
		69	Permanent Snow and Ices

Remote Sensing Data Collection and Mapping Units

A total of 170 Landsat TM scenes and 202 HJ-CCD scenes were collected respectively from the USGS EarthExplorer and the China Center for Resources Satellite Data and Application. These images were acquired in 2009, 2010 and 2011 (Figure 3). For each mapping unit, at least two free-cloud satellite images (cloud coverage below 10%), one acquired in leaf-on conditions (May to October) and the other acquired in leaf-off conditions (November to following April), were selected to form the multi-temporal satellite images.

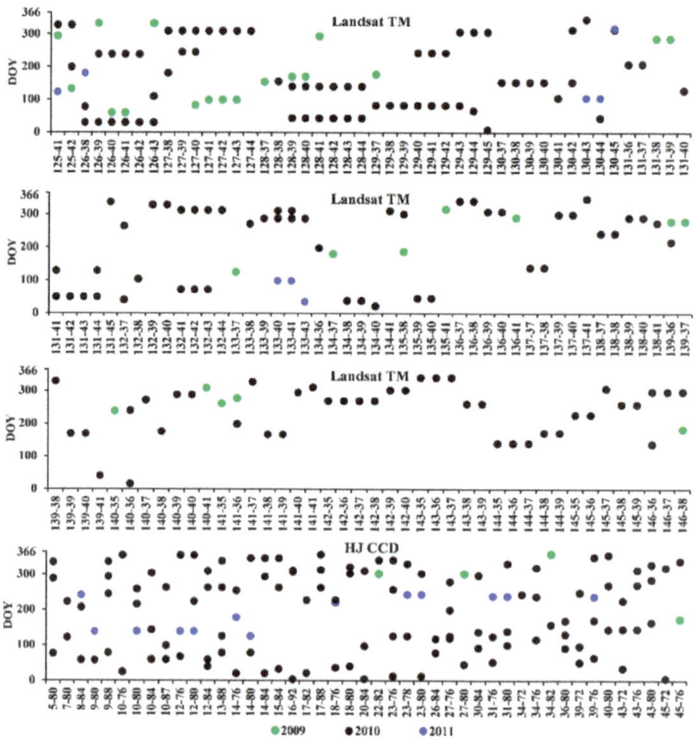

Figure 3. The temporal distribution (DOY, day of year) of Landsat TM images and HJ-CCD images for each tile in Southwestern China. The X-axis was the scenes (path-row) of the Landsat TM images and the HJ-CCD images.

The acquired Landsat TM images have been calibrated radiometrically and geometrically. The atmospheric correction for the Landsat images were processed by the Landsat Ecosystem Disturbance Adaptive Processing System (LEDAPS, [36]). Although systematic geometric correction has been conducted, there are still some residual geometric errors existing in the acquired HJ-CCD images. The precise geometric correction for these HJ-CCD images was conducted by Auto-Registration and Orthorectification Algorithm [23], which was specially designed for satellite images over mountainous areas. After processing, the geometric errors between HJ-CCD images and Landsat TM images were less than one pixel [37].

The topographical data were incorporated in the classification, including elevation, aspect and slope. Despite some geo-location errors (approximately 1–3 pixels) existing in the 30-m ASTER GDEM in this region [38], it was still used in this study having less missing data, higher spatial resolution and better topographic representation than other elevation data (Such as SRTM) [39]. Using the ASTER GDEM data, the slope and aspect were calculated by the ERDAS software. The high spatial resolution Google Earth images and existing thematic maps, including the Vegetation Map of the People's Republic of China (1:1,000,000) and the Chinese Land Use Database at 1:100,000 scale, were also used for generating the CLC-SW2010 product. Although these thematic maps and high spatial resolution images were not directly used in the automatically classification due to the scale and quality, they could provide lots of useful reference information for sample selection and classification error identification.

To conveniently and efficiently organize a large number of satellite images and relevant auxiliary data, the study area needs to be partitioned into a series of mapping units according to the image scenes. Because distinct differences have existed between the HJ image scenes and the Landsat image scenes, neither Landsat image scenes nor HJ image scenes, but rather the intersection regions between HJ and Landsat image scenes were chosen to divide Southwestern China into 190 mapping units. All satellite images and ancillary data were clipped by these mapping units. Meanwhile, all processes including segmentation, automatic classification and quality control were conducted in these mapping units. Notably, a part of the overlapping regions needed to be retained to meet the requirements for mosaicking the final classification results for each neighboring mapping unit.

Field Data Collection

Three collection modes of field land cover samples have been designed at the initial phase of the Carbon Special Program and the Ecological Decade Project to collect as many field samples as possible. A total of 35,792 field samples have been collected in this area. For each field sample, the predominant land cover class, the geographical position and the field photos were recorded.

The first mode was designed to collect the field samples located around 2 km from the main road and the distance of two neighbor field samples was less than 3 km. Based on this mode, 16,721 field samples were collected in 2011 and 2012 by the mapping group. Meanwhile, the independent validation group collected 4546 field validation samples in 2011 by the same mode, which were only used to assess the accuracy of the CLC-SW2010 product. All field samples collected by the first mode are shown by solid red dots in Figure 4.

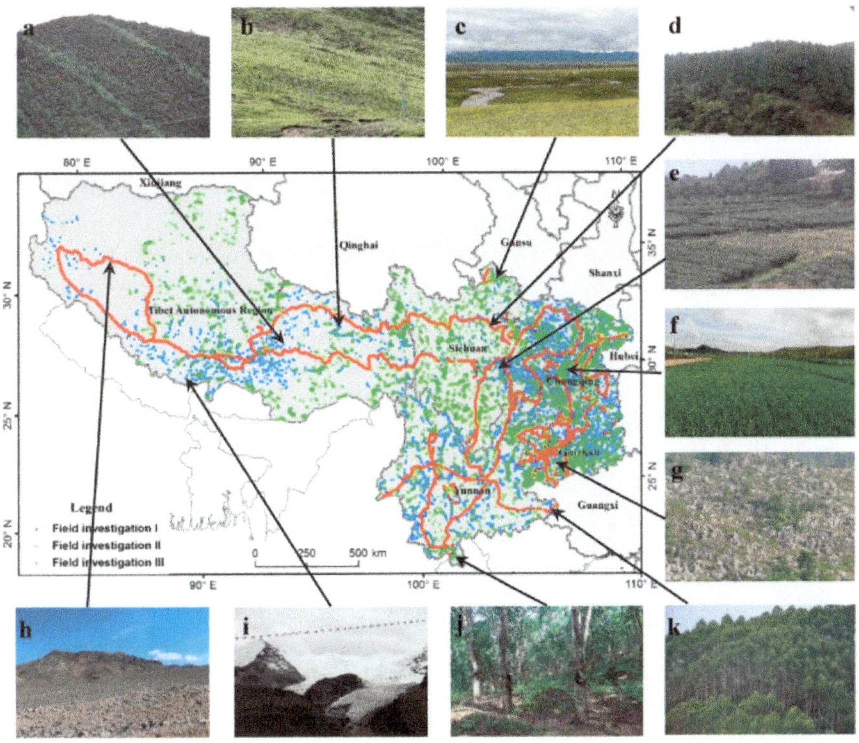

Figure 4. The spatial distribution of field samples used for generating the CLC-SW2010 product. The pictures (**a**–**k**) represent respectively the deciduous broadleaf shrubs, meadows, herbaceous wetlands, evergreen needleleaf forests, tea garden (shrub plantations), paddy fields, karst vegetation (herbosa), bare soils, permanent snow and ices, rubber plantations (woody plantations) and eucalyptus plantations (woody plantations).

The second mode was used to collect the field samples in the predefined plots, including forest plots, grassland plots, shrub plots and cropland plots. More than 10 field samples were randomly distributed in each sample plot. These field investigation tasks were conducted by scientific survey teams from the Carbon Special Program, including forest survey teams, grassland survey teams, remote sensing survey teams, *etc.* A total of 6772 field samples (shown by solid green dots in Figure 4) were provided to the mapping group, 1378 of which were not provided until the accuracy assessment was completed.

The purposes of the third mode for collecting field samples were neither to train rule sets nor to evaluate the accuracy, but rather to discover the classification errors in the preliminary classification results. In this mode, about 20 field verification points were randomly chosen by the stratified sampling method for each county-level administrative division. Supported by local governments, a total of 7553 verification points (shown by solid blue dots in Figure 4) were collected by hundreds of staff from the local environmental authorities.

Object-Oriented Multi-Resolution Segmentation

Image segmentation is a crucial step in object-oriented classification, which groups homogenous neighboring pixels into meaningful objects. In this study, a wide-used multi-resolution segmentation algorithm [40,41] embedded in the platform Cognition 8.7 was adopted to segment satellite images. The scale, shape and compactness are three critical parameters. The optimal values of these parameters were determined by the trial-and-error method and visual assessment of the segmentation results. For the highly heterogeneous landscape, the scale parameter was set as 25, the shape parameter was set as 0.1 and compactness as 0.7. All of the Landsat images and the HJ images were incorporated in the segmentation process, except the topographical data and available thematic maps.

Hierarchical Classification Trees Construction by Decision Tree Algorithms

The hierarchical classification improved the quality of the land cover product in mountainous areas by increasing the classification times and decreasing the number of land cover classes within a single classification process. The whole classification process was divided into multiple stages according to the conceptual hierarchical structure in this study. For each classification stage, an optimal classification tree was constructed from the given training sample sets and classification feature sets by using the decision tree classifier (See5 software, https://www.rulequest.com/see5-info.html). Then, it was conducted to assign a unique land cover class for each object. Inevitably, this process was executed iteratively by adjusting the input training sample sets and feature sets until a satisfactory classification result was achieved in each classification stage. The hierarchical classification trees of each mapping unit were constructed independently. Training sample sets, classification feature sets and the conceptual hierarchical structure are the critical parts for classification tree construction.

Training Sample Sets

A total of 22,115 field samples collected by field investigations (detailed description shown in Section 3.4) were conducted to build the classification trees. For each mapping unit, because the assigned field samples are usually limited and unevenly distributed in space, many supplementary training samples were appended. These samples came from the visual interpretation of selected satellite images by the experienced mappers. Meanwhile, the existing field samples and high spatial resolution Google Earth images also provided references for the visual interpretation.

Multi-Type Classification Feature Sets

In this study, more than 40 features, including spectral features, topographic features, texture features and shape features, were conducted for constructing classification trees and automatically classifying for each mapping unit, which are shown in Table 2. Some satellite-derived indices were also involved in classification feature sets, including the Normalized Difference Vegetation Index (NDVI), the Modified Normalized Difference Water Index (MNDWI, [42]), and the Normalized Difference Built-Up Index (NDBI, [43]). All of selected features were calculated in the eCognition 8.7 platform. For spectral features, the mean value and the standard deviation of each band of multi-temporal satellite images were involved.

Table 2. The multi-type features used in land cover classification. The *b1*, *b2*, *b3*, *b4* and*b5* denote respectively the blue, green, red, near infrared and mid-infrared band of satellite images. The GLCM is gray level co-occurrence matrix, which has a good ability to distinguish texture features of natural objects. The b_v is the border length of image object, and the p_v is the area of image object.

Types	Features	Calculation Method
Spectral features	The mean value of each band	
	The standard deviation of each band	
	Brightness	Brightness = $(b1 + b2 + b3 + b4)/4$ [44]
	NDVI	NDVI = $(b4 - b3)/(b4 + b3)$ [45]
	MNDWI	MNDWI = $(b2 - b5)/(b2 + b5)$ [42]
	NDBI	NDBI = $(b5 - b4)/(b5 + b4)$ [43]
Topographic features	Elevation	
	Slope	
	Aspect	
Texture features	GLCM-mean	Calculated in the eCognition 8.7 platform [44]
	GLCM-standard deviation	
	GLCM-entropy	
	GLCM-contrast	
Shape features	Shape index (SI)	$SI = b_v/p_v$ [44]
	Length/Width	

Conceptual Hierarchical Structure

Identifying the main land cover characteristics in each mapping unit and establishing a proper conceptual hierarchical structure are essential for the construction of hierarchical classification trees. Because the spectral signatures and the distinguishing capabilities of land cover class varied with the acquisition time of satellite images, the conceptual hierarchical structure needs to vary with mapping units. Figure 5 shows a representative conceptual hierarchical structure used for classification in a randomly selected mapping unit. It is noted that the structure in Figure 5 will not always be suitable for other mapping units.

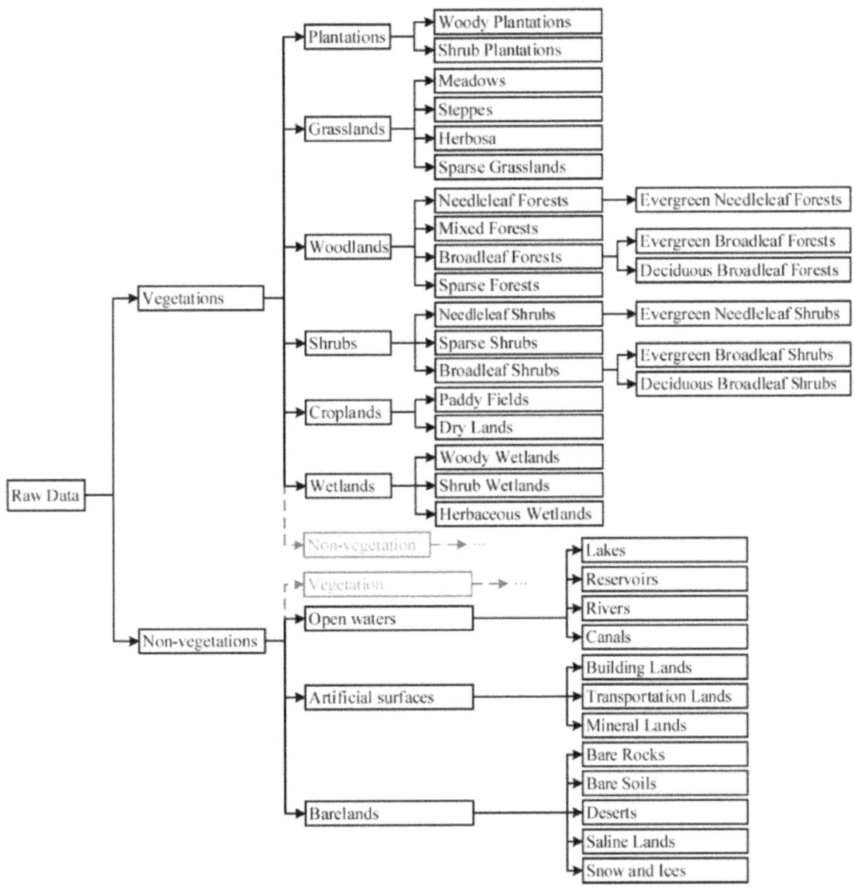

Figure 5. A representative conceptual hierarchical structure of the land cover classes used for generating the CLC-SW2010 product.

For each hierarchical structure in Figure 5, the classification errors were hard to avoid in the classification results, which could be introduced into the following hierarchical levels. Therefore, these errors in the following hierarchical level needed to be revised. For example, some non-vegetation might be confused with vegetation, so the misclassified non-vegetation needed to be extracted in the vegetation division stage as well as following stages (Dish line in Figure 5). Owing to spatial limitations, Figure 5 only shows the handling of non-vegetation and vegetation, and the handling of other land cover classes is the same.

Five-Step Interactive Quality Control Based on Knowledge

Step One: Interactive Quality Control Based on Geographical Rules

The geographical rules, including the single-temporal rules and the multi-temporal geographical rules (list in Table 3), were used for inspecting the obvious errors in preliminary classification results. If the assigned land cover class of an object broke these rules, it would be revised by an automatic or manual method. The single-temporal rules were mainly concerned with the relationships between the land cover classes and the relevant geographical environment. For example, open waters usually reside in relatively flat or low relief areas. The multi-temporal rules paid more attention to those land cover classes which changed frequently with seasons. For instance, if an object was assigned as deciduous vegetation, but its spectral signatures did not demonstrate an obvious change within a year, this object was obviously misclassified and needed further revision.

Table 3. The geographical rules used in the interactive quality control.

Types	Detailed Description
Single-temporal rules	Open waters and wetlands reside in a relatively flat or low relief areas
	The spatial distribution of each forest class commonly below the tree line
	The spatial distribution of glaciers and permanent snow usually under the snow line
	Croplands rarely distribute in the high-elevation region, such as above 4000 m
	Paddy fields locate close to the water source and relatively flat areas
Multi-temporal rules	The deciduous vegetation has obvious different spectral characteristics between leaf-on and leaf-off season, but the evergreen vegetation does not have this differences
	Open waters have great fluctuations in rainy season, the boundary of open water is the maximum boundary in rainy season
	Snow and ices have great fluctuation in winter, the boundary of ice and permanent snow is the minimum boundary in summer

Step Two: Interactive Quality Control Based on Available Thematic Maps

The available thematic maps provided a lot of useful reference information, such as the predominated land cover class, the area ratio and spatial distribution of each land cover class, which were useful for discovering potential classification errors in the preliminary classification result. To effectively use this information, a series of comparisons between the CLC-SW2010 product and available thematic maps were conducted. Once any inconsistencies occurred and were caused by the classification errors, the mapper would further revise classification errors by an automatic or manual method according to the area of classification errors, until any significant differences did not appear in the comparison results.

The location map of the township government (hereinafter referred to as township map) was used to discover the omitted residential land in the preliminary classification result in this study. It is difficult to extract automatically the residential land only relying on the spectral features, because the residential land in mountainous areas are usually composed by dispersed rural houses. The township governments are an important indicator for the residential land because they are usually situated in the concentrated regions of residential land. In this study, under the guidance of the township map and leaf-on satellite images, the omission residential lands were added artificially.

Step Three: Spatial Consistency Verification

The spatial discontinuity phenomena of land cover often occurred at the junction regions of the neighbor mapping units in classification results, which were mainly caused by the following differences: the acquisition time of selected satellite images in neighboring mapping units and the cognitive level of mappers. Therefore, the spatial consistency verification is indispensable to assuring the quality of the CLC-SW2010 product. To avoid any inspection omission, the whole of Southwestern China was divided into 24,130 grids with a side length of 10 km. The experienced mapper carefully checked and revised the spatial discontinuous phenomena grid by grid.

Step Four: Interactive Quality Control with Field Verification Points

To discover any further residual classification errors in the CLC-SW2010 product, a new round of field investigation was enforced to collect independent verification points (Section 3.4). The experienced mappers were carefully compared with the classification results of each verification point and its surrounding regions with the field records, photos and satellite images. Once the classification errors were determined, further revisions were needed.

Step Five: Interactive Quality Control with Statistics Reports

Some recognized statistical information, such as forest coverage, grassland area and cropland area, was recorded in government statistics reports and relevant documents, which were helpful for verification of the quality of the CLC-SW2010 product. Unfortunately, the differences of the definition for the same land cover classes have existed between the statistics report and the CLC-SW2010 product. Only when the differences of the area of a certain land cover class exceeded a predefined threshold value (10% of its area), should further revision be considered.

RESULTS

Hierarchical Decision Tree

For each mapping unit, a series of classification trees generated by the decision tree classifier (See Section 5) were conducted to produce the CLC-SW2010 product in this study. Owing to spatial limitations, this paper only shows three typical decision trees, which were conducted for distinguishing between vegetation and non-vegetation, vegetation divisions and forest divisions, respectively, in the junction zone (path/row: 138/38, WRS2) of the Sichuan province and Tibet autonomous region.

The Decision Tree for Distinguishing Vegetation and Non-Vegetation

Figure 6 shows a typical decision tree for distinguishing vegetation and non-vegetation. Generally, the differences of NDVI between vegetation and non-vegetation in the leaf-on season are obvious. Therefore, the mean value of NDVI derived from the HJ image in leaf-on season was used to distinguish between non-vegetation and vegetation. Then the MNDWI derived from the same image was used to identify open waters misclassified as vegetation, because it is sensitive to open waters.

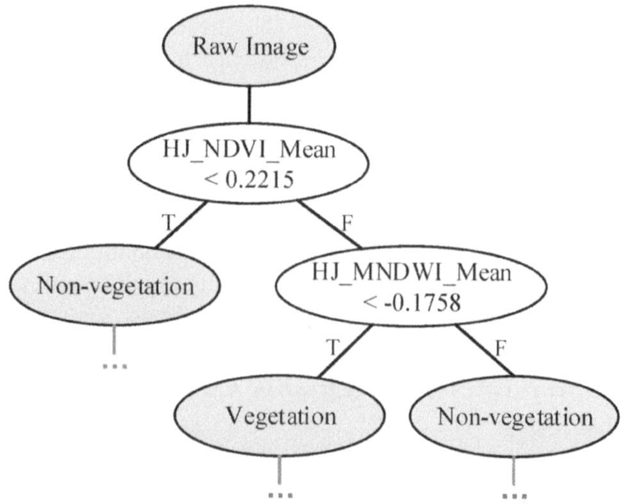

Figure 6. The decision tree for distinguishing vegetation and non-vegetation. The HJ-CCD image acquired on 24 July 2010. The ellipsis represents the land cover class needed for further division.

The Decision Tree for Vegetation Division

The vegetation extracted by above decision tree were further divided into forests, shrubs, grasslands, wetlands (covered with herbaceous and woody vegetation) and croplands by the decision tree shown in Figure 7. Obvious differences between the herbaceous vegetation (including withered grasslands and harvest croplands) and the woody vegetation (including shrubs and forests) existing in spectral reflectance of green band of leaf-on satellite images. Therefore, the feature of HJ_B2_Mean was used to distinguish them. In leaf-on season, the NDVI values (HJ_NDVI_Mean) between croplands and grasslands have differences, which were chosen to separate grasslands and croplands. The standard deviation of red and mid-infrared bands (TM_B3_Std and TM_B5_Std) was applied to reduce the commissions between grasslands and croplands because cropland objects usually have a higher degree of dispersion of spectral reflectance than grassland objects. Using the absorption feature of forest canopy in the red band (HJ_B3_Mean), the forests were extracted from the woody vegetation. The wetlands have relatively low reflectance in the near-infrared band (HJ_B4_Mean) due to the absorption feature of water, which was adopted to distinguish between wetlands and shrubs. The forests confused with the shrubs were extracted again by the spectral reflectance of blue band in leaf-off satellite images (TM_B1_Mean), and the grasslands confused with the croplands were extracted again by the spectral reflectance of mid-infrared band in leaf-off satellite images (TM_B7_Mean). Above all, each satellite image played a unique role and the multi-temporal satellite images contributed to effectively discriminating between each vegetation type.

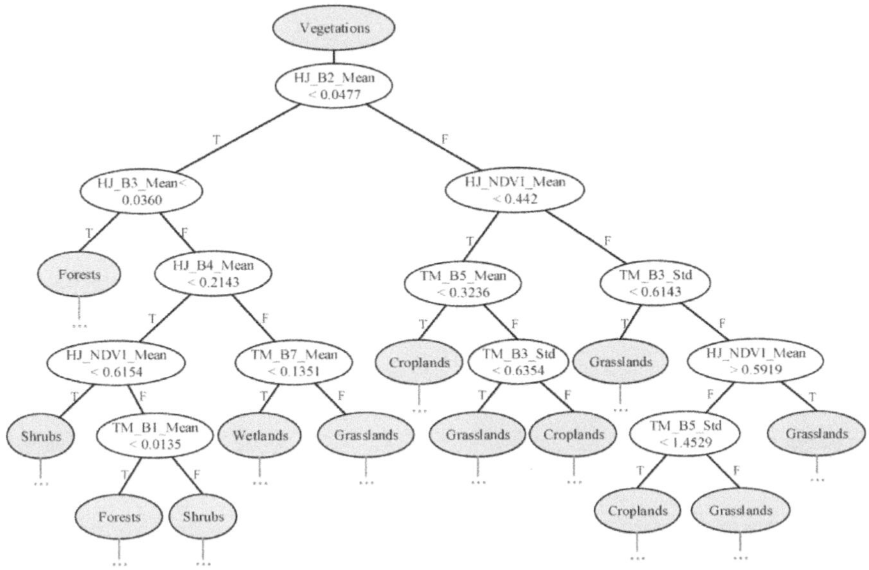

Figure 7. The decision tree for vegetation division. The Landsat TM image acquired on 28 September 2010. The Bx denotes x-th band of satellite image, the Std denotes the standard deviation.

The Decision Tree for Forests Division

Usually, the distinct differences existed in the near-infrared band of leaf-on satellite images among broadleaf forests, the needleleaf forests and the mixed forests, which were used to separate forests into broadleaf forests, needleleaf forests and mixed forests (Figure 8). The spatial distributions of each forest display distinct vertical stratification in mountainous areas, which was useful for discovering and revising classification errors of vegetation types. For example, the shrubs misclassified as needleleaf forests and broadleaf forests were extracted by the Elevation in Figure 8.

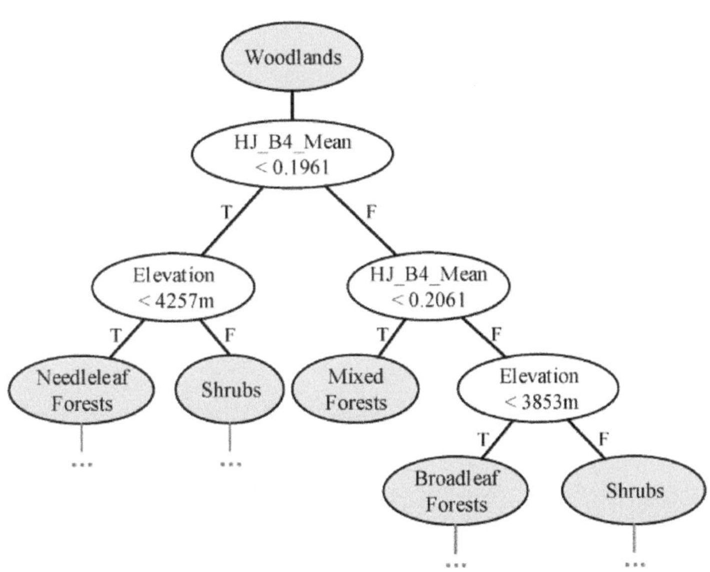

Figure 8. The decision tree for woodlands divisions.

Effectiveness of Interactive Quality Control

The five steps of knowledge-based interactive quality control were not conducted simultaneously. The steps based on geographical rules and the thematic maps were applied firstly in the automatic classification stage, and then the spatial consistency checking was conducted after forming a whole land cover map. The steps based on the verification points and statistics information were used in the last stage. Although all five steps were conducted to revise the classification errors contained in the CLC-SW2010 product, this section only shows two representative examples due to space limitations.

Figure 9 shows a typical example for revising residential land at high altitude by using a township map. The residential land in mountainous areas was often omitted by automatic classification approaches, but this omitted residential land was artificially added with the help of the township map and the high-resolution Google Earth images.

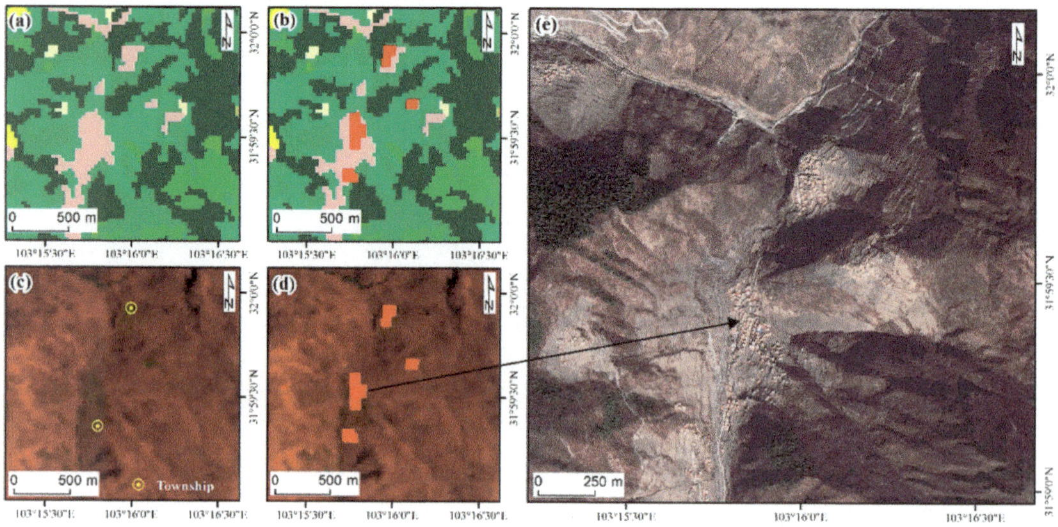

Figure 9. Interactive quality control based on the existing thematic map (township map). (**a**) and (**b**) are the results of unrevised and revised residential lands, respectively; (**c**) is the Landsat TM images and the locations of township; (**d**) is the additional residential lands with the help of the township map and high-resolution satellite images; (**e**) is the high-resolution Google Earth images.

A typical spatial discontinuous phenomenon is shown in Figure 10c,g, which mainly comes from confusing the evergreen shrubs with the deciduous shrubs. In this case, the classification trees used for distinguishing the evergreen shrubs and the deciduous shrubs needed to be reconstructed by adjusting training sample sets and feature sets for each mapping unit until a satisfactory result was obtained. After revision, the spatial discontinuous phenomenon was removed in the post-checking classification results (Figure 10d,h).

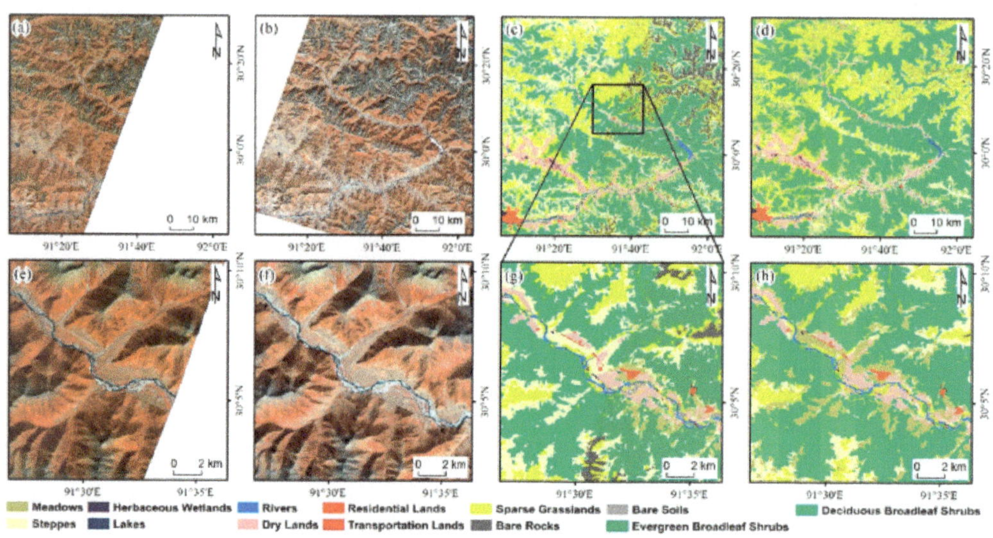

Figure 10. The results of the spatial consistency checking. (**a**) is the Landsat TM images (path/row: 138/39, acquired on 17 October 2009) used for classification of left mapping unit; (**b**) is the Landsat TM images (path/row: 137/39, acquired on 26 October 2009) used for classification of right mapping unit; (**c**) is the pre-checking classification result; (**d**) is the post-checking classification result; (**e**), (**f**), (**g**) and (**h**) are the enlarged images at black box in the (**a**), (**b**), (**c**) and (**d**), respectively.

The CLC-SW2010 Product and Product Accuracy

Taking the multi-temporal native HJ-CCD images and Landsat TM images as main input data, the 30 m-resolution CLC-SW2010 product (Figure 11) was produced by the proposed HC_MMK approach over the last three years.

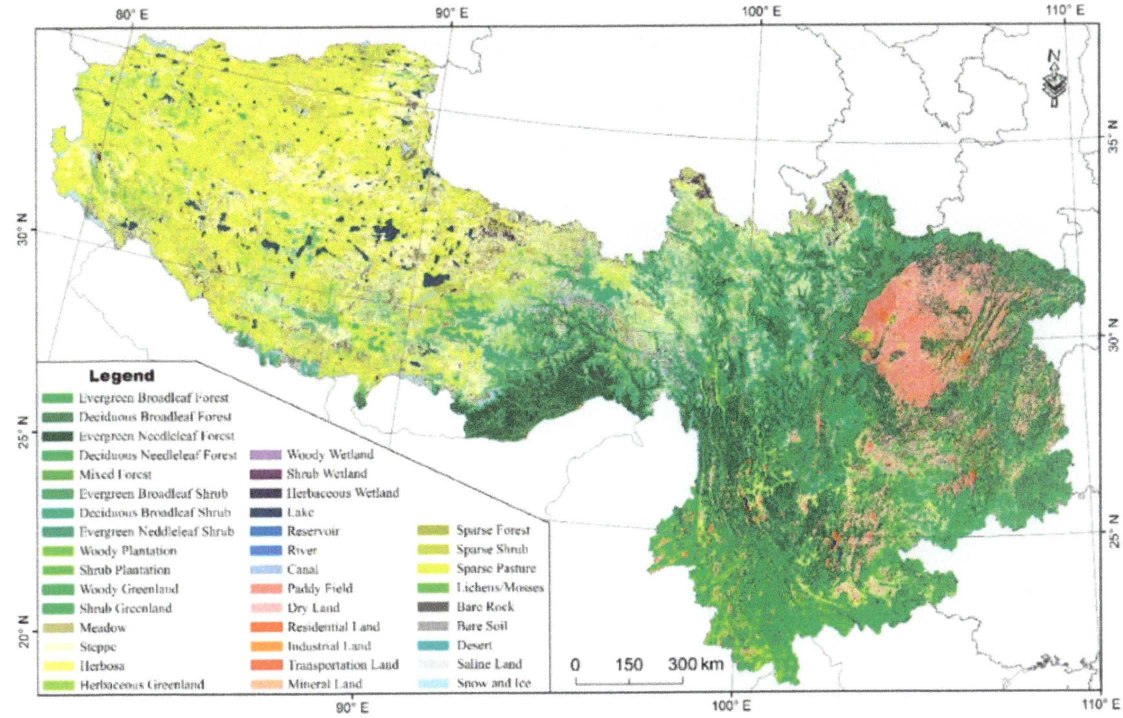

Figure 11. The CLC-SW2010 product.

In the CLC-SW2010 product, the grasslands and woodlands are the predominant land cover classes, accounting for 45.01% and 34.92% of the land area of Southwestern China, respectively. The areas of croplands and bare lands are less than that of grasslands and woodland, accounting for 10.45% and 6.10%, respectively. Artificial surfaces and wetlands are rarely found in Southwestern China, only accounting for 0.55% and 2.97%, respectively. The grasslands (including sparse grasslands) are mainly distributed on the Tibet plateau (81.06%), Western Sichuan plateau (11.21%) and the Karst region (7.73%). The woodlands are mainly found in Hengduan Mountain, Daba Mountain, Southeastern Tibet and Yunnan province. The cropland is mainly located in Sichuan Basin, Anning River valley and the "bazi" regions of Yunnan-Guizhou Plateau.

A total of 5924 validation points, which have been collected independently by a third-party group, were conducted to assess the accuracy of the CLC-SW2010 product. After independent validation, the accuracy of the primary categories and secondary categories of the CLC-SW2010 product reached 95.09% and 87.14%, respectively, and the kappa coefficients were 0.9345 and 0.8573 respectively. The detailed precision information of the CLC-SW2010 product is shown in Table 4 and Table 5.

Table 4. The confusion matrix of primary categories in the CLC-SW2010 product. The codes of each primary category are the same as the ones listed in Table 1.

Reference	Classification						Total	Producer's Accuracy (%)
	1	2	3	4	5	6		
1	2204	64	1	50	2	1	2322	94.92
2	42	744	2	27		1	816	91.18
3	6		665	1			672	98.96
4	35	13	3	1349	8	1	1409	95.74
5	7	3	1	20	474		505	93.86
6			1	2		197	200	98.50
Total	2294	824	673	1449	484	200	5924	
User's accuracy (%)	96.08	90.29	98.81	93.10	97.93	98.50		
Overall accuracy: 95.09%; Kappa coefficient: 0.9345								

Table 5. The confusion matrix of secondary categories in the CLC-SW2010 product.

Code	Area Ratio	Number of Samples	Producer's Accuracy	User's Accuracy	Code	Area Ratio	Number of Samples	Producer's Accuracy	User's Accuracy
101	5.30%	347	83.29%	88.38%	35	0.14%	142	95.77%	95.77%
102	0.80%	127	79.53%	82.11%	36	0.38%	195	95.90%	95.90%
103	15.49%	1303	90.82%	91.89%	37	0.00%	2	100.00%	100.00%
105	0.43%	36	66.67%	75.00%	41	2.64%	508	92.72%	91.99%
106	4.38%	266	81.20%	77.42%	42	7.81%	785	89.68%	88.33%
107	6.60%	243	75.72%	65.25%	51	0.47%	356	93.26%	94.05%
109	0.44%	74	87.84%	86.67%	52	0.01%	68	79.41%	87.10%
110	1.09%	42	85.71%	76.60%	53	0.06%	48	87.50%	93.33%
111	0.01%	3	100.00%	100.00%	54	0.01%	29	65.52%	95.00%
21	6.42%	316	73.42%	87.55%	63	17.39%	111	77.48%	67.19%
22	17.84%	217	62.21%	62.50%	65	2.03%	68	89.71%	98.39%
23	3.86%	173	84.39%	71.92%	66	2.42%	60	91.67%	88.71%
33	1.03%	117	99.15%	97.48%	68	0.18%	4	100.00%	57.14%
34	1.41%	216	99.54%	100.00%	69	1.32%	68	100.00%	98.55%

Overall accuracy: 87.14%; **Kappa coefficient:** 0.8573

The wetlands and bare lands have a higher accuracy (>98%) than other land cover classes, and the accuracy of the grasslands is in the end in CLC-SW2010 (Table 4). Confusing with croplands and shrubs is the main reason for grassland inaccuracy. Because the grasslands were usually accompanied by shrubs, especially in the Karst mountainous area, it is difficult to accurately distinguish the grasslands and the shrubs [46]. They have a similar spectral signature between the grassland and the returned farmlands with 15–25 degree slopes. Therefore, these returned farmlands were frequently classified as croplands rather than grasslands in the CLC-SW2010 product.

The evergreen needleleaf forests, herbaceous wetlands, rivers, lakes, reservoirs, paddy fields, residential lands and permanent snow and ice have a higher user's accuracy and producer's accuracy than other secondary categories (Table 5). The multi-source and multi-temporal satellite images, geo-knowledge, objected-oriented method and hierarchical classification played its roles in the accuracy of the CLC-SW2010 product. For instance, the spectral and shape signatures of the lakes were highly similar to the reservoirs. However, obvious differences existed in the spatial distribution between the lakes and the reservoirs in Southwestern China, the former mainly distributed in the plateau, and the latter mainly located in the agricultural region, which were useful for exactly distinguishing the lakes and the reservoirs.

DISCUSSIONS

Comparison with Existing 30 m-Resolution Land Cover Products

The CLC-SW2010 product were compared with other existing 30 m-resolution land cover products in 2010 over Southwestern China, including CLUDs (the third row of Figure 12), FROM-GLC (the fourth row of Figure 12) and GlobeLand30 (the fifth row of Figure 12). It was found that the quality of CLC-SW2010 was superior to these products according to the respective preliminary validation results, the accuracy of FROM-GLC and GlobeLand30 reached 71.54% [5] and 80.33% ± 0.2% [11], respectively. The lack of leveraging variability in phenological patterns is an important factor for the quality of GlobeLand30 and FROM-GLC [18]; however, it has been avoided in CLC-SW2010 by introducing multi-temporal satellite images and a variety of geo-knowledge. Although the total area of each land cover class is closer to reality, the boundary of land cover is vague and inaccurate in CLUDs (the third row of Figure 12).

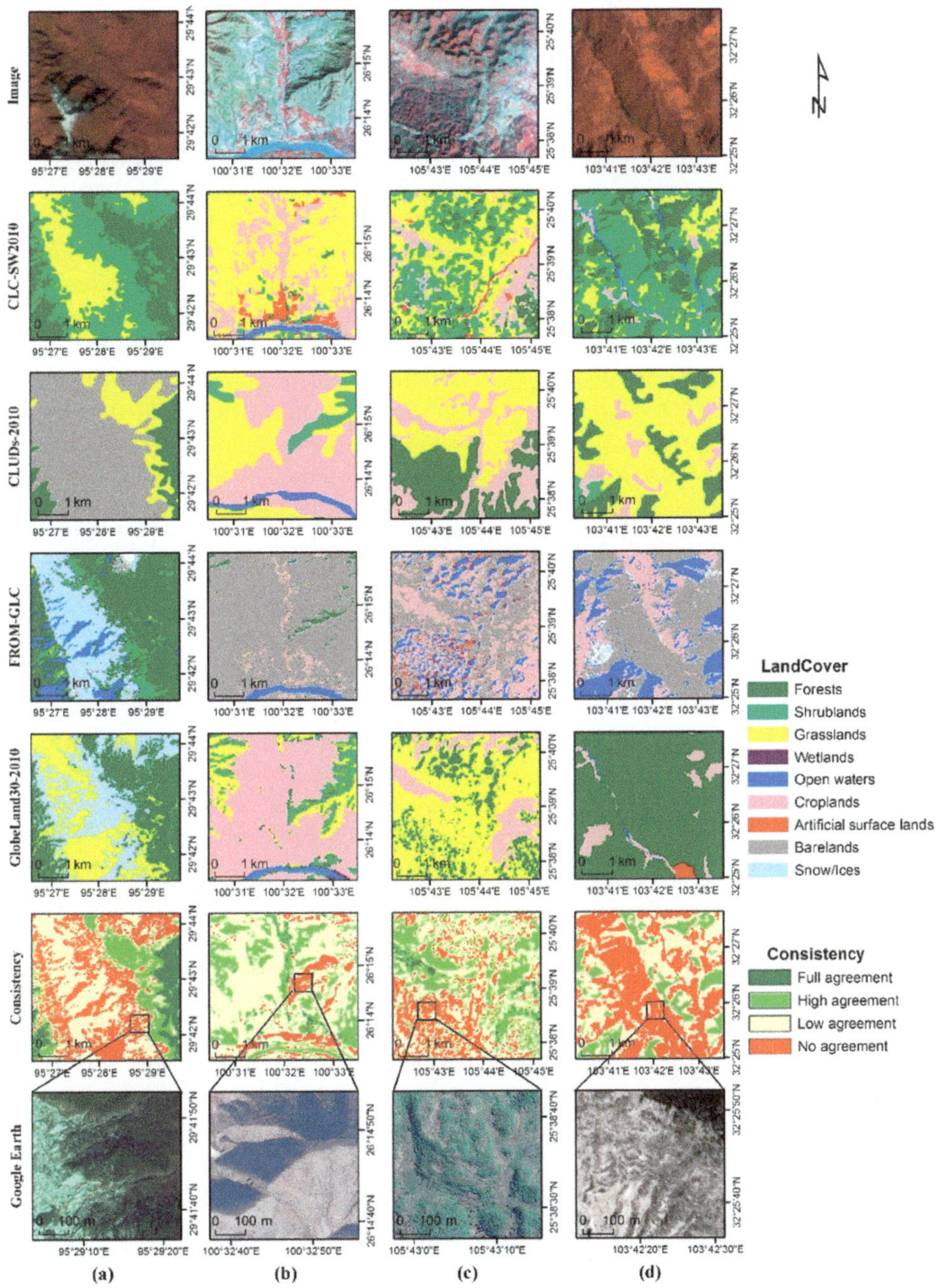

Figure 12. The comparison of classification results among four land cover products in four typical regions. The regions of **(a–d)** are located in Southeast Tibet, the middle part of Yunnan province, the Karst region of Guizhou province and the upper reaches of Minjiang River, respectively. For the consistency of pixels, the full agreement denotes all four products have a same land cover class, high agreement denotes there are two land cover classes among four products, low agreement denotes there are three land cover classes among four products and no agreement denotes there are a unique class in each product.

Four typical mountainous regions that are hard to map (a: Plateau regions host mostly grasslands; b: low mountains mainly consist of croplands; c: Karst regions with complex land cover and d: steep mountainous areas mostly covered with forests) were taken as examples (Figure 12) to present the comparison results of four land cover products in detail. The CLC-SW2010 product showed more distinctly the vertical stratification of vegetation (Figure 12a) and the fragmentized landscape (Figure 12b,c) in mountainous areas than the other three products, and the proposed HC-MMK approach especially designed for land cover mapping in mountainous areas was the main contributor. There was also many classification errors identified between mountain shadow and water in FROM-GLC (the fourth row of Figure 12), which were eliminated by some simple geographical rules in CLC-SW2010.

The Contribution of Each Component in the HC-MMK Approach

The Multi-Source and Multi-Temporal Data

In this paper, the native HJ-CCD images and Landsat TM images have been combined to provide more high-quality satellite images (Figure 3) for generating the CLC-SW2010 product. This combination can maximize the advantages of the native HJ-CCD images (wide-covered and short repeat cycle) and the Landsat TM images (plentiful spectral information and excellent quality). The multi-temporal satellite images were useful for distinguishing the evergreen vegetation and the deciduous vegetation, because they have differences in changes of spectral signature between leaf-on season and leaf-off season [19]. Meanwhile, the temporal signatures were used for identifying the dry lands, paddy fields and other land cover classes, and their spectral signatures frequently varied with season in this study.

Knowledge

The importance of knowledge for producing land cover products has been well acknowledged [11,16,47,48]. Subject to the availability, scale and quality of knowledge, at present, limited knowledge has been drawn upon to enhance the accuracy of land cover products, such as terrain knowledge [16,49], spatial distribution knowledge [11], and temporal signature [18,50,51]. In this paper, a variety of available knowledge, including geographical rules, existing thematic maps, field samples and statistical information, were applied together in the HC-MMK approach. Each type of knowledge played a unique role and all knowledge together contributed to quality improvement of the CLC-SW2010 product (Figure 9 andFigure 10). The attempt to draw upon of variety of knowledge in this study offered a recommendable method for other land cover mapping projects.

Hierarchical Classification

Hierarchical classification is an effective land cover mapping method for those areas with complex and heterogeneous landscapes by increasing classification time and decreasing the number of land cover classes in a single classification process [52,53,54]. The conceptual hierarchical structure is the core of this approach. A unique conceptual hierarchical structure was built in most research [53,54], which cannot meet the demands for land cover mapping over large regions with complex terrain, such as Southwestern China, because there are 38 land cover classes, 190 mapping units and various land cover classes. To solve this problem, a series of conceptual hierarchical structures aimed for each unique mapping unit were built in this study, which obviously improved the accuracy of the CLC-SW2010 product. This strategy also provided a useful reference for future research on land cover mapping over large areas.

Quality Control

Quality control is a critical step for producing a widely-used land cover product [11,55,56]. The procedures need to be applied to the whole classification process, including satellite image collection and preprocessing, field data collection, automatic classification and accuracy assessment. Only if each sub-step obtained a high-precision result could a high accuracy land cover product be achieved. This paper proposed a practical, interactive quality control method based on prior knowledge, which obviously eliminated many classification errors and spatial discontinuity phenomena (Figure 9 andFigure 10) and improved the quality of the CLC-SW2010 product. In addition, many strict criteria were also developed in the other stages; for instance, the accuracy of image-to-image co-registration was less than 2 pixels in mountainous areas for data preprocessing, and the total cloud cover was less than 10% for each selected satellite image.

The Merits and Limitations of the Current Work

The land cover classification system used in the CLC-SW2010 was more suitable for various ecological applications than other classification systems, because it gave adequate consideration to life form, vegetation height, leaf type, vegetation phenology, vegetation cover, and so on [35]. In addition, this classification system was easy to interoperate with multiple land cover classification systems, such as FAO LCCS, which will help to widen the application scope of the CLC-SW2010 product. A full set of automatic mapping processes, including satellite image preprocessing, classification and quality control, were proposed in this study. They were conducted to generate the CLC-SW2010 product, which obviously shortened the mapping cycle. The main users, including the local environmental authorities and the research teams of the Carbon Special Program, have been involved in the process of generating the CLC-SW2010 product. They not only provided the field land cover samples, but also discovered potential classification errors existing in the CLC-SW2010 according to their experiences.

Using the proposed HC-MMK approach, the overall accuracy of the CLC-SW2010 reached 87.14%. However, there are still some limitations in the current mapping work. Due to the lack of floodplain wetlands in land cover classification systems, the floodplain wetlands were classified as bare soils in the CLC-SW2010 product. They will limit its applications because distinct differences of ecological structure and functions existed between the floodplain wetlands and the bare soils. Meanwhile, some land cover classes have a lower producer's accuracy and user's accuracy in the CLC-SW2010 product, especially for mixed forests, steppes and meadows. The vague and non-quantitative definition and various spectral signatures are the main causes for the low accuracy of the mixed forests. The spectral signatures of the steppes and the meadows are easy to vary with the species, coverage, biomass and moisture content, which impacted distinguishing between steppes and meadows in the CLC-SW2010 product. Therefore, it is worth exploring how to further improve the accuracy of the mixed forests, steppes and meadows and optimize the land cover classification system according to application demands in the future.

CONCLUSIONS

This paper mainly introduced the proposed HC-MMK approach and the 30 m-resolution CLC-SW2010 product. The multi-temporal native HJ-CCD images, Landsat TM images and topographical data were taken as the main input data in this study. Hierarchical classification tree construction and a five-step knowledge-based interactive quality control were the key components of this proposed approach. Based on independent accuracy assessment by a third-party group, the overall accuracy of primary and secondary categories of the CLC-SW2010 product reached 95.09% and 87.14%, respectively, and the Kappa coefficients were 0.9345 and 0.8573, respectively.

To address critical issues for land cover mapping in mountainous areas, four key steps were taken in this study. Firstly, a combination of the native HJ-CCD images and Landsat TM images were taken as input data to construct the multi-temporal satellite images. Secondly, the satellite images with complex terrain were segmented with as much detail as possible in the image segmentation stage. Thirdly, for each mapping unit, a series of unique hierarchical classification trees were built to make the land cover classification as accurate as possible. Finally, a variety of prior geo-knowledge, including geographical rules, existing available thematic maps, verification samples and statistical information, was extensively used in the classification and quality control stage to improve the accuracy of the product.

The proposed HC-MMK approach can be used not only in mountainous areas, but also in plains, hills and other regions, because land cover mapping in mountainous areas is more complex than other regions. Overall, this study proposed an operable land cover mapping method and produced a 30 m-resolution CLC-SW2010 product. Most importantly, it provided an important reference point for other land cover mapping studies conducted over large areas or even the entire globe in the future.

ACKNOWLEDGMENTS

This research was funded jointly by the Strategic Leader Science and Technology project (XDA05050105), the "Ecology decade" specific project of MEP&CAS (STSN-01-04), the International Cooperation Key Project of CAS (GJHZ201320) and the National Natural Science Foundation of China (41271433, 41571373). We are especially grateful to Bingfang Wu, Yuan Zeng and Lei Zhang for their support and advice. We are also thankful to all the contractors, image providers and the anonymous reviewers for their valuable comments and suggestions.

AUTHOR CONTRIBUTIONS

All authors have made major and unique contributions. Ainong Li designed the framework of this research. Guangbin Lei drafted the preliminary version and Ainong Li finished the final version of the manuscript. Ainong Li, Guangbin Lei, Huaan Jin, Jinhu Bian, Xi Nan, Wei Zhao, Qiannan Liu, Guangbin Yang and Huan Yu took part in the field data collection. Jinhu Bian is in charge of satellite data collection and pre-processing. Guangbin Lei, Jinhu Bian, Zhengjian Zhang and Ainong Li completed the land cover mapping in Sichuan Province. Guangbin Lei, Huaan Jin, Jiyan Wang, Wei Zhao, Wenlan Feng, and Ainong Li completed the land cover mapping in Tibet autonomous region. Guangbin Yang, Guangbin Lei, Xiaomin Cao, Jianbo Tan and Ainong Li completed the land cover mapping in Guizhou province and Chongqing Municipality. Huan Yu, Xi Nan, Guangbin Lei and Ainong Li completed the land cover mapping in Yunnan provinces.

CONFLICTS OF INTEREST

The authors declare no conflict of interest.

REFERENCES

1. Yang, J.; Gong, P.; Fu, R.; Zhang, M.H.; Chen, J.M.; Liang, S.L.; Xu, B.; Shi, J.C.; Dickinson, R. The role of satellite remote sensing in climate change studies. *Nat. Clim. Chang.* **2013**, *3*, 875–883.
2. Shupeng, C.; van Genderen, J. Digital Earth in support of global change research. *Int. J. Digit. Earth* **2008**, *1*, 43–65.
3. Gross, J.E.; Goetz, S.J.; Cihlar, J. Application of remote sensing to parks and protected area monitoring: Introduction to the special issue. *Remote Sens. Environ.* **2009**, *113*, 1343–1345.
4. Wiens, J.; Sutter, R.; Anderson, M.; Blanchard, J.; Barnett, A.; Aguilar-Amuchastegui, N.; Avery, C.; Laine, S. Selecting and conserving lands for biodiversity: The role of remote sensing. *Remote Sens. Environ.* **2009**, *113*, 1370–1381.
5. Gong, P.; Wang, J.; Yu, L.; Zhao, Y.C.; Zhao, Y.Y.; Liang, L.; Niu, Z.G.; Huang, X.M.; Fu, H.H.; Liu, S.; *et al.* Finer resolution observation and monitoring of global land cover: First mapping results with Landsat TM and ETM+ data.*Int. J. Remote Sens.* **2013**, *34*, 2607–2654.
6. Hansen, M.C.; Potapov, P.V.; Moore, R.; Hancher, M.; Turubanova, S.A.; Tyukavina, A.; Thau, D.; Stehman, S.V.; Goetz, S.J.; Loveland, T.R.; *et al.* High-resolution global maps of 21st-century forest cover change. *Science* **2013**, *342*, 850–853.
7. Loveland, T.R.; Reed, B.C.; Brown, J.F.; Ohlen, D.O.; Zhu, Z.; Yang, L.; Merchant, J.W. Development of a global land cover characteristics database and IGBP DISCover from 1 km AVHRR data. *Int. J. Remote Sens.* **2000**, *21*, 1303–1330.
8. Friedl, M.A.; Sulla-Menashe, D.; Tan, B.; Schneider, A.; Ramankutty, N.; Sibley, A.; Huang, X.M. MODIS collection 5 global land cover: Algorithm refinements and characterization of new datasets. *Remote Sens. Environ.* **2010**, *114*, 168–182.
9. Arino, O.; Bicheron, P.; Achard, F.; Latham, J.; Witt, R.; Weber, J.L. GlobCover the Most Detailed Portrait of Earth. Available online: https://earth.esa.int/web/guest/-/globcover-the-most-detailed-portrait-of-earth-5910 (accessed on 30 March 2016).
10. Yu, L.; Wang, J.; Li, X.; Li, C.; Zhao, Y.; Gong, P. A multi-resolution global land cover dataset through multisource data aggregation. *Sci. China-Earth Sci.* **2014**, *57*, 2317–2329.
11. Chen, J.; Chen, J.; Liao, A.P.; Cao, X.; Chen, L.J.; Chen, X.H.; He, C.Y.; Han, G.; Peng, S.; Lu, M.; *et al.* Global land cover mapping at 30 m resolution: A POK-based operational approach. *ISPRS J. Photogramm. Remote Sens.* **2015**, *103*, 7–27.
12. Liu, J.Y. Study on national resources and environment survey and dynamic monitoring using remote sensing. *J. Remote Sens.* **1997**, *1*, 225–230.
13. Liu, J.Y.; Kuang, W.H.; Zhang, Z.X.; Xu, X.L.; Qin, Y.W.; Ning, J.; Zhou, W.C.; Zhang, S.W.; Li, R.D.; Yan, C.Z.; *et al.*Spatiotemporal characteristics, patterns, and causes of land-use changes in China since the late 1980s. *J. Geogr. Sci.***2014**, *24*, 195–210.
14. Giri, C.; Pengra, B.; Long, J.; Loveland, T.R. Next generation of global land cover characterization, mapping, and monitoring. *Int. J. Appl. Earth Obs. Geoinf.* **2013**, *25*, 30–37.
15. Kuenzer, C.; Leinenkugel, P.; Vollmuth, M.; Dech, S.

Comparing global land-cover products-implications for geoscience applications: An investigation for the trans-boundary Mekong Basin. *Int. J. Remote Sens.* **2014**, *35*, 2752–2779.

16. Li, A.N.; Jiang, J.G.; Bian, J.H.; Deng, W. Combining the matter element model with the associated function of probability transformation for multi-source remote sensing data classification in mountainous regions. *ISPRS J. Photogramm. Remote Sens.* **2012**, *67*, 80–92.
17. Jia, K.; Liang, S.L.; Zhang, N.; Wei, X.Q.; Gu, X.F.; Zhao, X.; Yao, Y.J.; Xie, X.H. Land cover classification of finer resolution remote sensing data integrating temporal features from time series coarser resolution data. *ISPRS J. Photogramm. Remote Sens.* **2014**, *93*, 49–55.
18. Yu, L.; Wang, J.; Gong, P. Improving 30 m global land-cover map FROM-GLC with time series MODIS and auxiliary data sets: A segmentation-based approach. *Int. J. Remote Sens.* **2013**, *34*, 5851–5867.
19. Lei, G.B.; Li, A.N.; Bian, J.H.; Zhang, Z.J.; Zhang, W.; Wu, B.F. An practical method for automatically identifying the evergreen and deciduous characteristic of forests at mountainous areas: A case study in Mt.Gongga Region. *Acta Ecol. Sin.* **2014**, *34*, 7210–7221.
20. Senf, C.; Pflugmacher, D.; van der Linden, S.; Hostert, P. Mapping rubber plantations and natural forests in Xishuangbanna (Southwest China) using multi-spectral phenological metrics from MODIS time series. *Remote Sens.***2013**, *5*, 2795–2812.
21. Turner, W.; Rondinini, C.; Pettorelli, N.; Mora, B.; Leidner, A.K.; Szantoi, Z.; Buchanan, G.; Dech, S.; Dwyer, J.; Herold, M.; *et al.* Free and open-access satellite data are key to biodiversity conservation. *Biol. Conserv.* **2015**, *182*, 173–176.
22. Bian, J.H.; Li, A.N.; Wang, Q.F.; Huang, C.Q. Development of dense time series 30-m image products from the Chinese HJ-1A/B Constellation: A case study in Zoige Plateau, China. *Remote Sens.* **2015**, *7*, 16647–16671.
23. Bian, J.H.; Li, A.N.; Jin, H.A.; Lei, G.B.; Huang, C.Q.; Li, M.X. Auto-registration and orthorecification algorithm for the time series HJ-1A/B CCD images. *J. Mt. Sci.* **2013**, *10*, 754–767.
24. Wang, Q.; Wu, C.Q.; Li, Q.; Li, J.S. Chinese HJ-1A/B satellites and data characteristics. *Sci. China Earth Sci.* **2010**, *53*, 51–57.
25. Yu, Q.; Gong, P.; Clinton, N.; Biging, G.; Kelly, M.; Schirokauer, D. Object-based detailed vegetation classification with airborne high spatial resolution remote sensing imagery. *Photogramm. Eng. Remote Sens.* **2006**, *72*, 799.
26. Ouyang, Z.T.; Zhang, M.Q.; Xie, X.; Shen, Q.; Guo, H.Q.; Zhao, B. A comparison of pixel-based and object-oriented approaches to VHR imagery for mapping saltmarsh plants. *Ecol. Inform.* **2011**, *6*, 136–146.
27. Huang, C.; Davis, L.S.; Townshend, J.R.G. An assessment of support vector machines for land cover classification. *Int. J. Remote Sens.* **2002**, *23*, 725–749.
28. Mountrakis, G.; Im, J.; Ogole, C. Support vector machines in remote sensing: A review. *ISPRS J. Photogramm. Remote Sens.* **2011**, *66*, 247–259.
29. Friedl, M.A.; Brodley, C.E. Decision tree classification of land cover from remotely sensed data. *Remote Sens. Environ.***1997**, *61*, 399–409.
30. Canty, M.J. Boosting a fast neural network for supervised land cover classification. *Comput. Geosci.* **2009**, *35*, 1280–1295.
31. Tseng, M.H.; Chen, S.J.; Hwang, G.H.; Shen, M.Y. A genetic algorithm rule-based approach for land-cover classification. *ISPRS J. Photogramm. Remote Sens.* **2008**, *63*, 202–212.
32. Lu, D.R.; Ding, Z.L. Climate change: Carbon budget and relevant issues. *Bull. Chin. Acad. Sci.* **2012**, *27*, 395–402.
33. Ouyang, Z.Y.; Wang, Q.; Zheng, H.; Zhang, F.; Hou, P. National ecosystem survey and assessment of China (2000–2010). *Bull. Chin. Acad. Sci.* **2014**, *29*, 462–466.
34. Deng, W.; Xiong, Y.L.; Zhao, J.D.; Qiu, D.L.; Zhang, Z.Q.; Wen, A.B. Enlightenment from international mountain research projects. *J. Mt. Sci.* **2013**, *31*, 377–384.
35. Zhang, L.; Wu, B.F.; Li, X.S.; Xing, Q. Classification system of China land cover for carbon budget. *Acta Ecol. Sin.* **2014**,*34*, 7158–7166.
36. Wolfe, R.; Masek, J.; Saleous, N.; Hall, F. Ledaps: Mapping North American disturbance from the Landsat record. *Int. Geosci. Remote Sens.* **2004**, *1*, 1–4.
37. Li, A.N.; Jiang, J.G.; Bian, J.H.; Lei, G.B.; Huang, C.Q. Experiment and accuracy analysis of automated registration and orthorectification for Landsat-like images based on AROP. *Remote Sens. Technol. Appl.* **2012**, *27*, 23–32.
38. Nan, X.; Li, A.N.; Bian, J.H.; Zhang, Z.J. Comparison of the accuracy between SRTM and ASTER GDEM over typical mountain area: A case study in the Eastern Qinghai-Tibetan Plateau. *J. Geo-Inf. Sci.* **2015**, *17*, 91–98.
39. Hayakawa, Y.S.; Oguchi, T.; Lin, Z. Comparison of new and existing global digital elevation models: ASTER G-DEM and SRTM-3. Geophys. *Res. Lett.* **2008**, *35*, 36–44.
40. Hou, Z.; Xu, Q.; Nuutinen, T.; Tokola, T. Extraction of remote sensing-based forest management units in tropical forests. *Remote Sens. Environ.* **2013**, *130*, 1–10.
41. Johnson, B.; Xie, Z. Unsupervised image segmentation evaluation and refinement using a multi-scale approach. *ISPRS J. Photogramm. Remote Sens.* **2011**, *66*, 473–483.
42. Trimble. *Ecognition Developer User Guide*; Version 8.7; Definiens: Munich, Germany, 2011.
43. Goward, S.N.; Markham, B.; Dye, D.G.; Dulaney, W.; Yang, J. Normalized difference vegetation index measurements from the advanced very high resolution radiometer. *Remote Sens. Environ.* **1991**, *35*, 257–277.
44. Gao, B.C. NDWI—A normalized difference water index for remote sensing of vegetation liquid water from space. *Remote Sens. Environ.* **1996**, *58*, 257–266.

45. Zha, Y.; Gao, J.; Ni, S. Use of normalized difference built-up index in automatically mapping urban areas from TM imagery. *Int. J. Remote Sens.* **2003**, *24*, 583–594.

46. Xu, E.Q.; Zhang, H.Q.; Li, M.X. Object-based mapping of karst rocky desertification using a support vector machine.*Land Degrad. Dev.* **2015**, *26*, 158–167.

47. Townshend, J.; Justice, C.; Li, W.; Gurney, C.; Mcmanus, J. Global land cover classification by remote-sensing—Present capabilities and future possibilities. *Remote Sens. Environ.* **1991**, *35*, 243–255.

48. Homer, C.; Dewitz, J.; Yang, L.M.; Jin, S.; Danielson, P.; Xian, G.; Coulston, J.; Herold, N.; Wickham, J.; Megown, K. Completion of the 2011 national land cover database for the Conterminous United States—representing a decade of land cover change information. *Photogramm. Eng. Remote Sens.* **2015**, *81*, 345–354.

49. Dorren, L.K.A.; Maier, B.; Seijmonsbergen, A.C. Improved Landsat-based forest mapping in steep mountainous terrain using object-based classification. *Forest Ecol. Manag.* **2003**, *183*, 31–46.

50. Wang, J.; Zhao, Y.Y.; Li, C.C.; Yu, L.; Liu, D.S.; Gong, P. Mapping global land cover in 2001 and 2010 with spatial-temporal consistency at 250 m resolution. *ISPRS J. Photogramm. Remote Sens.* **2015**, *103*, 38–47.

51. Mosleh, M.K.; Hassan, Q.K. Development of a remote sensing-based "boro" rice mapping system. *Remote Sens.* **2014**,*6*, 1938–1953.

52. Laliberte, A.S.; Fredrickson, E.L.; Rango, A. Combining decision trees with hierarchical object-oriented image analysis for mapping arid rangelands. *Photogramm. Eng. Remote Sens.* **2007**, *73*, 197–207.

53. Sulla-Menashe, D.; Friedl, M.A.; Krankina, O.N.; Baccini, A.; Woodcock, C.E.; Sibley, A.; Sun, G.Q.; Kharuk, V.; Elsakov, V. Hierarchical mapping of Northern Eurasian land cover using MODIS data. *Remote Sens. Environ.* **2011**, *115*, 392–403.

54. Gholoobi, M.; Kumar, L. Using object-based hierarchical classification to extract land use land cover classes from high-resolution satellite imagery in a complex urban area. *J. Appl. Remote Sens.* **2015**, *9*, 096052.

55. Vanselow, K.A.; Samimi, C. Predictive mapping of dwarf shrub vegetation in an arid high mountain ecosystem using remote sensing and random forests. *Remote Sens.* **2014**, *6*, 6709–6726.

56. Hansen, M.C.; Loveland, T.R. A review of large area monitoring of land cover change using Landsat data. *Remote Sens. Environ.* **2012**, *122*, 66–74.

CHAPTER 16

DESIGNING AN EXPERIMENT TO INVESTIGATE SUBPIXEL MAPPING AS AN ALTERNATIVE METHOD TO OBTAIN LAND USE/ LAND COVER MAPS

Yong Ge, Yu Jiang, Yuehong Chen, Alfred Stein, Dong Jiang and Yuanxin Jia

Keywords: Subpixel mapping, downscaling, land use/land cover, experimental design

INTRODUCTION

Various aspects of land use and land cover (LULC) are important in geo-information, environmental, and socioeconomic applications [1]. Remote sensing is a cost-effective and efficient means to obtain data for LULC monitoring over large land surfaces at a variety of spatial and temporal scales [2]. The presence of mixed pixels in remote sensing images, however, can cause difficulties in the extraction of accurate LULC information because their spectral characteristics reflect the composite signature of different LULC classes [3]. Soft classification (or spectral unmixing) methods have been proposed to address this problem by estimating the memberships of LULC classes within a pixel [4]. Unfortunately, such techniques might be unable to provide any indication of the spatial distribution of such LULC classes within each mixed pixel [5]. Alternatively, subpixel mapping (SPM) may be used as a post-processing step of soft classification to reduce the uncertainty in the spatial distribution of the subpixels within each mixed pixel. SPM, initially proposed by Atkinson [5], decomposes each mixed pixel into a fixed number of subpixels based on a zoom factor and it then assigns these subpixels to specific LULC classes [6,7]. It can be considered a classifier that transforms fraction images (*i.e.*, the output of soft classifications) into a hard classification map at the subpixel scales [8]. In contrast to common classifiers, SPM can produce a LULC map with finer spatial resolution than the original moderate- or low-spatial-resolution input image [8]. Thus, SPM offers a solution to the tradeoff between the spatial resolution of a sensor and its spectrum. It can process mixed pixels in low-, moderate-, and high-spatial-resolution imagery and thus save on the cost of obtaining images with higher spatial resolution [9,10]. In particular, in the analysis of LULC time series data, e.g., for change detection, SPM can be applied to low- or moderate-resolution images acquired in an earlier phase to produce LULC maps with spatial resolution consistent with maps produced from high-spatial-resolution images acquired during the current phase. It is therefore attractive to use SPM to derive finer-resolution LULC maps from coarser-resolution remote sensing images.

Since 1997, a number of techniques related to SPM have been developed. These include Hopfield neural networks [9,11], back-propagation neural networks [12], pixel swapping [13,14], spatial attraction models [15,16], indicator co-kriging [17,18], Markov random fields [19,20,21], advanced artificial intelligence-based algorithms [22,23], interpolation-based methods [24,25,26], and geometric methods [27,28]. To date, more than 200 articles on SPM methods have been published, a few of which have focused on applications for special classes, such as urban tree identification [29], urban building extraction [30], and lake area estimation [31]. Generally, however, these articles have focused on the development of SPM methods and performed relatively well. Little if any consideration has been given to whether SPM could feasibly provide alternative LULC

Citation: Yong Ge, Yu Jiang, Yuehong Chen, Alfred Stein, Dong Jiang and Yuanxin Jia, Designing an Experiment to Investigate Subpixel Mapping as an Alternative Method to Obtain Land Use/Land Cover Maps, doi:10.3390/rs8050360.

data in practical applications for large and complex regions where high-resolution LULC maps are lacking. Moreover, no research has examined how landscape heterogeneity [32], the different spatial distribution patterns (*i.e.*, areal, linear, or point patterns) of geographical objects [8,33], and zoom factors affect the performance of SPM in practical applications over large areas.

The objective of this study was to propose an experimental design to investigate the feasibility of using SPM to obtain LULC data in the absence of high-spatial-resolution LULC maps. Meanwhile, a combined index (CI) of landscape shape index (LSI) and areal pattern proportion (APP) was developed to explore the relationship between land surface complexity and the performance of SPM results. In the experimental design, the impact factors, such as zoom factor and land surface complexity, of SPM were explored to provide indications regarding practical applications. Different to most previous SPM studies that have concentrated on methodological development, this study was intended to provide practical guidelines for users to obtain LULC data using SPM. To assess the effectiveness of the experimental design, a case study of the Jingjinji region of China was implemented.

EXPERIMENTAL DESIGN

The experimental design is presented in Figure 1. First, SPM is implemented to obtain finer-resolution LULC maps. The specific steps include the extraction of fraction images and reference data from remote sensing images or existing datasets, setting more than three zoom factors, and selecting at least two representative SPM algorithms to generate SPM maps. Second, the performance of the SPM is evaluated by analysis of the results, including accuracy assessment indices, *i.e.*, overall accuracy (OA), Kappa coefficient (KP), producer's accuracy (PA), and user's accuracy (UA), spatial distribution patterns of geographical objects and landscape heterogeneity. Third, the impact of the SPM results is analyzed based on the zoom factors, different class levels, and SPM results in different subareas with different LSI, APP and CI.

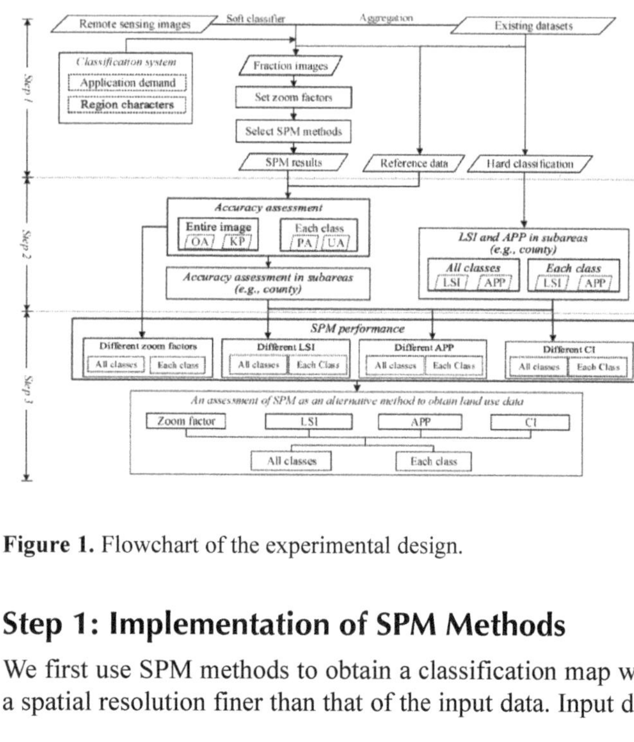

Figure 1. Flowchart of the experimental design.

Step 1: Implementation of SPM Methods

We first use SPM methods to obtain a classification map with a spatial resolution finer than that of the input data. Input data for SPM are the fraction images that express the proportion of each class within individual pixels. Fraction images can be used to determine the number of subpixels for each class within mixed pixels with a given zoom factor. The zoom factor is set to determine how many subpixels are in a pixel and to determine the spatial resolution of the SPM results.

Extracting Fraction Images and Reference Data

We distinguish two ways to obtain fraction images: soft classification from real remote sensing images and generation from existing LULC datasets such as upscaled data from high spatial-resolution classification data. For real remote sensing images, soft classification uses a classifier, such as a Bayesian or a support vector machine soft classifier, to directly generate fraction images with values ranging from 0 to 1 that indicate the proportion or possibility of pixels belonging to each class. Another way to extract fraction images is to aggregate existing classification data to simulate the output of soft classification, such as the Resources and Environmental Scientific Data Center of the Chinese Academy of Sciences [34], National Land Cover Data 1992 [35], and Global Land Cover datasets [36]. Usually, a mean filter (such as 3 × 3 pixels) or a regular polygon are used to calculate the class proportions within the mean filter or the polygon from the existing classification data to produce fraction images [7,13,17,18,33].

Reference data can be obtained in two ways: (1) validation points of ground truth for the SPM of real remote sensing images; and (2) reference maps for the SPM of synthetic fraction images degraded from existing classification data. Note that, for synthetic fraction images, each SPM map should have its corresponding reference map to ensure they have the same spatial resolution and size. The reference map is classified maps, which can be aggregated from existing classification datasets. More details for the generation of the reference maps can be discussed in Section 3.1.

Setting Zoom Factors

The zoom factor Z is a critical parameter that divides each coarse pixel in the fraction images into $Z \times Z$ finer subpixels. With the zoom factor of Z, the number of subpixels for each class within mixed pixels can be calculated by multiplying the class proportion by Z^2. Theoretically, it could be any integer value (>1) but an appropriate zoom factor should meet three requirements: (1) express the surface adequately; (2) retain the accuracy within an acceptable range; and (3) provide SPM output at the desired spatial resolution for practical applications, e.g., change detection requires maps with the same spatial resolution pertaining to different times. Therefore, selection of the optimal zoom factor could be considered the search for the optimal scale of SPM and several zoom factors should be tested and compared.

Selecting the SPM Algorithm

Different methods are suited to different situations because each has its own characteristics. Most SPM methods are based on spatial correlation, and therefore they are good at dealing with areal patterns. Some of these, such as pixel swapping [13] and Hopfield neural networks [9], involve iteration processes; thus, they require longer processing time. Others, such as geostatistics [37], are based on pattern prediction, and are more suited to dealing with point features. These methods use training images to predict subpixel distributions, and therefore auxiliary high-spatial-resolution data are essential for the training sample. Selection of any SPM method should meet the requirements of the application according to the specific situation, e.g., different land surfaces or types of LULC classes. To explore the potential of different SPM methods, two or more representative SPM methods should be used.

Step 2: Assessment of the SPM Output

SPM output is usually assessed using a confusion matrix and its statistical metrics. The reference data for the accuracy assessment are considered the correct classification of the surface; thus, the agreement between the SPM output and the reference data is considered to reflect the effectiveness of the SPM. In SPM, the spatial pattern of LULC is also significant in determining the classification results and to their accuracies. Thus, the SPM output is assessed in terms of accuracy assessments and the values of the landscape pattern index [38].

Accuracy Assessment

The accuracy of the SPM output is evaluated using a confusion matrix [2]. This matrix presents the differences between the SPM output and the reference data, where the diagonal elements show correct classifications and the others represent commission and omission errors for each class [2]. Four accuracy assessment indices (OA, KP, PA, and UA) are extracted from the confusion matrices to provide a quantitative assessment of the SPM results [2]. Both OA and KP measure the degree of agreement between the classification results and the reference data [2]. The difference between OA and KP is that KP excludes the possibility that a pixel might be classified correctly by randomness [2]. PA measures the ratio of pixels within a certain class in the reference map that are classified correctly in the classification map. Similarly, UA measures the ratio of pixels classified as a specific class in the classification map with those that are actually in that class in the reference map. In the experimental design, the accuracy of the SPM results with different zoom factors, at different class levels, and using different SPM methods are assessed and compared. Note, similar to traditional accuracy assessments for classified maps of remote sensing imagery, all subpixels within both pure and mixed pixels of fraction images can be used in the accuracy assessment, because the experimental design aims to investigate the overall performance of SPM for practical applications and because the LSI and the APP are calculated in an area including both pure and mixed pixels.

Landscape Heterogeneity Analysis

The LSI represents the measurement of spatial patterns, which could be the measurement of area/edge, shape, core area, and aggregation. It can be calculated at both the landscape and the class level, reflecting the measurement of the spatial pattern of the classification data over an entire area or for each class, respectively. Of the various indices, the LSI is chosen as a measurement of surface complexity because it has a direct interpretation for landscape heterogeneity [39]. It is defined as the total length of patch edges within a landscape, divided by the total area, and adjusted by a constant for a square standard. It is represented as

$$LSI = E/4\sqrt{A} \tag{1}$$

where E is the edge length and A is the patch area. The value of LSI begins at one and increases without limit as the landscape shape becomes increasingly irregular. Note that the calculation of LSI needs classified maps and each classified map can produce a LSI value by Equation (1).

Areal Pattern Proportion

The APP is an index with which to evaluate the areal spatial pattern, because most existing SPM methods are suited to areal features [16,18,24,25,40]. The APP is equal to the area of features with areal pattern, divided by the total area:

$$APP = \frac{A_{area}}{A_{area} + A_{linear} + A_{point}} \tag{2}$$

where A_{area} is the area of the areal features, A_{linear} is the area of features with linear pattern, and A_{point} is the area of features with point pattern [33]. The sum of A_{area}, A_{linear} and A_{point} is the total area. A_{area}, A_{linear} and A_{point} are extracted by a feature pattern recognition method [33]. First, all features are segmented by the seeded region-growing model from each fraction image. Second, the shape-density index is calculated for each extracted feature. Last, features are divided into the three types according to different values of shape-density index. More details can be found in [33].

Combined Index

Both the LSI and the APP affect the performance of the SPM output. A CI of LSI and APP is developed next to assess the performance of the SPM results:

$$CI = \alpha \times LSI + \beta \times APP \tag{3}$$

where α and β are control parameters that balance the contribution of LSI and APP. To control the range of parameters (−1 to 1) and of CI (0 to 1), the LSI is normalized into the range of 0 to 1.

Representative Spatial Pattern Classes

To investigate the performance of SPM in dealing with features with different spatial patterns, some LULC classes are selected as representatives of areal, linear, and point features. In Geographical Information Sciences research, geographical features are usually generalized and classified into three types, *i.e.*, polygon, polyline, and point, and these three patterns have been considered in many applications [33]. Areal features refer to high-resolution (H-resolution) cases where pixels are smaller than the features of interest, such as cropland and lakes, while point features refer to low-resolution (L-resolution) cases where pixels are larger than the features of interest [8], e.g., buildings. Linear features differ from both areal and point features because pixels within this pattern are both shorter and wider than the features of interest and they typically retain connectivity [33], e.g., roads.

Step 3: Analysis of SPM Performance

SPM Performance with Different Zoom Factors

Based on the accuracy assessment of all SPM output, the change of accuracy with different zoom factors is analyzed. The accuracy might decrease with increasing zoom factor [6,24,40,41], and therefore we are interested in establishing at which zoom factor the accuracy stabilizes.

SPM Performance of Different Classification Levels

Accuracies of SPM output at different classification levels are analyzed and compared. In a hierarchical classification system, different classification levels reflect the surface to differing degrees and they have different class amounts that also influence the SPM output.

SPM Performance of Different SPM Methods

The SPM output produced by different SPM methods is compared, and both similarities and differences are explored. The accuracy range and its change over the entire study area and for each class could also be discussed.

SPM Performance in Subareas with Different LSI Values

The spatial distribution of measurements calculated in subareas could be used to analyze SPM performance with different LSI values. The change of accuracy with LSI value can be illustrated using a scatterplot of measurements in subareas, where the LSI value is plotted as the x-axis and the accuracy assessment, such as OA and PA, is plotted as the y-axis. The accuracy assessment for different LSI ranges and the correlations between LSI and the accuracy assessment are calculated to explore the relationship between accuracy and LSI.

SPM Performance of Different Spatial Distribution Patterns

LULC classes with areal, linear, and point features are selected as representative features. Their accuracy assessments are compared to investigate the performance of SPM in dealing with different types of feature. Furthermore, the relationship between the APP values and accuracy assessments is discussed, as are the scatterplots of measurements in subareas for the APP values and accuracy assessments, and the correlation values between the APP and accuracy assessment.

Analysis of SPM Performance with CI Value

To analyze the relationship of SPM performance with both LSI and APP, the LSI and APP values are used as variables in a regression of accuracy assessments to obtain an estimation equation. Thus, when the LSI and APP values of a subarea are known, the equation could be used to estimate the performance of SPM and consequently, to determine whether the use of SPM is appropriate. Note, when exploring the relationship of the SPM results with the LSI, APP, and CI values, the LSI, APP, and CI indices and the accuracy of SPM results are calculated in county subareas.

CASE STUDY

Study Area and Data

The study area of the Jingjinji region, encompassing Beijing, Tianjin, and Hebei provinces, comprises an area of 220,000 km^2. The Jingjinji region continues to play a considerable role in the regional development of China because it contains the political and cultural center of Beijing and a significant port city (Tianjin). The area mainly comprises cropland, most of which is represented by dry land on the plains. The northern and western parts of the study area are characterized by hilly topography dominated by woodland and grassland. Built-up areas are dispersed widely throughout the plains but are more concentrated along the coast.

The fraction images are degraded from land use data with 30-m spatial resolution (Figure 2a,b), which are derived from Landsat TM imagery by means of human–machine interactive interpretation [42]. The input data for the SPM are synthetic land use fraction images with 1-km spatial resolution from 2005 (Figure 2c,d), provided by the Resources and Environmental Scientific Data Center of the Chinese Academy of Sciences. This land use data mainly focus on classes associated with human activities. In contrast to this, land cover data is more related to the physical and biological cover of the earth's land. To meet the requirements of many land resource management and ecological applications in China, the Chinese Academy of Sciences launched a project for generating land use datasets over China. This project was supported by many Chinese key technology research and development programs. The 1-km land use fraction images can be freely obtained from the Resources and Environmental Scientific Data Center, whereas the original 30-m spatial resolution classification data are not freely available. The steps for the generation of 1-km-resolution fraction images and reference data are as follows. (1) Polygons of the grids (such as the 1-km^2 grid) covering the Jingjinji region are produced; (2) Proportions of the classes within each polygon are determined by dividing the area of each class by the area of the polygon; (3) The estimated class proportions within each polygon are converted into fraction images; (4) Reference maps are obtained by transforming (or hardening) these fraction images with different spatial resolutions into labeled maps [16]. Additional details about the 1-km fraction images and the 30-m land used data can be found at [34].

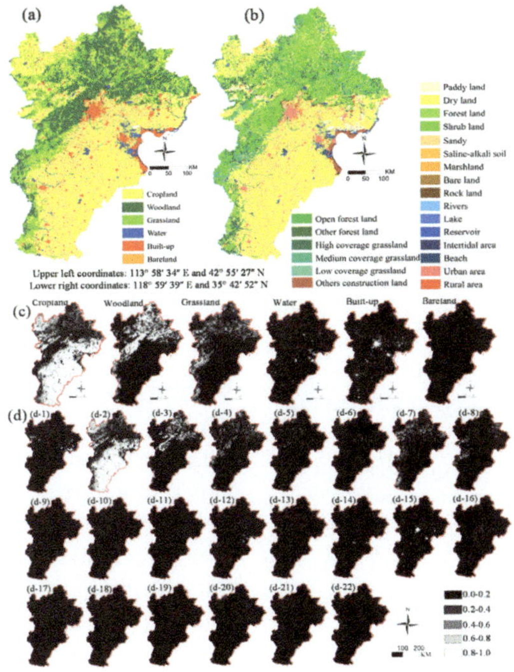

Figure 2. Original land use map with 30-m spatial resolution and fraction imagery with 1-km spatial resolution. (**a**) Original map at Level I; (**b**) original map at Level II; (**c**) fraction imagery at Level I; and (**d**) fraction imagery at Level II ((**d-1**) Paddy land; (**d-2**) Dry land; (**d-3**) Forest land; (**d-4**) Shrub land; (**d-5**) Open Forest land; (**d-6**) Other Forest land; (**d-7**) High-coverage grassland; (**d-8**) Medium-coverage grassland; (**d-9**) Low-coverage grassland; (**d-10**) River; (**d-11**) Lake; (**d-12**) Reservoir; (**d-13**) Intertidal area; (**d-14**) Beach; (**d-15**) Urban area; (**d-16**) Rural area; (**d-17**) Other construction; (**d-18**) Sandy; (**d-19**) Saline-alkali soil; (**d-20**) Marsh land; (**d-21**) Bare land; (**d-22**) Rock land).

In practice, 1-km-spatial-resolution land use datasets are often too coarse to meet the requirements of many geo-information applications [42]. Therefore, according to the experimental design, SPM methods are employed to produce land use maps at a scale finer than 1-km spatial resolution and they are freely provided for practical applications. The downscaling of these 1-km land use fraction images serves as an alternative means for sharing these data. Reference data used in the accuracy assessment are aggregated from the 30-m land use map and they have the same spatial resolution as the SPM output. Errors in the generation of the 30-m land use maps from the remote sensing imagery are not considered here, *i.e.*, we assume the 30-m land use map is error free.

The classification system used for these data consists of several land use categories grouped into a hierarchical nomenclature at two levels with names and codes, as shown in Table 1. The classification system was designed by the Resources and Environmental Scientific Data Center of the Chinese Academy of Sciences [34]. To explore the performance of SPM for different classification systems, the two classification levels are tested. Level I contains six land use categories: cropland, woodland, grassland, water, built-up, and bare land areas, each of which is separated into several subcategories at Level II. The study area incorporates all six land use categories at Level I and 22 of the 25 land use subcategories at Level II.

Table 1. Classification system used in this study at hierarchical levels I and II.

Level I	Level II
Cropland	11. paddy land, 12. dry land;
Woodland	21. forest land, 22. shrub land, 23. open forest land, 24. other forest land;
Grassland	31. high coverage grassland, 32. medium coverage grassland, 33. low coverage grassland;
Water	41. river, 42. lake, 43. reservoir, 44. glaciers and permanent snowfield *, 45. intertidal area, 46. beach;
Built-up	51. urban area, 52. rural area, 53. others construction land;
Bareland	61. sandy, 62. Gobi *, 63. saline-alkali soil, 64. marshland, 65. bare land, 66. rock land, 67. other unused land *.

Experiments and Methods

To apply the experimental design as described in Section 2, the 1-km land use fraction images as the input of SPM and reference data are generated by the four steps described in Section 3.1. The zoom factor of the SPM is set at values from 2 up to 10 to generate the SPM results with spatial resolutions of 500, 333, …, and 100 m. For each zoom factor, there is a reference map that has the same spatial resolution as the corresponding SPM result. For the accuracy assessment of the entire study area, confusion matrices are calculated by comparing the SPM results with the corresponding reference images. The

OA, KP, PA, and UA values are extracted as statistical indices from the confusion matrices. Note that, when exploring the relationships of LSI, APP, and CI to the accuracies of the SPM results, the LSI, APP and, CI are calculated in units of county from the 250 m reference map at Level I while OA and PA are also calculated in units of county by comparing the 250 m reference map with 250 m SPM maps at Level I.

To explore the potential of different SPM methods, two representative SPM methods based on spatial dependence are chosen: spatial-attraction-model-based subpixel mapping (SAMSPM) and vectorial-boundary-based subpixel mapping (VBSPM). SAMSPM was proposed originally by Mertens *et al.* (2006) and, subsequently, it has been applied successfully in several tests [6,16,22,23]. It is considered an efficient method for predicting the spatial distribution of LULC classes at subpixel scale. VBSPM was proposed by Ge *et al.* (2014) to improve the geometric SPM method [28]. Similar to SAMSPM, VBSPM is based on spatial dependence. These two SPM methods were chosen because neither involves iteration and they are both more efficient than other methods, e.g., pixel-swapping [13,14] for the generation of SPM maps with 22 classes over a large and complex area.

SAMSPM

SAMSPM uses the attraction between a subpixel and its neighboring pixels to represent the measurement of spatial dependence in estimating the spatial distribution of subpixels within a mixed pixel [15]. The attraction between a subpixel and its neighboring pixels is calculated using fraction values in the neighborhood as

$$p_{ab}(c) = Avg\left\{\frac{P_{ij}(c)}{d(p_{ab}, P_{ij})}|P_{ij} \in N_t\left[p_{ab}\right]\right\} \tag{4}$$

where $p_{ab}(c)$ is the raw attraction of subpixel p_{ab} ($a, b = 1, 2, \ldots, S$; S is the zoom factor) for class c ($c = 1, 2, \ldots, C$; C is the class number); $P_{ij}(c)$ is the fraction value of pixel P_{ij} ($i = 1, 2, \ldots, M, j = 1, 2, \ldots, N$; M and N are the width and length of the image, respectively) for class c in neighborhood $N_t[p_{ab}]$ of subpixel p_{ab}; $d(p_{ab},P_{ij})$ is the distance between pixel P_{ij} and subpixel p_{ab}. $Avg\{\}$ is the operator that gets the mean of its arguments. Further details about SAMSPM can be found in [15].

VBSPM

VBSPM first geometrically partitions a mixed pixel and then estimates the locations and lengths of the segments for each LULC class polygon within the mixed pixel using a vectorial boundary extraction model, as shown in Equation (5). A ray-crossing algorithm is then used to determine the LULC class of each subpixel within each vectorial boundary [27]:

$$V_{ic} = V_{ic}^{(0)} + \begin{cases} (1 - P_{ic})\frac{P_{(i+1)c}}{P_{(i+1)c}+P_{(i-1)c}}, & P_{(i+1)c} \neq P_{(i-1)c} \\ (1 - P_{ic})\frac{1}{2}, & P_{(i+1)c} = P_{(i-1)c} \end{cases} \tag{5}$$

where $V_{ic}^{(0)}$ and V_{ic} are the start and final locations of segment i ($i = 0, 1, \ldots, 7$) corresponding to pixel P_i ($i = 0, 1, \ldots, 7$) of class c ($c = 1, 2, \ldots, C$); P_{ic}, $P_{(i+1)c}$, and $P_{(i-1)c}$ indicate the proportions of class c in pixel P_i and to the left and right of neighboring pixel P_i, respectively. Further details can be found in [27].

Results and Analysis

SPM Output and Accuracies

Using the 1-km fraction images in Figure 2c,d as the inputs of SPM methods, representative output of SAMSPM and VBSPM with spatial resolutions of 500 m (*i.e.*, zoom factor = 2) and 100 m (*i.e.*, zoom factor = 10) at Levels I and II are shown in Figure 3. A similar spatial pattern is observed from the study area and more detail is obviously resolved with increasing spatial resolution. It indicates that SPM can produce maps at different spatial resolutions from the 1-km fraction images and that these maps are visually similar to the original classification map.

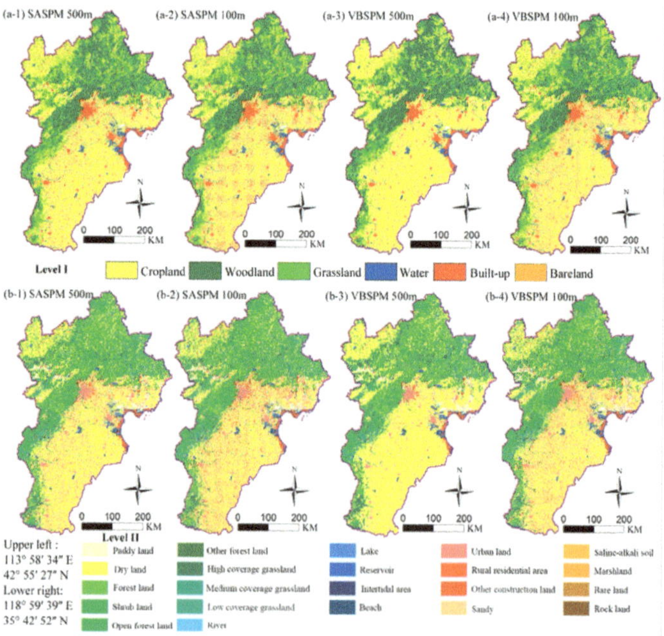

Figure 3. SPM output for spatial resolutions of 500 and 100 m: (**a-1**) and (**a-2**) SAMSPM output at Level I; (**a-3**) and (**a-4**) VBSPM output at Level I; (**b-1**) and (**b-2**) SAMSPM output at Level II; and (**b-3**) and (**b-4**) VBSPM output at Level II.

Entire Area

The OAs, KPs, PAs, and UAs of the SPM output obtained for Levels I and II are presented in Figure 4. It shows that OA decreases from 85% to 77% at Level I and from 83% to 75% at Level II, as the spatial resolution changes from 500 to 100 m. Similarly, KP decreases from 0.761 to 0.661 and from 0.751 to 0.656 at Levels I and II, respectively. Note, for a zoom factor greater than five, the accuracy assessments remain stable with a value of OA approximately equal to 78% at Level I and to 75% at Level II; generally, the OA is 2%–3% larger at Level I than at Level II.

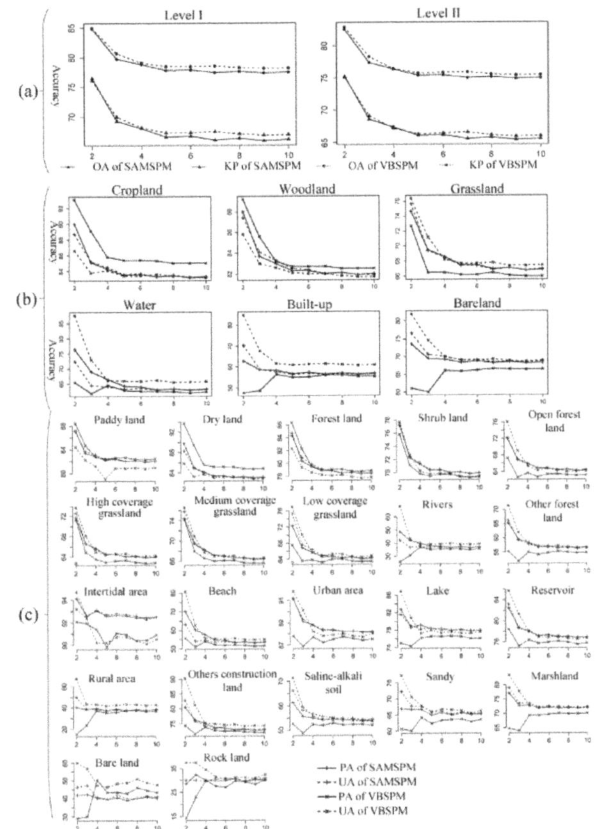

Figure 4. OA, KP, PA, and UA of SAMSPM and VBSPM at Levels I and II. (**a**) OA and KP values of SAMSPM and VBSPM at Level I and Level II; (**b**) PA and UA values of SAMSPM and VBSPM at Level I; (**c**) PA and UA of SAMSPM and VBSPM values at Level II.

Each Class

At Level I, the PAs of the classes in SAMSPM have the same rank as in VBSPM for all zoom factors. Cropland and woodland are highest, followed by grassland and water, and bare land and built-up areas are the lowest, as shown in Figure 4. Cropland, which is distributed mainly in plain areas, is the dominant land use class. It has the highest accuracy among the six classes at Level I with a range of 83%–93% because of the large proportion of areal features. Woodland is distributed mainly in the mountains and foothills. Its accuracy is the second highest among the six classes with a range of 82%–89%, which is also attributable to the relatively large proportion of areal features. Grassland is distributed mainly in the mountains, plateaus, and foothills. Its accuracy follows cropland and woodland with a range of 66%–76% because it has a relatively complex distribution. Water is distributed mainly in the plain and valley areas. It has relatively low accuracy with a range of 62%–88% because of the large proportion of linear and point features. The built-up area is distributed mainly in the plain areas. It has relatively low accuracy because of the large proportion of point features. Bare land has accuracy of up to 82% because of the various pattern features. Note, the influence of bare land on OA is limited because it accounts for only a very small proportion of the entire area.

At Level II, the PA and UA values vary much more than at Level I, as shown in Figure 4. Paddy land and dry land, subdivided from cropland at Level I, have accuracies ranging from 79% to 94%. The accuracies of forest, shrub land, and open forest land, subdivided from woodland at Level I, all range from 63% to 86% because they have similar areal features. The high, medium, and low coverages of grassland, subdivided from grassland at Level I, have similar patterns and accuracies ranging from 63% to 77%. Rivers, lakes, reservoirs, intertidal areas, and beaches, subdivided from water at Level I, have various accuracies. Intertidal areas have the highest accuracy because of the areal features distributed along the shoreline. Lakes and reservoirs also have relatively high accuracy ranging from 74% to 87% because of their areal patterns, whereas rivers have lower accuracy (<68%) because of their linear patterns. Urban areas, rural areas, and other construction land, subdivided from built-up areas in Level I, have varying accuracies. Urban areas have high accuracy ranging from 87% to 92% because most urban areas have a large patch size. Rural areas have relatively low accuracy (<70%) because the class patches are small, even though some are smaller or similar to the fraction pixel size (1 km^2). Other construction land has relatively high accuracy ranging from 72% to 90%. Classes subdivided from bare land at Level I also have various accuracies. For example, marshland and saline-alkali soil have relatively high accuracies of 71% and 65%, respectively, whereas rock land with linear pattern and bare and sandy land with complex patterns have relatively low accuracies at Level II of about 30%, 40%, and 53%, respectively.

Relationship between LSI and Accuracies within Subareas

Figure 5 shows the LSI values calculated in units of county from the reference map (Level I) at a spatial resolution of 250-m. Generally, the mountainous areas in the north and west have relatively high LSI values, whereas the plains in central and southeastern areas have relatively low LSI values. At the class level, the mountainous areas in the north and east have relatively low LSI values for cropland, woodland, and grassland. Woodland in central parts and water both have LSI values with only small differences.

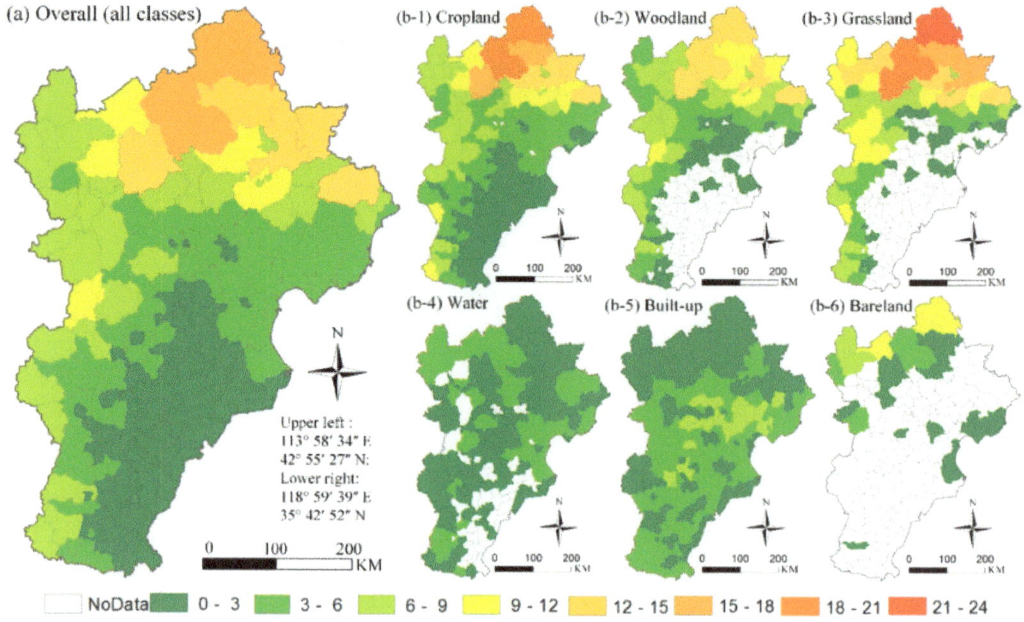

Figure 5. LSI values calculated for each county: (**a**) LSI for all classes; (**b-1**)–(**b-6**) LSI for each class.

To explore the relationship between LSI and the accuracy measurements in units of county, the OA and PA values (y-axis) are plotted against LSI (x-axis) in Figure 6. Apparently, the OA values show a declining trend as LSI grows. The same trend is found for the classes of cropland and built-up areas. Water has relatively low LSI values but its PA values are widely separated. The plotted dots for built-up areas can be separated into two groups: one group has the same trend as the OA (*i.e.*, the PA values decline as LSI grows), and the other group shows the PA values tending to increase with the increase of LSI. This may be due to the presence of two types of spatial pattern features: point features with LSI values of near one and low SPM accuracy, and areal features with high SPM accuracy. Figure 6 shows that OA has a relatively strong negative correlation of −0.75 with LSI, whereas PA has a low correlation with LSI in grassland, woodland and built-up areas. Less complex areas, with low LSI values, usually have high accuracy because of the stronger spatial correlation between pixels/subpixels and lower uncertainty in class allocation. Some classes, however, might contain many regular point features with low LSI values and low class accuracy. The reason for this may be that spatial correlation only works if the patch size is smaller than the pixel size. When predicting subpixel classes using neighboring pixel information, SPM may not perform well. This reveals that areas with low LSI values and classes with areal features usually have high accuracy.

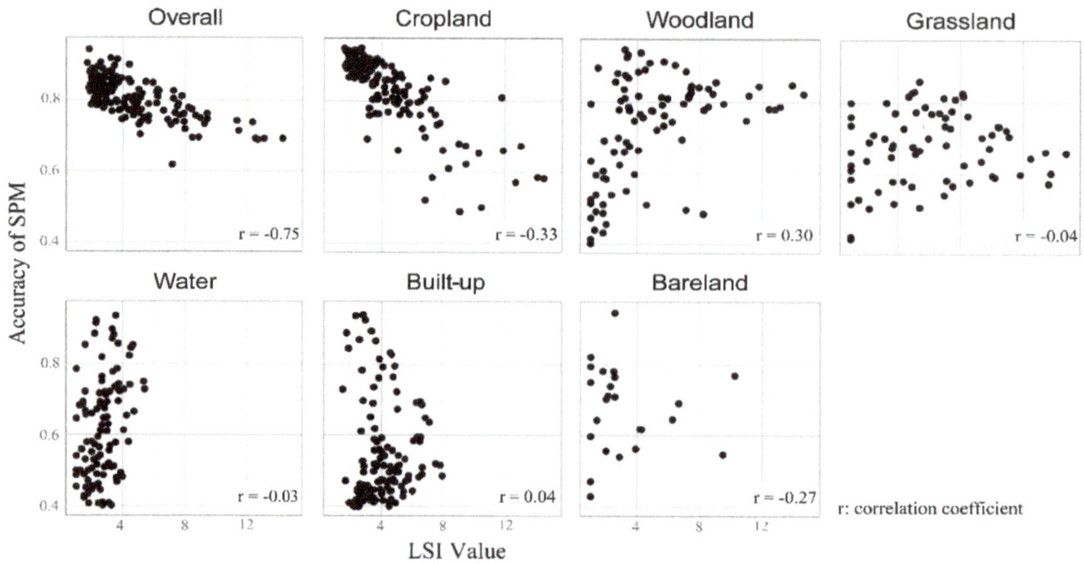

Figure 6. Correlation between LSI value (x-axis) and accuracy assessment (y-axis).

The average OA and PA values in units of county within the different LSI ranges are displayed in Table 2. It shows that units of county with an LSI range of 1–4 have an average OA of 85%, units of county with an LSI range of 4–7 have an average OA of 79%, units of county with an LSI range of 7–10 have an average OA of 75%, and units of county with an LSI value >10 have an average OA of 70%–71%. The PA values for cropland tend to decline with the increase of LSI. The PA values for grassland and water are relatively low for small LSI values but they remain relatively high for median LSI values. The average PA values for woodland are relatively low (61.48%) when the value of LSI is within the range 1–4, whereas they remain relatively high when the value of LSI is within the range 4–13. The PA values for built-up areas tend to increase with the increase of LSI. Furthermore, some average PA values show large jumps that could be caused by extreme values.

Table 2. Average OA and PA in units of county with different LSI ranges.

LSI Range	1–4	4–7	7–10	10–13	13–16	16–19	19–22
Overall	84.75%	79.09%	75.15%	71.52%	71.17%	71.58%	-
Cropland	88.52%	80.45%	70.56%	63.88%	62.82%	64.34%	64.52%
Woodland	61.48%	79.00%	78.42%	79.43%	81.86%	-	-
Grassland	56.56%	71.37%	70.41%	67.07%	61.78%	59.30%	61.63%
Water	55.07%	70.61%	-	-	-	-	-
Built-up	50.12%	52.98%	53.97%	-	-	-	-
Bareland	68.33%	64.33%	54.70%	76.94%	-	-	-

Three Spatial Distribution Patterns and APP Values within Subareas

A representative subarea containing all three spatial patterns is chosen to explore the SPM performance for different spatial patterns and the results and accuracy assessments of the subarea are shown in Figure 7. In this subarea, woodland is a typical areal feature with the highest accuracy. Cropland, distributed in a linear pattern along the valley, has relatively lower accuracy than woodland. The built-up area, representative of a point pattern class, has the lowest accuracy.

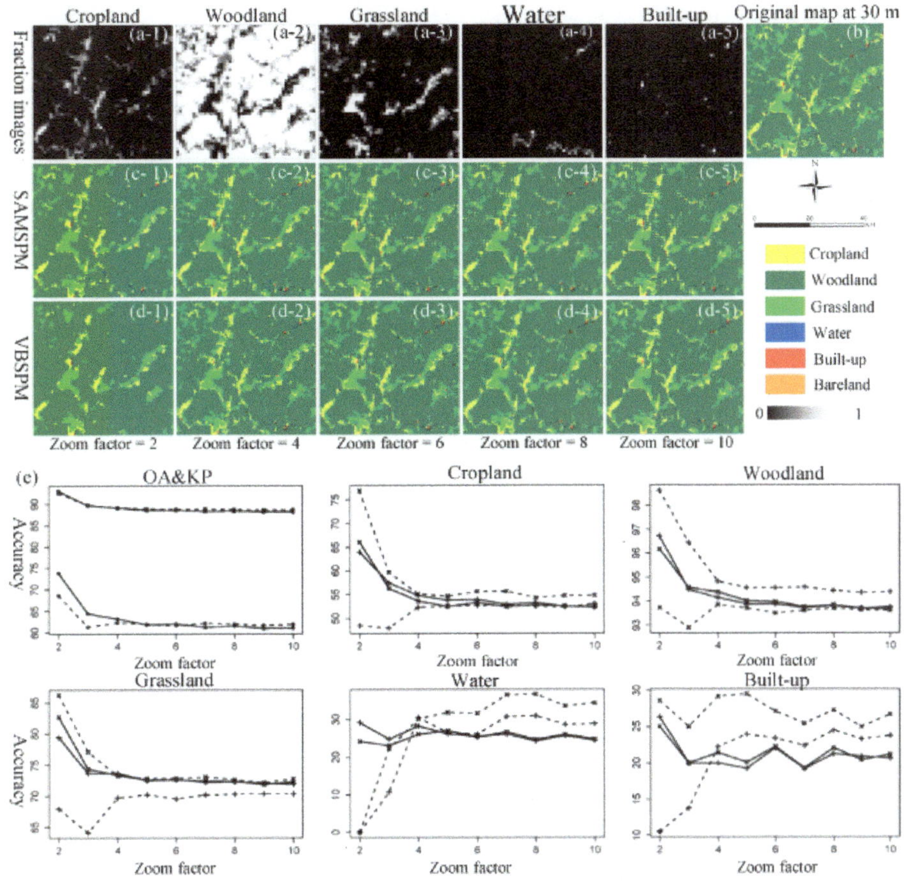

Figure 7. Fraction images, reference map, SPM results, and their accuracy assessments of a subarea containing features with areal, linear, and point patterns. (**a-1**)–(**a-5**) fraction images with spatial resolution of 1 km; (**b**) reference map with spatial resolution of 30 m; (**c-1**)–(**c-5**) upscaled reference map with zoom factor of 2, 4, 6, 8, and 10; (**d-1**)–(**d-5**) VBSPM results with zoom factor of 2, 4, 6, 8 and 10; and (**e**) OAs and PAs of subarea containing areal, linear, and point patterns.

Classes with areal patterns where class patches are much larger than the pixel size always have high accuracy, e.g., woodland at Level I with PA and UA values >94% (Figure 7). Classes with linear patterns, such as cropland (Figure 7) with PA and UA values ranging from 48% to 77%, have accuracies between those of areal and point features. The accuracy of classes with point patterns such as built-up areas (Figure 5), which are the so-called L-resolution cases [8], is the lowest among the three classes.

Figure 8 shows the APP values calculated in units of county. It can be seen that the spatial pattern of the APP values is similar to that of the LSI values.

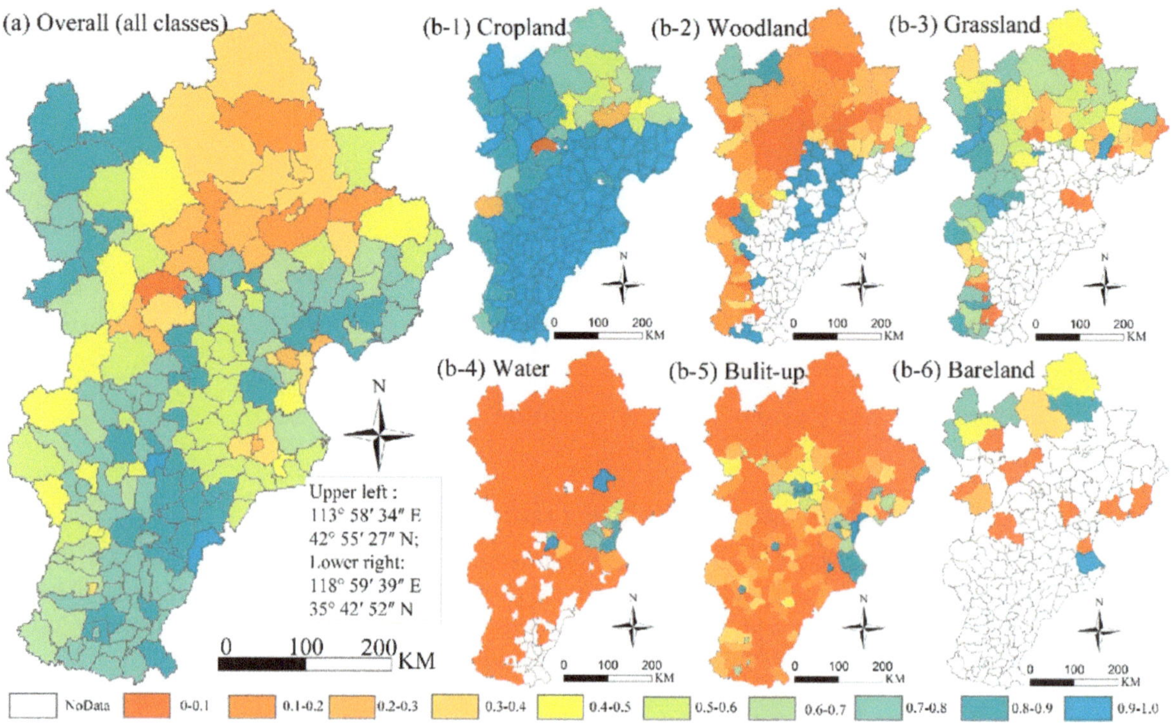

Figure 8. APP values calculated for each county. (**a**) APP for all classes; (**b-1**)–(**b-6**) APP for each class.

The OA and PA values (y-axis) are plotted against APP (x-axis) in Figure 9. For all classes, OA does not show a significant increasing trend with the increase of APP in Figure 9, but there is a slight increase from 78.52% to 84.43% in Table 3. However, the trend of increase of PA with the increase of APP is apparent in all of the classes, as shown in Figure 9. The reason for this increase in PA with APP value may be that higher APP values mean a greater proportion of areal features that usually have higher accuracy in SPM. Figure 9 shows PA has relatively strong correlation with APP, while the relationship between OA and APP is less obvious (correlation coefficient: 0.21). Less complex areas with low APP values usually have high accuracy because of their stronger spatial correlation between pixels/subpixels and lower uncertainty in class allocation. Some classes, however, may contain many regular point features with low APP values and low accuracy. The reason for this is spatial correlation works well for features where the patch size is smaller than the pixel size. This indicates that areas with low APP values and classes with areal features usually have high accuracy.

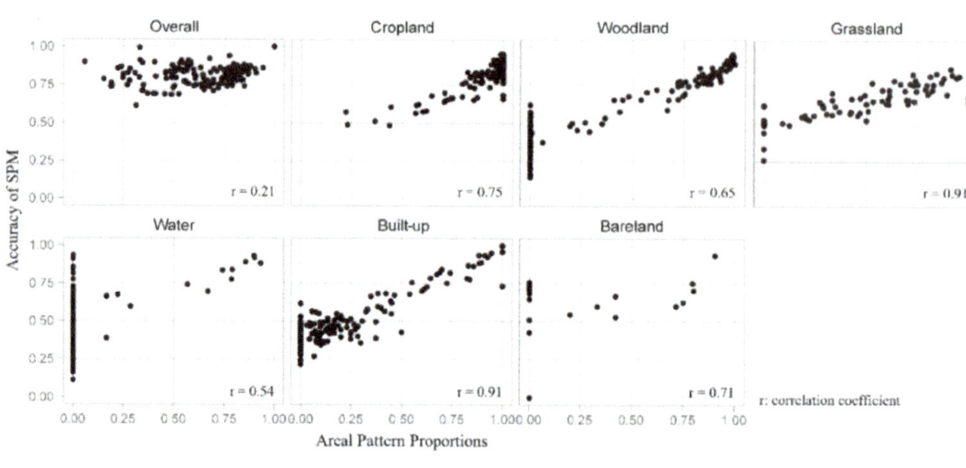

Figure 9. Correlation between APP value (x-axis) and accuracy assessment (y-axis).

Table 3. Average OA and PA in units of county with different APP ranges.

APP Range	0–0.2	0.2–0.4	0.4–0.6	0.6–0.8	0.8–1.0
Overall	78.52%	79.62%	82.00%	80.75%	84.43%
Cropland	-	52.55%	57.96%	64.70%	83.25%
Woodland	43.03%	48.91%	65.35%	71.35%	83.32%
Grassland	51.43%	59.88%	63.22%	72.01%	78.43%
Water	52.57%	62.50%	69.26%	75.45%	84.70%
Built-up	44.63%	50.13%	62.31%	78.54%	88.24%
Bareland	-	60.19%	60.19%	66.21%	93.65%

The average OA and PA values of subareas within different APP ranges are shown in Table 3. It can be seen that subareas with an APP range of 0–0.2 have an average OA of 79%, subareas with an APP range of 0.2–0.4 have an average OA of 80%, subareas with an APP range of 0.4–0.6 have an average OA of 82%, subareas with an APP range of 0.6–0.8 have an average OA of 81%, and subareas with an APP value >0.8 have an average OA of 84%. The PA values for cropland tend to increase with the increase of APP. For example, cropland with an APP value of <0.8 has an average PA value of <65%, whereas cropland with an APP value of >0.8 has an average PA value of about 83%. The PA values for woodland, grassland, water, bare land, and built-up areas are similar to cropland, showing an increase of the average PA value with increasing APP. Generally, the greater the proportion of the areal feature, as indicated by higher APP values, the higher the accuracy of SPM.

Impact of CI (Combined Index of LSI and APP) on SPM Results

To evaluate the impact of the CI on the SPM results, linear regression is used to determine the most suitable parameters for each CI of all classes and each class. The CI regression equations and plots of CI (x-axis) against OA and PA (y-axes) are shown in Figure 10.

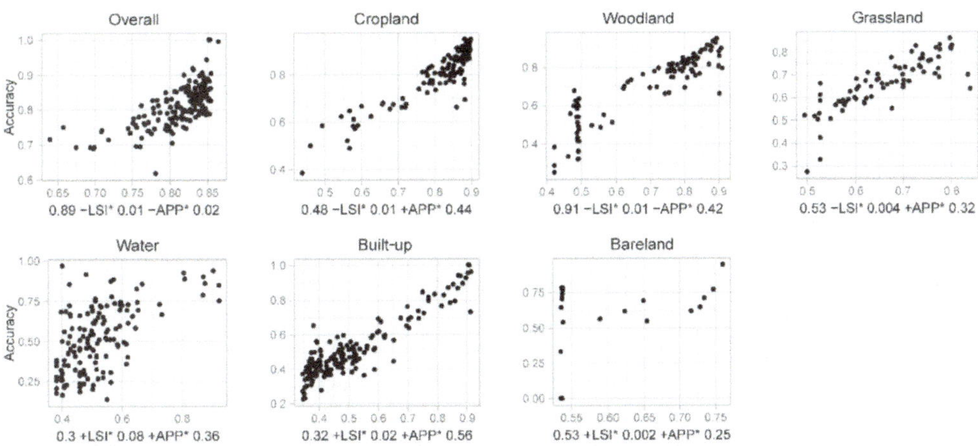

Figure 10. Correlation between CI value (x-axis) and accuracy (y-axis) at landscape level.

For all classes, LSI mainly affects the OA values, while the relationship between APP and OA is not significant. For each class, both LSI and APP are involved in the CI formulations of the six classes and the CI values have relatively good agreement with PA. It is noteworthy that some classes have low correlations between PA and either LSI or APP, while they have relatively high correlations with CI. For example, grassland and woodland do not show a declining trend in OA with increasing LSI values (see Figure 6), whereas they do show good linear relationships between PA and CI (see Figure 10). Thus, although LSI and APP may not effectively characterize the relationship with PA for some classes, CI is an effective solution for measuring the relationship between landscape heterogeneity and areal feature proportions with PA. Therefore, CI represents an alternative index of the relationship between PA and land surface complexity.

DISCUSSION

SPM Performance with Change of Zoom Factor

This study shows that OA and KP generally decrease as the spatial resolution of the SPM results becomes finer with the increase of the zoom factor. The reason for this can be attributed to the increase in the number of subpixels within a coarse pixel as the zoom factor increases; thus, greater uncertainty is generated regarding class allocation of the subpixels [25,40]. The reason for the stability of the accuracy assessment when the zoom factor is above five is that the proportion of mixed pixels within the 1-km fraction images becomes stable at this zoom factor. This is because the criterion defining mixed or pure pixels depends upon the zoom factor. As an illustration, it can be seen that the spatial resolution changes from 500 to 200 m when the zoom factor increases from 2 to 5, whereas it changes from 200 to 100 m when the zoom factor increases from 5 to 10.

SPM Performance at the Two Classification Levels

Comparing the accuracy assessments of the SPM results at the two classification levels reveals that classes with areal patterns at Level II, subdivided from Level I, have similar PA and UA values as those at Level I because of the similarities in their distribution patterns. For example, the PA of dry land (83%–94%), subdivided from cropland, is close to cropland (83%–93%). Classes with linear or point patterns, however, are different from classes with areal patterns. For example, the classes of intertidal areas and rivers at Level II, subdivided from water, have different accuracies. Intertidal areas have high accuracy (>90%) and rivers have low accuracy (<68%), whereas the accuracy of water is 62%–88%. Another example is built-up areas (58%–64%) at Level I, which has the subclass of rural areas (<40%) with a point pattern.

Performances of SAMSPM and VBSPM

Both SAMSPM and VBSPM perform well in the experiments and produce equivalent accuracies in different areas, but some differences remain in the detail (Figure 10). VBSPM produces slightly better results with greater OA (range: 0.3%–1.1%) than SAMSPM. VBSPM, however, generates results for built-up areas at Level I and rivers, rural areas, and rock land at Level II that are slightly less accurate (<1%) than SAMSPM with a low zoom factor (<4). This is because these classes do not represent H-resolution cases, for which VBSPM is most applicable [27]. The different performances between SAMSPM and VBSPM in these classes, however, have only minor influence on the OA.

SPM Performance in Subareas with Different Landscape Heterogeneity

A scatterplot is used to investigate the relationship between accuracy and LSI values. It shows that the OA and PA both decline with increasing surface complexity, which is measured by LSI. For the entire image, the less complex the surface, the better SPM performs. The complexity of the surface for the entire area and for each class can be measured separately by LSI at both the landscape (all classes) and the class level. For example, the experiment shows that subareas with LSI values in the range 1–4 have OA values of about 85%, subareas with LSI values in the range 4–10 have OA values of about 77%, and subareas with LSI values >10 have average OA values of about 70%. However, at the class level, PA is not always consistent with the LSI value. Usually, classes with H-resolution features (e.g., cropland and woodland) have relatively high PA values. Some classes with point features (e.g., water and built-up areas) have relatively low LSI and low PA values, because SPM does not perform well when dealing with point features. Areal features with lower LSI values have higher PA values than point features with low LSI values.

SPM Performance of Point, Linear, and Areal Pattern Classes

Among the three spatial patterns, classes with areal pattern have the highest accuracy. One explanation for this is that areal features have strong spatial correlation between pixels and subpixels; thus, those SPM methods based on the assumption of spatial dependence are well able to deal with such correlation, especially VBSPM. A second explanation is that the object size of areal features is generally much larger than the pixel size; thus, it contains a relatively large proportion of pure pixels that can lead to relatively high accuracy in the downscaled classification. Because the SPM methods used in this study are based on the assumption of spatial correlation, land use classes with linear patterns, such as rivers, often do not remain contiguous in the SPM results; thus, their accuracy is not high, especially for linear features with small widths. However, classes with point patterns have the lowest accuracy assessment among the three spatial patterns. This is because greater uncertainty in predicting the spatial locations of point features is generated because of the fewer constraints or lack of complementary information. Atkinson [8] stated that the goal of SPM for point features is to predict patterns rather than to generate an accurate prediction on a subpixel-by-subpixel basis. Thus, some SPM algorithms based on pattern prediction may be more suited to mapping land use classes with point patterns [33].

Analysis of SPM Performance with CI Value

The CI is a linear combination of the LSI and APP, which are the two indices related to SPM performance. The values of the regression parameters reveal that the contributions of the two indices may vary for different areas and different classes. For all classes, CI is similar to LSI, which indicates that the less complex the area, the less impact APP has and the better SPM may perform. At the class level, CI has different combinations of the LSI and APP. The SPM performance is better for smaller values of LSI and larger values of APP. Generally, CI could be used as an indicator to predict the performance of SPM prior to its use in practical applications.

Uncertainty of SPM Caused by Fraction Images of Soft Classification

The experiment uses the error free fraction images to avoid the impact of the uncertainty in soft classification on the SPM results. Actually, uncertainty or errors in soft classification are inevitable when classifying real remote sensing images [27]. As a result, uncertainty or errors from the soft classification of real remote sensing images would be propagated into SPM and the accuracy of SPM would be decreased. Ge [4] proposed a solution using the multiple-point simulation to reduce the uncertainty or errors in soft classification and it can be used in future SPM experiments on real remote sensing images.

Potential of SPM as an Alternative Method for Generating LULC Maps

Using the 1-km synthetic fraction images, the SPM methods based on the assumption of spatial dependence are able to produce land use maps at subpixel scales, with OA values of 77%–85% and 75%–83% for the six classes at Level I and 22 classes at Level II, respectively, over the large area of Jingjinji region in China. These accuracies decline with the increase of S, and stabilize at an OA of approximately 78% at Level I and 75% at Level II, when the zoom factor reaches five. The two SPM methods produce the highest accuracy for areal pattern classes (e.g., cropland, woodland, and grassland), medium accuracy for linear pattern classes (e.g., water), and the lowest accuracy for point pattern classes (e.g., bare land and built-up areas). In the subareas (*i.e.*, counties), less complex areas exhibit higher accuracy; *i.e.*, subareas with high LSI values have lower accuracy, and subareas with LSI values of <18 generate SPM results with OAs of >80%.

In summary, according to the SPM output, SPM has the potential to generate alternative LULC maps for regions where high-spatial-resolution maps are unavailable, especially in less complex areas with a large proportion of areal features. Thus, when using SPM to obtain finer-resolution LULC maps from real coarse remote sensing images or existing fraction images in future applications, the experimental design can be used. A primary assessment of the LSI and APP can be done to provide an indication of the most suitable method and to determine the appropriate zoom factor and classification system to use.

CONCLUSIONS

The objective of this study was to investigate the feasibility of using subpixel mapping (SPM) to obtain land use/land cover (LULC) data in the absence of high-spatial-resolution LULC maps. An experimental design was proposed to evaluate its feasibility for providing alternative LULC maps based on accuracy assessments and landscape measurements. In accordance with the experimental design, a case study was implemented in the Jingjinji region of China using 1-km land use fraction imagery as input data, the landscape shape index (LSI) as the measurement of surface complexity, and areal pattern proportion as the measurement of the spatial distribution pattern of geographical features. The results and analysis showed that overall accuracy (OA) was approximately 80% and that the accuracy declined with increasing zoom factor before stabilizing at a zoom factor of five. The accuracy of SPM was apparently associated with both the complexity and spatial pattern of the geographical features. Higher accuracy was obtained in less complex areas and when handling areal features, compared with handling linear and point features. Subareas with LSI values <4 showed OA values of approximately 85%, subareas with LSI values of 4–10 showed OA values of approximately 77%, and subareas with LSI values >10 had OA values of approximately 70%.

This experiment used an existing proportional land use dataset. In the future, soft classification results derived from remote sensing images should also be practicable. Moreover, other newly developed SPM methods, such as pattern-prediction-based SPM, might be worth consideration. Therefore, it would be interesting in the future to repeat this experiment using other SPM algorithms and actual remotely sensed images.

ACKNOWLEDGMENTS

This work was supported in part by the National Natural Science Foundation of China (Grant No. 41471296) and in part by the Key Technologies Research and Development Program of China (2012BAH33B01).

AUTHOR CONTRIBUTIONS

Yong Ge and Yuehong Chen conceived and designed the experiments, as well as analyzed the experimental results and wrote the paper; Yu Jiang performed the experiments and analyzed the experimental results; Alfred Stein contributed to edit the manuscript; Dong Jiang provided the experimental datasets; Yuanxin Jia contributed to data processing.

CONFLICTS OF INTEREST

The authors declare no conflict of interest.

REFERENCES

1. Lu, D.; Weng, Q. A survey of image classification methods and techniques for improving classification performance.*Int. J. Remote Sens.* **2007**, *28*, 823–870.
2. Foody, G.M. Status of land cover classification accuracy assessment. *Remote Sens. Environ.* **2002**, *80*, 185–201.
3. Fisher, P. The pixel: A snare and a delusion. *Int. J. Remote Sens.* **1997**, *18*, 679–685.
4. Ge, Y. Sub-pixel land-cover mapping with improved fraction images upon multiple-point simulation. *Int. J. Appl. Earth Obs. Geoinform.* **2013**, *22*, 115–126.
5. Atkinson, P.M. *Mapping Sub-Pixel Boundaries from Remotely Sensed Images, Innovations in GIS 4*; Kemp, Z., Ed.; Taylor and Francis: London, UK, 1997; pp. 166–180.
6. Ling, F.; Li, X.; Du, Y.; Xiao, F. Super-resolution land cover mapping with spatial-temporal dependence by integrating a former fine resolution map. *IEEE J. Sel. Top. Appl. Earth Obs. Remote Sens.* **2014**, *7*, 1816–1825.
7. Chen, Y.; Ge, Y.; Jia, Y. Integrating object boundary in super-resolution land cover mapping. *IEEE J. Sel. Top. Appl. Earth Obs. Remote Sens.* **2016**.
8. Atkinson, P.M. Issues of uncertainty in super-resolution mapping and their implications for the design of an inter-comparison study. *Int. J. Remote Sens.* **2009**, *30*, 5293–5308.
9. Tatem, A.J.; Lewis, H.G.; Atkinson, P.M.; Nixon, M.S. Multiple-class land-cover mapping at the sub-pixel scale using a hopfield neural network. *Int. J. Appl. Earth Obs. Geoinform.* **2001**, *3*, 184–190.
10. Wang, Q.; Shi, W.; Wang, L. Allocating classes for soft-then-hard subpixel mapping algorithms in units of class. *IEEE Trans. Geosci. Remote Sens.* **2014**, *52*, 2940–2959.
11. Li, X.; Du, Y.; Ling, F.; Feng, Q.; Fu, B. Superresolution mapping of remotely sensed image based on hopfield neural network with anisotropic spatial dependence model. *IEEE Geosci. Remote Sens. Lett.* **2014**, *11*, 1265–1269.
12. Wang, L.; Zhang, Y.; Li, J. Bp neural network based subpixel mapping method. In *Intelligent Computing in Signal Processing and Pattern Recognition*; Huang, D.-S., Li, K., Irwin, G.W., Eds.; Springer: Berlin, Germany, 2006; pp. 755–760.
13. Atkinson, P.M. Sub-pixel target mapping from soft-classified, remotely sensed imagery. *Photogramm. Eng. Remote Sens.***2005**, *71*, 839–846.
14. Su, Y.-F.; Foody, G.M.; Muad, A.M.; Cheng, K.-S. Combining pixel swapping and contouring methods to

enhance super-resolution mapping. *IEEE J. Sel. Top. Appl. Earth Obs. Remote Sens.* **2012**, *5*, 1428–1437.

15. Mertens, K.C.; de Baets, B.; Verbeke, L.P.C.; de Wulf, R.R. A sub-pixel mapping algorithm based on sub-pixel/pixel spatial attraction models. *Int. J. Remote Sens.* **2006**, *27*, 3293–3310.
16. Chen, Y.; Ge, Y.; Wang, Q.; Jiang, Y. A subpixel mapping algorithm combining pixel-level and subpixel-level spatial dependences with binary integer programming. *Remote Sens. Lett.* **2014**, *5*, 902–911.
17. Boucher, A.; Kyriakidis, P.C. Super-resolution land cover mapping with indicator geostatistics. *Remote Sens. Environ.***2006**, *104*, 264–282.
18. Wang, Q.; Shi, W.; Wang, L. Indicator cokriging-based subpixel land cover mapping with shifted images. *IEEE J. Sel. Top. Appl. Earth Obs. Remote Sens.* **2014**, *7*, 327–339.
19. Kasetkasem, T.; Arora, M.K.; Varshney, P.K. Super-resolution land cover mapping using a markov random field based approach. *Remote Sens. Environ.* **2005**, *96*, 302–314.
20. Li, X.; Du, Y.; Ling, F. Spatially adaptive smoothing parameter selection for markov random field based sub-pixel mapping of remotely sensed images. *Int. J. Remote Sens.* **2012**, *33*, 7886–7901.
21. Wang, L.G.; Wang, Q.M. Subpixel mapping using markov random field with multiple spectral constraints from subpixel shifted remote sensing images. *IEEE Geosci. Remote Sens. Lett.* **2013**, *10*, 598–602.
22. Wang, Q.; Wang, L.; Liu, D. Particle swarm optimization-based sub-pixel mapping for remote-sensing imagery. *Int. J. Remote Sens.* **2012**, *33*, 6480–6496.
23. Zhong, Y.F.; Zhang, L.P. Remote sensing image subpixel mapping based on adaptive differential evolution. *IEEE Trans. Syst. Man Cybern. Part B Cybern.* **2012**, *42*, 1306–1329. [PubMed]
24. Wang, Q.; Shi, W.; Atkinson, P.M. Sub-pixel mapping of remote sensing images based on radial basis function interpolation. *ISPRS J. Photogram. Remote Sens.* **2014**, *92*, 1–15.
25. Chen, Y.; Ge, Y.; Song, D. Super-resolution land-cover mapping based on high-accuracy surface modeling. *IEEE Geosci. Remote Sens. Lett.* **2015**, *12*, 2516–2520.
26. Ling, F.; Du, Y.; Li, X.; Li, W.; Xiao, F.; Zhang, Y. Interpolation-based super-resolution land cover mapping. *Remote Sens. Lett.* **2013**, *4*, 629–638.
27. Ge, Y.; Chen, Y.; Li, S.; Jiang, Y. Vectorial boundary-based sub-pixel mapping method for remote-sensing imagery. *Int. J. Remote Sens.* **2014**, *35*, 1756–1768.
28. Ge, Y.; Li, S.; Lakhan, V.C. Development and testing of a subpixel mapping algorithm. *IEEE Trans. Geosci. Remote Sens.***2009**, *47*, 2155–2164.
29. Ardila, J.P.; Tolpekin, V.A.; Bijker, W.; Stein, A. Markov-random-field-based super-resolution mapping for identification of urban trees in vhr images. *ISPRS J. Photogram. Remote Sens.* **2011**, *66*, 762–775.
30. Ling, F.; Li, X.; Xiao, F.; Fang, S.; Du, Y. Object-based sub-pixel mapping of buildings incorporating the prior shape information from remotely sensed imagery. *Int. J. Appl. Earth Obs. Geoinform.* **2012**, *18*, 283–292.
31. Muad, A.M.; Foody, G.M. Super-resolution mapping of lakes from imagery with a coarse spatial and fine temporal resolution. *Int. J. Appl. Earth Obs. Geoinform.* **2012**, *15*, 79–91.
32. Frazier, A.E. Landscape heterogeneity and scale considerations for super-resolution mapping. *Int. J. Remote Sens.* **2015**,*36*, 2395–2408.
33. Ge, Y.; Chen, Y.; Stein, A.; Li, S.; Hu, J. Enhanced sub-pixel mapping with spatial distribution patterns of geographical objects. *IEEE Trans. Geosci. Remote Sens.* **2016**, *54*, 2356–2370.
34. Resources and Environmental Scientific Data Center. Available online: http://www.Resdc.Cn/data.Aspx?Dataid=98(accessed on 13 December 2015).
35. Vogelmann, J.E.; Howard, S.M.; Yang, L.M.; Larson, C.R.; Wylie, B.K.; Van Driel, N. Completion of the 1990s national land cover data set for the conterminous united states from landsat thematic mapper data and ancillary data sources.*Photogram. Eng. Remote Sens.* **2001**, *67*, 650–662.
36. Bartholome, E.; Belward, A.S. Glc2000: A new approach to global land cover mapping from earth observation data.*Int. J. Remote Sens.* **2005**, *26*, 1959–1977.
37. Tatem, A.J.; Lewis, H.G.; Atkinson, P.M.; Nixon, M.S. Super-resolution land cover pattern prediction using a hopfield neural network. *Remote Sens. Environ.* **2002**, *79*, 1–14.
38. O'Neill, R.V.; Krummel, J.R.; Gardner, R.H.; Sugihara, G.; Jackson, B.; DeAngelis, D.L.; Milne, B.T.; Turner, M.G.; Zygmunt, B.; Christensen, S.W.; *et al.* Indices of landscape pattern. *Landsc. Ecol.* **1988**, *1*, 153–162.
39. Wang, Q.F.; Zhou, K.F.; Sun, L.; Chen, L.M.; Ou, Y.; Li, G.Y.; Qin, Y.F.; Wang, J.L. The study of landscape stability in yuli county by principal component analysis method based on rs and GIS technology. In Proceedings of the 2010 Conference on Piageng: Photonics and Imaging for Agricultural Engineering, Shanghai, China, 25–26 December 2010.
40. Chen, Y.; Ge, Y.; Heuvelink, G.B.M.; Hu, J.; Jiang, Y. Hybrid constraints of pure and mixed pixels for soft-then-hard super-resolution mapping with multiple shifted images. *IEEE J. Sel. Top. Appl. Earth Obs. Remote Sens.* **2015**, *8*, 2040–2052.
41. Zhong, Y.; Wu, Y.; Xu, X.; Zhang, L. An adaptive subpixel mapping method based on map model and class determination strategy for hyperspectral remote sensing imagery. *IEEE Trans. Geosci. Remote Sens.* **2015**, *53*, 1411–1426.
42. Liu, J.; Liu, M.; Tian, H.; Zhuang, D.; Zhang, Z.; Zhang, W.; Tang, X.; Deng, X. Spatial and temporal patterns of china's cropland during 1990–2000: An analysis based on landsat tm data. *Remote Sens. Environ.* **2005**, *98*, 442–456.

CHAPTER 17

AN AUTOMATIC PROCEDURE FOR EARLY DISASTER CHANGE MAPPING BASED ON OPTICAL REMOTE SENSING

Yong Ma, Fu Chen, Jianbo Liu, Yang He, Jianbo Duan and Xinpeng Li

Keywords: Disaster emergency mapping, optical image, automatic procedure, OptVM, SIFT-OFM

INTRODUCTION

Natural disasters, such as earthquakes, landslides, avalanches and debris flows, occur unexpectedly and suddenly, claiming huge losses of life and property and causing significant damage to the surrounding environment. A large number of cases show that rapid emergency response is effective in reducing the casualties and the loss caused by disasters. However, effectively making proper contingency plans is a troublesome problem faced by governments and experts [1].

Remote sensing technologies have the unique ability to help emergency managers streamline response and recovery by providing a backdrop of situational awareness which can be invaluable for assessing the impacts of the damage and to guide the disaster rescue [2]. Remote sensing plays an important role for the disaster rescue from the very early phase of a disaster, right through to long-term recovery [3]. Especially during the early stages of disaster response, a coordinated and reasonable plan for search and rescue (S&R) activities, logistics planning, and monitoring can save lives and property. However, the information (e.g., the geographical scope of the disaster areas, the magnitude and the spatial distribution of damage, and transportation infrastructures conditions) to support rapid response is limited. Remote sensing is a critical option for obtaining a comprehensive information about the disaster and for assessing the scope of damage, in particular to remote areas, where other means of assessment or mapping either fail or are of insufficient quality [4].

Various type of remote sensing sensors, platforms and techniques can be considered for emergency mapping [5]. The choice is mainly based on the type of disaster, the approximate extent of the affected areas and the requirement for monitoring the event. Generally, the main source of data for response activities is satellite remote sensing, as this gives the ability to monitor a wide footprint on the ground with limited or no access. In addition, modern satellite platforms can be triggered to change the acquisition angle to cover the affected areas in a short time for increasing the observation frequency of the regions of interest. The satellite imagery of disaster-hit areas can be obtained every day or even every few hours. Regarding the sensor type, the synthetic aperture radar (SAR) data and optical high-resolution (HR) or very high-resolution images (VHR) are generally obtained when disaster occur. SAR or InSAR systems are of great value for disaster response as it is a utility further enhanced by its all-weather capabilities, especially when the persistent of cloud cover over the affected areas make optical data unusable [6]. It has been achieved success in detecting surface displacement and offsets and height change of disaster areas by using the intensity, coherence or phase information of post-event SAR or InSAR data. However, for structural damage or some temporary change (e.g., damage buildings and the shelter sites), SAR and InSAR data are insufficiently sensitive and the final mapping results have been less conclusive, and marked by uncertainties. And most

Citation: Yong Ma, Fu Chen, Jianbo Liu, Yang He, Jianbo Duan and Xinpeng Li, An Automatic Procedure for Early Disaster Change Mapping Based on Optical Remote Sensing, doi:10.3390/rs8040272.

SAR-based change detection approaches suffer from a lack of archive data with the same acquisition parameters as the post-crisis imagery [7]. In addition, SAR imagery is less intuitive and is difficult to be interpreted by nonexperts, who must rely on sophisticated software to analyze the interferometry or amplitude coherence that require a longer processing time. Almost from the very onset of the disaster, optical satellite imagery is always available and provides the first glimpse of the disaster devastation. Optical high resolution imagery is the preferred choice to delineate the feature of interest (e.g., building, tents) and their current status (e.g., destroyed, burned, or moved) [5]. Generally, optical data can provide useful information to discriminate between damaged or non-damaged areas. Even though no further classification with respect to damage levels can be retrieved, the information derived from optical imagery can be effective [8]. The advantage of optical data is that their interpretation is intuitive for nonexperts, it can provide an overview of affected regions and sufficiently detailed information for decision-makers or aid workers to make S&R plan. However, the presence of cloud and shadows and variations in observation angles and geometric distortions limit the application of optical imagery for rapid emergency mapping [9].

This paper focuses on optical remote sensing image processing for the disaster mapping. One widely used approach for producing damage mapping is visual interpretation [10,11], which is tedious and labor intensive. Several automatic methodologies have been presented for disaster mapping, however, geometric distortions and improper co-registration between pre- and post-disaster imagery can result in a high false alarm rate in these automatic change detection approaches [7]. Therefore, a balance must be found between the product accuracy and the labor-cost, and a compromise must be found between the response time, the analysis depth, and the mapping accuracy. In this respect, we propose an automatic Optimizable Variational Model (OptVM) and scale-invariant feature transform (SIFT) constraint optical flow method (SIFT-OFM) to improve the accuracy of change mapping to produce the preliminary damage map, which usually will have a rather coarse character. In the early period of disaster, these initial maps could quickly provide some crucial information to relief teams and support the planning and coordination of rescue operations for emergency response. At the same time, these maps would give professional analysts or the public a guide to further checking whether a detected change is correct to improve the accuracy of the damage map by visual validation. With the availability of further earth observation data and more in-depth image analysis, the maps can be updated to incorporate this new information and provide a refined damage assessment [7].

Therefore, we propose an automatic optical remote sensing process to produce a cloud free product as a situation awareness map for identifying the overall damage and potential cascading effects, a change detection map for estimating the impacts of damage, and a temporal change map of key areas (such as the heavily damaged areas and temporary resettlement of victims) quickly in order to monitor the recovery process and guide the disaster recovery. This process can rely on the newest available optical data to rapidly produce the up-to-date maps, cover large parts of the affected area, and enable disaster managers to obtain an overview of the situation, assess the damage, and supply local logistic teams with reliable information on very short notice. The procedure and methods are introduced in Section 2. The experimental results and analyses, using the Nepal magnitude 8.1 earthquake on 25 April 2015, as an example, are given in Section 3. A discussion is presented in Section 4. The last section concludes the paper with the significance of this research and proposed future work.

THE PROCEDURE AND METHODS

The timeliness of receiving data and the length of time needed for data processing are key factors in rapid emergency mapping for disaster events. When a large disaster occurs (e.g., the Haiti earthquake, the forest fires in Australia, and the Nepal earthquake), many international organizations or commercial companies decrease the satellite revisit time to obtain the newest data for the affected areas by adjusting the acquisition angle [12,13]. This timely triggering provides enough data for disaster mapping. However, these agencies only offer the original satellite imagery, rather than processing or analysis results. The orderly organization and management for effective processing is scanty; and these multi-source and multi-temporal data are not sufficient for developing information applicable in humanitarian and natural crisis situations.

In this paper, a procedure is proposed to deal automatically with multi-source and multi-temporal remote sensing data, in order to produce a product that can serve in the early stages of disasters. The procedure is shown in Figure 1, which includes image fusion, image registration, and the production of maps. The procedure ultimately offers a cloudless remote sensing map, a change detection map, and a local temporal variation map for some typical areas during a short period of time. The proposed procedure is based mainly on the general rapid mapping framework proposed by Ajmar and Boccardo [14], but our proposed procedure is a specific implementation of a general framework except for map dissemination. In our procedure that uses some new techniques and methods, the accuracy of the products in each step has been greatly improved. Additionally, the whole procedure is fully automated without any artificial participation. The methods applied in each step of the procedure are described below.

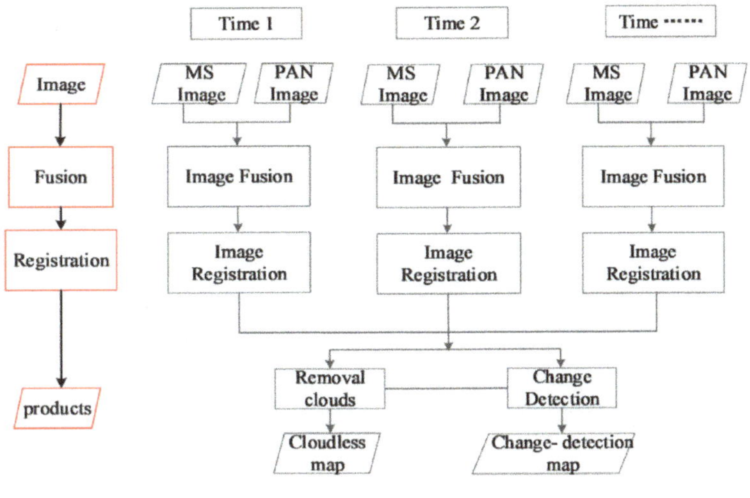

Figure 1. Procedure for dealing with multispectral and multi-temporal remote sensing images.

Image Fusion

Recently high spatial resolution images, such as QuickBird, WordView, GeoEye and GF-1, have been favored in many remote sensing applications, especially for the assessment of disasters [15]. Usually, high spatial resolution satellites provide image data with a low-resolution multispectral (LRM) image and a high-resolution panchromatic (HRP) image [16]. In order to obtain the high-resolution multispectral (HRM) images needed for detailed damage assessment, the image fusion process must be implemented. This is the first key step in our procedure.

There has been much research done on image fusion. The methods can be categorized into five types, including variable substitution (VS) methods, such as the IHS and PCA algorithms [17,18]; modulation fusion methods (MF), such as the Brovery algorithm [19]; multi-scale analysis (MSA) such as the wavelet fusion algorithm [20]; restoration based methods (RB), such as the adjustable model-based fusion algorithm [21]; and compressed sensing methods (CS) [16,22]. The first three types of methods can be viewed as traditional methods, which tend to cause color distortions. Additionally, spatial information weakens due to the wavelength extension of the new satellite HRP images. Inspired by the rapid development of super-resolution techniques, RB methods have been applied routinely to image restoration and fusion. These methods view the LRM image and the HRP image as the observations of the HRM image via an image degradation model. The HRM image is obtained by setting up a model and solving the model. The CS method is based on sparse representation, which utilizes the strong correlation between the LRM and HRP images to obtain a sparse coefficient, construct two coupled dictionaries, and reconstruct the fused image by the fused measurement [23]. This method usually needs a large collection of training images to obtain trained dictionary, requiring significant computing time.

In this paper, an improved fusion model, OptVM, is applied to produce high-quality HRM images automatically. The flowchart of the OptVM is shown in Figure 2. First, a PAN image is simulated by MS image, and the simulated PAN and original HRP image are matched. Next, the restored MS image is obtained by the relationship of spectrum and grayscale between the simulated PAN and LRM images; and finally, the model (Equation (1)) is optimized to get the HRM image. The model defines Pan-sharpening as an optimization of a linear over-determined system, which is based on gray-value and spectral sensitivity relationships between the original multispectral (MS), panchromatic (PAN), and fused images.

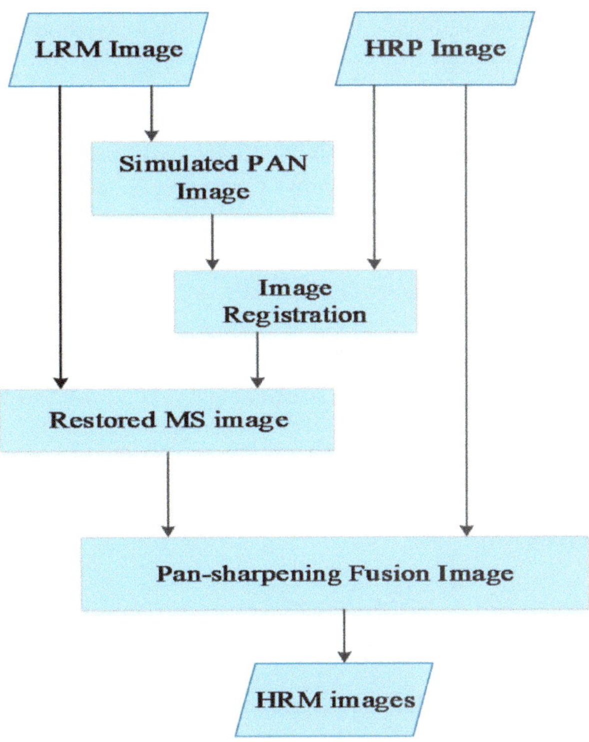

Figure 2. Flowchart of image fusion method.

The OptVM constructs a cost function to generate the HRM image based on three constrained hypotheses: (1) the radiation of an object in an LRM image should be equal to the integral average radiation of the same object in the HRM image; (2) spatial details of the HRP image and the HRM image (the target fused image) should be similar; (3) the values in the panchromatic band should be the summation of the respective multispectral band, if the spectral response range of the LRM image is nearly the same as the HRP image. According to the three assumption, the cost function can be defined as follows in Equation (1):

$$E(P_H, XS_H, XS_L) = \lambda_1 \sum_{n=1}^{4} \int (XS_L^n(x) - k \times XS_H^n(x))^2 dx$$
$$+\lambda_2 \sum_{n=1}^{4} \int \left(XS_H^n(x) - \sum_{S \in N(x)} W_{sx} XS_H^n(s) \right)^2 dx$$
$$+\lambda_3 \int \left(\sum_{n=1}^{4} W_n XS_H^n(x) - P_H(x) \right)^2 dx \qquad (1)$$

where XS_L, XS_H and P_H are the radiation of pixels in the LRM image, the HRM image and the HRP image respectively. The variables, λ_1, λ_2 andλ_3 represent the weight of the three models in the total energy function, which can be adjusted according the fusion result. Here, in next experiment, λ_1, λ_2 and λ_3 are set 1. The variables, n is number of fusion band. The parameter of k is based on the fact that the low-resolution pixels are formed from the high-resolution ones by low pass filtering followed by subsampling [24], and k represents the low pass filtering followed by subsampling. The express of k is shown as follows in Equation (2):

$$XS_L(x) = k \times \sum_{j \in M(x)} P_H(j) \qquad (2)$$

In the Equation (2), the pixel j is one of the pixels set M in HRP image which corresponds to the pixel x in LRM image. For example, if the resolution of LRM image is 8 m and HRP image is 2 m, each pixel in LRM image corresponds to 4 × 4 pixels set in HRP image in the same object. The variables in Equation (1), W_{sx} represents the weighted value of the pixel s which in the neighborhood window of pixel x in HRP image, the express is as follows Equation (2), including δ_x is the standard deviation.

$$W_{sx} = e^{-(P_H(x)-(P_H(s))^2/2\sigma_x^2} \qquad (3)$$

In the Equation (3), the size of neighborhood window is set 9 × 9 in the next experiment. The variables in Equation (1), Wn is the band weights of multispectral image in spectral response relationship. The express is shown as follows in Equation (4):

$$P_h(\lambda) = \sum_{n=1}^{4} W_n XS_L(\lambda) \qquad (4)$$

In the Equation (4), λ represent the Spectral value. The process of image fusion is the global minimum ofE(P_H,XS_H,XS_L),which is the linear constraint of the three basic models. The least square technique—LSQR [25] is applied to optimize the model. An important step of the processing is the image registration of the HRP image and the simulated panchromatic image by the LRM image to avoid fusion image distortion and spectral offset based on the optical flow and feature constraint method (introduced in Section 2.2).

Image Registration

Image registration is an important procedure in remote sensing image applications [26] and a crucial step in our chain. Previous studies of image registration methods can be divided into three categories, including area-based method, such as Cross-Correlation (CC) method [27], the mutual information (MI) [28]and the Sequential Similarity Detection Algorithms (SSDAs) [29]; feature-based method such as SIFT [30,31,32] and physically-based methods, such as multi-resolution elastic model [33] and the optical flow model [34,35]. The first two methods, which are based on the gray information and the spatial features of the images to register the images, cannot meet the real time requirements, as they require large computing capacity and highly complex algorithms, They are also not very robust. The physically based methods utilize physical models to describe the process of image deformation and solve the image physical model to achieve image registration. These methods have great advantages for speed and accuracy. In this paper, a physically-based method, the SIFT feature constraint optical flow method (SIFT-OFM), is proposed to register images. Figure 3 shows the flowchart of the algorithm. First, homologous points are obtained for the reference image and the sensed image using the SIFT feature matching method [36]. Then the affine transform coefficients are calculated to be a constraint for an optical flow model. Finally, the parameters of optical flow model are optimized through an iterative process to obtain high accuracy registration images.

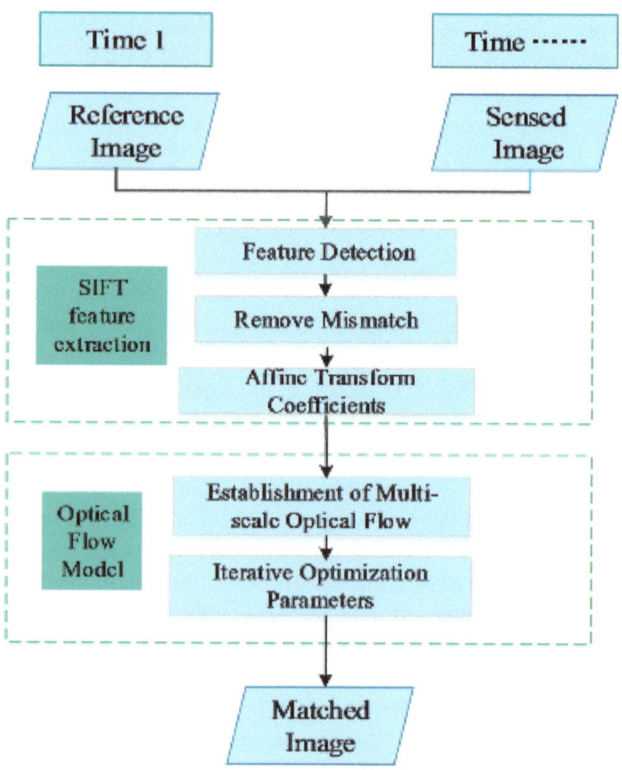

Figure 3. Flowchart of image registration method.

The optical flow model method is based mainly on the assumption of the local gradient model and the overall smoothness model of the pixel gray values, which regard the change of image gray scale as the change of optical flow [35]. The algorithm is able to get the high-precision matched image by considering the optical flow change as the displacement of the reference image and the sensed image to the register.

The optical flow model is based on the assumption of the constancy of brightness, which means that the gray value of the same object is not changing with time and the radiance consistent is in motion [37]. Thus, the spatial variation of the optical flow model can be expressed as follows in Equation (5):

$$E_{gray} = \int\int \varphi\left((I(x+u, y+v) - I(x,y))^2\right) dxdy \quad (5)$$

where I is the image intensity function of interval Ω (x, y), (x, y) ϵ Ω; u and v are the displacements in the x and ydirections, respectively, which is the offset between the reference image and the sensed image. The symbol, Ψ, represents a differentiable function on the interval Ω. To reduce the impact of light changes, the gradient energy equation is added into the total energy formula. The gradient energy equation is defined as follows in Equation (6):

$$E_{gradient} = \int\int \varphi\left(\left|\nabla I(x+u, y+v) - \nabla I(x,y)\right|^2\right) dxdy \quad (6)$$

Where $\nabla I = [I_x, I_y]^T$, and ∇I is the gradient vector of the image. For every pixel, there is only one equation (the sum of Equations (2) and (3)); but there are two variables, u and v, which can cause uncertain solutions and aperture problems. Thus, there is a need to add a regularization constraint. The regularization constraint plays an important role in the estimation intensity, the edge smoothness, and the edge retention of the optical flow field. In this algorithm, a global smoothness constraint proposed by Horn [37] is used to minimize the constraint term. The global smoothness constraint is described as follows in Equation (7):

$$E_{smooth} = \int\int \varphi\left(|\nabla u|^2 + |\nabla v|^2\right) dxdy \quad (7)$$

where ∇:=(∂x,∂y) is two dimensional differential operator.

Optical flow model is not suitable for matching large displacement images since the optical flow model can be expressed by expanding the Taylor series and omitting the high-order terms. Therefore, the SIFT-OFM adds the SIFT feature as a constraint into the optical flow model. The generation of the SIFT feature points includes four steps: extreme space detection, precise positioning of the SIFT feature point, determination of the main direction of the feature point, and the description of the building features. Based on these SIFT feature points, the affine coefficients are calculated as the initial value of the optical flow field to match the image. The total energy equation of SIFT-OFM is defined as follows in Equation (8):

$$\begin{aligned} E &= E_{gray} + \lambda E_{gradient} + \alpha E_{smooth} + \tau E_{SIFT} \\ &= \int\int \varphi\left((I(x+u, y+v) - I(x,y))^2\right) dxdy \\ &+\lambda \int\int \varphi\left(\left|\nabla I(x+u, y+v) - \nabla I(x,y)\right|^2\right) dxdy \\ &+\alpha \int\int \varphi\left(|\nabla u|^2 + |\nabla v|^2\right) dxdy \\ &+\tau \int\int \varphi\left((u-u_{SIFT})^2 + (v-v_{SIFT})^2\right) dxdy \end{aligned} \quad (8)$$

where λ, α and τ are the coefficient of the last three component. Based on previous experience, these are set to λ=1, α=50 and τ=1 in next experiment.

The minimum of u^* and v^*, are solved based on the energy function E, using Successive Over Relaxation (SOR) iterative method [38]. This is the displacement of the reference image and the sensed image. Then calculate the displacement based u^* and v^* and resample to get high-precision matched image.

Map Production

The procedure mainly provides two kinds of products. These are the cloudless maps for the disaster regions and the change detection maps for the early stage of the disaster. The flowchart for map production is shown in Figure 4. After obtaining the high precision matched disaster image, the clouds on the images are detected and removed to get cloudless remote sensing maps for determining the overall scope of the disaster and for automatically detecting the change between the pre- and post-disaster cloudless images. For the automated change map, artificial checking is needed to exclude the error detection, which is caused by the different types of sensors and the observed conditions between the two images. In cases of emergency, the automated coarse change map can be used directly to offer approximate information for managing crisis issues.

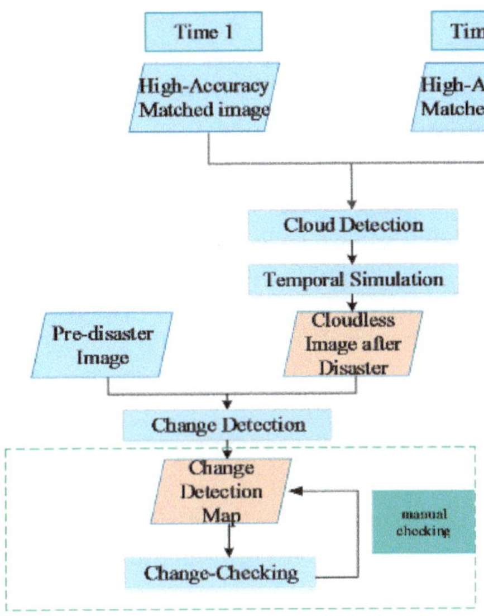

Figure 4. Flowchart of products processing.

Much research has been focused on the task of cloud removal to relieve the influences of clouds. In fact, the process of cloud removal is an information reconstruction process for missing information. The proposed approaches can be grouped into three categories [39,40]: (1) spatial-based methods without any other auxiliary information source; (2) spectral-based methods, which extract the complementary information from other spectra; and (3) temporal-based methods, which extract the complementary information from other data acquired at the same position and at different time periods. Spatial-based methods are often used for some small missing regions by interpolation, such as for the scan line (SLC) missing from Landsat-7 [41]. It is not appropriate for large-scale cloud removal. Spectral-based methods are based on the relationship of spectra in clear regions and restore the contaminated band of data by modeling a relationship between the contaminated band and the auxiliary band by methods such as the HOT method [42], and the STARFM method [43]. It is usually constrained by the spectral compatibility and has difficulty with thick clouds. Temporal-based methods utilize multi-temporal clear data for the same region to simulate the cloud cover area data for acquiring cloud-free images, such as local linear histogram matching (LLHM) approach [44], spatio-temporal MRF model [45] and etc. In this paper, we produce the cloud free maps by the temporal phases simulation method proposed by Gui [46], which utilize spectrum relationship, vegetation index and HOT parameters [47] to detect clouds on the images and calculate the feature difference between two temporal images to make up the cloud regions. The simulation method uses the minimum value of the feature difference and a multi-grid optimization algorithm. It converts the features of one image to the features of the reference image and then fills the areas of the clouds by the converted image that has no cloud in the same region. This generates a cloud free image in the disaster districts. In this study, time to acquire the difference between the images is very short, so the generated image can truly reflect the overall situation after the disaster by this method.

Several authors have presented semi- or fully- automatic methods to analyze images for damage change detection and assessment by optical data [3,48,49,50]. But geometric distortion, improper co-registration of pre- and post-disaster images can result in a high false alarm rate of automatic change detection methods. In this procedure, the new methods, OptVM and SIFT-OFM, are utilized to improve the accuracy of preprocessing results. And based on the high-precision preprocessing images, the Iteratively Re-weighted modification to the Multivariate Alteration Detection (IR-MAD) method proposed by Nielsen [51,52] is used for detecting change automatically in this procedure. IR-MAD is mainly based on the means of canonical correlation analysis (CCA). First, the maximum canonical correlation coefficient after linear transformation is determined for the pre- and post-disaster images [53]. The D-value and the variance of the linear transformed images are calculated. Finally, the pixel change is checked by the Chi-square distribution of the D-value and the variance. In order to highlight the changing information of the pixels to improve the accuracy of the change detection map, a weighted iteration method is used during the process of transformation. The iterations continue until the absolute value is less than the error limit (for this paper, the error limit is set as 0.001). The absolute value is the difference of correlation coefficient maximum value from two iterations, namely, before and the current conversion.

However, the change detection map has some error detection points because the map, which is generated by IR-MAD, could be influenced by the different conditions during the acquisition of the two images. Artificial checking is essential for removing the errors in the changed regions. Changes are totally possible during the early stages of disasters, since many people and agencies from other regions or countries have a great enthusiasm and willingness to help the victims and local government in relief and reconstruction activities. There are many public community remote sensing platforms, such as TomNod [54] and GIS Corps [55] which could add and update the information (not only in vector format but also by means of geotagged pictures and textual information), validate the information conveyed, and actively participate in the map production phase by coordinating the efforts of ordinary people and experts [56]. It could gather every tiny force to provide significant assistance for the victims and decision making department in the disaster region [57]. The automatic production change map can be used in on a community remote sensing platform to reduce labor costs and to provide a rapid response to the disaster.

EXPERIMENTS RESULTS AND ANALYSIS

Study Area and Data Sets

A powerful earthquake shook Nepal near its capital, Kathmandu, at 11:56 NST on 25 April 2015, killing more than 8000 people and flattening section of the city's historic center. Some of the oldest parts of the city, including seven UNESCO World Heritage sites, were severely damaged in the earthquake. This paper uses the Nepal earthquake as an example. The location of the study area, the center of Kathmandu, is shown in Figure 5. Four images from a GF-1 PMS sensor were selected, including a pre-earthquake image on April 11 and post-earthquake images on 27 April, 1 May, and 2 May, as shown in Figure 6. The specific observation parameters of the data set are shown in Table 1, and the parameters for the PMS GF-1 sensor are shown in Table 2. The proposed procedure was followed without any human intervention. The procedure will be discussed clearly, including the intermediate results of the image fusion and registration and the final products of the cloudless maps and the change map.

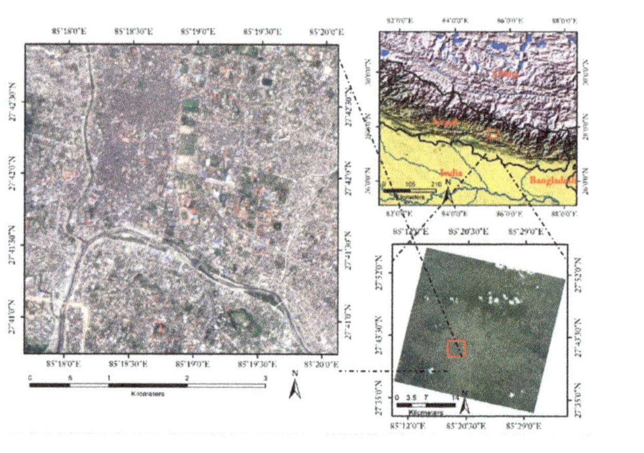

Figure 5. Spatial location of the study area (the resolution of left map is 2 m after fusion).

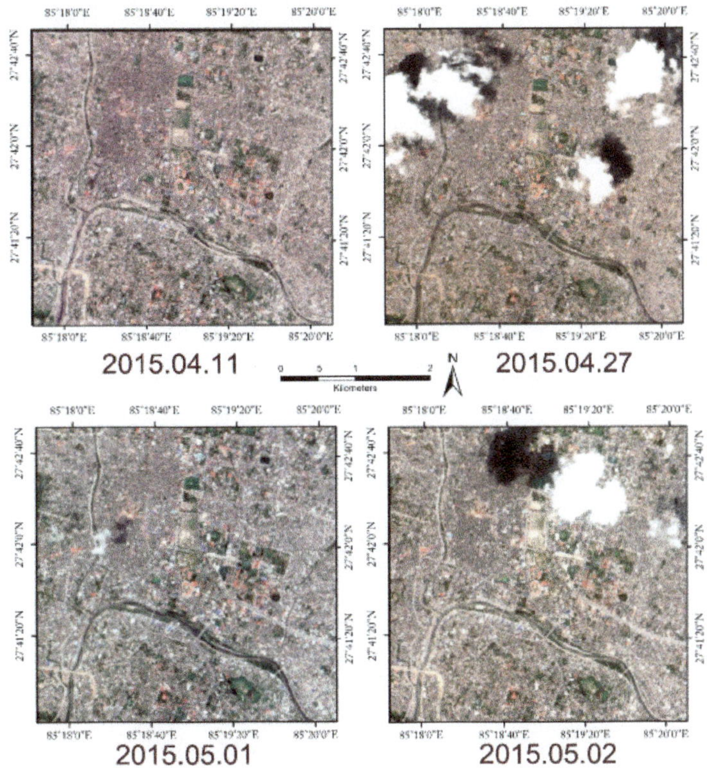

Figure 6. The four images of data set (all the four image is the original image of MS with 8 m resolution).

Table 1. The specific observations parameters of data set.

Acquisition Date	Sensor	Viewing Angle
11 April 2015	PMS-1	0°
27 April 2015	PMS-2	-24°
1 May 2015	PMS-1	-23°
2 May 2015	PMS-1	23°

Table 2. Parameters of GF-1 PMS sensor.

Parameter	PMS-1/PMS-2 Sensor	
Spectral range	PAN	0.45–0.90 μm
	MS	0.45–0.52 μm
		0.52–0.59 μm
		0.63–0.69 μm
		0.77–0.89 μm
Spatial resolution	PAN	2 m
	MS	8 m

Image Fusion Results and Analysis

In this section, the image from 27 April is taken as an example to show the accuracy of the image fusion results. Figure 7 shows part of the whole image to exhibit the image details clearly. The left image is multi-spectral data, and the right is panchromatic data. The upper left corner image in each view is an enlarged view of the small red box. Figure 8 shows the result of PCA fusion, GS-pansharpening fusion, wavelet transform fusion, and the proposed OptVM method. Visually, the four methods have achieved spatial resolution enhancement and keep a certain spectral fidelity; but the results of PCA, GS, and wavelet transforms have virtual phenomena in the edge of objects, especially the wavelet transform result being the worst. In detail, from the enlarged image of the small red box, the resulting color recovery of OptVM is finer than with the other three algorithms. For example, observe the red roof in the enlarged image in Figure 8. The OptVM method restored the red color of the roof regularly, while the color recovery results of the other three algorithms deviate from the red color, with some parts of roof being white

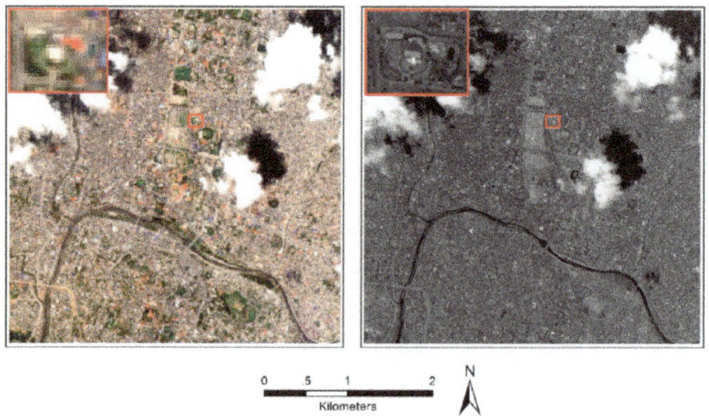

Figure 7. Multispectral (**left**) and panchromatic data (**right**) of the original image on 27 April (upper-left small image is the detail from the area of the red rectangle).

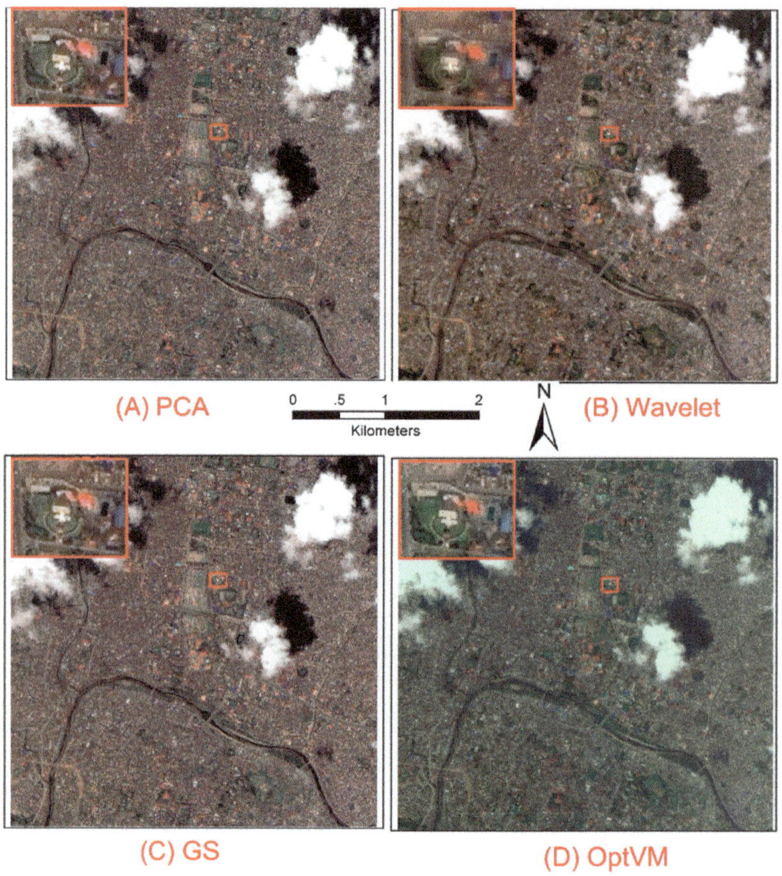

Figure 8. Result of various image fusion methods using (**A**) PCA; (**B**) wavelet; (**C**) GS; and (**D**) OptVM.

Six typical surface features are shown in Figure 9 which are sampled from the fused results (seen in Figure 8). These are sampled in relatively uniform areas to assess the spectral fidelity and the spatial location of the six points marked in Figure 9A. These pixels cover a large set of material in the scene and are uniform around their neighborhood. The spectra of the original MS image are used as ground truth for comparisons based on the reasonable assumption that the spectra around such uniform areas are most likely unchanged. The spectra results of the image fusion methods are shown in Figure 10; and the Euclidean distance difference and the spectral difference, expressed in spectral angle difference, are shown in Table 3. For uniformity of comparison, the values of the fusion results are normalized to 0–255. From Figure 10 and Table 3, it can be seen that the result of OptVM method can preserve the spectra of these pixels well, whereas the spectra from the PCA, the Gram–Schmidt (GS), and the wavelet fusion are very different from the truth. For example, the red roof of pixel #1 is shifted seriously in the result from wavelet fusion. For the ordinary roof (pixel #4) and bare land (pixel #6), the results of PCA and GS are lower than the truth, especially in red band, and the difference of the spectral angle is larger. In the vegetation area, the colors of the results based on the wavelet method (see Figure 9D) lack fidelity, while our algorithm performs very well in most regions (Figure 9F). In our method, a crucial step is that the simulated PAN and the original PAN need to be matched to reduce the distortion that is caused by the geometric deviation of the original MS and PAN images.

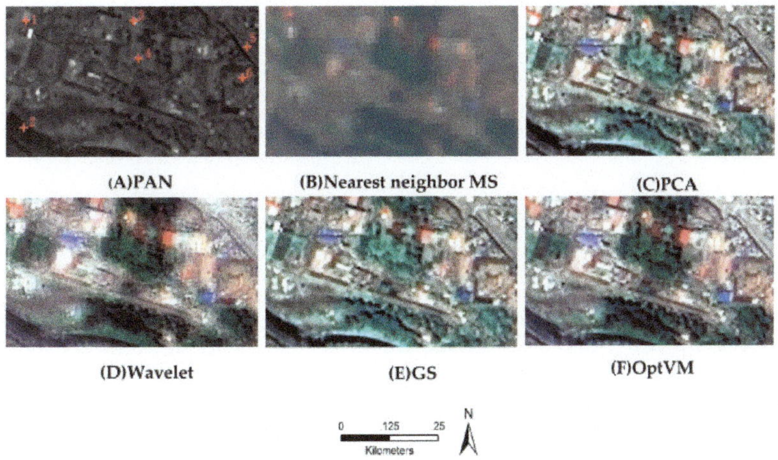

Figure 9. Comparison of the various algorithms on the urban area scene. (**A**) the PAN image; (**B**) the MS image after resampling using the nearest neighbor; (**C**) the PCA result; (**D**) the wavelet result; (**E**) the GS result; and (**F**) the OptVM result.

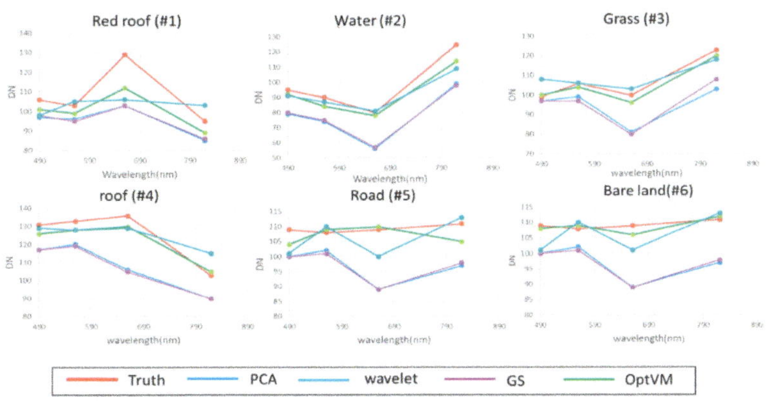

Figure 10. Spectral comparison between the various approaches in the urban area scene. The site of pixels marked in Figure 9A

Table 3. Spectral angle difference and Euclidean distance difference of marked pixels.

Category	Spectral Angle Difference				Euclidean Distance Difference			
	PCA	Wavelet	GS	OptVM	PCA	Wavelet	GS	OptVM
Red roof	29.75	25.71	30.10	19.13	0.066	0.108	0.065	0.044
Water	41.33	16.79	42.00	13.04	0.056	0.058	0.055	0.028
Grass	26.65	10.72	28.53	5.48	0.068	0.048	0.075	0.017
Roof	39.01	14.90	37.87	9.49	0.063	0.059	0.060	0.023
Road	26.44	12.37	26.70	7.94	0.051	0.050	0.054	0.030
Bare land	26.44	9.00	26.70	7.75	0.051	0.039	0.054	0.031

In order to comprehensively assess the quality of fused images, four objective evaluation metrics are selected, which include the spectral angle mapper (SAM), the error of relative dimensionless global in synthesis (ERGAS) metric, the Q4 index [58] and Quality with No Reference (QNR) metric [59]. The equations of four metrics are presented in the Appendix. SAM and ERGAS are measures of the global radiometric distortion of the fused images, and a smaller value indicates better fusion results. Q4 and QNR reflect the spectral and spatial quality of the fused images. Values closer to 1 denote the better result. The results of the four metrics are shown in Figure 11 and Table 4. In regards to the fidelity of the image spectrum and the image clarity, the fused images by GS and by our proposed algorithms appear better than those obtained by the PCA and the wavelet fusion methods. The QNR index obviously displays this difference, since replacing the original MS image information leads to the loss of information. Compared with the other fused methods, the SAM value of the GS fused image is very large, with the degree of sharpening being excessive in some parts of the results. The values for the four metrics for our method show that it is the best of the four methods. In fact, most of the existing image fusion algorithms are based on the band-to-band correlation, whereas our method synthetically utilize the relationship of spectrum, grayscale and spatial to produce an image with fewer distortions.

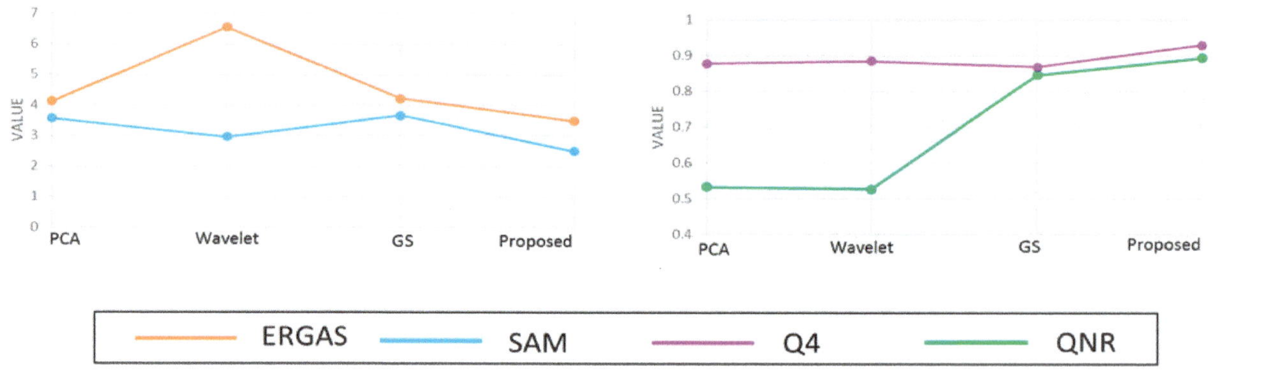

Figure 11. Line chart of the four metrics comparing fusion methods (ERGAS and SAM values are better when approaching 0, while Q4 and QNR values are better when approaching 1).

Table 4. Comparison of the value of above four index.

Index	PCA	Wavelet	GS	OptVM
ERGAS	4.118	6.551	4.214	3.485
SAM	3.566	2.957	3.656	2.492
Q4	0.533	0.527	0.846	0.894
QNR	0.878	0.886	0.86	0.931

Image Registration Results and Analysis

Image registration is based on the results of image fusion and select two images as the example (pre-earthquake on 11 April 2015, and post-earthquake on 27 April 2015). The pre-earthquake image is the reference image to match with the post-seismic image of 27 April. In Figure 12, the middle image is a combined image, which takes a 100 × 100 block from the reference image and also from the sensed image, alternately. Figure 12A–D are the detailed enlarged images, which show the position offset of the reference image and of the sensed image. In Figure 13, Figure 14, Figure 15 and Figure 16, the middle combined image is obtained in the same way as Figure 12. These figures are based on the reference image and the matched image using the optical flow model algorithm (Figure 13), the SIFT feature algorithm (Figure 14), the ENVI software registration workflow (Figure 15), and our proposed SIFT-OFM method (Figure 16).

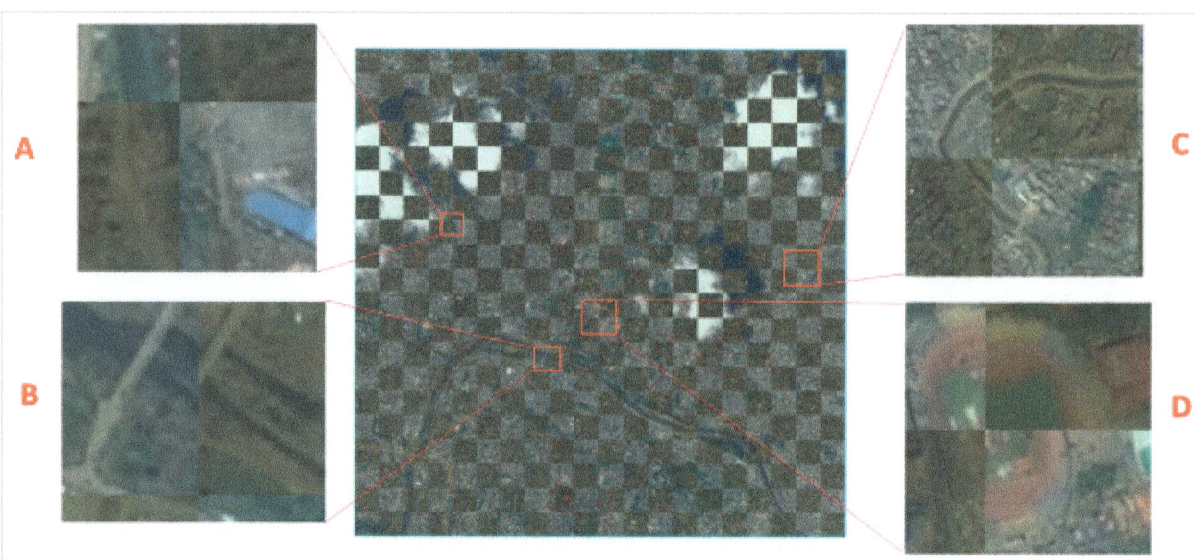

Figure 12. The combined image of the reference image and the sensed image (**A–D**show the detail of each red rectangle on the combined image).

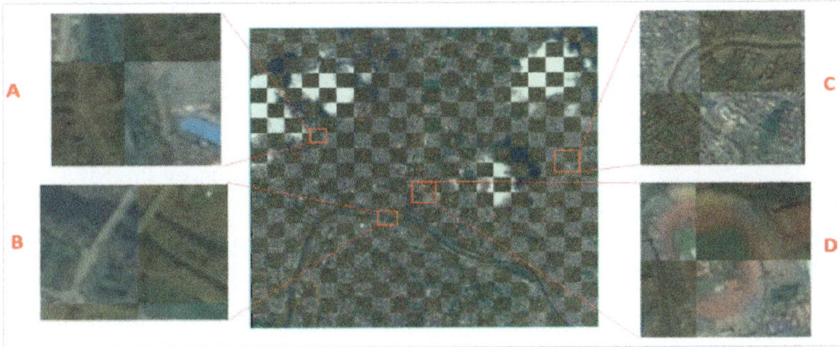

Figure 13. The combined image of the reference image and the matched image only using the optical flow model (**A–D** show the detail of each red rectangle on the combined image).

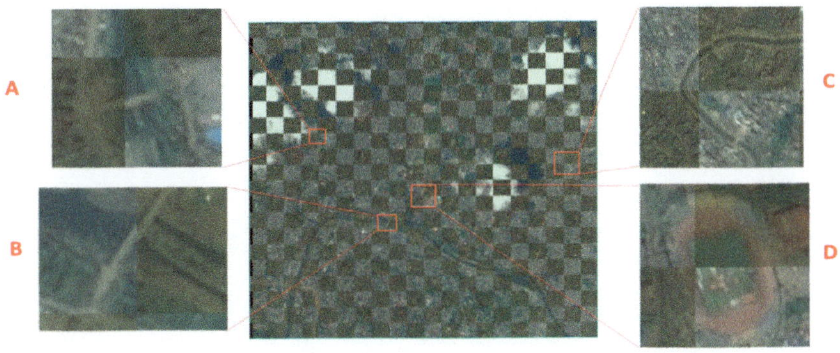

Figure 14. The combined image of the reference image and the matched image only using the SIFT algorithm (**A–D** show the detail of each red rectangle on the combined image).

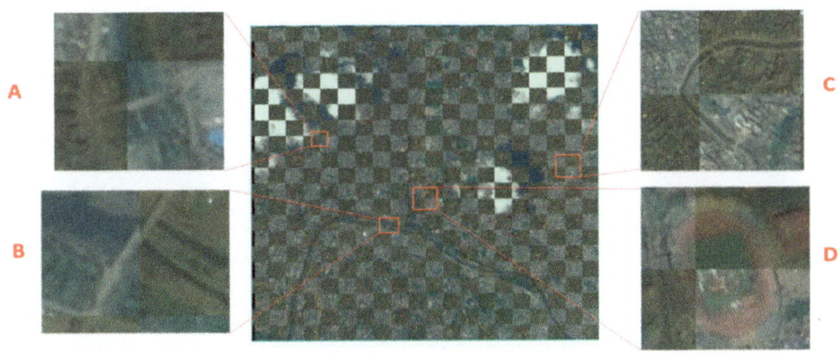

Figure 15. The combined image of the reference image and the matched image by using GCP in ENVI software (**A–D** show the detail of each red rectangle on the combined image).

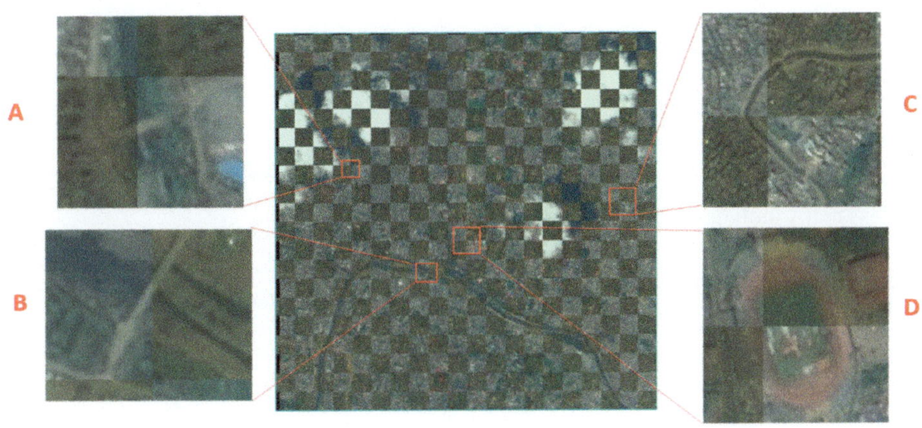

Figure 16. The combined image of the reference image and the matched image using SIFT-OFM (**A–D** show the detail of each red rectangle on the combined image).

From the detailed images in Figure 12, the dislocation between the reference image and the sensed image is very obvious, so it is necessary to register the two images for producing the products. From Figure 14 to Figure 16, the results are good. Figure 13 shows that the result that is based only on the optical flow model algorithm is the worst. It is because optical flow algorithm model is not suitable for the large displacement of two images, generally not more than two pixels; but in this study, the difference of the two images significantly exceeds at least 10 pixels. Furthermore, comparing the detailed image from Figure 14 to that in Figure 16, the SIFT algorithm performed well in Figure 14A,D; but in Figure 14B, the image of the dislocation of the bridges is very clear, while the road in the Figure 14C image is broken. The linear correction model of the affine transformation is utilized in SIFT feature algorithm and can cause some nonlinear distortions to be uncorrected. The result of Figure 15 is based on ENVI registration process, in which 34 ground control points (GCP) are marked by artificial selection and the overall error of these GCP is not more than 0.43 pixels. However, in both Figure 15A, B the discrepancy of the bridges is obvious and serious. This is related to the maldistribution of the GCP and the linear calibration model in the registration process. There is no obvious displacement or deviation in any of the images of Figure 16 produced by the SIFT-OFM method, which restricts the larger deviation by the SIFT feature and corrects the local linear or nonlinear distortion by the optical flow model.

The structural similarity index measurement (SSIM) [60] is the quantitative evaluation metric used to evaluate the precision of a registration image. The SSIM maps in Figure 17 are obtained by calculating the SSIM between the reference image and the matched image in a sliding window within 9 × 9. In the SSIM map, red indicates SSIM values close to 1; that is, the areas with high precision registration. It can be seen that the precision of the proposed method is higher for the SIFT feature and the ENVI automatic registration process. The existence of clouds and shadows in sensed images and the lack of clouds in the reference image leads to the low SSIM value in the clouds and shadows regions. Therefore, in the Table 5, the overall SSIM value in the whole image shows relatively low.

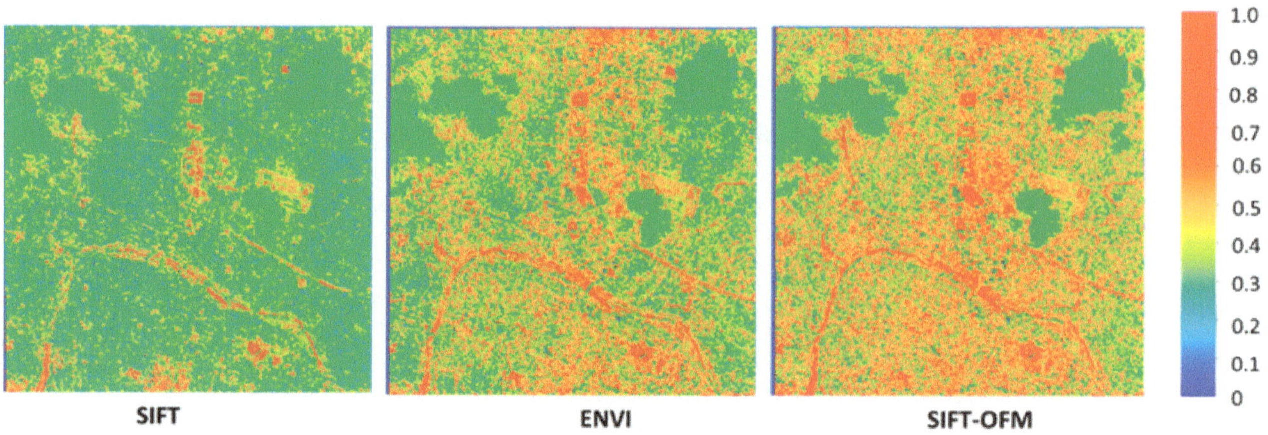

Figure 17. SSIM maps of three methods (the red area indicates SSIM values close to 1; that is, the more high precision registration regions).

Table 5. Comparison of SSIM value for whole subset image using three methods.

Index	SIFT	ENVI	SIFT-OFM
SSIM	0.6266	0.6415	0.7442

Cloudless Product and Analysis

The generation of cloudless maps includes cloud detection and temporal simulation based on the result of image fusion and image registration. The four images after image fusion and image registration are shown in Figure 18. From Figure 18, all the post-event images (images of 27 April, 1 May, and 2 May) included clouds and shadows, which affect the ability of the decision makers and analysts to estimate and grasp the overall situation. Therefore, cloud and shadow removal and temporal simulation for producing a cloud free image is indispensable.

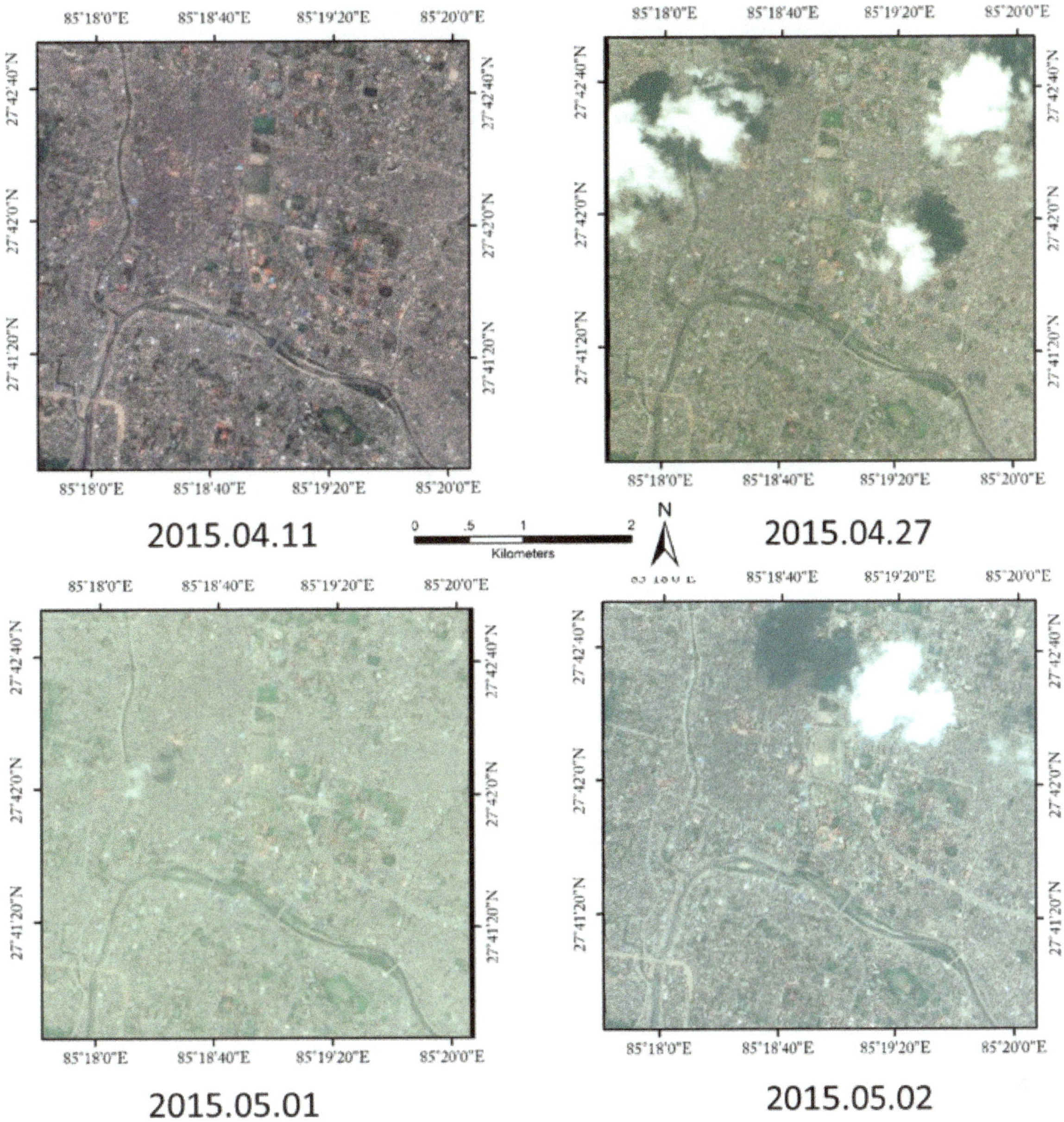

Figure 18. The data set of four images after image fusion and registration (compared with Figure 6, the resolution and the registration accuracy have been improved greatly).

The cloud detection results by our method introduced in Section 2.3 are shown in Figure 19, in which red areas represent cloud regions and green areas represent shadow regions. By visual comparison, most of the cloud and shadow have been detected, while only a small portion of the thin clouds have not been detected. The cloud and shadow based on the result in Figure 19 is used to produce the cloud free image by temporal simulation based on the adjacent time image after the earthquake. The cloudless image is shown in Figure 20, in which Figure 20A is the image of 1 May and Figure 20B is the image of 2 May. Figure 20C is the cloudless map. The image, Figure 20C, is based mainly on the Figure 20B image, with the cloud and shadow regions made up from the Figure 20A image. Figure 21 show the detailed enlarged images of the rectangle in Figure 20. In Figure 21, the scenes in the cloud and shadow regions are recovered by the temporal simulation method and keep consistency in color and texture using neighbors. It provides a high quality cloudless image and identifies the overall damage and potential cascading effects accurately. Figure 22 shows the cloudless maps of the post-event on 27 April, 1 May, and 2 May. Comparing with the original image of Figure 18, the cloudless maps eliminate the influence of clouds and shadows and can be used easily as backdrop maps or situation maps. The cloud and shadow regions of the image on 27 April (seen Figure 18 right-up image and Figure 19 right-up image) is replaced by the pre-event cloudless image on 11 April (seen Figure 18 left-up image) to obtain a cloudless map (seen Figure 22 left image), as there is no post-event image before 27 April to make up the cloud and shadow regions. The left cloudless image of Figure 22 contains some pre-event information, which may make a confusion by users if they only use the single cloudless map, so some red vectors of cloud and shadow regions are added to distinguish the pre and post event data. In the internal regions of the vectors are the pre-event information, while the external region are the post-event information. This kind maps is only used in the short time after the event when there are few post-event images and the images still have overlapping cloud-covered areas.

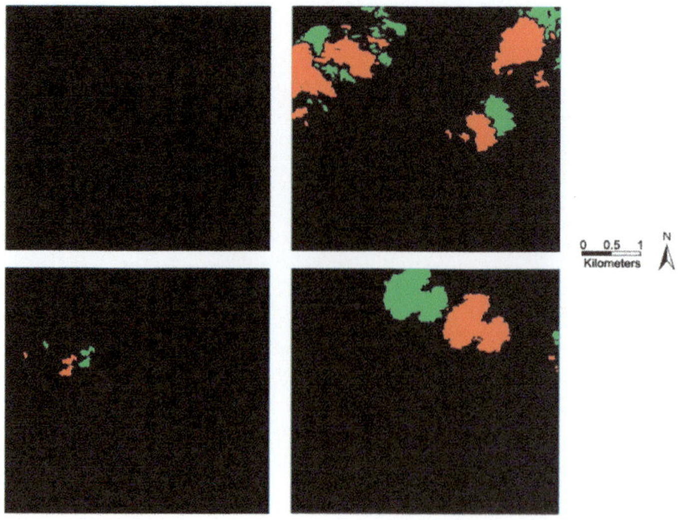

Figure 19. The cloud detection products of the four image in Figure 18 (red areas represent cloud regions and green areas represent shadow regions).

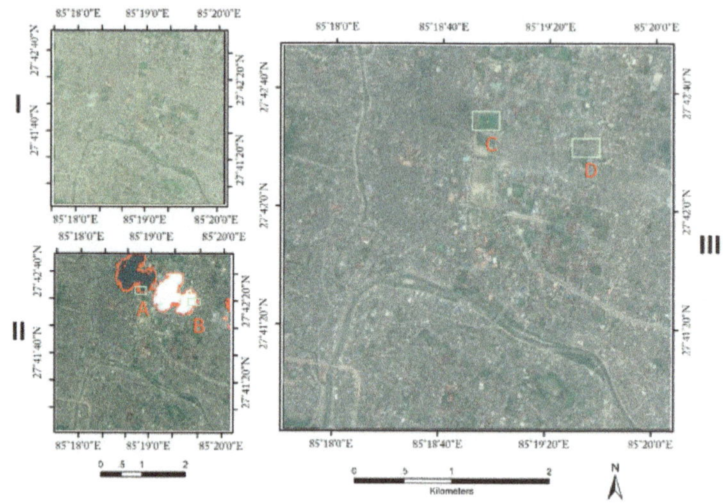

Figure 20. The cloudless map of 2 May. (**I** and **II** are the original images of 1 May and 2 May, **III** is the cloudless map after removing the cloud and shadow). (**A**) and (**B**) are the shadow and cloud regions in the original image; (**C**) and (**D**) are the same areas after removing the influence of cloud and shadow corresponding to (**A**) and (**B**) regions. The detail of **A–D** are shown in Figure 21).

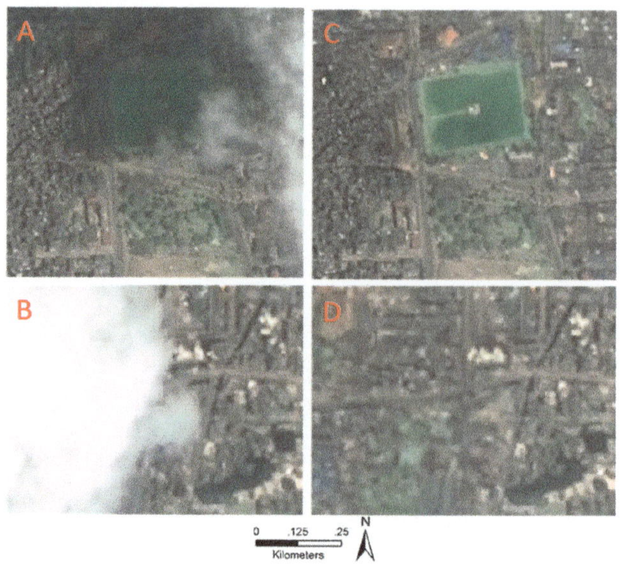

Figure 21. The detailed enlarged images of the rectangle in Figure 20. (**A** and **B** are the original images from 2 May, **C** and **D** are the simulated cloudless images).

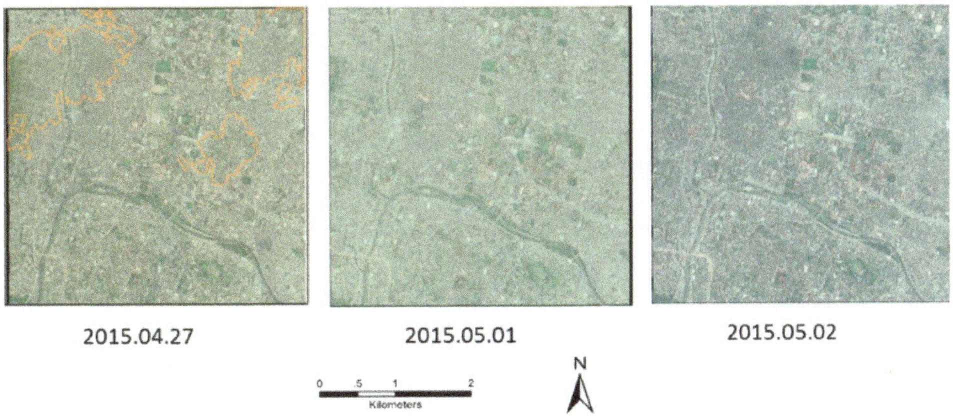

Figure 22. The cloudless maps for the post-event images from 27 April, 1 May and 2 May.

Change-Detection Product and Analysis

The change-detection map produced by IR-MAD in this paper is a coarse detection map and needs further manual checks to improve the accuracy of the product. In this section, the four images of Figure 18 are tested. The image in Figure 23I is the change-detection map between 11 April and 27 April, in which clouds and shadows are considered as the change, andFigure 23II is the change-detection map after removing the cloud and shadow regions based on the Figure 23I image. Figure 23A–D give the detailed enlarged images of the red rectangle in the large maps. Figure 23A illustrates the change of open grass land before and after earthquake, where some tents have been put up as temporary refuge. Figure 23B indicates the landmark building of the Kathmandu, Bhimsen Tower, completely collapsed after the earthquake. Figure 23C represents an error change detection result, as the different imaging angle of two images result in the different positions of tall buildings and their shadows in the images. Though the images are matched by the method proposed in Section 2.2, the differences caused by the viewing angle is very difficult to be corrected. Figure 23D also shows an error change-detection result, because part of the shadow area is missing from detection. From the above analysis, the change-detection map produced by our proposed procedure is a coarse result without distinguishing the type and grade of change and many errors are detected, which require manual checking to refine the work.

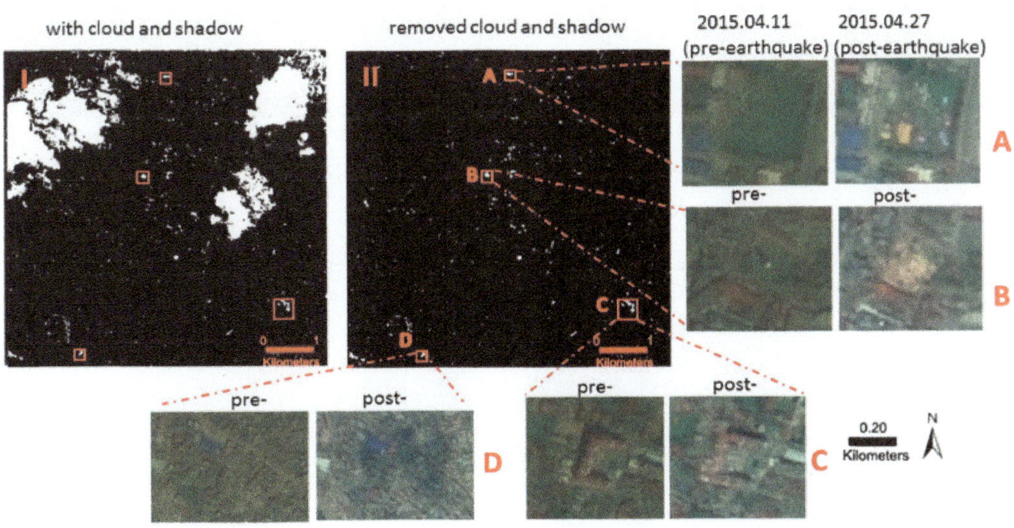

Figure 23. The change-detection map of image between 11 April and 27 April. (**I** image contains the cloud and shadow, **II** image remove the impact of cloud and shadow based on I image, **A–D** are the different detected changing scenes, A and B are the right detected scenes, C and D are false detected scenes).

The image in Figure 24I is the same image shown in Figure 23II, which is the change map between 11 April and 27 April. The image in Figure 24II is the change between 1 May and 11 April and the image in Figure 24III is the change between 27 April and 1 May. All of these eliminate the impact of cloud and shadow. Figure 25 shows the detailed enlarged images of the red rectangles of Figure 24. In Figure 25A, the Durbar Square of Kathmandu, which damaged severely in the earthquake, is shown, but this change (the damage) is not discovered because the clouds and shadows exist in the image of 27 April (seen in Figure 25A). In Figure 24III, the change is observed, since the image of 1 May is cloudless in this region. For B sense in the Figure 24, the reason of discovering the change in II, not in I and III map is the same with the A sense as the influence of shadow (seen the detail in Figure 25B). Figure 25C shows the Kal Mochan Temple of Kathmandu, which also collapsed after the earthquake. In Figure 24I,II, the change area is the same large, but small in Figure 24III map. This is because the feature on the ground has no major changes in the post-earthquake images (27 April and 1 May), while the feature changed obviously between pre- and post-earthquake images (Figure 24I is the result of 11 April and 27 April, Figure 24II is the result of 11 April and 1 May). There are just some rocks from the collapsed temple that have been consolidated during the period from 27 April to 1 May to clean the area of the ruins (shown in Figure 25C). By these comparisons, the progress of the rescue will be learned. The same reason applies to the Figure 25D,E scenes. From these images, we see that many tents were put up in some open space nearby the houses of victims as temporary shelters because of the constant large aftershocks. The temporary shelters were migrated or revoked on 1 May as the rescue proceeded. So some temporary facilities would disappear on the image of 1 May (seen in Figure 25D,E); and no or few change areas would be detected in image III of Figure 24. The results of this paper mainly provide a quick change detection map to determine the urgent requirements, as well as offer guidance to check the change regions precisely through a community remote sensing platform for experts or the public, which is possible to reduce labor costs and improve efficiency in the maximum extent. Figure 26 show the detected changes vector superposition on cloudless maps of 27 April. The vector of the left hand map is the final result, which was extracted from the image I of Figure 26; and the vector of the right hand map removes the error regions of the final result by visual checking. The right vector map is the combination of visual interpretation by four independent people. This vector map can be considered the relative correct result for comparing. The simple statistic is shown in Table 6. The number of change regions in the original vector map (shown in Figure 26 left hand map) is 758, while the number in the checked vector map (shown in Figure 26 right hand map) is 556, and the error detected change regions are 202. The percentage of truth change regions and original change detected regions is 73.4%. This can meet the application requirements to a certain degree. In the true change regions, most of them are tents or shelters (shown Figure 23A, Figure 25D,E, and Figure 27A), some are cars on the street or other temporary features, and a few are the collapsed or damaged buildings (shown in Figure 25C and Figure 27B). In the error change regions, as analyzed above, most of the error is caused by the changes in different view angles between pre-event maps and post-event maps, and some are caused by undetected clouds, thin clouds, shadows, or other unknown errors.

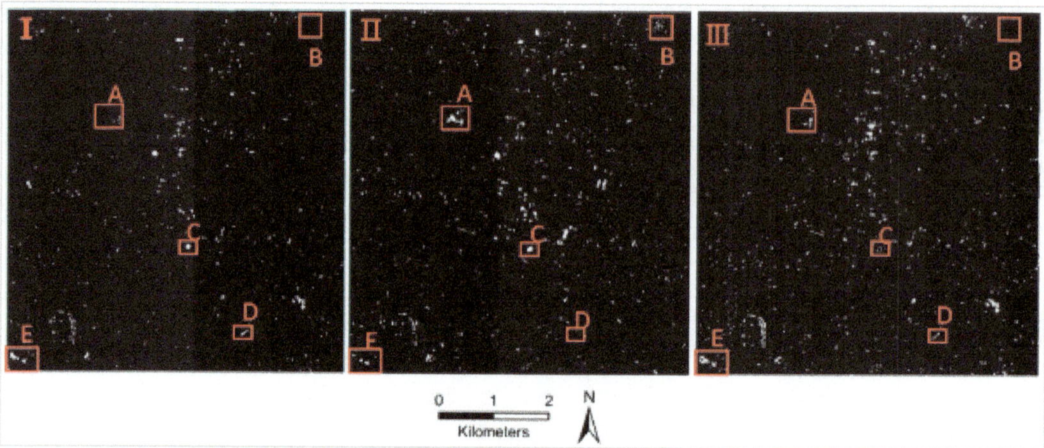

Figure 24. The change-detection map of image between 11 April, 27 April and 1 May after removing the cloud and shadow (**I** is the change map of 11 April and 27 April, **II**is the change map of 11 April and 1 May, **III** is the change map of 27 April and 1 May.**A–E** region are the different sense and the detail show in Figure 25).

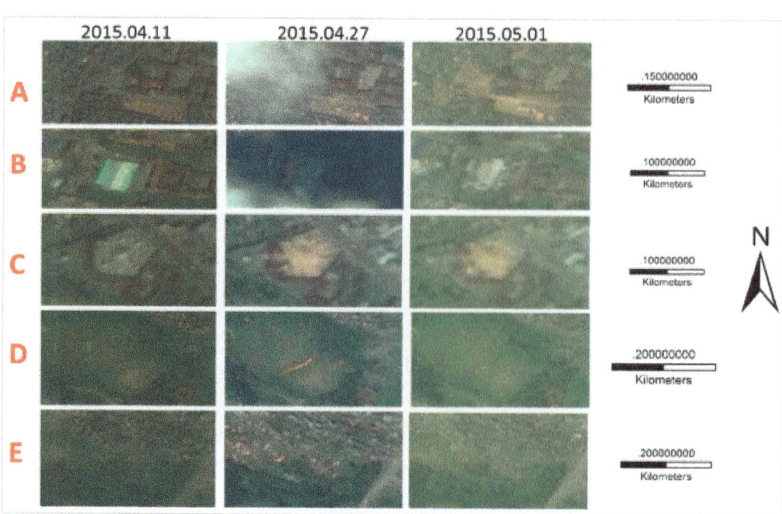

Figure 25. The detailed enlarged images of the red rectangle in Figure 24 (**A** and **B** are influenced by clouds and shadows; **C** is a collapsed temple; **D** and **E** are the open spaces where many tents are being set up).

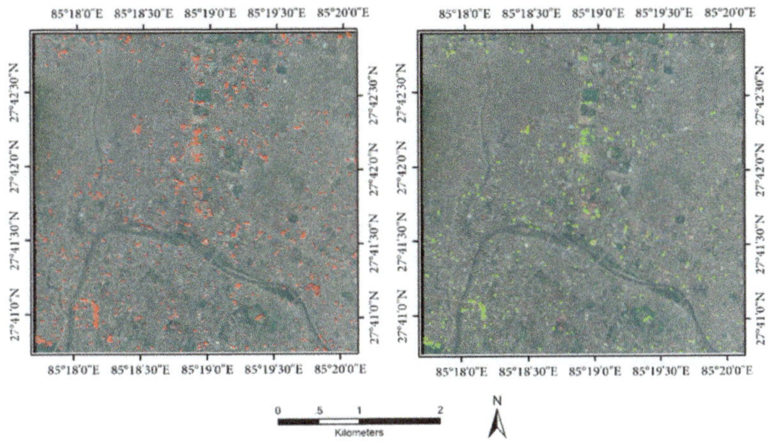

Figure 26. The detected changes vector superposition on the cloudless map of 27 April. (the vector of the left hand map is extracted from the I image of Figure 24, and the vector of the right hand map has removed the error regions of the left hand vector map by visual checking).

Figure 27. The temporal change detail map of two scenes (A shows a series of changes in Center Park from 27 April to 2 May and B shows the change of a collapsed tower).

Through the change detection map, the rescue progress can be observed. An example is shown in Figure 27. The image inFigure 27A shows the difference over four days in the open space of a park in the center of Kathmandu. The image from 27 April, the day after earthquake, indicates that the situation in this open space is in chaos, with a variety of temporary residences set up. However, by the sixth day after the earthquake, on the 1 May, the temporary settlements are orderly. This phenomenon is more obvious and more and unified residences and relief tents are set up by 2 May. The image in Figure 27B shows the series of changes of Bhimsen Tower. From 27 April to 2 May, the bricks and masonry from the ruins are cleaned up effectively and the situation changed from one of chaos to order. These temporal change maps indicate the progress of the disaster rescue, allowing for more objective guidance to formulate the next plan of rescue and allocate rescue resources reasonably.

Discussion

The goal of this work was to solve the problems related to the use of optical remote sensing data in disaster emergency mapping. The main problems are the impact of clouds in obtaining a clear situation backdrop map after the disaster and the contradiction between obtaining the poor precision change detection maps by automated methods and the relatively labor intensive process by visual interpretations. Through a series of new methods proposed in this paper, the accuracy of optical data pre-processing results have been improved greatly and the produced change detection maps can fulfill the fundamental requirement of disaster S&R activities, which is the information regarding the disaster in terms of the number, location, distribution, arrangement, extent, and impact of the devastation. These rather coarse change maps also can be used as a guideline and the basis for visual observation to check the correctness of the detected change in objects. This can then generate more refined damage assessment maps and up-to-date maps for evaluation.

Map Accuracy and Timeliness

Reliability of the results and timeliness of the process are the key factors in emergency mapping. In previous studies, the biggest problem of optical data used in damage mapping by automatic change detection method was the calibration stage, which is a time consuming procedure with poor accuracy of results. In this paper, OptVM for image fusion and SIFT-OFM for image registration are proposed for producing a high accuracy fused image and matched image to obtain a rather high precision change detection map. Comparing the results based on traditional image fusion and registration methods, the results based on our proposed methods are the best overall (see the analysis in Section 3.2 and Section 3.3). The false alarm rate of the final change detection scenes, which are caused by improper co-registration, can be reduced by relying on the rather high accuracy of the fused image and the matched image. Timeliness is also an issue. The time for processing the images of 11 April, 27 April, and 1 May from the Kathmandu example was only about 46 min to obtain the cloudless map shown in Figure 20 and the change map shown in Figure 24 using a PC with 8GB RAM and a 64-bit operating system. This is well worth the effort to obtain a cloudless situation backdrop map and a rather coarse change map. The time can be shortened if cluster or GPU parallel computing is used.

Limitations

Although a series of methods or strategies are used in this study to reduce the effects of optical imagery limitations, some limitations are still presented in the maps produced by this study. First, cloudless situation maps are available only when cloud cover is less than 50% of the image taken on the same day; otherwise, the cloudless maps contain a lot of information from the previous images. When the cloud coverage in the image is more than 80%, the image is completely unavailable, and only the SAR data is available to determine the situation of the cloud and shadow areas. The other limitation is that the generation of cloudless maps requires at least two post-event images with non-overlapping cloud covered areas. Post-event data are a bit more challenging due to cloud cover during the time of satellite passes. In some cases, getting a post-event cloudless map of the entire damaged area requires a great deal of time. However, multiple satellites passes and multiple data increase the possibility of obtaining cloudless images. Generally, a single post-event imagery is available in the first hours after the event. If it has cloud covered areas, there is no method other than to just replace the cloud areas with the nearest pre-event clear data when we want to get cloudless maps. Our proposed method can produce a cloudless map in this case. It is also useful for rescue teams to learn the situation of the affected areas using cloudless maps before they start search and rescue (S&R) activities within a few hours after the disaster. In parts of the country without internet access, the professional rescue teams need to take a cloudless printed map. It is important to provide cloudless maps even though cloud cover areas are replaced by pre-event clear data. In this kind cloudless map (seen Figure 22 left image), the vectors of detected cloud and shadow regions are provided to distinguish where is the pre-event or post-event data to avoid confusion for users.

Next, undetected clouds and shadows, variations in solar illumination, different off-nadir angles, and some temporary objects (e.g., car traffic), which could influence the results of change detection, still exist and reduce the accuracy of change maps in this example (as shown in Figure 23C,D). These errors are difficult to exclude from automatic change detected maps. It is only by manual visual checking that errors can be removed to obtain more precise change maps. That is why we emphasize the important of manual participation for further checking the change maps in this study. These change maps are just the coarse result, which provides some basic main information for the affected regions. In addition, the detected changes in the product maps are not divided into categories and degrees, which also rely on manual interpretation to measure.

Then, a main limitation is the data resolution. As the policy of data acquisition and fund problems, the highest data resolution obtained was only 2 m with GF-1. This is not sufficient for detecting and assessing the damaged buildings, except those that are completely collapsed (shown in Figure 27B) or partially collapsed (shown in Figure 25C). The main purpose of the change detection maps in this study is to detect the scene after the earthquake to determine whether it was changed or not. The change detection maps could provide guidance of the changed regions for further checking as mentioned above.

Lastly, though image-based mapping has played an important role in rapid post-disaster damage assessment, the detail and accuracy of assessment has not reached the level of ground-based surveys [9]. The limitation of image-based damage assessment is not only related to the spatial resolution of the sensors, but also the reflected information reflected by image itself. Generally, most of the satellite images for disaster mapping are acquire by vertical angle (or small angle offset vertical), and the provided information of images, such as for building, is the situation of roofs. The roof information is well suited for the identification of extreme damage states, i.e., completely destroyed structures [61]. But for some not collapsed buildings with cracks or inclined walls, the results of image-based mapping are largely missed and lead to assessment deficiency of this type damage in the final results. Recently oblique or multi-perspective data, which provide information of both roofs and facades, has been identified as a potential data source for a more comprehensive building damage assessment [62,63].

Conclusions and Future Work

Using remote sensing for emergency mapping is now a common operational approach for disaster risk management. Both optical and radar data are the main input data when disasters occur. However, for damage assessment or analysis purposes, a number of results have been presented after every event mostly based on optical data. These can be easy to read and analyze to generate fast maps showing the location, scale, or the extent of disaster or crisis situation. In this study, a rapid procedure for transforming optical remote sensing data is proposed to produce cloudless situation backdrop maps and change detection maps for the emergency needs in the early phases of disaster recovery. On the one hand, the precision and reliability of the products has improved significantly by using the new methods, which include OptVM and SIFT-OFM. On the other hand, the automatic processing time is shortened by proposed procedure. It is useful to provide some rapid and actionable information for the governmental agencies, multilateral organizations, and NGOs in a timely fashion. While only the GF-1 temporal data were tested in this study as the limitation of data acquisition, the proposed procedure is also applicable to the other remote sensing data. It can even provide up-to-date situation awareness, loss estimation, and recovery monitoring maps to support decision making by processing multi-source and multi-temporal optical remote sensing data together.

The accuracy of the results from the automatic methods is obviously not as good as the manual visual interpretation by good cooperation of professional image analysts or skilled operators [63]. Manual visual interpretation, is always a rather labor intensive task. An ideal solution is to minimize the impact of restrictions to improve the accuracy of automatic methods results continuously. These automatic methods results can save the labor-cost to the maximum extent and provide some guideline for further refine the results by manual visual checking, though the results is rather coarse. This is also our main purposes of this study. International cooperation and volunteer contributions play an important role in mapping for disaster impact assessment at the urban level. International agencies and volunteers should be coordinated properly to ensure that common mapping guidelines and proper communication mechanisms are in place, for more effective cooperation among the involved actors especially during major emergencies.

Remote sensing is an effective tool at the disposal of a disaster manager. It can supply authentic and objective information for various rescue resources and the general public, especially in remote or inaccessible areas during the early period of the disaster. It would be strongest when allied with other tools, such as catastrophe risk models or GIS tools [64], to provide a powerful guidance to streamline evacuation, issue disaster declarations, coordinate S&R, coordinate logistics, conduct damage assessment, distribute relief supplies, and aid in long-term recovery. Some emerging technologies and initiatives will provide updated change and geospatial information as well as the resources distribution and demand data to analyze it in a time frame suitable for emergency response purposes. In the future, integrating different remote sensing data based on emergency mapping and other GIS information or analysis tools will be the popular trend to reduce economic losses, ensure public safety, and meet the basic subsistence needs of the people affected in an emergency response context.

ACKNOWLEDGMENTS

This work was supported by the civil aerospace pre-research project during the "12th Five-Year Plan", the satellite special project of National Development and Reform Commission: the system of remote sensing satellite data real-time active service and the 135 Strategy Planning of Institute of Remote Sensing and Digital Earth, CAS.

Author Contributions

Yong Ma developed the methods, carried out the experiments and analyzed the results. Fu Chen and Jianbo Liu supervised the research. Yang He, Jianbo Duan and Xinpeng Li prepared the data. Yong Ma and Fu Chen wrote the manuscript.

Conflicts of Interest

The authors declare no conflict of interest.

Appendix

The four indices used in this paper are defined as follows

(1) The Spectral Angle Mapper (SAM): The SAM measures the absolute angle between the spectral vectors between the reference and fused images, which reflects the spectral distortion. The best value is 0, and the equation is shown as follow

$$SAM(A(i,j), B(i,j)) = \arccos\left(\frac{\langle A(i,j), B(i,j)\rangle}{\|A(i,j)\|_2 \cdot \|B(i,j)\|_2}\right)$$

where A(i, j) and B(i, j) are the value of pixel (i, j) on the reference and fused images.

2) The error of relative dimensionless global in synthesis (ERGAS): ERGAS is a comprehensive index, which gives a global depiction of the radiometric difference between the reference and fused images. The best value is 0 and it is defined as

$$ERGAS(A_k, B_k) = \frac{100}{ratio}\sqrt{\frac{1}{N}\sum_{k=1}^{N}\left(\frac{RMSE(A_k, B_k)^2}{Mean(B_k)}\right)}$$

where A is the fused image and B is the reference image, ratio is the ratio of the spatial resolutions of the PAN image and the MS image, RMES is the root mean square of image A and B, Mean represents the average value of the image.

3) The Q4 Index: the Q4 index [34] is an extension of the quality index, which is a product of the correlation coefficient, mean bias, and contrast variation. The best value is 1 and it is described as follow:

$$Q4_{D\times D} = \frac{\delta_{z_1 z_2}}{\delta_{z_1}\delta_{z_2}} \cdot \frac{2\delta_{z_1}\cdot\delta_{z_2}}{\delta_{Z_1}^2+\delta_{Z_2}^2} \cdot \frac{2\left|\overline{Z_1}\right|\cdot\left|\overline{Z_2}\right|}{\left|\overline{Z_1}\right|^2+\left|\overline{Z_2}\right|^2}$$

$$Z_1 = X_1 + i\times X_2 + j\times X_3 + k\times X_4$$

$$Z_2 = Y_1 + i\times Y_2 + j\times Y_3 + k\times Y_4$$

$$Z_1^* = X_1 - i\times X_2 - j\times X_3 - k\times X_4$$

$$\left|\overline{Z_1}\right| = \sqrt{\overline{Z_1}\times\overline{Z_1^*}} = \sqrt{(\overline{X_1^i})^2 + (\overline{X_2^i})^2 + (\overline{X_3^i})^2 + (\overline{X_4^i})^2}$$

where $\left|\overline{Z_2}\right|$ is the same way with $\left|\overline{Z_1}\right|$, X_1 and Y_1 are the first band of image A and B respectively, and so are the other bands. σ_{z1z2} is the covariance of z_1 and z_2, andσ_{z1} and σ_{z2} is the variance of z_1 and z_2. The first term of the equation is the modulus of the hyper-complex CC between z_1 and z_2, and the second and third terms measure contrast changes and the mean bias on all bands, respectively. D is block size of image and in this paper the size is 32*32.

4) Quality with No Reference (QNR): QNR measure the quality of fused image and reference image from the spectral and spatial distortion. The best value is 1 and it is defined as follow:

$$QNR = (1 - D_\lambda)^\alpha (1 - D_s)^\beta$$

$$D_\lambda = \sqrt[p]{\frac{1}{L(L-1)}\sum_{l=1}^{L}\sum_{\substack{r=1\\ r\neq 1}}^{L}\left|Q\left(\widehat{G_l},\widehat{G_r}\right) - Q\left(\widetilde{G_l},\widetilde{G_r}\right)\right|^p}$$

$$D_S = \sqrt[q]{\frac{1}{L(L-1)}\sum_{l=1}^{L}\left|Q\left(\widehat{G_l},P\right) - Q\left(\widetilde{G_l},\widetilde{P}\right)\right|^q}$$

where L is the number of bands, G^ is fused HRM image ,and G~ is LRM image. p is integer index of amplifying the spectrum distortion, generally taking p = 1. P is HRP image and P~ is panchromatic image after processing by low-pass filter. q is integer index of amplifying of spatial distortion, generally taking q= 1,and α=β=1.

REFERENCES

1. Dou, M.; Chen, J.; Chen, D.; Chen, X.; Deng, Z.; Zhang, X.; Xu, K.; Wang, J. Modeling and simulation for natural disaster contingency planning driven by high-resolution remote sensing images. Future Gener. Comput. Syst. **2014**, 37, 367–377.
2. Huyck, C.; Verrucci, E.; Bevington, J. Remote Sensing for Disaster Response: A Rapid, Image-Based Perspective. InEarthquake Hazard, Risk, and Disasters; Academic Press: Cambridge, MA, USA, 2014; pp. 1–24.
3. Joyce, K.E.; Belliss, S.E.; Samsonov, S.V.; McNeill, S.J.; Glassey, P.J. A review of the status of satellite remote sensing and image processing techniques for mapping natural hazards and disasters. Prog. Phys. Geogr. **2009**, 33, 183–207.
4. Voigt, S.; Kemper, T.; Riedlinger, T.; Kiefl, R.; Scholte, K.; Mehl, H. Satellite image analysis for disaster and crisis-management support. IEEE Trans. Geosci. Remote Sens. **2007**, 45, 1520–1528.
5. Boccardo, P.; Tonolo, F.G. Remote Sensing Role in Emergency Mapping for Disaster Response. In Engineering Geology for Society and Territory; Springer International Publishing: New York, NY, USA, 2015; Volume 5, pp. 17–24.
6. Arciniegas, G.A.; Bijker, W.; Kerle, N.; Tolpekin, V.A. Coherence- and amplitude-based analysis of seismogenic damage in Bam, Iran, using ENVISAT ASAR data. IEEE Trans. Geosci. Remote Sens. **2007**, 45, 1571.
7. Voigt, S.; Schneiderhan, T.; Twele, A.; Gähler, M.; Stein, E.; Mehl, H. Rapid damage assessment and situation mapping: Learning from the 2010 Haiti earthquake. Photogramm. Eng. Remote Sens. **2011**, 77, 923–931.
8. Baiocchi, V.; Dominici, D.; Giannone, F.; Zucconi, M. Rapid building damage assessment using EROS B data: The case study of L'Aquila earthquake. Ital. J. Remote Sens. **2012**, 44, 153–165.
9. Kerle, N. Satellite-based damage mapping following the

2006 Indonesia earthquake—How accurate was it? Int. J. Appl. Earth Obs. **2010**, 12, 466–476.

10. Ehrlich, D.; Guo, H.; Molch, K.; Ma, J.; Pesaresi, M. Identifying damage caused by the 2008 Wenchuan earthquake from VHR remote sensing data. Int. J. Digit. Earth **2009**, 2, 309–326.
11. Chini, M. Earthquake Damage Mapping Techniques Using SAR and Optical Remote Sensing Satellite Data; INTECH Open Access Publisher: Rijeka, Croatia, 2009.
12. Dell'Acqua, F.; Bignami, C.; Chini, M.; Lisini, G.; Polli, D.A.; Stramondo, S. Earthquake damages rapid mapping by satellite remote sensing data: L'Aquila 6 April 2009 event. IEEE J. Sel. Top. Appl. Earth Obs. Remote Sens. **2011**, 4, 935–943.
13. Yamazaki, F.; Kouchi, K.I.; Kohiyama, M.; Muraoka, N.; Matsuoka, M. Earthquake damage detection using high-resolution satellite images. In Proceedings of the IEEE International Geoscience and Remote Sensing Symposium, Anchorage, Alaska, 20–24 September 2004; IEEE: Piscataway, NJ, USA, 2004; Volume 4, pp. 2280–2283.
14. Ajmar, A.; Boccardo, P.; Disabato, F.; Tonolo, F.G. Rapid Mapping: Geomatics role and research opportunities. Rend. Lincei **2015**, 26, 63–73.
15. Zhang, Y. Understanding image fusion. Photogramm. Eng. Remote Sens. **2004**, 70, 657–661.
16. Jiang, C.; Zhang, H.; Shen, H.; Zhang, L. A practical compressed sensing-based pan-sharpening method. IEEE Geosci. Remote Sens. Lett. **2012**, 9, 629–633.
17. Shahdoosti, H.R.; Ghassemian, H. Spatial PCA as a new method for image fusion. In Proceedings of the 16th CSI International Symposium on Artificial Intelligence and Signal Processing (AISP), Shiraz, Iran, 2–3 May 2012; IEEE: Piscataway, NJ, USA, 2012; pp. 90–94.
18. El-Mezouar, M.C.; Taleb, N.; Kpalma, K.; Ronsin, J. An IHS-based fusion for color distortion reduction and vegetation enhancement in IKONOS imagery. IEEE Trans. Geosci. Remote Sens. **2011**, 49, 1590–1602.
19. Zhang, Y. Problems in the Fusion of Commercial High-Resolution Satelitte as well as Landsat 7 Images and Initial Solutions. Int. Arch. Photogramm. Remote Sens. Spat. Inf. Sci. **2002**, 34, 587–592.
20. Pradhan, P.S.; King, R.L.; Younan, N.H.; Holcomb, D.W. Estimation of the number of decomposition levels for a wavelet-based multiresolution multisensor image fusion. IEEE Trans. Geosci. Remote Sens. **2006**, 44, 3674–3686.
21. Zhang, L.; Shen, H.; Gong, W.; Zhang, H. Adjustable model-based fusion method for multispectral and panchromatic images. IEEE Trans. Syst. Man Cybern. B Cybern. **2012**, 42, 1693–1704.
22. Li, S.; Yang, B. A new pan-sharpening method using a compressed sensing technique. IEEE Trans. Geosci. Remote Sens.**2011**, 49, 738–746.
23. Ghahremani, M.; Ghassemian, H. Remote sensing image fusion using ripplet transform and compressed sensing. IEEE Geosci. Remote Sens. Lett. **2015**, 12, 502–506.
24. Ballester, C.; Caselles, V.; Igual, L.; Verdera, J.; Rougé, B. A variational model for P+ XS image fusion. Int. J. Comput. Vis. **2006**, 69, 43–58.
25. Paige, C.C.; Saunders, M.A. LSQR: An algorithm for sparse linear equations and sparse least squares. ACM Trans. Math. Softw. (TOMS) **1982**, 8, 43–71.
26. Le Moigne, J.; Netanyahu, N.S.; Eastman, R.D. Image Registration for Remote Sensing; Cambridge University Press: England, UK, 2011.
27. Avants, B.B.; Epstein, C.L.; Grossman, M.; Gee, J.C. Symmetric diffeomorphic image registration with cross-correlation: Evaluating automated labeling of elderly and neurodegenerative brain. Med. Image Anal. **2008**, 12, 26–41.
28. Dame, A.; Marchand, E. Second-order optimization of mutual information for real-time image registration. IEEE Trans. Image Process. **2012**, 21, 4190–4203.
29. Yu, X.; Chen, X.; Jiang, M. Motion detection in moving background using a novel algorithm based on image features guiding self-adaptive Sequential Similarity Detection Algorithm. Optik-Int. J. Light Electron Opt. **2012**, 123, 2031–2037.
30. Lowe, D.G. Distinctive image features from scale-invariant keypoints. Int. J. Comput. Vis. **2004**, 60, 91–110.
31. Mikolajczyk, K.; Schmid, C. Scale & affine invariant interest point detectors. Int. J. Comput. Vis. **2004**, 60, 63–86.
32. Sima, A.A.; Buckley, S.J. Optimizing SIFT for matching of short wave infrared and visible wavelength images. Remote Sens. **2013**, 5, 2037–2056.
33. Bajcsy, R.; Kovačič, S. Multiresolution elastic matching. Comput. Vis. Graph. Image Process. **1989**, 46, 1–21.
34. Zhang, Z.; Li, J.Z.; Li, D.D. Research of automated image registration technique for infrared images based on optical flow field analysis. J. Infrared Millim. Waves **2003**, 22, 307–312.
35. Liu, J.; Yan, H. Phase correlation pixel-to-pixel image co-registration based on optical flow and median shift propagation. Int. J. Remote Sens. **2008**, 29, 5943–5956.
36. Huo, C.; Pan, C.; Huo, L.; Zhou, Z. Multilevel SIFT matching for large-size VHR image registration. IEEE Geosci. Remote Sens. Lett. **2012**, 9, 171–175.
37. Horn, B.K.; Schunck, B.G. Determining optical flow. In 1981 Technical Symposium East; International Society for Optics and Photonics: Bellingham, WA, USA, 1981.
38. Allahviranloo, T. Successive over relaxation iterative method for fuzzy system of linear equations. Appl. Math. Comput.**2005**, 162, 189–196.
39. Shen, H.; Li, H.; Qian, Y.; Zhang, L.; Yuan, Q. An effective thin cloud removal procedure for visible remote sensing images. ISPRS J. Photogramm. Remote Sens. **2014**, 96, 224–235.
40. Lin, C.-H.; Tsai, P.-H.; Lai, K.-H.; Chen, J.-Y. Cloud removal from multitemporal satellite images using information cloning. IEEE Trans. Geosci. Remote Sens. **2013**, 51, 232–241.

41. Chen, J.; Zhu, X.; Vogelmann, J.E.; Gao, F.; Jin, S. A simple and effective method for filling gaps in Landsat ETM+ SLC-off images. Remote Sens. Environ. **2011**, 115, 1053–1064.

42. Zhang, Y.; Guindon, B.; Cihlar, J. An image transform to characterize and compensate for spatial variations in thin cloud contamination of Landsat images. Remote Sens. Environ. **2002**, 82, 173–187.

43. Gao, F.; Masek, J.; Schwaller, M.; Hall, F. On the blending of the Landsat and MODIS surface reflectance: Predicting daily Landsat surface reflectance. IEEE Trans. Geosci. Remote Sens. **2006**, 44, 2207–2218.

44. Helmer, E.; Ruefenacht, B. Cloud-free satellite image mosaics with regression trees and histogram matching. Photogramm. Eng. Remote Sens. **2005**, 71, 1079–1089.

45. Cheng, Q.; Shen, H.; Zhang, L.; Yuan, Q.; Zeng, C. Cloud removal for remotely sensed images by similar pixel replacement guided with a spatio-temporal MRF model. ISPRS J. Photogramm. Remote Sens. **2014**, 92, 54–68.

46. Zhengke, G.; Fu, C.; Jin, Y.; Xinpeng, L.; Fangjun, L.; Jing, Z. Automatic cloud and cloud shadow removal method for landsat TM images. In Proceedings of the 10th International Conference on Electronic Measurement & Instruments (ICEMI), Chengdu, China, 16–19 August 2011; IEEE: Piscataway, NJ, USA, 2011; Volume 3, pp. 80–84.

47. Zhu, Z.; Woodcock, C.E. Object-based cloud and cloud shadow detection in Landsat imagery. Remote Sens. Environ.**2012**, 118, 83–94.

48. Chini, M.; Pierdicca, N.; Emery, W.J. Exploiting SAR and VHR optical images to quantify damage caused by the 2003 Bam earthquake. IEEE Trans. Geosci. Remote Sens. **2009**, 47, 145–152.

49. Turker, M.; Sumer, E. Building-based damage detection due to earthquake using the watershed segmentation of the post-event aerial images. Int. J. Remote Sens. **2008**, 29, 3073–3089.

50. Miura, H.; Midorikawa, S.; Kerle, N. Detection of building damage areas of the 2006 central Java, Indonesia, earthquake through digital analysis of optical satellite images. Earthq. Spectra **2013**, 29, 453–473.

51. Canty, M.J.; Nielsen, A.A. Automatic radiometric normalization of multitemporal satellite imagery with the iteratively re-weighted MAD transformation. Remote Sens. Environ. **2008**, 112, 1025–1036.

52. Nielsen, A.A. The regularized iteratively reweighted MAD method for change detection in multi-and hyperspectral data. IEEE Trans. Image Process. **2007**, 16, 463–478.

53. Marpu, P.R.; Gamba, P.; Canty, M.J. Improving change detection results of IR-MAD by eliminating strong changes.IEEE Geosci. Remote Sens. Lett. **2011**, 8, 799–803.

54. Qiang, L.; Yuan, G.; Ying, C. Study on disaster information management system compatible with VGI and crowdsourcing. In Proceedings of the 2014 IEEE Workshop on Advanced Research and Technology in Industry Applications (WARTIA), Ottawa, ON, Canada, 29–30 September 2014; IEEE: Piscataway, NJ, USA, 2014; pp. 464–468.

55. Resor, E.E.L. The Neo-Humanitarians: Assessing the Credibility of Organized Volunteer Crisis Mappers. Ph.D. Thesis, Massachusetts Institute of Technology, Cambridge, MA, USA, 2013.

56. Boccardo, P. New perspectives in emergency mapping. Eur. J. Remote Sens. **2013**, 46, 571–582.

57. Barrington, L.; Ghosh, S.; Greene, M.; Har-Noy, S.; Berger, J.; Gill, S.; Lin, A.Y.-M.; Huyck, C. Crowdsourcing earthquake damage assessment using remote sensing imagery. Ann. Geophys. **2012**, 54.

58. Alparone, L.; Baronti, S.; Garzelli, A.; Nencini, F. A global quality measurement of pan-sharpened multispectral imagery. IEEE Geosci. Remote Sens. Lett. **2004**, 1, 313–317.

59. Khan, M.M.; Alparone, L.; Chanussot, J. Pansharpening quality assessment using the modulation transfer functions of instruments. IEEE Trans. Geosci. Remote Sens. **2009**, 47, 3880–3891.

60. Amintoosi, M.; Fathy, M.; Mozayani, N. Precise image registration with structural similarity error measurement applied to superresolution. EURASIP J. Adv. Signal Process. **2009**, 2009, 1–7.

61. Fernandez Galarreta, J.; Kerle, N.; Gerke, M. UAV-based urban structural damage assessment using object-based image analysis and semantic reasoning. Nat. Hazards Earth Sys. Sci. **2015**, 15, 1087–1101.

62. Vetrivel, A.; Gerke, M.; Kerle, N.; Vosselman, G. Identification of damage in buildings based on gaps in 3D point clouds from very high resolution oblique airborne images. ISPRS J. Photogramm. Remote Sens. **2015**, 105, 61–78.

63. Kerle, N.; Hoffman, R.R. Collaborative damage mapping for emergency response: The role of cognitive systems engineering. Nat. Hazards Earth Syst. Sci. **2013**, 13, 97–113.

64. Eguchi, R.T.; Huyck, C.K.; Ghosh, S.; Adams, B.J.; McMillan, A. Utilizing New Technologies in Managing Hazards and Disasters Geospatial Techniques in Urban Hazard and Disaster Analysis; Springer: Houten, The Netherlands, 2010; pp. 295–323.

CHAPTER 18 DEVELOPMENT OF MOBILE MAPPING SYSTEM FOR 3D ROAD ASSET INVENTORY

Nivedita Sairam, Sudhagar Nagarajan and Scott Ornitz

Keywords: Mobile mapping system, transportation asset inventory, LiDAR

INTRODUCTION

The lifetime of Civil Engineering infrastructure projects are generally made up of three phases—design and planning, construction and maintenance. The use and maintenance phase has the longest duration and involves most of the project cost. Different construction engineering projects, such as development of roads, bridges, parks and other utilities are expected to be in a usable state for several years. However, damage and deterioration are unavoidable. The amount of money spent on repair or replacement of the assets are also very high. In order to make the process of maintenance and improvements more manageable, it is necessary to document the assets and inventories in an appropriate format. The recent advancements in digital maps and GIS (Geographic Information System) technology has improved the efficiency of asset and inventory management. GPS (Global Positioning System) is an instrument that is widely used to locate and map the positions of assets. The most important task in the development of an asset management system is the collection of accurate spatial data and its related attributes. Manual data collection using field data loggers with GPS and traditional survey methods are commonly seen for asset inventory data collection. Though this methodology provides accurate results, it involves more time and manpower. Several remote sensing products, such as aerial imageries, terrestrial photographs and laser point clouds with high accuracies also contribute to the development of an asset inventory database. This paper focuses on development and usage of a mobile mapping system for documenting transportation asset inventories.

A typical mobile mapping system is equipped with a laser scanner, panoramic cameras, GPS and IMU (Inertial Measurement Unit) positioning setup, on-board computer and a storage device for data logging. Airborne LiDAR data and aerial images are highly suitable for extracting features, such as building footprints, roads and vegetation cover. Mapping of infrastructure facilities may require much higher level of detail including accurate positions of road signs and signals, curb height and width of pavements, availability of ramps on the sidewalks, etc. Mobile mapping system tends to be more advantageous in these scenarios. More recent mapping products with up-to date data is preferred over highly accurate outdated data for most cartography applications. Since the field techniques are largely subjective, it calls for a technology that is both economical and expedient. With the introduction of low cost and portable laser scanners in the market, most of the mobile mapping systems are equipped with laser scanners in addition to video cameras. Thus, a combination of georeferenced pictures and three-dimensional point cloud provides a better scope for extracting information from the scene. The paper is divided into following sections: The first section discusses the background for the study which includes a detailed description about existing methods of data collection and literature references supporting various advancements in

Citation: Nivedita Sairam, Sudhagar Nagarajan and Scott Ornitz, Development of Mobile Mapping System for 3D Road Asset Inventory, doi:10.3390/s16030367.

applications of mobile mapping systems in the domain of asset management. The second section describes the methodology involved which includes the task of building the system, different components and their communication networks, co-registering multi-sensor data, basic processing of the data and integrating multi-sensor data into a single GIS system. Specific details about the experiments conducted, discussion about the proposed framework and concluding remarks.

BACKGROUND

Spatial data about roadside asset inventories and other municipal utilities are generally collected using traditional land survey methods. There are two types of information recorded—inventory (structures and road signs) and condition of the inventories [1]. The field survey methods use GPS and total stations to record the location of various assets. GPS can be easily used to map street furniture like lamp posts, sign boards, etc. [2]. Though manual survey methods are highly accurate in two-dimensional space, it is difficult to obtain the third dimension of points with high accuracy. These methods are time consuming and cumbersome, as well [3]. There are several other data management and data integrity challenges associated with using traditional survey methods for replenishing asset inventory database. The data might not be consistent since a number of manual tasks are involved [4]. There are also cases where the data is incoherently spread across different systems, in different formats, depending on the survey crew and temporal aspects of data collection. There is a definite need to integrate the inventory data as a single system with a common database structure in order to add relevant attribute data that is relevant during supervision, maintenance and replacement jobs. A good comparison of techniques by manual and mobile mapping data collection is explained in [5]. The introduction of field data collectors into the market has improved the consistency of survey data and also resulted in a less cumbersome and systematic way of data collection. Field data collectors are devices with inbuilt GPS to collect and store details of features. These devices log the location and attribute data directly to a server or have a storage inbuilt [6–8]. Mobile GIS technology is also increasingly used for asset inventory data collection. Mobile applications with spatial analysis capabilities are increasingly found in the market which can be used to collect data about features using the internal GPS of the phone. Though the process of data collection is direct and simple, the accuracy of a phone GPS is low and the features are coarsely mapped. When asset inventory data is collected using field survey methods, most of the time and resources are spent on field work. The office work is limited to data migration from the data collector to the servers and entry of attribute data [5]. While using field data collectors, it is not always possible to verify the details without visiting the field again. However, newer instruments are equipped with cameras which aids in review and verification of the collected data.

Though these field data collection techniques are widely used, the procedures are labor intensive and time consuming. There is also a considerable level of difficulty in three dimensional modelling of the assets. These limitation factors motivated technological advancements and eventually led to the development of a mobile mapping system which directly collects 3D data of the vicinity. The concept of a mobile mapping system which integrates multiple sensors on a moving platform has a history dating back to 1970s. A photo logging instrument was attached to a vehicle to collect images of transportation asset inventories. Though satellite positioning technology like GPS were unavailable during that time, a combination of accelerometers, gyroscopes and odometers were used to determine the course of the vehicle movement and direction. The photos were eventually georeferenced based on the recorded vehicle positions [9]. After the introduction of GPS, a combination of GPS, IMU and other sensors were used for positioning. Video cameras with time-stamped frames were used to record images. Urban development and growth in transportation infrastructure encouraged the development of more efficient and improved techniques of mobile mapping. Further, precise positioning using GPS in kinematic mode led to the possibility of direct geo-referencing. Combination of direct geo-referencing concepts and digital imaging technology led to lower costs, better accuracy and increased flexibility of MMS [10]. Cameras capture images of the scene. Though it is possible to get the third dimension by using multiple overlapping stereo images, the process is indirect and tedious. On the other hand, laser scanners determine the three-dimensional coordinate of each point in the scene directly by calculating the range and direction of the laser pulse. The rise of LiDAR technology led to compact, light-weight and inexpensive laser scanners in the market which could be easily mounted on the mobile mapping system. Thus, integrating laser scanners along with this system provides the coordinates of points in three dimensional space. Fusing these geo-referenced multi-sensor data provides a better opportunity to find solutions for specific problems in geospatial domain [11]. Mobile mapping systems reduce the data collection time significantly. Literature relating to the applications of mobile mapping system in the field of transportation are explained in detail by [12]. With an appropriate set up of the sensors, the efficiency of data collection can be maximized. In addition to imaging cameras and laser scanners, depending on the application, the system can also be equipped with thermal cameras, hyper spectral scanners or ground penetrating radars. The laser scanner typically rotates in a 2-D plane and collects data over 360 degrees. A vertical inclination of the laser scanner would lead to better coverage as it collects points from overhead structures that are even perpendicular to the vehicle movement [13]. For approaches aiming to detect the road markings and features on the road surface, it is advantageous to orient the cameras/laser scanners facing downward [14]. A pair of stereo-cameras in the place of normal cameras for capturing asset information improves the accuracy of third dimension [15]. It provides a possibility to measure feature points directly from images. The major limitation of this approach is inaccurate measurement of relative distances between objects. This can be attributed to the different challenges leading to inaccuracies in calibrating the cameras. In addition to the vehicle based systems, the mobile mapping system can also be mounted on different platforms, such as, backpacks, UAVs, boats, carts or balloons depending on the application. There are light-weight laser scanners which helps to build a system weighing less than 30 pounds paving way for multi-modal data collection [16]. The positioning components—GNSS and IMU were used to determine the quality of a newly proposed wheelchair tires in a novel research [17]. Assessment of the condition of the pavements turns out to be a byproduct of the system setup thereby, making it a viable technology for condition assessment for assets. The survey grade mobile mapping systems are basically equipped with a commercial laser scanner, panoramic cameras, GPS-INS, dead-reckoning devices, such as odometers and a dedicated post processing software. An average absolute accuracy of 0.5 m is expected from most commercial mobile mapping systems [18]. An important aspect to be considered while migrating to mobile mapping technology from traditional survey methods is the high initial cost of the system. Several approaches [19–21] explain the development of a low cost mobile mapping system with laser scanners and cameras on-board. It should be noted that, a low cost mobile mapping system provides data accurate enough to detect tree canopy from it [22]. The required accuracy of positioning components for a mobile mapping system is much lesser than that of an airborne LiDAR. Hence, the overall cost can be reduced by choosing a less accurate GPS-IMU system. Though mobile mapping systems reduce the data collection time tremendously, an equivalent amount of time and resources are generally spent on extracting required information from the collected data. Thus, most time and resources for creating an asset inventory is spent digitizing features and entering attribute details. Due to the intensive office work involved during data digitization, research involving semi-automatic and automatic methods of feature extraction from mobile mapping data gained importance. Different algorithms and methodologies utilizing multi-sensor data (LiDAR point clouds and images) have been proposed for road infrastructure modelling. There are also several proposed methodologies for automatic/semi-automatic detection of some urban scene features from mobile mapping data. The most prominent features include road centerlines, road boundaries, lanes, median, sidewalk, curbs, poles, signboards and roadside vegetation/trees. Depending on the type of the scene and the sensors used for data collection, there are some algorithms and concepts that are commonly used to automate feature extraction. RANSAC—RANdom SAmpling Consensus method of plane detection is used to detect and classify road planes from LiDAR data [23]. A method of classification of urban features using multi-level segmentation and then connecting coherent planar components is explained by [24]. This paper also describes region growing, mean-shift and connected component algorithms. An innovative method of detecting curbstones from high density LiDAR data prove that LiDAR point clouds from mobile platform provides better extraction accuracy when compared with airborne LiDAR data [25]. Experiments prove that automated detection of road markings by using a range dependent thresholding function on the rasterized LiDAR data can be used for detecting lane markings. Several other approaches for detecting road corridor features from multi-sensor data are given in [26–30]. An ideal solution for asset inventory data collection would be a fast and cost effective technology providing an easy method of data collection. Though most existing mobile mapping systems cater to the current requirements, there is a high cost involved. Time consuming manual feature extraction also adds to the disadvantage. This paper describes the process of building a cost-effective mobile mapping system using mobile LiDAR and cameras for data collection. The paper also discusses various semi-automatic and automatic techniques of feature extraction for detecting transportation assets

METHODOLOGY

Transportation assets mainly consists of roads and pavements. However, other assets such as sign boards, signal lights, poles and electric, telecommunication and water utilities which are found above and below the roads and pavements also belong to the major components of transportation assets. These assets help to improve the efficiency of the roadway network thereby adding safety of transportation [31]. Additionally, it should be possible to monitor the assets regularly, which would help detect overshooting trees, condition of roads/pavements and roadside vegetation changes. Mobile mapping data can also be used for specific purposes, such as determining the presence of ramp in the sidewalk and adherence with other ADA (Americans with Disabilities) policies. Since it is important to collect and maintain a detailed inventory of these assets in an appropriate format, the mobile mapping system which helps in collecting images and three-dimensional point cloud data of the scene can be used to document the position and condition of the above mentioned assets.

Development of MMS That Meets the Requirements

Required Sensors

The primary components of the mobile mapping system being built for the purpose of transportation asset inventory data collection includes:

- One or more Cameras—Nikon 3200, 3300: Cameras capture pictures/video frames of the scene, thereby providing the asset managers with digital pictures portraying the conditions of assets.
- One or more laser scanners—Velodyne HDL-32E: LiDAR records 3D point data of the vicinity in the mapping frame, which, helps create 3D model of the scene and extract features.
- One or more GPS receivers, Gyros, Accelerometers—Geodetics, Geo-iNav: GPS-IMU integrated solution improves the frequency of recording positions from 1 Hz (GPS) to 125 Hz. The integrated navigation solution includes the position as well as the orientation of the vehicle (X, Y, Z, heading, roll and pitch).
- On-board computer—Brix: The software for controlling the functioning of the sensors are installed on the computer. It is important to Wi-Fi enable the computer, in order to connect it remotely using a laptop/tablet.
- External storage device—Samsung 1TB Solid State drive (SSD): The laser scanner can record up to 700,000 points per second. The navigation data is also quite voluminous. Since, the velocity of data recording is high, it is necessary to use a Solid state drive which provides high speed data logging.

Velodyne HDL-32E

The Velodyne laser scanner is widely used for mapping applications. There are a few survey grade laser scanners, such as Optech Lynx and Riegl VQ, which have accurate and high precision GPS-IMU solution. While, these scanners are very expensive, there are several inexpensive laser scanners that are cheaper than Velodyne. For instance, Hokoyu and ProtoX2D are some examples of cheap laser scanners. These scanners have just one laser diode and are mostly used for obstacle detection in robotics applications. They have a small range and insufficient accuracy for mapping applications. Velodyne has 32 laser channels covering a vertical swath of +10.67 to ′30.67 degrees. The LiDAR head can rotate 360 degrees in the two-dimensional plane. The sensor works on infra-red band with wavelength of 905 nm. The sensor is light weight and compact. The HDL-32E Lidar collects up to 700,000 points per second. The time of flight method is used to determine the range of these points. The sensor typically has a range of 1 m to 70 m. i.e., points from the scene within a range of 70 m from the sensor are measured. These attributes make Velodyne an obvious choice for developing low-cost mapping solution. The points are stored as frames. Each frame corresponds to a 360 degree rotation of the laser scanner in the 2D plane.

Nikon Cameras—3200, 3300

Cameras have been deployed in asset mapping. Charge Coupled Device (CCD) cameras were used initially which were soon replaced by SLR cameras. The large CCD arrays and pixel size of DSLR cameras when compared to the point and shoot digital ones makes it more apt for mobile mapping applications. Nikon D3200 has a fixed focal length of 35 mm and an angle of view which is 1.5 times the format equivalent of the focal length. Nikon D3300 has a variable focal length of 15 mm to 55 mm. The angle of view is also 1.5 times the format equivalent of the focal length. Both cameras capture either 30 or 50 frames per second in the video mode depending on the chosen resolution. The duration per video is 20 min and requires about 1.3 GB of memory. Technologically advanced cameras, such as GoPro, are also widely discussed about their role in mobile mapping. However, the fish-eye lens of GoPro is prone to create distortions and decrease the quality of the cartography product.

Geodetics—Geo-iNav

The Geo-iNav instrument from Geodetics provides a GPS-IMU integrated solution. It consists of a GPS which provides positions with a horizontal accuracy of 5 cm and vertical accuracy of 10 cm on L1/L2 real time kinematic mode and horizontal accuracy of 1.5 m and vertical accuracy of 2.5 m on L1. Standalone mode. The Inertial measurement unit consists of accelerometers and gyros. The IMU has the following specifications,

- Gyroscope Dynamic Range: ˜150″/s.
- Gyroscope Bias In-run stability (1 σ): 3″/h.
- Accelerometer Dynamic Range: ˜3 g.
- Accelerometer Bias In-run stability (1 σ): <0.1 mG

For a mapping application, the expected absolute accuracy of feature points is <20 cm. Hence, it is necessary to deploy an IMU with less than 3″/h of gyro bias. The Geo-iNav unit weighs 567 gram and has size of 550 cubic cm which makes it portable and apt for mobile mapping applications. It provides navigation solution at a frequency of 125 Hz. The navigation and geodetic data provided by Geo-iNav includes, navigation solution (position, velocity and attitude) and raw GPS/IMU data which could be used for post processing.

Since the mobile mapping system could be mounted on a variety of mobile platforms from trucks to boats, it is essential to create a robust system on which these sensors could be assembled. The first important aspect to consider is the stability of the platform. The IMU, GPS antenna, cameras and the laser scanner are required to be in a stable position throughout the data collection. Additional care should also be taken to ensure that the IMU does not receive undesired vibrations due to the vehicle's internal mechanics. The second aspect of consideration is the position of the GPS antenna. It is important to mount the GPS antenna in a position such that the signal from the GPS satellites are not occluded. The orientation of the laser scanner and cameras also play a major role in the efficiency of data collection. By changing the vertical orientation of the laser scanner, it is possible to collect points from the roads as well as overhead structures. The laser scanner is mounted on top of an arm which could rotate about the horizontal axis as shown in the Figure 1. In addition to the sensors and navigation instruments, the mobile mapping system also needs a laptop/tablet computer which would connect to the brix thru Wi-Fi or wired network and allows the operator to control the sensors.

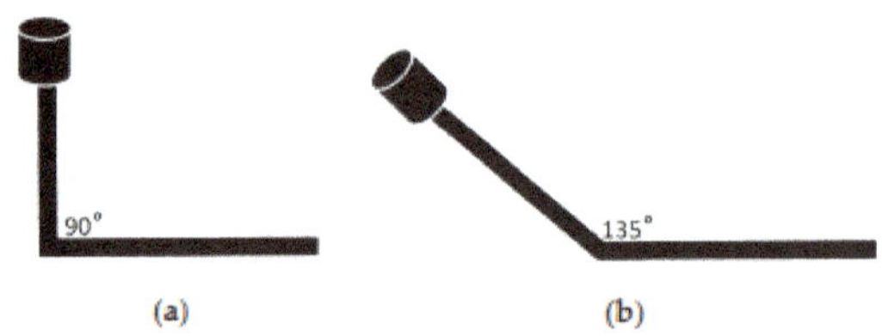

Figure 1. Different vertical orientations of the laser scanner. Practically, an angle of 130–140 degrees provides a good data coverage for typical urban environments.

Once the components are put together on the platform, the sensors must be calibrated before they are deployed. Amongst a number of systematic errors that reflect on the data, the error due to boresight calibration is very significant. Calibration is the process of compensation for the misalignment between different sensors belonging to the system [32–34]. In general, the misalignment involves both translation and rotation components. This misalignment is measured between the centers of all other sensors and the IMU center. Translation and rotation between the centers of all other sensors and the IMU center. Translation and rotation between the following pairs are measured.

1. GPS antenna center to the center of IMUThe origin and axes of the IMU are marked on its box. Since, the centers of IMU and GPS antenna are physically separated and identifiable, the distance between their centers can be manually measured along the X, Y and Z directions of the IMU as shown in Figure 2. The GPS antenna does not have any orientation. Hence, there are only translation components (dx, dy, dz). The body frame of the sensor platform has its positive X-axis aligned in the direction of the vehicle, positive Y-axis extends out on the side (right hand) and positive Z-axis pointing down. Based on current setup of the IMU, the negative X- axis of IMU is aligned along the forward direction of the body-frame, the Z- axis of IMU is set along the vertical axis of the body-frame but point up and the Y-axes of both body frame and IMU frame are

aligned. Based on the specified mounting parameters, the GPS is located in IMU's positive X direction. Hence, the displacement between IMU origin and GPS antenna along IMU's X-axis should be subtracted from the sensor measurements. Thus, while computing the boresight misalignment parameters, the value of dx is negative. For similar reasons, dy and dz are positive. These translation values are used by the IMU while integrating the GPS-IMU data. The measurement should be accurate to 10 cm or less in order for the post-processing software to accurately determine the offset.

Sensors 2016, 16, x 7 of 18 mounting parameters, the GPS is located in IMU's positive X direction. Hence, the displacement between IMU origin and GPS antenna along IMU's X-axis should be subtracted from the sensor measurements. Thus, while computing the boresight misalignment parameters, the value of dx is negative. For similar reasons, dy and dz are positive. These translation values are used by the IMU while integrating the GPS-IMU data. The measurement should be accurate to 10 cm or less in order for the post-processing software to accurately determine the offset.

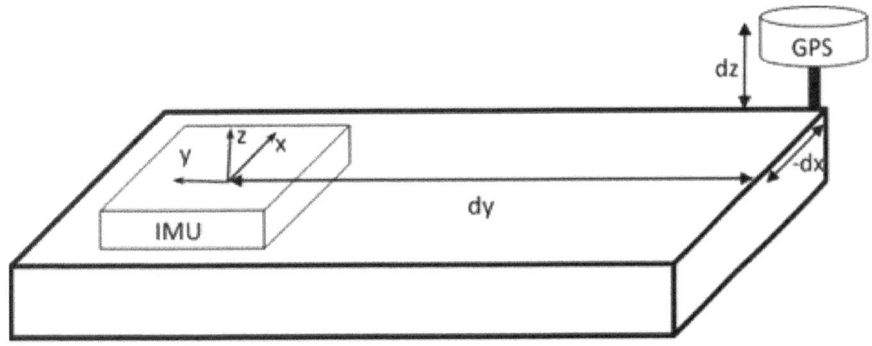

Figure 2. Boresight misalignment between the Inertial Measurement Unit (IMU) and GPS.

2. LiDAR (origin from which laser pulse is triggered) to the center of IMUThe origin and axes of the IMU are marked on the box. On the other hand, it is difficult to manually identify the axes of the laser scanner. Hence, the translation and rotation between the two origins are determined by using a Terrestrial Laser Scanner (TLS). The TLS and the mobile mapping unit are placed on a levelled plane. Both the mobile laser scanner and the TLS scans the vicinity. The mobile scanner is fine scanned by the TLS such that the IMU markings are clearly visible on the point cloud. Points from the TLS point cloud belonging to the IMU center and IMU axes are picked. The origin and axes of the TLS point cloud are shifted to the IMU center and axes of the mobile mapping system. and axes of the mobile mapping system. Since, the scale component is fixed, 3-D Helmert rigid body transformation is applied [35]. Now, the translation and rotation between the point clouds from mobile laser scanner and TLS (centered at IMU) are measured using Iterative Closest Point (ICP) methodology introduced by [36,37]. One implementation of ICP as an open source software-cloud Compare is used. The translation and rotation values determine the boresight misalignment between IMU and Laser Scanner.

CAMERA CALIBRATION [39–41]

a. Interior Orientation—Interior orientation is a part of camera calibration where the measurements relating to the camera, such as the perspective center and focal length are determined. It also involves finding the scaling, skew factors and lens distortion. The interior orientation is performed in lab conditions by clicking pictures of a regular grid from different angles. By transforming the image pixels to real world lengths, the metrics of imaging can be determined

b Exterior Orientation—Exterior orientation parameters change for every picture. It is the position and orientation of the camera with respect to a coordinate system, while capturing each photo. Space Resection is a conventional method used to determine exterior orientation parameters. It involves the measurement of ground control points and digitizing them on overlapping images to determine the camera position (X, Y, Z) and orientation (omega, phi, kappa).

Co-Registration of Multi-Sensor Data

The mobile mapping system consists of two cameras, a laser scanner and a GPS-IMU instrument. Once the data collection is complete, the trajectory needs to be smoothed and the data from different sensors should be synchronized. The trajectory appears rugged mostly at places where there is a GPS signal outage. We smooth the trajectory, by creating a piecewise polynomial function.

Trajectory Interpolation

The frequency of Geo-iNav is 125 Hz. The combined solutions of GPS and IMU records a position for every 1/125th of a second. However, the Velodyne laser scanner collects up to 700,000 points per second. In order to geo-reference the points, ideally, we would need a position/orientation of the vehicle for approximately every 1/20,000th of a second. Since, the frequency of the laser scanner is much higher than that of the INS, the trajectory curve is interpolated using piecewise polynomial function in the shape of a cubic Spline (B-Spline) with timestamp (t) as its parameter as shown in Equation (1). Piecewise cubical Spline is preferred over linear interpolation because, it provides better continuity and also considered as the best fitting curve for the road trajectory [42,43].

$$f(x,y) = \bigcup_{i=0}^{n} (a_i t^3 + b_i t^2 + c_i t + d_i, e_i t^3 + f_i t^2 + g_i t + h_i) \tag{1}$$

Time Synchronization

Coordinates of points on the trajectory, images and LiDAR point clouds are uniquely identified by their timestamp. The GPS time from Geo-iNav is considered as the standard and the clocks of all other sensors are corrected for GPS time. The Velodyne laser scanner records two timestamps: System time in microseconds and GPS time in seconds. The system time is the number of microseconds calculated from the beginning of the hour in UTC (Universal Time Coordinated). Thus, offset between the system time of Velodyne and GPS time is corrected using the GPS timestamp recorded by the Velodyne laser scanner. A video file has the following properties that can be used to timestamp the image frames –

1 Modified date (timestamp)—Timestamp recorded when the video was stored to the SD card (Assumed as the finishing time of the video)
2 Duration of the video
3 Frames per second (fps)

The SLR cameras record the system time when the videos are saved to the SD card. This is arbitrarily assumed as the finishing time of a video. The beginning timestamp of the video is calculated from the duration and finishing time. Using the value of fps, the timestamp for every image frame is interpolated. The timestamp of the image frames are calculated based on the camera's system time which is prone to have an offset from the GPS time. In order to determine the offset between camera's clock and the GPS time, a picture of the GPS log (Latitude, Longitude, Altitude, GPS Week/Second) from the visualization screen is captured before the data collection. The difference between the system time of the camera and the GPS time on the captured picture gives the offset between the two clocks. The timestamps of the video frames are corrected for this measured offset.

Direct Geo-Referencing

Direct Geo-referencing is the process of determining the position and orientation of the laser scanners and cameras (sensors) of the mobile mapping system by using the navigation data from the GPS-IMU device. The LiDAR and the cameras capture points in the scene which are external of the platform. The position and orientation of these sensors determined through direct geo-referencing techniques help in determining the coordinates of the points captured by these sensors in the mapping coordinate system [44]. The point clouds were registered by applying the interpolated rotation values (heading, roll, pitch) obtained from the INS, corresponding X, Y, Z position from GPS-IMU integrated navigation data and the boresight rotation and translation between IMU and the laser scanner as shown in Equation (2).

$$P_G = p_{ECEF} + R_{IG} R_{ba} (R_{sb} P_s + dT_{sb}) \tag{2}$$

where,

PG—Coordinate of the captured point in ECEF system, p_{ECEF}—GNSS sensor position in global ECEF system, R_{IG}' Rotation matrix from origin of WGS system to origin of local frame,

$$R_{iG} = \begin{bmatrix} -\sin(\text{long}_i) & -\sin(\text{lat}_i) * \cos(\text{long}_i) & \cos(\text{lat}_i) * \cos(\text{long}_i) \\ \cos(\text{long}_i) & -\sin(\text{lat}_i) * \sin(\text{long}_i) & \cos(\text{lat}_i) * \sin(\text{long}_i) \\ 0 & \cos(\text{lat}_i) & \sin(\text{lat}_i) \end{bmatrix} \tag{3}$$

Longi , lat_i are geodetic coordinates of p_{ECEF},

R_{ba}' Rotation matrix from body frame to local frame,

R_{sb}—Rotation matrix from scanner to IMU (b-frame)—boresight rotation matrix

P_s—Coordinate of the point in scanner frame (as recorded by Laser Scanner)

dT_{sb}—Offset between scanner frame and IMU center (b-frame)—boresight translation.

In order to improve the accuracy of direct geo-referencing, GCPs (ground control points) are collected and used to rectify any errors in the geo-referenced point cloud. It is important to choose control points that are clearly identifiable on the point cloud as well as images. The points ought to be well-distributed in the area where data is collected. The survey is carried out using GPS and total stations. The coordinates are recorded in WGS System. The measured points are identified on the registered point clouds. The point clouds are corrected by applying 3D Helmert rigid body transformation using the GCPs as parameters.

Geo-Referencing Images

The mobile mapping system captures images/video frames continuously. A scene is captured by multiple images with significant overlap between the frames. Hence, the first step in geo-referencing is to connect the consecutive images using common points—tie points [45]. The traditional aero-triangulation (AT) methodology is used to orient and connect the images. The indirect method of exterior orientation is considered very accurate and commonly used for registering images. It involves the use of ground control points to geo-reference the images. Once AT is performed, the measured values of the control points are used to geo-reference the images. With more accurate GPS-IMU navigation sensors, direct geo-referencing of images using the position of the sensor and calibration parameters are increasingly used in several mobile platforms, such as UAVs, mobile mapping systems etc. In order to use the direct method of geo-referencing, it is important to accurately determine the boresight misalignment parameters. That is, the separation of the sensor origin (camera) from the origin of the IMU. Since, it is unlikely to measure this separation with centimeter level of accuracy, a combination of direct and indirect geo-referencing techniques could be used in this case. Once bundle block adjustment is performed on the set of images using the GPS-IMU position data, the GCPs can be used to correct for the errors in calibration [46,47].

Quality Check Methodology

It is important to assure the mapping quality of the collected data before populating the asset inventory database or using them for feature extraction. The measures of quality for the survey data involves determining the relative and absolute accuracies. A particular feature might be recorded in multiple image or point cloud frames. This may be due to overlapping frames or multi-path (repeat trajectory) data. Relative accuracy is the measure of drift between the same features measured on different frames. The major cause for inconsistency between consecutive frames are systematic or random errors of the sensors involved. Positioning error will reflect in the relative accuracy of multi-path data. Features like sign boards, pavement boundaries and building corners can be used to quantify relative errors. Absolute accuracy is the closeness of the collected data to the true position of the features. In order to check absolute accuracy of the collected data, a combination of the set of GCPs used for geo-referencing and a new set of GCPs are used. The coordinates of points from the geo-referenced point cloud and images are checked against the GCPs collected thru field survey. The root mean square deviation of these two sets of values determines the absolute accuracy of the laser scanning data and images. Once, the quality of data is assured, the data is processed so that, outliers are removed and only the relevant data is used for the experiment.

Processing the Data

Filtering of Data

Laser scanning data may contain noise and erroneous returns. The first step is to filter these erroneous points. Different errors due to laser scanner and the ways they affect the quality of data are explained by [48]. Sparse returns which fall out of the typical range of the laser scanner may be considered as noisy returns and can be removed from the data. Further, unrealistic elevation values might also be the result of inconsonant returns or incorrect range calculation. Noisy points are filtered from the data by using a basic range threshold. Points not belonging to the range are treated as outliers and are removed. Secondly, it is important to narrow down the entire dataset to the regions related to the domain of interest and eliminate unrelated parts of the data. A transportation asset inventory database would typically contain roads and roadside features. In order to localize the point cloud, road centerline and road width are extracted from the TIGER (Topologically Integrated Geographic Encoding and Referencing) dataset [49]. The TIGER shapefile shows the centerline of the roads, the attribute data holds the type of the road which would provide a fair idea about the width of the road. The extent is determined by providing a buffer to the prescribed road width, in order to accommodate the roadside landscape, pavement and other features of interest. This process of localizing the point cloud to contain only the features of interest is very helpful in avoiding misclassification of features and also significantly reduces the processing and computation time.

Automatic Bare-Earth for Cross Section

The raw LiDAR data consists of returns from all features in the vicinity which includes both ground and non-ground points. In a road scene, there could be ground points, such as the road surface, sidewalk, median and there are also points from non-ground features like trees, electric lines, cars, poles, sign boards and other roadside structures. The easiest and straight-forward method to classify ground points from non-ground points is by exploiting their planarity property. Almost all ground points lie on well-defined planes. For example, the roads and pavements are typically planar. RANSAC method determines the best fit plane on a set of points. RANSAC algorithm, when run repeatedly on the outliers of each iteration, results in a number of significant planes that fits the point cloud. The points belonging to these planes are classified as the ground points (bare-earth) and the outliers of the final iteration of RANSAC are non-ground points. Thus, ground points can be differentiated from non-ground points using this procedure.

Creation of GIS Database for Asset Management

The concept of using GIS technology for Asset Management is well-established. GIS provides a concrete mapping between spatial data and its corresponding attribute details, which helps in making well-informed decisions about infrastructure management and maintenance [31]. Displaying an asset with its location on a scaled map provides better visualization and helps in decision making. Applying spatial relationship constraints like connectivity, containment and adjacency helps maintain data integrity and consistency. A number of commercial software tools with rich Computerized Maintenance Management System (CMMS) capabilities are very popular in the GIS industry for almost a decade. Beginning from electric and water up to transportation and public safety, GIS asset management systems are being implemented for almost every utility domain. Most of these frameworks have a customized database structure with internal relationships mapped between different components of the asset management system. Largely, they are exclusive for a particular utility. As discussed in the previous sections, most transportation asset management frameworks use the traditional methods of data collection using GPS and other field data collectors. The data is then imported into CAD format which is later converted into GIS compatible formats. The pioneers in asset management software support almost all data formats belonging to common field data collection methods. However, the best practice of integrating data from mobile mapping systems directly into a GIS database is a state-of-the-art research domain. The structure of databases and spatial relationships between the assets do not depend on the data collection methodology. However, mobile mapping system is not a direct method of data collection. It involves a number of processing tasks before documenting the actual assets of interest. It involves tasks that range from geo-referencing to feature extraction. An ideal framework for managing asset data collected using a mobile mapping system should include these processing tasks along with the implementation of a GIS database for asset inventory.

Characteristics of an asset management framework using mobile mapping data:

- Ability to manage voluminous data efficiently—large point clouds
- Maintain consistency between multi-sensor data—images, LiDAR point clouds, thermal/ hyperspectral data etc.
- Appropriate versioning of collected datasets.
- Implementation of basic automatic feature extraction algorithms and interface to manually digitize features.
- Relational database design to store assets and their spatial relationships.
- Visualization, rendering and fluid user interface.

Experiments

The mobile mapping system with the described components are assembled as shown in Figure 1. The GPS antenna and IMU are fixed upright. The system is mounted on a uniformly levelled surface. The arm of the laser scanner is adjusted depending on the mounting of the mobile mapping system so that maximum points from the scene are collected. The system also requires a battery to supply power for the components to function. Once, the system is set up, boresight misalignment of each sensor with respect to the IMU center is measured. The center of the IMU and its axes are marked on its cover. The x, y and z misalignment values between GPS and IMU center are fed into the IMU processing software so that the integrated navigation solution is automatically corrected for this misalignment. The second set of calibration values which includes the rotation and translation between IMU and laser scanner/cameras are applied to the data during the process of geo-referencing. The system is powered and individual components are tested for their functionality. The SSD is also tested for proper data logging and completeness of data. Once, the data collection is complete, the trajectory is post-processed to apply differential corrections and to integrate GPS-IMU using "GeoRTD" from Geodetics. The data is converted into appropriate format for synchronizing with data from other sensors. The GPS-IMU data which is originally in binary format is converted into ASCII (American Standard Code for Information Interchange). The point cloud data from laser scanner is converted from PCAP (Packet Capture) to CSV (Comma Separated Values) format. Further, the videos recorded by the camera in MP4 (MPEG-4 AVC – Advanced Video Coding) format are converted to individual frames which are images in JPEG (Joint Photographic Experts Group). The data from different sensors are synchronized based on the recorded timestamp. The system time of all sensors are measured in UTC system. The data is compensated for the 17 s offset measured between UTC and GPS time. However, there can also be a possible drift between the internal system time recorded by the cameras and GPS time. A drift of 24.5 s and 28.5 s are measured between the system times of the two cameras and the GPS time respectively. At the end of this synchronization step, each return from the laser scanner, each frame recorded by the camera and each position recorded by the GPS-IMU sensors will be timestamped based on GPS time. A piece-wise polynomial, spline function is fitted for X, Y, Z, heading, roll and pitch values with timestamp as the parameter. Thus, the navigation data corresponding to each point from the laser point clouds/each image can be determined from the spline functions based on their respective timestamps. Additionally, the rotation and translation values corresponding to boresight misalignment/sensor calibration are also applied. Once the data is geo-referenced, the point clouds that are initially in sensor's coordinate system were converted to a global coordinate system. Ground Control Points are required to determine the exterior orientation (EO) parameters of images and also to check the quality of geo-referenced point clouds. Points that are well-defined on the ground, on images and on point clouds are considered for GCPs. GPS and Total Stations are used to measure GCPs. Based on rough reconnaissance of the area, GPS stations are determined. There are locations from which maximum number of GCPs lie in the vicinity. The GPS instrument is set up on these locations in static mode to record the station coordinates. A total station is then set up at these locations to determine the coordinates of the GCPs. The EO parameters of the images are determined by using the ground control points in addition to calibration values and navigation data. The quality of geo-referenced point clouds is determined by calculating the root mean square deviation between the measured GCPs and the corresponding points on the point cloud. The data collection was carried out by mounting the system on a truck (Figure 3a) and a golf cart (Figure 3b). The former was primarily used to collect asset inventories for the City of Dania Beach—a larger area with high speed vehicles and traffic. The latter was used for data collection within the FAU, Boca Raton Campus. Since, the golf cart is limited to be used on pavements, it is unlikely that it covers wider roads in a single run. Additionally, golf carts are not equipped with shock absorbers. Hence, the undulations in the pavement reflect in the navigation data as noise. The vehicles were driven at an average speed of 10 mph in order to have good coverage of data. The laser scanning data is stored as number of frames with each frame having a time range of 0.2 s. The density of the point cloud depends on the number of returns. Since, the scanner is oriented at an approximate angle of 135 degrees as shown in Figure 1, the road surface is clearly captured. There are larger number of returns from the road surface because it is closer to the scanner. In an urban road side environment, an average of 100,000 points (returns) are collected in a frame. Pictures are collected at the rate of 30 frames per second by the Digital SLR camera. The pictures are extracted from the video at an interval of 0.2 s which is the same as the duration of a LiDAR frame.

(a) (b)

Figure 3. Mobile mapping system mounted on a truck (a) and a golf cart (b) for data collection.

The LiDAR frames are loaded in Leica Cyclone environment to digitize the features of interest. The assets of interest for transportation inventory management includes, sidewalks, medians, guardrails, fencing, misc. concrete structures, lighting, landscape areas, delineators, striping, symbols and messages, crosswalks, stop bars, raised pavement markers, attenuators and highway signs. In addition to the spatial component, the attributes for these assets are also populated. The typical database structure for transportation asset inventory data management includes attributes, e.g., condition and dimensions. However, road signs should also be tagged with the symbol represented by the sign. The spatial component of the assets may be points, lines or polygons depending on its geometry and may also require some customized attributes. Hence, each asset is stored in an individual table with attributes as shown in Table 1. The Asset Id is unique for every asset. In addition to the process of manual digitization of assets, there are a few established automatic and semi-automatic methods of feature extraction that can be implemented in future.

Asset	Attributes (Format/Type)
Sidewalk	Width (Double) Curb Height (Double) Length of the segment (Double) Availability of ramp (Boolean—true/false) Condition (Integer; range—1–10) Comments (Text) Geometry (Polygon)
Median	Width (Double) Height (Double) Length of the segment (Double) Condition (Integer; range: 1–10) Comments (Text) Geometry (Polygon)
Guard Rail	Height (Double) Length of the segment (Double) Condition (Integer; range: 1–10) Comments (Text) Geometry (Line)
Fencing	Height (Double) Length of the segment (Double) Condition (Integer; range: 1–10) Comments (Text) Geometry (Line)

Asset	Attributes (Format/Type)
Lighting	Height (Double) Type (Text) Condition (Integer; range: 1–10) Comments (Text) Geometry (Point)
Landscape Areas	Area of landscaping (Double) Condition (Integer; range: 1–10) Comments (Text) Geometry (Polygon)
Delineators	Height (Double) Type of Delineator (Text) Condition (Integer; range: 1–10) Comments (Text) Geometry (Point)
Lanes	Type of striping (Text) Condition (Integer; range: 1–10) Comments (Text) Geometry (Line)
Road Markings	Type of striping (Text) Condition (Integer; range: 1–10) Comments (Text) Geometry (Line)
Road Signs/Boards	Message (Text) Type of sign (Text) Condition (Integer; range: 1–10) Comments (Text) Geometry (Point)

Extraction Methodology

As discussed in the previous section, mobile mapping reduces the data collection time greatly. However, digitizing individual features from the geo-referenced point clouds and images requires huge amount of human effort and is time-consuming. Since, we have cameras and a laser scanner onboard, it improves the scope for automatic/semi-automatic feature extraction. Algorithms involving geometry of feature points can be used in combination with image processing algorithms to get better results. There are a number of different methods being proposed by researchers in geomatics and data mining scientists for automatic feature detection and extraction. Some common methods used in feature detection that can be used to extract different transportation asset inventories are described in this section.

Automatic Road Marking Extraction

The major advantage of a mobile mapping system with multiple sensors onboard is that the data from each sensor complements the other and helps extract maximum information about the scene. Implementing a combination of image processing and geometry algorithms on images and point cloud data helps in detecting the road markings with almost nil manual intervention. The edges of the sidewalk detected from the images are used to define the boundary of the road and the bare-earth LiDAR points within the boundary are considered as road points. In order to ensure clear visibility for the drivers, road lane markings have a much higher reflectance when compared to the road surface. The boundaries of the road markings can be extracted from the images [27] as well as by classifying the LiDAR point cloud based on the intensity of the return pulse [14]. An example of such road marking extraction is illustrated in Figure 4. Once the parts of the image corresponding to the road markings are localized (Figure 4a), the image is segmented. The segmented image is a smoothened form of the original image with less noise (Figure 4b). Finally, edges are detected from the segmented image using Canny's edge detection algorithm [50]. These edges are processed to extract road markings (Figure 4c).

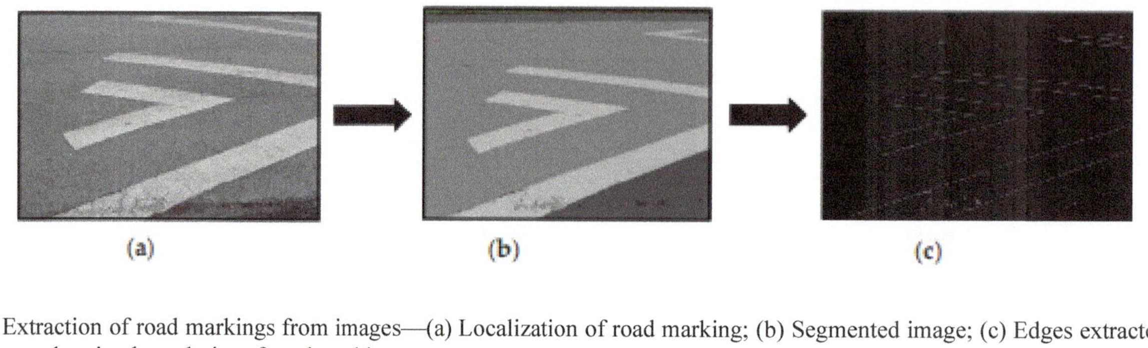

Figure 4. Extraction of road markings from images—(a) Localization of road marking; (b) Segmented image; (c) Edges extracted from segmented image showing boundaries of road markings.

Road Sign Extraction

Road signs are important components of the transportation asset inventory database. Road signs are non-ground points and do not lie on a definite plane. They are made up of aluminum which is highly reflective and also clearly distinguishable on the point cloud. The shape of the sign board can be determined from the boundary of clusters of high intensity returns [29,51]. Signboards are of fixed dimension range, height and shape. Hence, by using the height and dimension constraints, points that do not belong to road signs can be eliminated and the precise shape of the sign board can be obtained (Figure 5a). It is also important to know the symbol on the road sign to add attribute information to the extracted spatial data. Since the content on the sign board cannot be well identified from the point cloud, geo-referenced images are used for this purpose. The location of the signboard can be determined on the set of referenced images by using the coordinate corresponding to the centroid of the sign board extracted from the point cloud (Figure 5b).

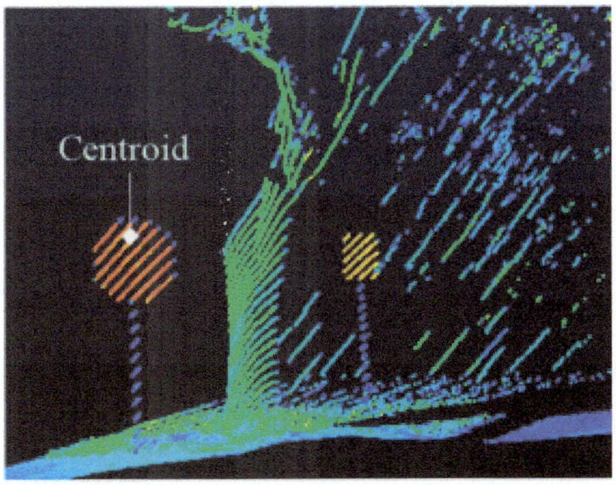

Figure 5. Extraction of road signs from Lidar point clouds and images. (a) Points from point cloud whose intensity >100 (high reflectance of aluminum); height from curb > 5 ft.; dimension between 0.4 m and 1.6 m; (b) Locating the signboard on the image using the coordinate of its centroid.

Extraction of Other Assets

Several methods utilizing the geometry of point clouds and also the intensity of returns are used to extract assets like poles and lamp posts. Considering a high density LiDAR point cloud, the poles and lamp posts can be modeled as cylinders with a height and volume threshold [30]. However, in a sparse point cloud, poles and lamp posts are seen as upright lines. RANSAC Line Fit can be used to distinguish poles from lamp posts [23]. Other features of interest include building faces and road side landscape. Building faces can be extracted by using their planar property because they are always planes perpendicular to the ground surface [26]. Landscape on the sides of the road are typically ground points reflecting green. Combining intensity values from image pixels with the classified LiDAR ground points, roadside landscape can be extracted.

DISCUSSION

Field data collection methods possess a number of potential challenges such as high data collection time, increased manual effort, safety constraints and limitation of information collected. Additionally, several data management aspects like data transformation, storage and integrity are not efficiently streamlined when field data collection methods are used. For instance, the workflow currently adopted by the FDOT (Florida Department of Transportation) involves field data collection through traditional survey methods. The data is converted into CAD (Computer-Aided Design) format; the images captured are converted to appropriate GIS raster format and are used as an overlay on as-built drawings. These methods often involve multi-step processes which are difficult to automate for an organized work-flow. It is also challenging to integrate newly collected data with existing data. A mobile mapping technology overcomes almost all of these limitations. Though the initial cost of the system is high, once the system is built, it makes frequent data collection easier and helps procure up-to-date information. The experiment developed a very economic system whose cost sums up to 65,000 USD. Considering the recurring cost of man power involved in traditional field survey methods, the initial cost involved with a mobile mapping system providing mapping grade accuracy turns out to be optimally advantageous. Major challenges involved in using a mobile mapping system is the processing of voluminous data and laborious digitization of individual features. These limitations can be overcome using improved data management techniques and automatic feature extraction methodologies. The use of field data collection methods using improved positioning instruments and hand-held mapping devices is a huge leap from traditional surveying methods in the last decade and now, mobile mapping system is the newer mapping technique owing to the number of advantages over its predecessors

Advantages of a mobile mapping system:

Less laborious data collection process.

- Possibility of frequent data collection for constantly changing scenarios

- Data can be reviewed for condition assessment at the comfort of office
- Minimum field work involved
- Direct and accurate elevation information from mobile LiDAR data
- Easy to conduct survey over inaccessible areas.
- Minimum/nil hindrance to commuters and traffic.

CONCLUSIONS

Considering the importance of a detailed asset inventory database and the limitations of traditional surveying methods to achieve that, there is a need for an economic solution with improved capabilities. Eventually, a cost-effective mobile mapping system was built to provide a solution to the limitations posed by the previous techniques. The geo-referenced data from the mobile mapping system provides a three-dimensional model of the entire scene. This provides a direct method to accurately determine the three-dimensional coordinates of road assets in the corridor. Though, there is no single well-established methodology for automatic extraction of features from the point clouds and images, a number of proven techniques to detect individual assets have been demonstrated with the experimental data. Additionally, the advantages of mobile mapping system over other techniques makes it an important technology for asset inventory mapping. The current framework includes geo-referencing tools and quality check methods for point clouds and images. The current research involve building a UAV-based mobile mapping system to monitor the condition of overhead road infrastructure. The future work will also include developing more automated feature extraction methods to further improve office processing cost in asset inventory management.

REFERENCES

1. Olsen, M.J. NCHRP 15-44 Guidelines for the Use of Mobile LIDAR in Transportation Applications; National Cooperative Highway Research Program: Washington, DC, USA, 2013; p. 243.
2. Lange, A.F.; Gilbert, C. Using GPS for GIS data capture. In Geographical Information Systems, 2nd ed.; Longley, P.A., Goodchild, M.F., Maguire, D.J., Rhind, D.W., Eds.; John Wiley and Sons: New York, NY, USA, 1998; pp. 467–479.
3. Laflamme, C.; Kingston, T.; McCuaig, R. Automated mobile mapping for asset managers. In Proceedings of the 23th International FIG Congress, Munich, Germany, 8–13 October 2006.
4. He, G.; Orvets, G. Capturing road network data using mobile mapping technology. Proc. Int. Arch. Photogramm. Remote Sens. 2000, 33, 272–277.
5. Findley, D.J.; Cunningham, C.M.; Hummer, J.E. Comparison of mobile and manual data collection for roadway components. Transp. Res. C Emerg. Technol. 2011, 19, 521–540.
6. Donatis, M.D.; Bruciatelli, L. MAP IT: The GIS software for field mapping with tablet pc. Comput. Geosci. 2006, 32, 673–680.
7. Zingler, M.; Fischer, P.; Lichtenegger, J. Wireless field data collection and EO–GIS–GPS integration. Comput. Environ. Urban Syst. 1999, 23, 305–313.
8. Pundt, H. Field data collection with mobile GIS: Dependencies between semantics and data quality. GeoInformatica 2002, 6, 363–380.
9. Tao, V.C. Mobile mapping technology for road network data acquisition. J. Geospat. Eng. 2000, 2, 1–14.
10. Klaus, P.; Schwarz, N.; El-Sheimy, N. Mobile mapping systems—State of the art and future trends. Int. Arch. Photogramm. Remote Sens. Spat. Inf. Sci. 2012, 35, 759–768.
11. Schwarz, K.P.; El-Sheimy, N. Kinematic Multi-Sensor Systems for Close Range Digital Imaging. Int. Arch. Photogramm. Remote Sens. 1996, 31, 774–785.
12. Williams, K.; Olsen, M.; Roe, G.; Glennie, C. Synthesis of Transportation Applications of Mobile LIDAR. Remote Sens. 2013, 5, 4652–4692.
13. Cahalane, C.; Lewis, P.; Mcelhinney, C.; Mccarthy, T. Optimising Mobile Mapping System Laser Scanner Orientation. ISPRS Int. J. Geo Inf. 2015, 4, 302–319.
14. Kumar, P.; Mcelhinney, C.P.; Lewis, P.; Mccarthy, T. Automated Road Markings Extraction from Mobile Laser Scanning Data. Int. J. Appl. Earth Observ. Geoinf. 2014, 87, 125–137.
15. Van Geem, C.; Gautama, S. Mobile mapping with a stereo-camera for road assessment in the frame of road network management. In Proceedings of the 2nd International Workshop "The Future of Remote Sensing, Antwerp, Belgium, 17–18 October 2006.
16. Kukko, A.; Kaartinen, H.; Hyyppä, J.; Chen, Y. Multiplatform Mobile Laser Scanning: Usability and Performance. Sensors 2012, 12, 11712–11733.
17. Meruani, A.; Woodruff, G.W. Tweel™ Technology Tires for Wheelchairs and Instrumentation for Measuring Everyday Wheeled Mobility. Master's Thesis, Georgia Institute of Technology, Atlanta, GA, USA, 2006.
18. Hunter, G.; Cox, C.; Kremer, J. Available online: http://www.isprs.org/proceedings/XXXVI/1-W44/papers/Hunter_full.pdf (accessed on 15 November 2015).
19. Madeira, S.; Gonçalves, J.A.; Bastos, L. Sensor Integration in a Low Cost Land Mobile Mapping System. Sensors 2012, 12, 2935–2953.
20. Silva, J.F.C.D.; Camargo, P.D.O.; Gallis, R.B.A. Development of A Low-Cost Mobile Mapping System: A South American Experience. Photogramm. Rec. Photogramm. Rec. 2003, 18, 5–26.
21. Ellum, C.M.N.; El-Sheimy, N. A mobile mapping system for the survey community. In Proceedings of the 3rd International Symposium on Mobile Mapping Technology (MMS 2001), Cario, Egypt, 3–5 January 2001.
22. Jaakkola, A.; Hyyppä, J.; Kukko, A.; Yu, X.; Kaartinen, H.; Lehtomäki, M.; Lin, Y. A Low-Cost Multi-Sensoral Mobile Mapping System and Its Feasibility for Tree Measurements. ISPRS J. Photogramm. Remote Sens. 2010, 65, 514–522.
23. Lam, J.; Kusevic, K.; Mrstik, P.; Harrap, R.; Greenspan, M. Urban scene extraction from mobile ground based

lidar data. In Proceedings of the 5th International Symposium on 3D Data, Processing, Visualization & Transmission (3DPVT 2010), Paris, France, 17–20 May 2010.

24. Vosselman, G. Point Cloud Segmentation for Urban Scene Classification. ISPRS Int. Arch. Photogramm. Remote Sens. Spat. Inf. Sci. 2013.

25. Zhou, L.; Vosselman, G. Mapping Curbstones in Airborne and Mobile Laser Scanning Data. Int. J. Appl. Earth Observ. Geoinf. 2012, 18, 293–304.

26. Manandhar, D.; Shibasaki, R. Auto-extraction of urban features from vehicle-borne laser data. ISPRS J. Photogramm. Remote Sens. 2002, 34, 650–655.

27. Sun, H.; Wang, C.; El-Sheimy, N. Automatic Traffic Lane Detection for Mobile Mapping Systems. In Proceedings of the 2011 International Workshop on Multi-Platform/Multi-Sensor Remote Sensing and Mapping, Xiamen, China, 10–12 January 2011.

28. Tao, C.; Li, R.; Chapman, M.A. Automatic reconstruction of road Centerlines from mobile mapping image sequences. Photogramm. Eng. Remote Sens. 1998, 64, 709–716.

29. Pu, S.; Rutzinger, M.; Vosselman, G.; Elberink, S.O. Recognizing Basic Structures from Mobile Laser Scanning Data for Road Inventory Studies. ISPRS J. Photogramm. Remote Sens. 2011, 66, S28–S39.

30. Brenner, C. Extraction of Features from Mobile Laser Scanning Data for Future Driver Assistance Systems. In Advances in GIScience; Lecture Notes in Geoinformation and Cartography; Springer: Berlin, Germany; Heidelberg, Germany, 2009; pp. 25–42.

31. Schultz, A.J. The Role of GIS in Asset Management: Integration at the Otay Water Distict. Ph.D. Thesis, University of Southern California, Los Angeles, CA, USA, 2012.

32. Habib, A.; Bang, K.I.; Kersting, A.P.; Chow, J. Alternative Methodologies for LiDAR System Calibration. Remote Sens. 2010, 2, 874–907.

33. Mostafa, M.M.R. Boresight calibration of integrated inertial/camera systems. In Proceedings of the International Symposium on Kinematic Systems in Geodesy, Geomatics and Navigation (KIS 2001), Banff, AB, Canada, 5–8 June 2001.

34. Rieger, P.; Studnicka, N.; Pfennigbauer, M.; Zach, G. Boresight Alignment Method for Mobile Laser Scanning Systems. J. Appl. Geod. 2010, 4, 13–21.

35. Watson, G. Computing Helmert Transformations. J. Comput. Appl. Math. 2006, 197, 387–394.

36. Chen, Y.; Medioni, G. Object Modeling by Registration of Multiple Range Images. In Proceedings of the IEEE International Conference on Robotics and Automation, Sacramento, CA, USA, 9–11 April 1991.

37. Besl, P.J.; McKay, D.N. A method for registration of 3-D shapes. IEEE Trans. Pattern Anal. Mach. Intell. 1992, 14, 239–256.

38. CloudCompare—Open Source Project. CloudCompare—Open Source Project. Available online: http://www.danielgm.net/cc/ (accessed on 10 November 2015).

39. Zhang, Z. Flexible Camera Calibration by Viewing a Plane from Unknown Orientations. In Proceedings of the IEEE 7th International Conference on Computer Vision, Kerkyra, Greece, 20–27 September 1999.

40. Scott, P.J. Close Range Camera Calibration: A New Method. Photogramm. Rec. 1976, 8, 806–812.

41. Bouguet, J.-Y. Camera Calibration Toolbox for Matlab. Available online: http://www.vision.caltech.edu/bouguetj/calib_doc/ (accessed on 14 December 2015).

42. Gontran, H.; Skaloud, J.; Gilliéron, P.Y. A mobile mapping system for road data capture via a single camera. In Advances in Mobile Mapping Technology; Taylor & Francis Group: London, UK, 2007; pp. 43–50.

43. Atkinson, K. Modelling a Road Using Spline Interpolation. Reports on Computational Mathematics. Available online: http://homepage.math.uiowa.edu/~atkinson/ftp/roads.pdf (accessed on 15 October 2015).

INDEX

A

Active contour model (ACM) 97
Advanced Spaceborne Thermal Emission and Reflection Radiometer (ASTER) 26, 53
Analytic hierarchy process (AHP) 1, 2
areal pattern proportion (APP) 224
Areas of interest (AOI) 85
Area under curve (AUC) 33
artificial neural network (ANN) 153
Artificial neural network (ANN) 1, 13, 17
Asset Management 265, 271
Automatic subpixel water mapping method (ASWM) 112

B

Bivariate statistical analysis (BSA) 1

C

Calibration dataset 57
Canadian Hydrographic Service (CHS) 194
canonical correlation analysis (CCA) 242
Cation-exchange capacity (CEC) 54
Changjiang River basin 107
Civil Aviation Administration of China (CAAC) 153
Co-efficient of Variation (CV) 70
Complex adaptive system (CAS) 19
Computerized Maintenance Management System (CMMS) 265
Constrained energy minimization (CEM) 116
constraint optical flow method (SIFT-OFM) 238, 240

D

days of the year (DOY) 165
Decision tree (DT) 1, 14
Digital elevation mode (DEM) 25
Digital elevation model (DEM) 5
Digital Elevation Model (DEM) 139
Digital Elevation Models (DEMs) 191
Discriminant analysis (DA) 1

E

Ecological Decade Project 206, 207, 208
Electrical Conductivity (EC) 46
Endmember signature 115
Enhanced Thematic Mapper 47
Environment for Visualizing Images (ENVI) 69

F

Fast Line-of-sight Atmospheric Analysis of Spectral Hypercubes (FLAASH) 113
Flood inundation 137, 143
Flood simulations 140
Food and Agriculture Organization (FAO) 70

G

Genetic Algorithm (GA) 17
Geographical Information System (GIS) 175
Geographically weighted regression (GWR) 17
Geographic information system (GIS) 1
Geographic Information Systems (GIS) 137
Global Monitoring (GMM) mode 160
Global positioning system (GPS) 69
GPS (Global Positioning System) 259
Gravimetric moisture 88, 93
gray-level co-occurrence matrix (GLCM) 146
Gray-level co-occurrence matrix (GLCM) 98
ground control points (GCP) 248

H

high-resolution panchromatic (HRP) 239
hydrodynamic program 140
Hydrologic Engineering Centers River Analysis System (HEC-RAS) 137

I

Increased visible reflectance 50
Information content model (ICM) 1, 2
Interactions between Soil Biosphere and Atmosphere (ISBA) 85
International Panel on Climate Change (IPCC) 195

Iterative Closest Point (ICP) 263

J

Jianghan Basin downstream 24
Jinsha River belt 4
Jinsha River flow 3

K

Kappa coefficient (KP) 224

L

Landsat Thematic Mapper 53, 54
Landslide susceptibility map 18, 33, 40
Landslide susceptibility mapping (LSM) 1
Land surface water (LSW) 97
Land use and land cover (LULC) 223
Light Detection and Ranging equipment (LiDAR) 146, 153
Linear spectral mixture analysis (LSMA) 115
Lipschitz function 101
local linear histogram matching (LLHM) 242
Logistic regression (LR) 17
low-resolution multispectral (LRM) 239

M

Marine Environmental Service (MEDS) 194
Markov random field (MRF) 97
maximum likelihood (ML) 145, 153
Mean absolute error (MAE) 116
mean sea-level (MSL) 194
Meteorological analysis system 82
Model Sensitivity 39
Moderate Resolution Imaging Spectroradiometer (MODIS) 53
multiple endmember spectral analysis (MESMA) 181
Multiple endmember spectral mixture analysis (MESMA) 116
multi-scale analysis (MSA) 239
Multispectral Scanning (MSS) 47

N

Nearest optical image 87
Near-infrared (NIR) 68, 69
Near infrared reflectance (NIR) 50
Normalized difference salinity index (NDSI) 45
Normalized Difference Vegetable Index (NDVI) 26
Normalized difference vegetation index (NDVI) 45
Normalized Difference Vegetation index (NDVI) 56
Normalized Difference Vegetation Index (NDVI) 210
Normalized Difference Water Index (NDWI) 26

O

Object Based Information Analysis (OBIA) 154
Operational Land Imager (OLI) 53
Optimizable Variational Model (OptVM) 238
Ordinary cokriging (OCK) 57
out-of-bag (OOB) 149

P

panchromatic (PAN) 239
Particle swarm optimization (PSO) 17
Principal Components Analysis (PCA) 68
Probability density function (PDF) 100

R

Receiver operation characteristic (ROC) 33
Root-mean-square error (RMSE) 113
Root Mean Square Error (RMSE) 57, 70

S

Salinity index (SI) 45, 46
Salinity Index (SI) 69, 75
Saudi Arabia 45, 50, 51, 52
scale-invariant feature transform (SIFT) 238
Silty clay classes 55
Soil moisture 81, 82, 83, 84, 85, 86, 87, 88, 89, 90, 91, 92, 93, 94, 95
Soil Salinity Information Extraction (SSIE) 45
Soil Salinity Map 73
Soil Surface Characteristic (SSC) 81
Soil unit 54, 61
Soil-Vegetation-Atmosphere-Transfer model (SVAT) 82
Solar radiation 46
Spatial Variation 73, 76
Spectral mixture analysis (SMA) 111
structural similarity index measurement (SSIM) 248
support vector machine (SVM) 176
Support vector machine (SVM) 1, 17, 19
Synthetic aperture radar (SAR) 94
Synthetic Aperture Radar (SAR) 157, 172
Systematic error (SE) 116

T

Tasseled Cap Transformation (TCT) 68
Terrestrial Laser Scanner (TLS) 263
Textural map 86, 90, 93
Thematic Mapper (TM) 53
Topography bedding intersection angle (TOBIA) 26
Transformed Normalized Vegetation Index (TNDVI) 68
Triplet Markov Field 97

U

Universal Transverse Mercator (UTM) 69
unmanned aerial vehicle (UAV) 146

V

vertical stratification 205, 206, 213, 217

W

Water class 101, 102
World Geodetic System (WGS) 69